The 71F Advantage:
Applying Army Research Psychology
for Health and Performance Gains

Edited by Paul T. Bartone, Ross H. Pastel, and Mark A. Vaitkus

Foreword by Major General David A. Rubenstein, USA

PUBLISHED FOR THE
CENTER FOR TECHNOLOGY AND NATIONAL SECURITY POLICY
BY NATIONAL DEFENSE UNIVERSITY PRESS
WASHINGTON, D.C.
2010

Opinions, conclusions, and recommendations expressed or implied within are solely those of the contributors and do not necessarily represent the views of the Defense Department or any other agency of the Federal Government. Cleared for public release; distribution unlimited.

This edition published by Books Express Publishing
Copyright © Books Express, 2010
ISBN 978-1-907521-65-2
To purchase copies please contact info@books-express.com

In memory of the 248 Soldiers of Task Force 3–502 who perished
when their plane crashed on December 12, 1985, in Gander, Newfoundland,
as they were returning home from peacekeeping duty in the Sinai.
And in honor of all of those who continue to fight and sometimes die
in the cause of world peace and security.

Contents

Part III: Psychology of Leadership

Part IV: Soldier Psychological Health

Part V: Applying Psychology to Chemical and Biological Defense

Part VI: Psychology and Military Families

Part VII: Perspectives on Army Research Psychology

Illustrations

Figures

Tables

Foreword

The book you hold in your hands is a powerful tool in the arsenal of our Soldiers' well-being and, in some cases, very survival. Since our nation began putting Soldiers in harm's way, our forces have experienced the psychological trauma of war. In some cases, the number of casualties due to psychological injury approached the rates of physical injury. As leaders, we spend countless days training our Soldiers to survive the actual battle. Combat training, battle drills, physical fitness, and medical readiness help ensure their physical survival and ultimate victory in the combat zone. Now, we dedicate ourselves to helping ensure their psychological survival and victory when the battle is over.

This book grows out of paper presentations and discussions at the Army Research Psychology Biennial Conference held April 24–25, 2008, in Bethesda, Maryland. It provides a snapshot of groundbreaking research currently being done by Army research psychologists around the globe. The book illuminates the many contributions of Army research psychologists to supporting Soldiers and their families, as well as enhancing their performance and well-being. Research breakthroughs and updated treatment methods in the areas of human performance, psychological and environmental stressors, military and family wellness, personality and organizational factors, and leader and cognitive performance are outlined within this volume. The book also serves as a valuable historical document, providing a vivid picture of the 71F Army research psychology specialty and including a number of personal career stories. It should appeal to a wide range of audiences, including operational commanders, academics, students, and anyone curious about what military research psychology is all about.

Today, research psychologists remain true to the fundamental principles of Army medicine: to care for those who have borne the perils of battle in the service of the Nation. There is no higher calling.

Major General David A. Rubenstein
Chief, U.S. Army Medical Service Corps
Deputy Surgeon General of the U.S. Army

Preface

This is a book by and about military research psychologists—behavioral scientists who apply the principles and methods of psychology toward the goal of understanding and improving human health and performance in our defense forces. This group of specialists, along with their counterparts in the Army's sister Services and the many civilian research psychologists throughout the Department of Defense (DOD), forms a critical and essential core asset for our current and future national security. Since World War II, we have relied increasingly on research psychologists within DOD to design better and more efficient selection and classification tools and strategies, more effective training programs for military personnel, and improved leader development programs.

In addition, these specialists have applied rigorous research methods to understanding the limits and capacities of human mental and physical performance in a range of demanding operational environments. Their work has provided much of the knowledge that has allowed senior national security leaders to design systems and policies that appreciate and capitalize on our human talent, rather than ignore or frustrate it. Some of this research relies upon the use of animal models, an approach that has proven essential for understanding and developing treatments for chemical defense, stress, and traumatic brain injuries. More recently, military research psychologists have applied their expertise to understanding the causes and cures of stress-related injuries in troops, such as post-traumatic stress disorder, traumatic brain injury, drug and alcohol problems, and even suicide. This work being done by research psychologists not only helps with operational effectiveness, but it also helps the individuals who are put under great stress during military operations to cope with that pressure; it can save lives and enhance well being. Military research psychologists are responsible for significant advances in understanding human resilience and hardiness under stress, knowledge that is being applied in the design of more effective training and prevention programs.

As the military operations that we are engaged in as a nation and as allies grow in complexity, understanding the human dimension is more essential than ever. No matter how good our technologies may be, technology alone will not win the day in the challenges before us. It is the quality, strength, intelligence, and morality of our human systems that ultimately will allow us to realize the shared goal of a world in which security and prosperity are the normal state for all inhabitants. The contributions of research psychologists such as those in this volume are critical for moving us toward that goal. The Center for Technology and National Security Policy is pleased to publish this volume in the interest of sharing these critical conclusions with all Services, with our allies whose personnel face similar situations, and with those dealing with comparable situations in civilian life.

Hans J. Binnendijk,
Vice President for Research and Applied Learning
National Defense University

Introduction

71F, or "71 Foxtrot," is the AOC (area of concentration) code assigned by the U.S. Army to the specialty of Research Psychology. Qualifying as an Army research psychologist requires, first of all, a Ph.D. from a research (not clinical) intensive graduate psychology program. Due to their advanced education, research psychologists receive a direct commission as Army officers in the Medical Service Corps at the rank of captain. In terms of numbers, the 71F AOC is a small one, with only 25 to 30 officers serving in any given year. However, the 71F impact is much bigger than this small cadre suggests. Army research psychologists apply their extensive training and expertise in the science of psychology and social behavior toward understanding, preserving, and enhancing the health, well being, morale, and performance of Soldiers and military families. As is clear throughout the pages of this book, they do this in many ways and in many areas, but always with a scientific approach. This is the 71F advantage: applying the science of psychology to understand the human dimension, and developing programs, policies, and products to benefit the person in military operations.

This book grew out of the April 2008 biennial conference of U.S. Army Research Psychologists, held in Bethesda, Maryland. This meeting was to be my last as Consultant to the Surgeon General for Research Psychology, and I thought it would be a good idea to publish proceedings, which had not been done before. As Consultant, I'd often wished for such a document to help explain to people what it is that Army Research Psychologists "do for a living." In addition to our core group of 71Fs, at the Bethesda 2008 meeting we had several brand-new members, and a number of distinguished retirees, the "grey-beards" of the 71F clan. Together with longtime 71F colleagues Ross Pastel and Mark Vaitkus, I also saw an unusual opportunity to capture some of the history of the Army Research Psychology specialty while providing a representative sample of current 71F research and activities. It seemed to us especially important to do this at a time when the operational demands on the Army and the total force were reaching unprecedented levels, with no sign of easing, and with the Army in turn relying more heavily on research psychology to inform its programs for protecting the health, well being, and performance of Soldiers and their families.

Everyone present at the 2008 meeting was invited to submit a paper for inclusion in the proceedings, and we soon extended this invitation to every 71F we could reach. The response was so great that the project grew into the present book, with contributions from nearly all 71Fs. It took longer to complete than we planned, but we think the result was worth the wait. This book is the first of its kind, providing a detailed overview of research projects and activities of Army research psychologists, as well as several personal and historical accounts. It should serve as a valuable resource for anyone curious about research psychology in the U.S. Army, and those wanting to know more about the kinds of insights and products generated by this group.

This book is primarily the work of the individual chapter contributors. However, it would never have been finished without the dedication and diligence of my two co-editors, Ross Pastel and Mark Vaitkus. Both are not only top-flight behavioral scientists in their own right, but also dedicated public servants. They worked on this project mostly nights and weekends, with no ambition for personal gain, but rather with the hope of providing something beneficial to the field, to senior leaders, and to junior and future 71Fs.

Having Center for Technology and National Security Policy (CTNSP) Research Assistant Marek Rewers on the project was almost like having another editor. In addition to substantive writing contributions, Marek took responsibility for tracking and organizing hundreds of files, keeping the different versions straight, and fixing what seemed at times like an endless supply of formatting and reference problems. The book could not have been completed without Marek's fine work. Also, absolutely stellar editing and review assistance was provided by CTNSP Research Assistants Yong-Bee Lim, Simone Erchov, and Anna Nelson (whom we borrowed from the Industrial College of the Armed Forces).

We owe a special debt of gratitude to Hans Binnendijk, Director of CTNSP, for his unflagging support of the project since it was first proposed. Other colleagues at CTNSP have helped in many ways, especially Bill Berry, Director of the Life Sciences Division, Debra Cagan, a Senior Research Fellow, and Bill Bode, Chief Editor. Paul Strohl and Craig Zugschwerdt of NDU Audiovisual services generously helped with the illustrations. The professional staff at NDU Press, especially Jeff Smotherman and Lisa Yambrick, handled the many details of finalizing the manuscript and getting it ready for print with characteristic efficiency and good grace.

We also wish to thank our current Medical Service Corps Chief, Major General David Rubenstein, and the Deputy Corps Chief, Colonel Dawn Smith, for their ongoing encouragement and support on this project, and for the 71F AOC more broadly. Former Corps Chiefs Brigadier General (Ret.) Sheila Baxter and Brigadier General (Ret.) Richard Ursone also have our collective gratitude for their leadership and support along the way. Finally, the Medical Service Corps Historian, Colonel (Ret.) Richard Ginn, helped not only with key interviews of 71Fs, but also with his regular upbeat reminders that it is important for all of us to remember and learn from the past.

Paul T. Bartone

Part I

What Is Army Research
Psychology?

Chapter 1

U.S. Army Uniformed Research Psychologists: Making a Difference Yesterday, Today, and Tomorrow

Gerald P. Krueger

The United States emerged from the two World Wars as a global economic and military power. During the 65 years since the end of World War II, the Nation has deployed its forces around the world to serve in numerous smaller, more regionally based military actions ranging from intense regional wars to peacekeeping and to offer humanitarian assistance (Grimmett, 2009). During that time, the weapons available to the U.S. military have become both more precise and lethal, and many new technologies have been developed for countering enemy attacks. The military has embraced electronics and computers, developed unmanned weapons platforms, military robots, and smart bombs, and selected armed systems with longer standoff ranges. It has also determined means to interconnect distributed forces with network-centric communication systems, permitting electronic interactions throughout the chain of command. Such technological innovations contribute to an increasingly complex and stressful working environment for Soldiers both during training and in military operations.

Superior technological systems and weapons may help win battles, but they do not ensure winning wars. Machines must be directed, operated, controlled, mastered, maintained, and repaired—*by people*. To master the battlefield, the Army continually retains a sizeable manpower-intensive force structure. Despite its thousands of systems, the Army is still very much a people-rich organization—a Soldier-centric combat force. Since World War II, there have been several large conventional but regional battlefield wars: Korea (1950–1953), Vietnam (1967–1973), the Persian Gulf War in Iraq and Kuwait (1991), and war in Iraq again in 2003. However, the more contemporary military missions have involved smaller skirmishes and operations such as peacekeeping and nation-building (for example, in Bosnia, Haiti, and Somalia, among others). Since 2001, heavy participation in antiterrorist missions has produced paradigmatic shifts in military tactics. In particular, counterinsurgency (COIN) warfare (initially experienced in Vietnam) has returned to the forefront for conducting current military operations in Iraq and Afghanistan (U.S. Army Field Manual 3–24, *Counterinsurgency*, 2006).

Lower intensity asymmetric battlefields involve both high- and low-tech weapons, requiring ever-changing alternative tactics. These contemporary skirmishes necessitate more, not less, direct human involvement in fluid battlefield operations in which combatants and support personnel alike are exposed not only to the traditional battlefield stressors, but also to new threats conjured up from centuries-old tribal animosities that return cruel, crude, but sophisticated guerrilla and terrorist tactics to the forefront (Krueger, 2008).

To help ensure a human-oriented balance in a military milieu seemingly driven by technologists, the U.S. Army engages sizeable numbers of behavioral science researchers, among other specialists. Since the 1950s, without much fanfare, a relatively small coterie of Army Medical Service Corps (MSC) uniformed research psychologists has consistently played significant roles; and they continue to be called upon in increasingly important ways as military forces restructure to meet the challenges of the new millennium. Currently categorized into the Army's Military Occupational Specialty (MOS) 71F, these uniformed psychologists do research to advance knowledge and develop useful products for enhancing Soldier performance and to prevent, reduce, or cope with neuropsychiatric injuries, including post-traumatic stress disorder (PTSD) and traumatic brain injury (TBI). Today's uniformed psychologists accomplish laboratory-based research in neuroscience and do studies on human performance, sleep management, psychosocial and environmental stressors, personality and social/organizational factors affecting Soldier health and performance, leader development, and other human factors issues. These research psychologists are keenly focused on applying the tools and methods of science to find solutions to the most pressing human problems facing U.S. military forces today, and they communicate research findings quickly and effectively to leaders and decision-makers throughout the Department of Defense (DOD) (Bartone, 2007).

As social-political climates and the nature of warfare both continually change, uniformed research psychologists face exciting but tough new problems. For the past several years, General Robert Scales, USA (Ret.), has argued that our forces are already involved in World War IV—the social scientists' war—in which he contends that social scientists must assist military leadership in developing smaller teams of Soldier-warriors who understand cultural context by training them to be skilled in governance, statesmanship, and diplomacy so they can thrive in an alien environment and capture the psychocultural rather than the geographical high ground (Scales, 2006, 2009). Given that for decades the mantra of Army Soldiers has been to "make noise, break stuff, blow things up, and kill people," General Scales presents both the Army and social scientists with some very big challenges.

Background

A historian I am not. I write this chapter based on memorable impressions stemming from 40 years of involvement, including my 25-year Active-duty career as a uniformed U.S. Army research psychology officer from 1969 to 1994, and on my 15-year military human factors psychology consultant services for various elements of DOD. This chapter presents some of my personal recollections from the exciting research laboratories, programs, and projects I participated in, and from other programs my colleagues carried out. All the research was directed toward preserving the health and performance of Army Soldiers, but the efforts were also meant for Sailors, Airmen, Marines, and even Coast Guard personnel. This brief history recalls only some of the many contributions made by U.S. Army MSC uniformed research psychologists—colleagues I had the honor of working for or with, or in some cases officers I might have mentored. They are now the next generation of military psychologists proudly wearing the uniform of the U.S. Army.

Personal Experience

While I was an Active-duty graduate student at Johns Hopkins University from 1972 to 1975, several professors asked what kind of training I expected to obtain to prepare me to do research in an Army laboratory. Their faces portrayed quizzical wonderment at my answer: I aimed not only to become well versed in experimental design, statistics, and quantitative analysis, but also to learn the fundamentals for doing research in perception, cognition, psychological and physiological performance, neuroscience, work-related stress, human factors engineering, ergonomics, information processing, and decisionmaking. The professors were accustomed to students specializing in basic research in a particular topic, not a panoply of applications of behavioral science. I told them military research psychologists needed to be agile enough to do good work on a myriad of Army Soldier performance topics at several military research labs.

Why was my answer to their question so broad? Because in the mid-1970s, those Army Medical Department *buzzwords* described job and career opportunities for uniformed Army research psychologists participating in the U.S. Army's MSC officers' allied sciences research career program. After attaining a doctorate, psychology officers were expected to serve in a sequence of 3- to 5-year research assignments in the above topical areas at different Army laboratories. Soon after graduation, that was how many Army MSC psychologists began their careers as uniformed researchers. Some more than others may have had a vision as to how our stints on Active duty would play out; for example, some officers intended to fulfill one 4-year assignment as payback for their education and then depart Active duty. However, for many of us it seemed that the winds of fate and the Army frequently changed; and with the changes, our research careers evolved as opportunities to serve in different arenas cropped up one after another. Like many MSC colleagues, my own Army career evolved. I gained a superb education at Hopkins, and I frequently was called upon to employ my university training in most of those topical areas, as well as in additional ones not originally envisioned.

What Is Uniformed Army Research Psychology?

Military research psychology might be conceptualized as a specialized sliver of behavioral science in that it deals with topics of human behavior and human performance revolving around the work and lives of Soldiers, Sailors, Marines, and Airmen. Uniquely, it also involves the social psychology of Soldiers' families. Today, in the post-conscription era, Soldiers work in an all-volunteer force—a unique job setting. Those Soldiers and uniformed psychologists who devote a career to military service participate in a rather select profession in our society—one not well understood by psychologists who have not served in the military. This, then, speaks volumes about how different the profession of being a uniformed military research psychologist is from that of other researchers, such as our colleagues in academia. The focus of military psychology is with Soldiers. This story could therefore be about the psychology of soldiering, but it is more about a group of special uniformed research psychology officers who are Soldiers themselves. It is about those who work hard to preserve the psychological health and performance of our troops in everything they do.

As a profession, *military psychology* is not limited to those wearing the uniform of the U.S. Army. It entails decades of work by thousands of behavioral scientists in the uniformed and civil service sectors. Contributions have come from not only the Army, but also the Navy, Air Force, DOD, and countless university professors and graduate students, all quietly doing research—a team of scientists fulfilling important roles for the U.S. military over at least seven decades. Modern-day military psychology research took on a high level of importance during World War II when it was estimated that more than 2,000 psychologists (researchers and practitioners, uniformed and civilian) worked on the war effort (Parsons, 1972). For more complete descriptions of what *military psychology* is and what it has been about since World War II, readers are referred to publications such as Wiskoff (1997, 1998), Gal and Mangelsdorff (1991), Cronin (1998), Mangelsdorff (2006), Kennedy and Zillmer (2006), and the 21 annual volumes of the American Psychological Association's Division 19 journal, *Military Psychology*.

Understanding the Military Milieu

The military services are keen on training people to acquire and accentuate skill sets, so the Army pushes opportunities for advanced education in the belief that smarter Soldiers and officers are better members of the force. Accordingly, shortly after graduation with a Ph.D. in psychology, new Army officers participate in several months of basic MSC officer training, which helps them to assimilate military culture and Army values necessary to their jobs as behavioral scientists studying "what makes Soldiers do what they do." After completing a first or second research assignment at an Army laboratory, career progression officers receive 5 to 6 months of advanced military officer training. Expectation of additional mid-career training encourages them to take correspondence courses in the Army's intermediate level of education (formerly the Command and General Staff College) and other developmental courses to prepare them for higher leadership positions. Very occasionally, some psychology officers at higher rank participate in senior service schooling at such institutions as the U.S. Army War College to prepare for the highest levels of military leadership.

Soldier Psychologists: Ready for Action

One of the basic tenets of the Army's uniformed research psychology program is that psychologist officers are expected to be ready to deploy with Soldiers to any environment where the forces are engaged in field training or military operations. After studying and doing research in academic settings, uniformed psychology officers deem it important to prepare themselves as Soldiers by taking additional military skills training. These endeavors include the physically demanding 3-week Airborne (parachutist) course, 2 weeks of Air Assault training, or other specialized training such as that leading to earning the skill-based Expert Field Medical Badge. After completing such training, it is customary to wear on the Army uniform not only the medical caduceus of the Medical Service Corps, but also the military skill badges earned.

Such skills training rounds out the Soldier scientist, giving these officers a better understanding of what Soldiers undergo in training. It provides insights into the stresses entailed in soldiering, and it vividly demonstrates what skill sets and capabilities Soldiers are expected to bring to the battleground—all grist for the psychologist's study of Soldier behavior. Just as important, wearing the specialized skill badges gives these psychologists instant rapport with the Soldiers they often must entice to volunteer for psychological research projects, or with Soldiers they must survey and interview in an attempt to grasp the complex nature of unique soldiering tasks and missions being studied. Many among us can recall Soldiers in the 82^d Airborne Division or in the Ranger Regiment claiming that a researcher, as a non–badge wearer, is not likely to understand the nuances of soldiering jobs because "he's just a leg" (not airborne qualified) and "just doesn't get it."

To survey, study, and assist Soldiers, many of our uniformed research psychologists have accompanied large numbers of troops to far-forward areas, and some have even deployed with them into combat theaters. Such field data collection missions usually last several weeks to several months, but they rarely match the lengthy overseas deployments of other Soldiers. Participation by research psychologists in deployments overseas became prevalent throughout the 1990s, and this tendency holds true now as research psychologists accompany Mental Health Advisory Teams (MHATs) repeatedly deploying to Iraq, Afghanistan, and other conflicts (Gifford et al., 1992; Gifford, 1995; Bartone, Adler, and Vaitkus, 1998; Bartone, 2006).

The Army Uniformed Research Psychology Career Path

While acknowledging General Scales's challenge, this chapter presents a reflective snapshot—an outline of many efforts of the Army's Medical Service Corps research psychologists. Since about the mid-1950s, the number of Army uniformed research psychologists at any point in time has ranged from 20 to 40 officers. These psychologists have had a significant impact by shaping the military force; informing Army decisionmaking for establishing personnel policies; adjusting Army doctrine, troop force planning, and training; and influencing Soldier performance during actual military operations. In an abbreviated accounting of how far this elite group of military psychologists has come since the 1950s, this chapter projects the role uniformed research psychologists might play in the social scientists' war—World War IV and beyond.

The U.S. Army Medical Research and Materiel Command

For almost 60 years, the U.S. Army's MSC uniformed research psychologists served as key research members, principally at one of five research laboratories belonging to the U.S. Army Medical Research and Materiel Command (USAMRMC). There has been a smattering of other important assignment opportunities, about which more will be said. Since the mid-1970s, the laboratories employing uniformed research psychologists have been the Walter Reed Army Institute of Research (WRAIR), the U.S. Army Research Institute of Environmental Medicine (USARIEM), the U.S. Army Aeromedical Research Laboratory (USAARL), the U.S. Army Medical Research Institute of Chemical Defense (USAMRICD), and the Letterman Army Institute of Research (LAIR), which was deactivated in 1995.

Each of the latter four medical laboratories generally had from three to five uniformed research psychologists assigned, but WRAIR usually had 10 or more. For decades, WRAIR has also maintained a small behavioral research lab in Heidelberg, Germany, where typically four to six uniformed research psychologists were employed. Army research psychologists, from two to six at a time, also served as research program staff officers at Headquarters, USAMRMC, at Fort Detrick, Maryland.

Additionally, two Army materiel development laboratories traditionally employed two or three uniformed research psychologists:

the U.S. Army Human Engineering Lab (HEL) at Aberdeen Proving Ground, Maryland, and the Natick Research, Development, and Engineering Center (Natick RDEC), a materiels research lab at Natick, Massachusetts.

The Planned Career Pattern

Typical career patterns for Army Medical Service Corps officers have changed over the years. A current generic career path description can be found in Department of the Army Pamphlet 600-4, "Army Medical Department Officer Development and Career Management" (U.S. Department of the Army, 2007). Here, the career field of 71F, Research Psychologist, is described on pages 74–75 and portrayed graphically on page 96. Within the MSC and U.S. Army Medical Department (AMEDD), 71F Research Psychologists are classified as "Medical Allied Science" officers, together with 71A, Microbiologists; 71B, Biochemists and Research Physiologists; and 71E, Laboratory Science officers.

Whereas most officer assignments in the larger branches of the "big Army" are for only 2 to 3 years, the MSC uniformed Ph.D.-level psychologist scientist historically has had at least a 4- to 5-year stabilized tour of duty at a particular lab. There are several reasons for this difference. The longer tour permits the officer to:

- become established in a viable lab research program
- prepare and obtain approved research proposals
- acquire funding support
- set up lab equipment and experimental apparatus and recruit participants
- conduct data collection on sizeable projects
- analyze collected data and document the results
- present findings at Army decisionmaking forums and scientific and professional meetings en route to publishing the results in open literature journals.

After the initial 3- to 4-year assignment, psychology officers were generally expected to rotate to another medical research lab, where they would take on slightly different research problems, perhaps studying other research populations. For example, an officer might be expected to alternate between studies on issues confronting armor forces or artillery personnel—or infantry Soldiers, airborne troops, aviators, Rangers, or Special Operations forces. Or he might be assigned to a basic science laboratory to conduct experiments using animal species as models, where the research is oriented on determining basic mechanisms of bodily defense against disease or chemical, biological, or environmental threats. Of course, there were exceptions to that general rotational assignment pattern. For example, when an officer was identified as having particular expertise related to a hot research topic, by agreement he might stay on at that lab for longer than 4 or 5 years—particularly at WRAIR, where some officers moved about the Institute, in some cases into higher managerial roles. Many officers remained affiliated with WRAIR for several consecutive tours and some even for their entire military careers.

Staff Officer Work

Like other Army allied scientist officers, research psychologists were expected to spend a 2- to 3-year tour in a staff officer role at places like the USAMRMC Headquarters at Fort Detrick. There the staff work included assisting in headquarters-level management and administration of large, Federally funded research programs in the labs, and enacting and overseeing the command's extensive extramural research contract program, generally spread out among numerous prestigious universities around the country. This meant staff officers regularly prepared and defended budget outline documents to the Army Headquarters at the Pentagon, and to DOD offices where liaison work with Congress was carried out to negotiate critical defense research appropriation funding decisions. Such staff work placed young officers in the role of shaping the research programs, determining priorities for the Army and the labs, and influencing decisions about how millions of precious research dollars would be spent to ensure the Army, the taxpayer, and our Soldiers got the best payoff for the country's investments.

Initially, many young research psychologists looked at being selected for a staff officer assignment as something to be avoided, as it normally took a scientist away from the research bench for several years at a time, usually at mid-career—just when he was ready to make some important research advancement. Thus, a staff assignment could be an intrusion. However, unlike no other group of allied science specialists, research psychologists learned to grow into staff officer roles.

Army research psychologists excelled in their careers because of their staff officer experiences.

Staff officer psychologists gained invaluable understanding of the bigger picture—the inner workings of how the Army and DOD do business. They learned what works and often what does not. Importantly, they learned how to improve processes, make things work better, increase the likelihood of the labs achieving their research missions, and obtain better productivity from research programs for taxpayers and Soldiers. Subsequently, when those research psychologists who had grown in their staff officer roles returned to the laboratories, they had groomed themselves to become the chiefs of multidisciplinary research branches, taking on increasing levels of program and personnel management responsibility. Some officers became deputy directors of the labs themselves, and a few were even selected as laboratory scientific and technical directors, taking command of the institutes and laboratories, attaining perhaps the highest levels of management afforded uniformed scientists in the Army and in DOD.

Army Research Psychology: Often Multidisciplinary

Army research generally is multidisciplinary and involves much teamwork. Perhaps as a young Army captain or a junior major, assignment to a research laboratory employing from 85 to 1,100 research staff members often meant one had arrived as "the new kid on the block." Initially, that meant "latching on" to extant research programs of senior, more established officers, or working for and with seasoned civil service research program directors already entrenched for decades in managerial roles. It follows that few uniformed research psychologists assigned to laboratories accomplished their high levels of productivity alone. Army research experiments, studies, or projects are usually accomplished through group effort.

Depending on the laboratory and its programs, much of the project work involved multidisciplinary studies that called for collaboration among researchers from many scientific disciplines, including physiology, neuroscience, biology, chemistry, audiology, anthropology, biomechanics, biodynamics, biophysics, bio-optics, nutritional science, microbiology, immunology, and other biological or behavior-related sciences. In some labs, doing research projects meant interacting with military equipment designers, engineers, and automotive and aviation platform specialists; designers of individual Soldier equipment, protective clothing, specialized uniforms, parachutes, tents, and night-vision systems; or military training and doctrine developers and other parties. One of the thrills for research psychologists was recruiting, training, and working with Soldier volunteers, who agreed to be subjected to behavioral studies in hopes the test results would benefit all Soldiers.

Often it was the military psychologist whose expertise in experimental design and statistics determined the direction taken by research programs. Since the goals of large military research projects frequently revolve around predictions of equipment operator performance, psychologists were best equipped to guide multidisciplinary research teams in accounting for the trickiest of human performance issues: individual differences, motivation, and learning and skill development. Performance studies predicting people's behavior as it related to experimental design issues became the forte of applied military psychologists. The most effective and persuasive officer psychologists repeatedly rose to the challenge as they took on key leadership roles in designing, running, managing, and directing such studies (Krueger, 1998, 2006).

Conducting Research in Military Laboratories

As mentioned above, Army research psychologists conducted the bulk of their work at five medical research and two materiel development laboratories. These are outlined here mostly according to the organizational missions the labs held and are followed by descriptions of assignments at a few other organizations that employed uniformed psychologists. The names of many of the MSC psychologists who worked in each lab are mentioned, and where possible I listed dates of their assignments. No doubt I missed some names because I retired from Active duty in 1994 and could not keep fully attuned to activities since then. I apologize to those Army psychologists I have not mentioned. Additionally, providing a short sketch of the contributions of each psychologist would make for a very lengthy chapter. Perhaps that goal can be fulfilled in a later effort, as the numerous accomplishments of so many colleagues probably warrant a book to do them all justice.

WRAIR Division of Neuropsychiatry

Auspicious beginnings. In 1951, the U.S. Army formally established a medical psychology research unit at WRAIR in Washington, DC. This unit eventually became known as the Division of Neuropsychiatry

(Div. NP). Any history of the Div. NP must include the significant productivity and influence of Joseph V. Brady. After serving as a combat infantry platoon leader in World War II, Joe received his Ph.D. from the University of Chicago in 1951, and his service as an MSC officer at Walter Reed began that same year and lasted two decades. A foundation investigator in the Div. NP, he identified himself as a neurobehavioral scientist. From 1951 to 1963, he was Chief of the Department of Experimental Psychology and was Deputy Director of Div. NP from 1963 to 1970.

Among his many achievements at WRAIR, Joe Brady pioneered research into the ties between drug effects and emotional behavior (Brady, 1956) and performed the often-cited "ulcers in executive monkeys" experiment, which suggested a link between stress and peptic ulcers (Brady, 1958). That research was significant in establishing the notion that stress can lead to physical illness. His collaborative work with M. Sidman, J.W. Mason, and W.J. Nauta produced a classic series of interdisciplinary studies of the relationships between the limbic system, the endocrine system, and behavioral stress. Among the many important discoveries of that project was that psychological stress could be far more damaging than physical stress. Joe Brady's work in the 1950s, along with that of his colleagues at WRAIR, cast important influences into the development of psychology and neuroscience themselves.

In his lengthy, productive stay at WRAIR, Joe Brady was the epitome of a role model. For 20 years, he exuded behavioral science leadership for the Army. His work at WRAIR constitutes a cornerstone for the MSC psychology program. He set the pace for many MSC research psychologists who followed him in doing neuroscience and behavioral biology research at WRAIR and elsewhere. If Joe Brady was not the key person in formulating the nucleus of what became known as the MSC research psychology program as we know it today, then he certainly was the most emulated scientist on whose model career the tenets of the program were crafted. Joe retired from the Army in 1970 when he left WRAIR to become Professor and Director of Behavioral Biology Research at the Johns Hopkins School of Medicine in Baltimore. Always the productive scientist and educator, as recently as June 15, 2008, at age 86, Joe Brady received the prestigious Mentorship Award from the College on Problems of Drug Dependence.

Over the years, Div. NP traditionally employed about 20 percent of the entire WRAIR work force, including as many as 120 full-time research and staff members. One of the more prominent MSC psychologists was Frank J. Sodetz, who, in picking up his mentor Joe Brady's mantle, did superb basic neuroscience research, first in the Department of Experimental Psychology, and then as Director of Div. NP. Being recognized for his managerial leadership skills, Frank also served for several years as the Deputy Director of the entire WRAIR while he was simultaneously the Research Psychology Consultant to the Army Surgeon General. Before his Army retirement, Frank also spent several years as Commander of the WRAIR's Armed Forces Institute of Medical Sciences, a biological research lab for infectious diseases in Bangkok, Thailand. In that assignment, he was the first MSC research psychologist to command a USAMRMC laboratory.

Division NP Headquarters Office. Due to the size of the Division of Neuropsychiatry and the number of personnel assigned, a number of MSC uniformed research psychologists served in staff officer positions in the Div. NP main office at WRAIR, adjacent to the Walter Reed Army Medical Center. Some of the administrative work entailed budgeting and preparation of research progress and annual reports. In numerous cases, these officers also simultaneously continued their active involvement in research projects sponsored by Div. NP. Among the many officers who served in that office are John H. McDonough, Robert H. Stretch, Kenneth Zych, G. Rufus Sessions, and Thomas G. Raslear. After fulfilling leadership roles in the Department of Medical Neuroscience, Frederick J. "Rick" Manning had a tour at the Heidelberg research unit and, on his return, became Deputy Director of Div. NP (1981–1987), then Director (1987–1992), and for a year Executive Officer of WRAIR (1992). Robert K. Gifford also served a tour as Executive Officer of WRAIR.

Department of Medical Neuroscience. Historically, one-third of WRAIR's sizable Div. NP program was dedicated to studies of basic neuroscience research (DOD's 6.1 program funding), including a mix of animal and human experiments done to elucidate mechanisms of human stress reactions and responses to a wide array of stresses encountered by military personnel. Rick Manning served as Chief of the Department of Experimental Psychology from 1976 to 1978 before it was combined with the Department of Neuroendocrinology and Neurochemistry to become the Department of Neuroscience.

An important era in Div. NP started around 1971 and lasted until 1976. Congress mandated a large investment in drug abuse research in response to many Vietnam soldiers becoming addicted to heroin and facing discharge with an opiate dependence. Div. NP was a center for behavior and physiology research on this problem, and many uniformed research psychologists joined the division during that era, including Timothy Elsmore, George Rufus Sessions, Steven R. Hursh, Thomas G. Raslear, George Koob, Jeffery Witkin, and David M. Penetar. The sudden termination of the program in 1976 led to a reduction in force of many civilian employees as well as a restructuring of the careers of many uniformed psychologists. Nevertheless, this drug research program had a lasting influence on the scientific careers of many of these psychologists.

Steven R. Hursh served as Chief of the Physiology and Behavior Branch of Neurosciences (formerly the Department of Experimental Psychology) and Chief of the Medical Neurosciences (1981–1987). He was Deputy Director of Div. NP (1987–1992) and succeeded Rick Manning as Director (1992–1995). Rick and Steve each served as the Research Psychology Consultant to the Army Surgeon General. Steve Hursh's basic research on substance abuse and animal behavior was the foundation for what became known as Behavioral Economics, a current theoretical framework for translational research on drug abuse. Subsequent studies culminated in a new theory of behavioral motivation. Eventually that work, which descended from his early work at Walter Reed with primates, was highlighted in a *Psychological Review* article (Hursh and Silberberg, 2008). Steve later directed his research efforts to development of Soldier and transportation operator fatigue models used in risk management of work schedules in both military and civilian work settings.

George Koob became a leader in the discovery of the neurochemistry of drug addiction. His first experiments on reinforcement by electrical brain stimulation in a large and productive animal laboratory in Medical Neurosciences became the basis for his highly influential work on dopaminergic mechanisms of behavioral reinforcement. After working on the drug research program at WRAIR, Jeff Witkin became a staff scientist in the intramural research program at the National Institutes for Drug Abuse.

G. Rufus Sessions was a long-time member of the Experimental Psychology Department and subsequently the Department of Medical

Neurosciences. Later in his career, Rufus returned as the Chief, Physiology and Behavior Branch of Medical Neurosciences (1990–1993). He then joined Frederick Hegge, a Department of the Army civilian (DAC), in a Headquarters USAMRMC–sponsored Tri-service Joint Working Group on Drug Dependent Degradation of Military Performance, a program that subsequently funded the research of numerous psychologists and physiologists around the command.

David M. Penetar served in Medical Neuroscience and became Chief of the Physiology and Behavior Branch, before he served in the Department of Behavioral Biology, also in Div. NP. Thomas G. Raslear did basic research on animal signal detection and later collaborated with Hursh on behavioral economics. Raslear moved to the Department of Microwave Research in Div. NP and conducted studies on the behavioral toxicology of electromagnetic radiation. After leaving Army research, he became the head of human factors research for the Federal Railroad Administration, where he sponsored studies that validated Hursh's fatigue model for prediction of human error accident risk in railroad operations. Other MSC uniformed psychologists who spent at least a tour doing research in the Department of Medical Neurosciences included Ross H. Pastel and Bruce Hamilton.

Department of Psychophysiology. The second part of WRAIR Div. NP constituted applied psychology studies that were conducted in the Department of Psychophysiology at WRAIR's Forest Glen annex near Silver Spring, Maryland. Laboratory studies were configured to do psychophysiological assessments of Soldier performance in sustained duration missions (command and control simulations) that caused Soldiers to experience significant loss of sleep and rest. In the 1950s and 1960s, the department was headed by Harold L. Williams, who gave it fame as a center of excellence in sustained performance and sleep loss research. It was Williams's seminal work on the effects of sleep deprivation that first described the Walter Reed Sleep Lapse (microsleeps) phenomenon (Williams, Lubin, and Goodnow, 1959).

The department continued sleep deprivation studies for another decade under the direction of Dr. Frederick W. Hegge, a well-known DAC in our field. In about 1985, Dr. Hegge took a career sidestep to work for the Headquarters USAMRMC, where he provided senior staff leadership for numerous programs in two premiere research areas: Drug Dependent Degradation of Performance, and Military Operational Medicine. These large research programs both funded the work and challenged the numerous MSC research psychologists and other allied scientists in several of the MRMC labs to take on new and changing roles in psychological research applications.

In 1983, under the leadership of Lieutenant Colonel Gregory Belenky, a research psychiatrist, the department was renamed the Department of Behavioral Biology. During the 1980s and 1990s, highly productive studies continued on Soldier performance during sustained military operations, accomplishing both significant sleep deprivation studies in the lab and naturalistic studies in the field while the troops participated in extensive sustained operations training exercises or in large military unit combat readiness test scenarios. Some of that work was codirected by this author and included the team leadership work of John R. Leu and Harold R. Smith (Krueger 1989, 1991). This field test program also included important collaborative work with other MRMC uniformed psychologists in efforts that illustrated that wearing chemical protective clothing for extended periods has debilitating effects on Soldier performance (Krueger and Banderet, 1997). With novel test instrumentation, including dynamic wrist-activity monitoring technology, members of the department configured many of the significant results from the sleep and fatigue research into Soldier alertness/ fatigue and work-rest modeling algorithms. This work, pioneered by Colonels Gregory Belenky and Daniel Redmond, USMC, was elaborated on with the involvement of Steven R. Hursh at Div. NP headquarters.

Under the tutelage of research psychiatrists Colonels Gregory Belenky and Paul Newhouse, uniformed psychologists Vincent M. O'Donnell and David M. Penetar, and a number of DAC researchers (Balkin, Thorne, Wesensten et al.) in the Department of Behavioral Biology accomplished 25 years (1984–2009) of significant in-house exploratory research on stimulant drugs, hypnotic sleep aids, and other pharmaceuticals for use in Soldier fatigue countermeasure programs. That work continues today under the direction of Balkin and Wesensten.

Results from decades of research in this highly productive department shaped the Army's doctrinal planning and operational policies on Soldier and unit work-rest scheduling; sleep discipline and fatigue management and countermeasures, including napping; and use of sleep-promoting compounds, stimulant drugs, alertness enhancers, physiological status monitoring, fatigue and performance modeling, and other workforce planning measures. For a list of research publications from the first 30 years of the Department of Psychophysiology (1958–1986) identifying the productivity of numerous MSC research psychologists and their colleagues, see Krueger (1986).

Department of Military Psychiatry. The remaining third of WRAIR Division NP's work was carried out in the Department of Military Psychiatry, also housed at the Forest Glen Annex. For decades, those research programs were directed by Dr. David H. Marlowe, a noted Army civilian social anthropologist. Marlowe employed numerous uniformed research psychologists in examination of more holistic stresses facing Soldiers in training and especially while they were deployed to various regions of the world. He and his staff pioneered research in stress syndromes associated with the harshness of military occupations and lifestyle. Studies were carried out on the effects of good or poor leadership, unit cohesiveness, stress hardiness and resilience, and regimental unit replacement policies (Vaitkus and Griffith, 1990); operational stress and mental/behavioral effects of stress on Soldiers and their families; prevention of mental health problems and performance decrements in Soldiers; post-traumatic stress in response to significant events (for example, the Gander Newfoundland airliner crash, Bartone et al., 1989); mental health assessments in combat theater deployments; identifying factors that impede Soldiers from seeking medical treatment for psychological problems; and related topics.

Some of the department's work is done collaboratively with military psychologists stationed at WRAIR's Heidelberg research unit, and with other scientists throughout the Army. A contemporary example of such work is the support provided in the Army's deployable MHATs to Afghanistan and Iraq; officers from both Forest Glen and Heidelberg have participated in each of the MHATs deployed to combat theaters. Among others, Sharon A. McBride, Paul D. Bliese, Jeffrey Thomas, Dave Cotting, Dennis McGurk, and Michael Wood participated in MHAT missions.

Uniformed MSC psychologists who did research with the Department of Military Psychiatry reads like a who's who of the MOS for research psychologists. Some worked in the department itself at Forest Glen, some in WRAIR's Heidelberg field research office, and others assigned elsewhere working in collaboration. Some of the names associated with the Department of Military Psychiatry are Lawrence H. Ingraham, Frederick "Rick" J. Manning, Paul T. Bartone, Robert K. Gifford, Mark A. Vaitkus, Carl A. Castro, Kathryn Knudson, Robert H. Stretch, J. Edward McCarroll, James A. Martin, James E. Griffith,

Sharon A. McBride, Melba C. Stetz, Paul D. Bliese, Scott Killgore, Dave Cotting, Paul Wright, Edward Edens, Michael Wood, and Jeffrey Thomas. Robert K. Gifford and Carl A. Castro served stints as Chief of the Department of Military Psychiatry.

U.S. Army Medical Research Unit–Europe at Heidelberg

Since its founding in September 1977, the WRAIR field research lab at Heidelberg, the U.S. Army Medical Research Unit–Europe (USAMRU–E), has been closely allied with the Department of Military Psychology at Forest Glen. The principal focus in the USAMRU–E has been on human dimensions research in the forward-deployed environment. This lab unit, commanded by a research psychologist, provides a highly responsive and agile research capability to field commanders and the European Theater Surgeon on human aspects of mobilization, training, and leadership that promote Soldier psychological readiness and effective unit performance and aid in prevention and treatment of psychiatric casualties.

Initially, USAMRU–E was charged to examine issues of Soldier stress in a large deployed force in Europe (as many as 300,000 troops and their families). This forward-deployed research lab collects psychological and biomedical data with overseas military units to determine the nature and extent of stressors on Soldiers; ascertain the health and performance consequences of those stressors; understand factors that diminish or enhance morale, cohesion, and psychological well-being in Army forces in Europe; and identify mediating factors such as stress hardiness that build resiliency in individuals and military units to decrease vulnerability to stress.

The psychologists in this highly productive small research unit often completed quick response assessments of human dimensions survey data on U.S. Army Europe Soldiers deployed to theaters such as Iraq and Afghanistan. The findings identified critical pre- and post-combat stressors for Soldiers and determined their relation to psychiatric symptomatology, health, and well-being. From these endeavors, the Army's MHATs were formulated and are used extensively today. As a result of work done in support of these teams, Carl A. Castro led this unit's efforts to develop innovative and effective treatment/training programs for returning troops known as *Battlemind training* that have now been adopted Army-wide (Castro, Hoge, and Cox, 2006).

While the mission of the European research unit has changed over the years, there is a continuing focus on applying psychological research and knowledge to protect and preserve the health of the force. Prominent MSC psychologists who had one or more tours of duty at the Heidelberg lab, and most of whom also served as commander of USAMRU–E, include Lawrence H. Ingraham, Frederick J. Manning, Robert K. Gifford, Paul T. Bartone, Mark A. Vaitkus, Jeffrey Thomas, Carl A. Castro, James Ness, Dennis McGurk, Paul D. Bliese, and Oscar Cabrera.

WRAIR Field Office at Fort Bragg

After a stint directing field work on sustained operations and Soldier performance research at WRAIR's Department of Behavioral Biology, John R. Leu rekindled and developed the influential medical research liaison assignment at Fort Bragg, North Carolina. Officers serving in this position, which was first held by Stanley H. Holgate (1983–1986), brought about much interchange and collaborative field research in the arenas of 18th Airborne Corps and 82d Airborne Division. For a time, representing the U.S. Army Research Institute of Environmental Medicine (USARIEM), Lolita Burrell directed this work at Fort Bragg as she accomplished multidisciplinary field research with the military units located there.

U.S. Army Research Institute of Environmental Medicine

The Army Surgeon General opened USARIEM at Natick, Massachusetts, in 1961. The lab was formed as a merger of elements of the Harvard Fatigue Laboratory's renowned lab of human physiology (Horvath and Horvath, 1973), the Army's Climatic Research lab at Lawrence, Massachusetts, the Quartermaster's Environmental Protection Research and Earth Sciences Divisions at Natick, and portions of the Army Medical Research Lab formerly located at Fort Knox, Kentucky. USARIEM's mission includes basic and applied science in environmental medicine, physiology, military nutrition, and Soldier performance research. It features integrated cellular, tissue, animal, and human research programs. Multidisciplinary scientists at USARIEM have especially been known for their cutting-edge physiological and psychological studies examining Soldier performance and health variables in harsh working environments such as intense heat, extreme cold, and high terrestrial altitudes. Over time, USARIEM also became recognized as one of the world's leading military research laboratories on physical fitness, biomechanics, the biophysics of protective clothing, and the health and performance effects of military field nutrition.

USARIEM supports a generalized tri-Service research mission in preventive, occupational, and environmental medicine, especially as it pertains to unique military work tasks and missions. Research psychologists have been an integral part of USARIEM's research programs every step of the way. An extensive list of 30 years of USARIEM publications (1961–1992) sheds light on the high productivity of the many scientists who worked there (Krueger, Cardinal, and Stephens, 1992).

Researchers have continually worked in world-class climatic and environmental research chambers and biomechanics facilities at Natick. But USARIEM is also renowned for carrying out numerous large operational and training data collection exercises with hundreds of Soldiers in the field. Its research findings provide products to enhance Soldier capabilities and reduce health risks. The recommendations of the lab's scientists have led to improvements in Army and DOD personnel training policies and preventive medicine guidance, decreasing Soldier injuries, lost duty time, and medical costs. USARIEM materiel development support programs produce recommendations for design specifications and product improvements for individual Soldier clothing and equipment, the nutrition content of military field rations, and pharmaceuticals used as prophylactics for battlefield environmental stressors. USARIEM identified strategies for personal physiological and psychological status monitoring systems and developed a variety of algorithms to prevent or detect decrements in Soldier health and performance. In terms of health hazard assessment of Army developmental materiel systems, USARIEM works closely with the Army's Center for Health Promotion and Preventive Medicine (CHPPM) to provide expertise on thermal and hypoxic conditions.

Research at USARIEM aims to maximize Soldier health and performance in extreme environments and under occupational stress. Over the decades, many MSC uniformed psychologists have been assigned to USARIEM's Human Performance Research Directorate, where they conducted studies on the cognitive and behavioral decrements related to environmental and operational stressors. Typically, the research talents of these psychologists were cross-matrixed to work collaboratively with other scientists in the lab and field on numerous multidisciplinary studies of the effects of environmental stressors on Soldier performance in high heat or extreme cold (Kobrick and Johnson, 1991) and at high altitude (Banderet and Burse, 1991), in sustained military operations, or while Soldiers wore chemical protective clothing (Krueger and Banderet, 1997). Psychologists examined the effects of various pharmaceuticals and nutritional add-ons for inclusion in military food and field rations; measured rifle marksmanship as affected by environmental stressors and other independent variables; and assessed effects of arduous military training—such as Ranger and Special Forces selection programs (IOM Committee on Military Nutrition, 1999), injury countermeasures for the airborne training program, and U.S. Military Academy at West Point fitness programs. USARIEM scientists conducted special studies of female Soldiers in training and operations (Vogel and Gauger, 1993), and so on.

One of the best illustrations of effective transfer of important lab research findings to Soldiers in the field was USARIEM's series of occupational, preventive, and environmental guidance manuals, prepared to help deployed Soldiers and commanders to preserve Soldier health and performance in harsh environments. These manuals were produced for a sequence of overseas deployments beginning with Operations *Desert Shield/Storm* in Southwest Asia in 1990–1991 (Glenn et al., 1990), and followed with similar preventive medicine guidance for deployments to Somalia, Haiti, Bosnia, Rwanda, and other locales. That highly successful series of manuals (distributed overseas by the tens of thousands) was spearheaded predominantly by USARIEM MSC research psychologists. After USARIEM set the pace for this program for several years, the mission for technology transfer of such preventive medicine guidance to Soldiers in the field was assumed by the CHPPM at Aberdeen Proving Ground, Maryland.

MSC psychologists effectively served in upper level management positions at USARIEM as well. Mary Z. Mays served as Chief of the Military Performance Division; J. Frazier Glenn was Deputy Commander of USARIEM from 1988 to 1992; Gerald P. Krueger served as Scientific Technical Director and military Commander from July 1990 to June 1994; and David M. Penetar served as Science Director and Commander of USARIEM from October 1997 to August 2000. Following the lead of Frank Sodetz as lab Commander in Thailand, Colonels Krueger and Penetar were the second and third MSC psychologists to command a USAMRMC lab.

Additional MSC psychologists who worked at USARIEM over the decades include R. Curtis Graeber, Donald R. Sweeney, Dennis M.

Kowal, Jared B. Jobe, Terry M. Rauch, Harold E. Modrow, Kathryn Popp, Sharon A. McBride, Gina E. Adam, Lolita Burrell, and Lynee D. Murray.

Liaison at Canadian Defence and Civil Institute of Environmental Medicine

In 1980–1981, the MRMC established a medical research liaison and exchange scientist position with the armed forces of Canada. This amounted to stationing a uniformed MSC research scientist (initially Captain David O. Cote, a human factors specialist) at the Defence and Civil Institute of Environmental Medicine (DCIEM) at Downsview, Toronto. The purpose was to facilitate exchange of scientific information between the USAMRMC laboratories and DCIEM, whose research missions most closely resemble those of USARIEM and HEL. From 1986 to 1991, the second person to fill the position at DCIEM was Robert H. Stretch, a research psychologist who served as an investigator on studies of combat stress, readjustment of Canadian Vietnam veterans, and small group interactions of command post personnel during sleep deprivation and sustained performance studies.

U.S. Army Aeromedical Research Laboratory

The U.S. Army Aeromedical Research Laboratory was established as a research unit at Fort Rucker, Alabama, in October 1962. It was redesignated as a laboratory in 1969. Collocated with the Army Aviation Center and School, USAARL's predominant mission was initially to accomplish aviation medicine research for the growing Army air force, which during and after the Vietnam conflict included thousands of helicopter aviators and crews. USAARL is also positioned across the street from the Army Aeromedical Center, where aviation medicine is key, Army flight surgeons are trained, and guidelines for crewmember flight physicals are determined, prompting much cross-fertilization among the respective organizational missions. In 1974, the research staffs of the vision and auditory sciences sections of the Army Medical Research Lab at Fort Knox, Kentucky, relocated to Fort Rucker and were incorporated into the USAARL aviation medicine mission base. Several uniformed psychologists from that lab were among the newcomers.

In 1970–1971, USAARL established an Aviation Psychology Division to examine helicopter pilot workload and flight performance issues. Research projects included extensive use of instrumented helicopters and simulators for in-flight and lab assessments of Army helicopter pilot performance under various stressful flight conditions. Some of those studies included sustained flight operations involving aviator fatigue; extended flight periods while using night vision goggles; flight while wearing chemical protective clothing; flight performance of aviators taking drugs such as chemical warfare agent prophylaxis compounds, antihistamines, and even stimulants; copilot navigation workload during nap-of-the-Earth flight; assessments of pilot workload during helicopter transatlantic self-deployment flights; workload during helicopter air-to-air combat scenarios; verification of developmental aeromedical evacuation equipments; and so on.

The first uniformed MSC psychologist to establish and direct USAARL's Aviation Psychology Division was Mark A. Hofmann (1971–1972). Mark left Active duty soon thereafter to become a DAC and went on to represent human factors in Army aviation development programs in St. Louis, Missouri; from there he moved on to become Deputy Director of the Human Engineering Lab at Aberdeen, where he continued to employ MSC research psychologists. Kent A. Kimball succeeded Mark as Aviation Psychology Chief at USAARL. Kent devoted considerable effort to completing the design and construction of the Laboratory's Helicopter In-flight Monitoring System (HIMS), an airborne data capturing system capable of simultaneously measuring pilot and helicopter performance aboard a JUH–1H Iroquois research helicopter. The HIMS provided Army aviation psychologists with a unique tool that was used for several decades to examine pilot performance in actual flight, simulating military air operations. After completing his initial obligated Active-duty tour in 1975, Dr. Kimball took on the role of Aviation Psychology Division Chief as a DAC. In keeping with a broadened mission including performing human factors evaluations of developmental aviation systems, he changed the name of the division to the Biomedical Applications Research Division (BARD), which he directed for over a decade. Gerald P. Krueger, who initially served as Chief of the Manned Systems Branch (1976–1980), returned to USAARL to serve as the BARD Division Director from 1988–1990.

MSC psychologists who worked in either the Aviation Psychology or the BARD Division at USAARL included Andrew Martin, James Bynum, Michael G. Sanders, Michael A. Lees, Aaron W. Schopper, Bruce Hamilton, George Mastrianni, and Charles A. Salter.

MSC psychologists also worked in other divisions of the USAARL, typically doing vision or auditory science work. These included

Franklin F. Holly, Richard R. Levine, and Bruce C. Leibrecht. James H. Patterson, Jr., who served in the Fort Knox lab, was among those in 1974 who relocated to USAARL, where Jim worked in auditory research before he too converted to a DAC position to direct the blast overpressure research program for another two decades. Along the way, Aaron W. Schopper served as the Biodynamics Research Division Chief; and Robert M. Wildzunas served as Chief of the Visual Sciences Branch (1996–1997) and later as Research Director of the Aviation Human Performance Division (2005–2008).

As part of the Army's Manpower-Personnel Integration (MANPRINT) Program, USAARL gradually broadened its behavioral science studies to involve assessments of health hazards associated with operation of large developmental Army weapons and materiel systems in accordance with Army Regulation 40–10 (see Krueger, 1983; Ranadive et al., 1983). Assessments included multidisciplinary studies of the mechanical threats to the body by impact, shock, acceleration, vibration, acoustical noise, and visual enhancement via night vision systems. MSC psychologists involved in some of the most prominent of these studies at USAARL included Bruce C. Leibrecht, Gerald P. Krueger, Aaron W. Schopper, Franklin F. Holly, and Jim Patterson while he was a DAC.

More recent assignments of MSC military psychologists who worked at USAARL include James M. King (1991–1993), Robert M. Wildzunas, (1993–2000 and 2005–2008), Lawrence C. Katz (1997–2000), Gina E. Adam (2000–2004), Michael W. Boye (2003–2005), Melba C. Stetz (2005–2008), and Michael N. Dretsch (2007–present).

Research psychologists who regularly performed U.S. Army Reserve Individual Mobilization Augmentee duty with USAARL include James H. Patterson, Michael A. Lees, Elmar T. Schmeisser, William D. Springer, and Christopher P. Long.

U.S. Army Medical Research Institute of Chemical Defense

The U.S. Army Medical Research Institute of Chemical Defense traces its origins to 1915 when the War Department gave the Army Medical Department responsibility for designing protective equipment against chemical agents. The Medical Department was tasked to provide the first gas masks to the Army in 1917. In 1922, the Medical Research Division was organized at Edgewood Arsenal, Maryland, and charged to study the pharmacological action of chemical warfare agents. In 1940, increased emphasis was placed on development of methods for prophylaxis and treatment of chemical warfare casualties. At the end of World War II, the United States learned that Germany had stockpiled organic phosphate compounds far more deadly than those in the Allied arsenal. Russia had seized German stocks of some chemical agents (for example, tabun) and reassembled a chemical warfare agent factory in Russia. Basic and applied medical research continued on these colorless, odorless threat compounds, among others, in the search for medical defense against chemical agents.

On Thanksgiving Day 1969, President Richard Nixon declared that the United States would unilaterally withdraw from researching and producing chemical and biological weapons (which the United States had begun in earnest in 1941). It took almost a decade to disengage some Army labs from that sort of work. Washington ratified the Biological Weapons Convention in 1975 but stubbornly declined to do the same with the Chemical Weapons Convention through the 1990s. One prominent laboratory affiliated with chemical weapons research (both offensive and defensive work) was the Army Biomedical Research Lab at Edgewood Arsenal. Others in biological and infectious diseases were at Fort Detrick.

During the 1970s, the following MSC research psychologists worked at the Army Biomedical Research Lab: George Crampton, Stanley H. Holgate (1970–1977), Ronald Peterson (1972–1976), Thomas Bennett (1973–1977), Agu Pert (1973–1976), Thomas Bevin (1974–1977), James M. King (1977–1981), John H. McDonough (1975–1985), Daniel L. Rickett (1976–1982), David M. Penetar (1977–1981), Donald B. Headley (1978–1983), and James A. Romano (1978–1982). These officers worked on animal and human experiments studying various chemical threats, key pharmacological countermeasures, and behavioral research issues. This was prior to the Biomedical Research Lab being taken over by the Army Medical Department, circa 1979, with some assignment overlap extending into the time when the Medical Research Institute of Chemical Defense came into being.

In a major shift of emphasis (July 1979 through the early 1980s), DOD reestablished the Biomedical Research Lab as a part of the Office of the Army Surgeon General. It became a subcommand of USA-MRMC, with a directive to serve as the lead laboratory for research in medical chemical defense for DOD. This initiative was part of the Army's transition from development of offensive chemical weapons, mostly under the auspices of the Army Materiel Command's Chemical

Systems Laboratory, also located at Edgewood, to significant research into *medical defense* against chemical battlefield threat agents. Thus, the mission and tenor were altered in favor of medical research to protect our forces against enemy use of chemical threat agents. In May 1981, the lab was redesignated the U.S. Army Medical Research Institute of Chemical Defense.

The main thrust at the institute is first to develop antidotes and medical countermeasures to chemical warfare agents, but also to inform military medical policymaking regarding the preparation of combatants for a fight in a chemically contaminated battlefield. Additionally, the lab is tasked to train medical personnel in the management of chemical casualties. Whereas most of the research findings immediately go into DOD medical decisionmaking, some published references documenting work done at USAMRICD labs are scattered through the biochemical literature, but they are not commonly found in the psychological research literature per se.

As a key part of multifaceted medical research programs at USAMRICD, MSC research psychologists join other colleagues in both government and nongovernment agencies to conduct numerous programs and experiments using either animal or human models to develop or evaluate pretreatment prophylaxis, treatment drugs, and medical antidotes for administration to Soldiers who might be exposed to battlefield chemical threats. The uniformed research psychologists who worked at USAMRICD tend to be experimental physiological or pharmacological psychologists, or neuroscientists, with backgrounds in comparative behavioral biology and biological toxicology.

Memorable MSC psychologists who worked at the institute after it was formed in 1979–1980 include these officers, some of whom served a second or third tour at this laboratory: John Frazier Glenn (1981–1985), Charles Burdick (1982–1985), Mary Z. Mays (1984–1987), Carl A. Castro (1988–1992), Daniel L. Rickett, Harold E. Modrow (1983–1986), James A. Romano, James M. King (1995–1998), and Maurice Sipos (1997–present). James A. Romano was Commander of USAMRICD from 1995 to 2000, making him the fourth MSC research psychologist to serve as an MRMC laboratory Commander.

Letterman Army Institute of Research

At the Presidio of San Francisco, the Army Surgeon General formed the Western Medical Research Laboratory in 1966 to carry out research in tropical medicine, nutrition, surgery, blood replacement, pathology, and psychiatry. From 1971 to 1974, the Army opened four new laboratory buildings at the Presidio. The facility was renamed the Letterman Army Institute of Research, as it adjoined the Letterman Army Medical Center. LAIR was primarily a medical research center doing specialized experimentation on artificial blood, laser physics, and treatment of trauma.

A few MSC research psychologists worked at LAIR with Colonel Edwin S. Beatrice, MC, who headed up a vision research program doing studies involving ocular hazards. In particular, Dr. Beatrice's studies examined electromagnetic threats (that is, lasers) risking damage to the eyesight of Soldiers or to visual performance. In 1974, when the Army Medical Research Lab at Fort Knox closed, at least three of its uniformed MSC psychologists, Andre J. Lloyd, Peter A. O'Mara, and Bruce C. Leibrecht, relocated to LAIR to continue doing visual research in Beatrice's division. Other MSC research psychologists who worked on assignment at LAIR included George Luz, James H. Patterson, Samuel Shiflett, Eugene Grossman, and Gregory Lewis.

Both the Letterman hospital center and the research institute were deactivated in 1995 as part of the Base Realignment and Closure. Portions of LAIR's visual science program were moved to the Brooks City Base, Texas, where they were made part of a tri-Service research program on ocular hazards. More recently, Michael W. Boye, assigned to the detachment there, has been developing and validating new ocular diagnostic tools for field use to permit rapid battlefield triage of eye injuries, including those from lasers.

U.S. Army Human Engineering Laboratory

The HEL was founded at Aberdeen Proving Ground in 1951 as a part of the Army's extensive materiel development community, known most recently as the Army Materiel Command. For the past decade, HEL, now known as the Human Research Engineering Directorate, has been a subcommand of the Army's corporate lab, the Army Research Laboratory headquartered at Adelphi, Maryland. From the beginning, under the directorship of Dr. John D. Weisz, a DAC research psychologist, HEL devoted one-third of its efforts to field experiments measuring the performance of Soldiers operating new developmental hardware systems. Another third was devoted to ensuring appropriate application of human engineering design guidance for development and testing of hundreds of Army materiel systems over the decades. Importantly for this treatise, one-third of HEL programs consisted of basic psychological research (program 6.1 funding) on human sensory input and handling of information. It is in this later section of work at HEL where two and sometimes three MSC uniformed research psychologists were assigned to work.

After their initial 4-year tours at the lab, George R. Price and Lynn C. Oatman stayed on at HEL as DACs, where they continued to manage and conduct highly productive basic science programs in auditory and vision research. A third officer, Murray Levine, did stress research with monkeys in the mid to late 1960s, leaving after one tour. Other uniformed psychologists fulfilled a single 4-year assignment at HEL before leaving Government service, and still others went on to military assignments where they established their own programs. Such was the case with Lawrence E. Symington, a specialist in psychological reaction time studies, who worked as an MSC research psychologist at HEL (1969–1972), doing work on computer display symbology for application in Army missile command and control centers. After his tour, Dr. Symington moved to the Natick RDEC, where after a 4-year tour, he stayed on as a DAC and became a prominent human factors researcher and member of the Natick Army Labs management staff.

James M. King was assigned at HEL from 1986 to 1991 and worked in applied research, focusing on military performance in simulated operational conditions. For a time King served as Director of HEL's basic research program, and for almost 2 years he was Deputy Director of HEL.

Gradually, during the 1980s and early 1990s, the Army Medical Department assigned fewer MSC research psychologists directly to HEL. However, the lab continued an active collaborative relationship with uniformed psychologists in the USAMRMC labs largely because of the Army's need for health hazard assessment data inputs on major developmental systems (tanks, helicopters, anti-aircraft gun systems, missiles, and so forth) as part of the MANPRINT program and for providing medical inputs into the system acquisition decisionmaking forums at the Department of the Army staff level. This medical research relationship with HEL was solidified by Headquarters, USAMRMC, when, circa 1979, MRMC began assigning a medical liaison staff officer to work part-time at HEL to ensure input of medical research findings into the Army Medical Department's portion of the health hazard assessments of developing the Army's major weapons systems. John D. LaMothe, MSC, a pharmacist, was the first such MRMC medical liaison to HEL.

The second and third MRMC medical liaison officers to the Army's human engineering programs, and therefore to HEL, were MSC research psychologists Bruce C. Leibrecht and Gerald P. Krueger. They teamed with occupational and preventive medicine specialists in the Army Surgeon General's office and at the Army Environmental Hygiene Agency (now the Center for Health Promotion and Preventive Medicine) to shape the health hazard assessment program, formalize it, and institutionalize it by crafting and writing Army Regulation 40–10 for the Army's Health Hazard Assessment Program for Developmental Materiel Systems. That formal program and the regulation, published first in 1983 (Ranadive et al., 1983), remain largely unchanged and stand as a key and effective domain within the Army MANPRINT program.

Natick Army Labs–Soldier Systems Center

The Army Materiel Command's Natick Research, Development, and Engineering Center (NRDEC) at Natick, Massachusetts, is a center for materiel development of Soldier clothing, individual protective equipment, parachutes, tentage, and field kitchen equipment. It is also home to DOD's entire food and nutrition research and development program. This NRDEC is called the Soldier Systems Center, but due to numerous name changes since the 1950s, this lab, the host organization at the Natick Army facility, is often simply referred to as the Natick Army Labs. Since USARIEM always has been collocated on the Natick facility, it too is often identified as one of the Natick Army labs.

Over the early decades of the MSC research psychology program, two uniformed MSC psychologists were usually assigned to the Natick RDEC. They worked as applied researchers, conducting what would be labeled as human factors testing, or in some cases they spearheaded experimental designs for field tests of Natick developmental items such as clothing, individual equipment, chemical protective gear, and other equipment items as well as participating in food preference tests for newly developed field food rations (for example, Meals Ready to Eat) and evaluations of field kitchen kits. Sometimes these officers acted as military coordinators for contractor work. Such was the case for Stanley H. Holgate, who in the late 1980s coordinated most of the work done on the extensive, well-known anthropometry survey of several thousand men and women to establish the Army's anthropometry database used to set the tariffs for most of the military clothing in DOD. This anthropometry database is also used in countless commercial applications both in the United States and abroad.

MSC psychologists who worked at NRDEC include Richard F. Johnson (1970–1972), Lawrence E. Symington (1972–1975), Earl S. Stein (1975–1977), Robert H. Stretch (1985–1986), and Stanley H. Holgate (1986–1992). Among uniformed psychologists at the Natick Army labs in the 1960s were Frederick Isgrig, Philip Friedman, Douglas Bloomquist, and Harold O. Kiess, with J. Bradley Swanson, Laurence G. Branch, Thomas L. Nichols, and F. Thomas Eggemeir in the 1970s.

Headquarters, USAMRMC Research and Development Staff Officer Positions

At least 30 uniformed MSC psychologists served as Research and Development (R&D) staff officers at Headquarters (HQ), USAMRMC (Fort Detrick). Some served two or more such tours as R&D staff officers. Most of these assignments were to act as R&D programs officers in one of the five to six Research Area Director (RAD) staffs. These personnel staffs supplied the Commanding General of USAMRMC with effective span of control over Army medical research lab programs, and over extensive intramural and extramural contract research programs in such diverse topical areas as infectious disease research and defense against biological agents, whether from naturally occurring environmental exposures or from weaponized threats; research in combat casualty care; military operational medicine research and applications; field and combat dentistry and maximal facial injury research; and medical research to protect against chemical agents. There were occasionally other select areas that warranted a singular identifiable title by virtue of either the substantial funding being invested or a high medical research priority in Pentagon circles. Additional staff positions filled by research psychologists were in the USAMRMC Research Plans and Programs office and elsewhere. Some of these MSC R&D staff officers worked in various capacities for the Headquarters, with duty in the USAMRMC's Pentagon or Surgeon General offices in the Washington, DC, area.

In these multifaceted roles, psychology officers assisted the command in the programmatic oversight of millions of dollars in research in the nine USAMRMC laboratories. The important work carried out in all of these staff officer positions involved almost daily coordination with senior Army staffers in the Pentagon, the Army Surgeon General's office, senior medical research staffers in the sister Services, and the USAMRMC subordinate laboratories' commanders, directors, and management staffs. Work included preparation of the command's short-, mid-, and long-range research plans in programmatic areas, preparation of budget estimates, formulation of requests for research funding, justification of resource requests, preparation of descriptive explanations of ongoing research programs, and preliminary results and progress reports.

As indicated earlier, many uniformed psychologists excelled at fulfilling headquarters staff officer roles. The command, the nine research laboratories, and the officers themselves all benefited greatly from their work. A chart listing many psychologists who served in R&D staff officer positions appears as an appendix to this chapter. Following are a few who left significant memories if only because they reached the pinnacle of staff officer roles within the command.

James A. Romano not only rose to the executive managerial level as Commander of USAMRICD at Edgewood (1995–2000) but subsequently served as Deputy Commander at HQ, USAMRMC at Fort Detrick, and for 6 months in 2006 as the Commander of all of USAMRMC.

John Frazier Glenn, who served a tour at USAMRICD at Edgewood and for several years as Deputy Commander of USARIEM at Natick, took on an assortment of senior staff officer positions at HQ, USAMRMC, where he spent the last dozen years of his Active-duty career. Ultimately, in his last assignment before retiring, Frazier acted as Deputy Assistant Commander for Research and Technology Programs, a position he assumed after retirement and now holds as a DAC.

Robert K. Gifford served as Research Area III Director for the Systems Hazards Research Program (1993–1995) at HQ, USAMRMC, renaming it the Military Operational Medicine Program. More recently, Carl A. Castro took on the RAD III leadership role in September 2007. As senior R&D staff officers, the position of RAD entails carrying out programmatic and budgeting oversight for many of the research efforts conducted in three to four of the labs in which MSC research psychologists work.

After one assignment to the Academy of Health Sciences, Terry M. Rauch worked as a staff officer in RAD III in the 1980s, followed by an eventful research assignment to USARIEM. For me, Terry receives the award for perfecting the role of consummate staff officer, for subsequent to his USARIEM stint, he spent his entire career in the Washington, DC, area working as a medical research staff officer in various positions at HQ, USAMRDC, and lastly in the office of the Under Secretary of Defense for Health Affairs–Force Health Protection. In these capacities, Terry served as a Pentagon front-runner for many of the research programs in which our psychologists were involved. At one point, he was Commander, U.S. Army Science and Technology Center, Europe. For years, he coached the Army marathon team and personally ran numerous marathons.

Other Lab Assignments

The mainstay common career path of alternating assignments through the USAMRMC labs or working at the two materiel development labs (HEL and Natick RDEC) was flexible enough to permit occasional assignments to other exciting organizations and locations. There have always been several other Army postings where one or two MSC research psychologists (rarely more than three at a time) were assigned.

Among the numerous opportunities for assignments during my Active-duty days were chances for uniformed MSC research psychologists to work at the Army's Combat Development Experimentation Center (CDEC) at Hunter-Liggett, Fort Ord, California, or at the Armed Forces Radiobiological Research Institute (AFRRI) at Bethesda, Maryland, and even a pair of assignments at the National Aeronautics and Space Administration (NASA) Ames Research Center in California. On occasion, one could be assigned as an exchange scientist in the sister Service labs—for example, the Navy's Aerospace Medical Research Lab at Pensacola, Florida, or the Armstrong lab collocated with the U.S. Air Force School of Aerospace Medicine at Brooks Air Force Base, San Antonio, Texas.

Research Psychology Assignments to Academia

Academic research and teaching assignments were also possible for uniformed psychologists at places like the Army's Academy of Health Sciences at Fort Sam Houston, Texas; the U.S. Army School of Aviation Medicine at Fort Rucker, Alabama; the U.S. Military Academy at West Point, New York; the U.S. Army War College at Carlisle, Pennsylvania; the Air Force Academy at Colorado Springs, Colorado; the Uniformed Services University for the Health Sciences (USUHS) at Bethesda, Maryland; and the National Defense University and the Industrial College of the Armed Forces, both at Fort Lesley McNair, Washington, DC.

Some of these special assignments involved placing an officer in an organization without an existing personnel authorization (a slot or billet) for a research psychologist. In many cases, such nontraditional assignments resulted in creation of permanent MSC research psychologist billets where none existed before. For example, due to their proven value as researchers, teachers, and leaders, new MOS 71F permanent billets were created at the U.S. Military Academy at West Point, the National Defense University, and the Army Physical Fitness Research Center at the U.S. Army War College.

Several MSC uniformed research psychologists held prominent teaching, research, and staff management positions at military academic institutions where they mentored and taught students and accomplished superb research or program management as well.

Paul T. Bartone and Mark A. Vaitkus both served in the Department of Behavioral Sciences and Leadership at the U.S. Military Academy. They not only worked with countless cadets but also found time to do research and publish their works on such topics as "Combat Exposure and Post Traumatic Stress Disorder" (Adler, Vaitkus, and Martin, 1996), "Cognitive and Personality Predictors of Leader Performance" (Bartone, Snook, and Tremble, 2002), and "Factors Influencing Small-unit Cohesion in Cadets" (Bartone, Johnsen, Eid, Brun, and Laberg, 2002). More recently, Lolita Burrell and Paul Lester have carried on the tradition at the academy, where current studies center on developing more effective mentors and leaders.

Colonel Paul T. Bartone continued this upper level academic research tradition by carrying out a Fulbright Fellowship, working predominantly with Norwegian forces in Europe while a staffer at the Industrial College of the Armed Forces. Bartone is now assigned at the National Defense University's Center for Technology and National Security Policy. For some of his works there, see his timely treatise on "Positive Meaning in Military Operations: Reflections on Abu Ghraib" (Bartone, 2005, 2008); and his thorough outline of factors involved with leaders influencing *stress hardiness* and *resilience* in small units (Bartone, 2006).

Mark A. Vaitkus served as acting Director of Research at the Army Physical Fitness Research Institute at the U.S. Army War College, where he conducted studies with leading researchers in the fields of heart disease, stress management, physiology, prostate cancer, and nutritional health as they applied to senior military leaders over 40. After graduating with honors from the Industrial College of the Armed Forces in 2007, Mark became an Associate Professor of Behavioral Science on its faculty, where he teaches the Strategic Leadership course and conducts research to improve the college's strategic leadership and executive development assessments.

A number of MSC psychologists, particularly while they were assigned at WRAIR, served in an adjunct psychology professorship role at USUHS, usually in the Department of Medical Clinical Psychology. Ross H. Pastel garnered the first full-time assignment (2008–present) for a MSC research psychologist to do research and teaching there.

After postdoctoral cross-training in clinical psychology, Robert H. Stretch served as a liaison scientist and professor from WRAIR's Department of Military Psychiatry to the USUHS Department of Psychiatry (1995). More recently, Charles A. Salter was Director of the Congressional Programs Office and then the Scientific Director of the Office of Grants and Contracts Management at USUHS; and Robert K. Gifford acted as Director of Admissions for the USUHS School of Medicine.

Mary Z. Mays was named the Air Force Academy's professor of the year (circa 1998) while she was assigned as a professor of military psychology in the Department of Behavioral Sciences and Leadership at the Academy in Colorado Springs, Colorado. G. Rufus Sessions and George Mastroianni also served several years as psychology professors on staff there.

James M. King served in the Health Care Studies and Clinical Investigations Activity at the Army's Academy of Health Sciences (1982–1986). The focus was on evaluation studies addressing stress, chemical-biological defense, patient data capture, physical fitness, and so on. At one point, Robert K. Gifford served as Chief of the Staff and Faculty Development Division at the academy. More recently, Sandie Escolas was assigned to the Office of the Dean at the academy, where she addressed institutional challenges from evaluating effectiveness of various curriculum approaches to identifying causes of early attrition from AMEDD training programs at the academy.

Sandie has also deployed to Iraq as part of an AMEDD research and field assessment team.

Anticipating the Future

The introduction to this chapter suggested that projections would be made concerning the role of uniformed research psychologists in the "war of the social scientists and beyond." While this treatise could proceed in many directions, this chapter limits discussion to just two thrusts: a simple listing of some areas where more military psychology research is needed, and some changes that might be predicted.

Areas for Subsequent Research

Traumatic brain injury. In 2007, at a Human Sciences Research Conference I coordinated for the Assistant Secretary of the Navy, then Navy Surgeon General, Vice Admiral Donald C. Arthur, described suffering his own brain injury in a motorcycle crash. In assessing his recovery, Admiral Arthur indicated he could "pass any neuropsychological test" administered to him, but he also knew somehow that his cognitive functioning had still not returned to normal. He challenged psychologists and other behavioral scientists to explain that. Secondly, he pointed out that TBIs suffered through exposure to explosive blasts of the type our forces encounter in Iraq and Afghanistan are not quite like those resulting from motorcycle crashes or football. Admiral Arthur's experience and his quest for answers to understanding TBIs provide challenges for neuropsychological research. We must determine how to assess the extent of cognitive impairment resulting from the kind of TBIs Soldiers and Marines experience from exposure to single or repeated blasts from explosions brought about by improvised explosive devices. Suitable diagnosis and recovery treatment regimens, as well as functional tests for recertification of cognitive wellness, are needed.

Augmented cognition. For 5 years, the Defense Advanced Research Projects Agency devoted significant funding to push the state of the art of research on augmented cognition (Aug Cog). Aug Cog work to date directly touches many facets of military psychology, invoking the possibility of impacting Soldier performance in a select number of operational scenarios, especially those of equipment operators faced with cognitive overload in stressful operating conditions (Schmorrow and Stanney, 2008). The Aug Cog concept is to configure an array of psychophysiological sensors to monitor the cognitive state of an operator, and then, based on those monitored signals, the machinery will make decisions to automatically provide augmentation to that operator (that is, to unburden the operator from excessive workload). While that may sound attractive, more psychological research is needed to ensure that such machine augmentation is better than human decisionmaking, including whether Aug Cog mechanisms should supersede the operator's choice to be augmented or not by automated machinery. As technological advances are made, the military applications must be validated and confirmed as providing "value added assistance" to human operators of systems in military settings.

PTSD and prevention of military suicides. These two topics are the hot issues of the day and no doubt are covered adequately elsewhere. Predictably, much more psychological research should be done on these two military stress-related maladies, which in some cases might even be linked.

Psychological aspects of combat. Robert K. Gifford (2006) reported that although we have learned much about the positive and negative psychological consequences of serving in war, there is even more we do not understand. He suggests three critical areas for future research: long-term (longitudinal) consequences for those who have been in combat, positive and growth-enhancing reactions to having been in combat, and psychological reactions to having killed other people. All three charges are right on target and need to be researched.

Preparing Soldiers for combat stress. In the "Military Life" book series, the volume on *Operational Stress* (Adler, Castro, and Britt, 2006) identifies psychological research needs that still confront us. One is identifying the best methods for preparing military personnel psychologically for operational stress. As Adler, Castro, and Britt point out, there is little empirical evidence regarding the best method for providing that support. Should training be geared toward the individual or the unit? Can military personnel benefit from better preparation for the psychological stressors encountered in combat?

Human dimensions as promulgated by U.S. Army Training and Doctrine Command. The Army's future direction takes its cue primarily from the development of doctrine at the U.S. Army Training and Doctrine Command (TRADOC)—essentially the staff cells at the Army's schools charged with predicting where, what, and how the Army will fight and envisioning what weapons and personnel will be needed. Recent developments at Headquarters, TRADOC, have witnessed a rekindling of the notion of Human Dimensions being a critical aspect of all Army doctrinal planning (see TRADOC Pamphlet 525–3–7, June 2008). TRADOC's "doctrine" defines *human dimension* as the:

> moral, physical, and cognitive components of Soldier, leader, and organizational development and performance essential to raise, prepare, and employ the Army in full spectrum operations. Army concepts acknowledge the Soldier as the centerpiece of the Army, but none, individually or collectively, adequately addresses the human dimension of future operations. This concept provides an integrating and forcing function that draws on other joint and Army concepts to describe those aspects of a highly nuanced human dimension interacting at all levels.

It is important for uniformed research psychologists and their Army civilian counterparts in the laboratories to engage and participate with Army doctrinaires at TRADOC as they flesh out those concepts and plans for the future Army.

Human diversity research for the military. Three areas of human diversity in the force call for more behavioral science research. Due to current total force strength limitations, and in part due to the fast tempo of recent military encounters overseas, all three Services have seemingly curtailed their plans of significantly increasing the number of women in the force. Nevertheless, those plans will likely come to fruition eventually. In addition to the other variables behavioral scientists study about all Soldiers, women undergo additional gender-related stressors, raising psychological and physiological issues that call for additional biobehavioral research.

Secondly, increasing numbers of Soldiers for whom English is a second language are joining the ranks. They join the Army not only for patriotic and economic reasons, but also to speed up the immigration and naturalization process as a reward for their Active-duty service. Additional psychological research needs to be done concerning integration of minority Soldiers into the Army.

Integration of Soldiers with alternative sexual preferences (gays and lesbians) into the total force has been inadequately addressed since the Revolutionary War (Shilts, 1993; Herek, Jobe, and Carney, 1996). While this is not a new concern, suggestions of impending change to the current military policy regarding Soldiers' individual sexual preferences ("don't ask, don't tell") also foretell that more psychological research is called for in topical areas pertaining to integration of Soldiers with alternative sexual orientation.

Military family stresses during deployment. Research is needed on ways to enhance the resilience of military families (see *Military Families*, Volume 3, in "Military Life" series, edited by Castro, Adler, and Britt, 2002). Such research should provide better understanding of how military spouses and children cope throughout the deployment cycle, particularly when the Servicemember is deployed to a combat zone.

Psychology of serving in the military Reserves. Volume 4 of the "Military Life" Series (Britt, Adler, and Castro, 2006) echoes Wisher and Freeman's (2006) call for a thorough understanding of the impact of increased activation on the psychology of serving in the Reserves. This includes the stresses civilian communities encounter when Reservists deploy on Active duty and when they return. Better methods are needed to facilitate reintegration of Reservists into their home and civilian work lives.

Adopting an overall health and wellness philosophy. While much psychological research has focused on military operational fitness (both physical and mental), a glaring need for new research ideas concerns the *overall health and wellness* of the troops and their families. The medical treatment communities of the three Services are spending countless dollars on medical diagnosis and treatment issues to work with not only Servicemembers, but also their dependent families, on such issues as smoking, obesity, hypertension, diabetes, and stress. If surveys of the poor health of new recruits and many military members' dependent teenagers are any indication, there are opportunities for big gains in the overall fitness of the force and additional gains through applying preventive medicine to save huge sums in military medical funding. More behavioral research is needed on health and wellness topics to prompt program Health and Wellness program actions in our total force.

General Scales's 9-point challenge. In his article, "Clausewitz and World War IV" (2006, 2009), General Scales states that we are already in World War IV, the war of the social scientist, and he addresses nine elements for victory. Each element contains specific challenges for military psychologists. Primarily, they involve a need for more psychological research. Briefly, they are:

- cultural awareness: Soldiers need empathy for foreign cultures
- building alien armies and alliances: Soldiers need to build foreign armies
- perception changing as art, not science: shaping opinions of foreign peoples
- inculcate knowledge and teach wisdom: take time to immerse in alien cultures
- tactical intelligence: leverage human sciences to determine whom to kill or not
- psychological and physiological tuning: provide psychological conditioning for close combat
- develop high-performing Soldiers and small units: selection for bonding and psychological preparation of small tactical units
- leadership and decisionmaking: develop intuitive decision-making in young, ambidextrous leaders
- intuitive battle command: teach commanders how to think and intuit rather than what to think; anticipate how the enemy will act.

71F study is needed. This chapter depicts a variety of assignments, mentions myriad research careers, and lists opportunities for uniformed research psychologists to contribute in meaningful ways. However, memories are still fresh of the early 1990s personnel downsizing threats to our Medical Service Corps and specifically to our MOS as research psychology officers. See, for example, the article by Ruck and Mitchell (1993) as they grappled with post–*Desert Storm* era questions about the need for, and the role of, uniformed behavioral scientists in the Air Force. Recall that Army personnel downsizing anticipated a so-called peace dividend after the fall of the Berlin wall in 1989, and that the dividend (cuts in troop strength) was postponed until after *Desert Storm* (early 1991). Subsequent personnel cutbacks and laboratory consolidations under Base Realignment and Closure Commission recommendations have put additional stresses on research psychology billets.

Thus, if we scan General Scales's list of challenges for World War IV, it suggests that it may be time for another analytical review of what the Army should have in the way of uniformed research psychologists going forward. As we approach what may be another downsizing of U.S. forces in both Iraq and Afghanistan, it might behoove us to reexamine which are our strongest skill sets, and we should look toward the mid- to long-range future to envision where our group's members can best meet the needs of the Army and the overall military. In a strategic planning exercise, we might want to predict where and how many of our uniformed research psychologists ought to be assigned, and what kind of work they should be prepared to conduct.

Changes for Military Research Psychologists

Jointly sponsored and conducted research projects. In the future, we will likely witness more joint laboratory cooperation in research programs and projects. There will be more cross-Service collaborative studies among the Air Force, Army, Navy, and Marines. Tri-Service–sponsored and –funded research programs will become more common. The Army's movement of armor forces from Fort Knox, Kentucky, to join the infantry at Fort Benning, Georgia, provides synergistic opportunities to conduct forward-looking performance studies combining several major combat arms communities of how the Army should prepare to fight in subsequent wars (note the merging of tank and infantry forms of warfare in counterinsurgency battles in Iraq). Our psychologists are encouraged to participate in such tri-Service and combined arms field test endeavors.

International networking and multinational studies. There will be an enhanced international network of military psychologists involved in research. More military psychologists will collaborate on multinational experiments, joint projects, and international research programs. There will be more sharing of resources, instrumentation, and data. Joint multinational studies are more difficult to coordinate and carry out, but the rewards of getting to know one's allies help to maintain working relationships that often pay off when joint military maneuvers are called for to meet operational commitments. Use of the Internet helps to facilitate some of this activity. Military psychologists will continue to be in the forefront of joint military service and multinational research programs.

Sharing results internationally. Countries with formal alliances agreeing to cooperate will sponsor more professional military psychology meetings at which researchers will present technical papers on their work and intermingle with foreign counterparts. Whereas the North

Atlantic Treaty Organization (NATO) used to be a major sponsor of cooperative research panels, boards, international meetings, and exchange forums, the prediction is that there will be more such activity from countries outside NATO that are also friendly with the United States.

Electronic communication and publications media. Omniaccess to the Internet and other forms of electronic telecommunications will make international and even cross-country collaborations more likely. More publishing of experimental and research test results will be done on the Internet. Everyone seems to want things in an easily stored digital format these days. Consequently, print media publications will likely become more scarce.

Conclusion

This chapter only scratches the surface of what could be written about our coterie of Army MSC uniformed research psychologists. Since I write with a memory bank that reflects on my departure from Active duty as far back as 1994, no doubt I have short-changed many of the more contemporary MSC psychologists in MOS 71F. Would space permit, many other vignettes concerning our psychologists could be mentioned or elaborated. There are indeed many more accomplishments of uniformed research psychologists that have received scant attention in this chapter. Nevertheless, I hope the reader now has a sense of the remarkable history and contributions of this specialty. The remaining chapters in this volume will serve as further contemporary examples of how Army research psychologists are working to protect and enhance the health and performance of our military personnel and their families.

References

Adler, A.B., Castro, C.A., and Britt, T.W. (eds.) (2006). *Military life: The psychology of serving in peace and combat. Vol. 2. Operational stress.* Westport, CT: Praeger Security International.

Adler, A.B., Vaitkus, M.A., and Martin, J.A. (1996). Combat exposure and posttraumatic stress symptomatology among U.S. soldiers deployed to the Gulf War. *Military Psychology* 8 (1), 1–14.

Banderet, L.E., and Burse, R.L. (1991). Effects of high terrestrial altitude on military performance. In Gal, R. and Mangelsdorff, A.D. (eds.). *Handbook of Military Psychology,* 233–254. New York: John Wiley and Sons.

Bartone, P.T. (2005). The need for positive meaning in military operations: Reflections on Abu Ghraib. *Military Psychology* 17 (4), 315–324.

———. (2006). Resilience under military operational stress: Can leaders influence hardiness? *Military Psychology* 18 (Supplement), 131–148.

———. (2008, April). Army uniformed research psychologists: Making a difference. *The Military Psychologist* 24, 17–22.

———. (2008). *Lessons of Abu Ghraib: Understanding and preventing prisoner abuse in military operations.* Defense Horizons 64. Washington, DC: National Defense University Press.

Bartone, P.T., Adler, A.B., and Vaitkus, M. (1998). Dimensions of psychological stress in peacekeeping operations. *Military Medicine* 163, 587–593.

Bartone, P.T., Johnsen, B.H., Eid, J., Brun W., and Labert, J.C. (2002). Factors influencing small-unit cohesion in Norwegian navy officer candidates. *Military Psychology* 14 (1), 1–22.

Bartone, P.T., Snook, S.A., and Tremble, T.R. (2002). Cognitive and personality predictors of leader performance in West Point cadets. *Military Psychology* 14 (4), 321–338.

Bartone, P.T., Ursano, R.J., Wright, K.M., and Ingraham, L.H. (1989). The impact of a military air disaster on the health of assistance workers. *Journal of Nervous and Mental Disease* 177, 317–318.

Brady, J.V. (1956). Assessment of drug effects on emotional behavior. *Science* 8, 123 (3206), 1033–1034.

———. (1958). Ulcers in executive monkeys. *Scientific American* 199 (4), 95.

Britt, T.W., Castro, C.A., and Adler, A.B. (eds.) (2006a). *Military life: The psychology of serving in peace and combat. Vol. 1, Military performance.* Westport, CT: Praeger Security International.

Britt, T.W., Adler, A.B., and Castro, C.A. (eds.) (2006b). *Military life: The psychology of serving in peace and combat. Vol. 4, Military culture.* Westport, CT: Praeger Security International.

Castro, C.A., and Adler, A.B. (2005). Operations Tempo (OPTEMPO). *Military Psychology, Special Issue* 17 (3), 131–246.

Castro, C.A., Adler, A.B., and Britt, T.W. (eds.) (2006). *Military life: The psychology of serving in peace and combat. Vol. 3, The military family.* Westport, CT: Praeger Security International.

Castro, C.A., Hoge, C.W., and Cox, A.L. (2006). Battlemind training: Building soldier resiliency. In *Human Dimensions in Military Operations—Military leaders' strategies for addressing stress and psychological support,* 42-1–42-6. Meeting proceedings RTO–MP–HFM–134, Paper 42. Neuilly-sur-Seine, France, NATO Research and Technology Organization.

Committee on Military Nutrition Research, Food and Nutrition Board, Institute of Medicine (1999). *Committee on military nutrition research: Activity report 1994–1999.* Washington, DC: The National Academies Press.

Cronin, C. (ed.) (1998). *Military psychology: An introduction.* Needham Heights, MA: Simon and Schuster Custom Publishing.

Department of the Army (2007). Army Medical Department Officer Development and Career Management. (Pamphlet 600-4). Washington, DC: Headquarters Department of the Army. Available at <www.army.mil/usapa/epubs/pdf/p600_4.pdf>.

Gal, R., and Mangelsdorff, A.D. (eds.) (1991). *Handbook of Military Psychology.* New York: John Wiley and Sons.

Gifford, R.K. (1995). Military field research in Somalia: Findings and implications for military psychology. *The Military Psychologist* (Summer 1995).

———. (2006). Psychological aspects of combat. In Britt, T.W., Castro, C.A., and Adler, A.B. (eds.). *Military life: The psychology of serving in peace and combat. Vol. 1, Military performance,* 15–30. Westport, CT: Praeger Security International.

Gifford, R.K., Marlowe, D.H., Wright, K.M., Bartone, P.T., and Martin, J.A. (November/December 1992). Unit cohesion in Operations *Desert Shield/Desert Storm. Journal of the U.S. Army Medical Department,* PB 8–92–11/12, 11–13.

Glenn, J.F., Burr, R.E., Hubbard, R.W., Mays, M.Z., Moore, R.J., Jones, B.H., and Krueger, G.P. (eds.). (1990). *Sustaining health and performance in the desert: Environmental medicine guidance for operations in southwest Asia* (USARIEM Technical Notes No. 91–1 and 91–2, pocket version). Natick, MA: U.S. Army Research Institute of Environmental Medicine.

Grimmett, R.F. (2009). Instances of use of United States armed forces abroad, 1798–2008 (Congressional Research Service Report for Congress, No. 7–5700). Washington, DC: Congressional Research Service.

Headley, D.B. (1982). Effects of atropine sulfate and pralidoxime chloride on visual, physiological, performance, subjective and cognitive variables in man: A review. *Military Medicine* 147, 122–132.

Herek, G.M., Jobe, J.B., and Carney, R.M. (eds.) (1996). *Out in force: Sexual orientation and the military.* Chicago: The University of Chicago Press.

Horvath, S.M., and Horvath, E.C. (1973). The *Harvard Fatigue Laboratory: Its history and contributions.* Englewood Cliffs, NJ: Prentice-Hall, Inc.

Hursh, S.R., and Silberberg, A. (2008). Economic demand and essential value. *Psychological Review* 115 (1), 186–198.

Ingraham, L.H., and Manning, F J. (1981). Cohesion: Who needs it, what is it, and how do we get it to them? *Military Review* LXI (6), 2–12.

———. (1984). *Boys in the barracks: Observations on American military life.* Philadelphia, PA: Institute for the Study of Human Issues.

Kennedy, C.H., and Zillmer, E.A. (eds.) (2006). *Military psychology: Clinical and operational applications.* New York: The Guilford Press.

Kobrick, J.L., and Johnson, R.F. (1991). Effects of hot and cold environments on military performance. In Gal, R., and Mangelsdorff, D. (eds.). *Handbook of Military Psychology,* 215–232. New York: John Wiley and Sons.

Krueger, G.P. (1983). The role of the behavioral scientist in assessing the health hazards of developmental weapon systems. In Schopper, A.W., and Nowak, U.V. (eds.). *Proceedings of the Army Medical Department Behavioral Sciences R&D Conference.* Fort Rucker, AL: U.S. Army Aeromedical Research Laboratory.

———. (1986). *Publications of the Department of Behavioral Biology: 1958–1986, Walter Reed Army Institute of Research* (WRAIR–BB–Technical Report No. 86–1). Washington, DC: Walter Reed Army Institute of Research.

———. (1989). Sustained work, fatigue, sleep loss and performance: A review of the issues. *Work and Stress* 3 (2), 129–141.

———. (1991). Sustained military performance in continuous operations: Combatant fatigue, rest and sleep needs. In Gal, R., and Mangelsdorff, D. (eds.). *Handbook of Military Psychology,* 255–277. New York: John Wiley and Sons.

———. (1998). Psychological research in the military setting. In Cronin, C. (ed.). *Military psychology: An introduction,* 15–30. Needham Heights, MA: Simon and Schuster Custom Publishing.

———. (2006). U.S. Army research in human performance. In Mangelsdorff, A.D. (ed.). *Psychology in the service of national security,* 71–81. Washington, DC: American Psychological Association.

———. (2008). Contemporary and future battlefields: Soldier stresses and performance. In Hancock, P.A., and Szalma, J.L. (eds.). *Performance under Stress,* 19–44. Aldershot, UK: Ashgate Publishing Co.

Krueger, G.P., and Babkoff, H. (1992). Stimulants to ameliorate sleep loss during sustained operations. *Military Psychology, Special Issue* 4 (4), 189–287.

Krueger, G.P., and Banderet, L.E. (1997). Effects of chemical protective clothing on military performance. *Military Psychology, Special Issue* 9 (4), 251–415.

Krueger, G.P., Cardinal, D.T., and Stephens, M.E. (1992). *Publications and technical reports of the United States Army Research Institute of Environmental Medicine (1961–1991)*. Natick, MA: U.S. Army Research Institute of Environmental Medicine.

Mangelsdorff, A.D. (ed.) (1999). Military Cohesion: Special Issue. *Military Psychology* 11 (1), 1–128.

———. (ed.) (2006). *Psychology in the service of national security*. Washington, DC: American Psychological Association.

McDonough, J.H. (2002). Performance impacts of nerve agents and their pharmacological countermeasures. *Military Psychology* 14 (2), 83–177.

Parsons, H.M. (1972). *Man-machine system experiments*. Baltimore, MD: The Johns Hopkins University Press.

Pastel, R.H. (2001). Proceedings of the international conference on the operational impact of psychological casualties from weapons of mass destruction. July 25–27, 2000. Supplement to *Military Medicine* 166 (12), 1–91.

Penetar, D.M., and Beatrice, E.S. (1986). Effects of atropine sulfate on human pursuit tracking performances. *Aviation, Space, and Environmental Medicine* 57, 654–658.

Ranadive, M.W., Fitz, R.J., Krueger, G.P., Wolfe, H., and Jones, L. (1983). *U.S. Army Regulation 40-10: Health hazard assessment program in support of the Army materiel acquisition decision process*. Washington, DC: Headquarters Department of the Army.

Romano, J.A., and King, J.M. (2002). Chemical warfare and chemical terrorism: Psychological and performance outcomes. *Military Psychology*, Special Issue 14 (2), 83–177.

Ruck, H.W., and Mitchell, J.L. (1993). Perceived need for and roles of uniformed behavioral scientists in the United States Air Force. *Military Psychology* 5 (4), 219–233.

Scales, R.H. (2009). Clausewitz and World War IV. *Armed Forces Journal*, July 2006, 15–24.

Schmorrow, D.D., and Stanney, K.M. (2008). *Augmented cognition: A practitioner's guide*. Santa Monica, CA: Human Factors and Ergonomics Society.

Shilts, R. (1993). *Conduct unbecoming: Lesbians and gays in the U.S. military, Vietnam to the Persian Gulf.* New York: St. Martin's Press.

U.S. Army Combined Arms Center and the U.S. Marine Corps Combat Development Command (2006). *Counterinsurgency* (U.S. Army Field Manual 3-24). Fort Leavenworth, KS: U.S. Army Combined Arms Center. Available at <http://usacac.army.mil/CAC/Repository/Materials/COIN-FM3-24.pdf>.

U.S. Army Training and Doctrine Command (2008). The U.S. Army concept for the human dimension in full spectrum operations: 2015–2024 (TRADOC Pamphlet 525-3-7, June 11, 2008). Fort Monroe, VA: U.S. Army Headquarters, Training and Doctrine Command.

Vaitkus, M., and Griffith, J. (1990). An evaluation of unit replacement on unit cohesion and individual morale on the U.S. Army all-volunteer force. *Military Psychology* 2 (4), 221–239.

Vogel, J.A., and Gauger, A.K. (1993). *An annotated bibliography of research involving women: Conducted at the U.S. Army Research Institute of Environmental Medicine* (USARIEM Technical Note No. 93–5). Natick, MA: U.S. Army Research Institute of Environmental Medicine.

Williams, H.L., Lubin, A., and Goodnow, J.J. (1959). Impaired performance with acute sleep loss. *Psychological Monographs* 484, 73 (14), 1–26.

Wisher, R.A., and Freeman, M.W. (2006). The U.S. Reserve component: Training strategies for adapting to deployment. In Britt, T.W., Adler, A.B., and Castro, C.A. (eds.), *Military life: The psychology of serving in peace and combat. Vol. 4, Military culture* (81–96). Westport, CT: Praeger Security International.

Wiskoff, M.F. (1997). Defense of the nation: Military psychologists. In Sternberg, R.J. (ed.), *Career paths in psychology: Where your degree can take you* (245–268). Washington, DC: American Psychological Association.

Wiskoff, M.G. (1998). Careers in military psychology. In Cronin, C. (ed.), *Military psychology: An introduction* (257–267). Needham Heights, MA: Simon and Schuster Custom Publishing.

Part II

Protecting and Enhancing Cognitive Performance

Chapter 2

Army Research Psychology: Moving from Science to Solutions

Steven R. Hursh

This is a story about one person's career as an Army research psychologist. The theme of this story is not the person but the opportunity —how an opportunity at the right time with the right team can turn possibility into probability, and probability into product.

Upon graduation from Wake Forest University, I was commissioned as a product of the Army Reserve Officers' Training Corps. I was an Armor Corps lieutenant at the height of the Vietnam War in 1968. However, I wanted to go to graduate school in psychology and applied for a delay of Active duty to pursue further study. Perhaps because I was less dangerous in the laboratory than on the battlefield, my request was granted. I attended the University of California, San Diego, and specialized in animal learning and conditioning under Edmund Fantino. After finishing 4 years later, I was granted a transfer to the Medical Service Corps.

Fantino knew Fred Hegge of the Division of Neuropsychiatry at the Walter Reed Army Institute of Research (WRAIR), and with the help of Captain Frank Sodetz, I found a congenial assignment to the Division of Neuropsychiatry, Department of Experimental Psychology, where I joined the animal research laboratory looking at problems of drug abuse and treatment. The Department of Experimental Psychology was founded by Joseph V. Brady around 1950, probably making him the first person to be designated as an Army research psychologist, a separate pursuit from clinical psychology. His organization at WRAIR has the distinction of having been home to many noted research psychologists, including Murray Sidman, Richard Herrnstein, Philip Hineline, John Boren, Larry Stein, Elliot Hearst, Hal Williams, Elliot Valenstein, William Hodos, Norman Krasnegor, Dennis Kelly, Joe Sharp, Sol Steiner, Ed Schein, Bill Stebbens, Bernie Beer, and George Koob, to name a few. For me, it was an opportunity to use the tenets of basic conditioning and choice to address a problem of pressing concern to the military—namely, the many Vietnam veterans returning home with an addiction to heroin.

My next 18 years were dedicated to related research with primates and the development of what has become known as behavioral economics, a framework for understanding the strength of reinforcement and the conditions that lead to behaviors of excess. But this essay will focus on the last 5 years of my Army career and the development of a product that has become the basis of a solution to cognitive fatigue, a common problem facing millions of civilians around the world. For me, there is nothing better than being part of an organization that is driven not simply by a thirst for knowledge but also by a thirst for solutions.

The Medical Research and Development Command Modeling Initiative

During the latter stages of my Army career, I was asked to represent the Army Medical Research and Development Command—now the Medical Research and Materiel Command (MRMC)—not as a research psychologist, but as a career scientist with a general interest in promoting the products and achievements of the command. I was not alone in this type of assignment, and several other research psychologists had paved the way, notably Fraser Glenn and Dan Rickett. During one of three tours in those hectic halls, I was tapped to lead a project seeking to model the effects of medical products as force multipliers—aids to Soldier health and performance that have a material effect to improve the performance capabilities of military units. The rationale for this initiative was to put medical products on a similar footing with other materiel, which nearly always enjoys the benefits of battlefield modeling to illustrate its impact on lethality, sustainability, or force protection. In May 1989, a generous budget was provided to develop these various models in conjunction with a Science Applications International Corporation (SAIC) team led by Rich McNally and supported by statistician Joe Fanzone.

Perhaps the largest share of our budget was dedicated to modeling the improvements in unit effectiveness afforded by the use of chemical and biological medical defense products. We modeled protection offered by nerve agent pretreatments and therapies and biological threat agent vaccines (Hursh and McNally, 1993). However, of greater interest to me as a psychologist were products coming from the Military

Operational Medicine program. Specifically, heat stress prediction and management products from the Army Institute for Environmental Medicine and sleep and fatigue prediction and management products from WRAIR caught my attention (Hursh and McNally, 1993). Both programs were seeded with many influential research psychologists, including Jerry Krueger, Curt Graeber, Dave Penetar, and later Ross Pastel, Paul Bliese, and William Killgore.

The initial modeling platform for all these studies was the Army Unit Resilience Analysis (AURA) program developed by the Ballistics Research Laboratory at Aberdeen Proving Grounds. The code was massive and could only be executed on a supercomputer. My role was therefore supervisory, with all the coding done by SAIC. The very first model of the effects of sleep deprivation on Army unit effectiveness came into being under that contract, building on some initial fatigue code developed by Terry Klopcic in 1989 that was subsequently translated by Rich McNally into a FORTRAN computer program in 1993 for ease of experimentation. This code allowed us to demonstrate that the productivity of artillery units was maximized when they had 8 hours of sleep a day—enough to be fully rested. These gains in productivity outweighed the losses caused by devoting extra time to sleep. This model was later abandoned based on certain fundamental contradictions with laboratory research results from WRAIR on chronic sleep restriction (Belenky, Wesensten, Thorne, Thomas, Sing, Redmond et al., 2003). However, the success of that initial model as a tool to translate basic fatigue research results into operational productivity metrics laid the foundation for additional research and modeling that continues today.

The WRAIR Sleep Research Program and Modeling of Sleep and Performance

I retired from Active duty 2 years after the AURA fatigue modeling work was completed and in 1995 joined SAIC to lead their biomedical modeling and analysis program. We continued to conduct modeling in support of many of the MRMC research areas, but my technical involvement focused on fatigue modeling. Of particular importance to this story was the long tradition of sleep research at WRAIR going back to the early 1950s, with studies stimulated by reports of sleep deprivation torture during the Korean War. Williams, Lubin, and Goodnow (1959) reported on studies of sleep deprivation at WRAIR and the resulting acute impairment of performance, which they characterized as due to lapses or "brief periods of no response accompanied by extreme drowsiness and a decline in EEG alpha amplitude." This portrayal of the physiological impact of sleep loss remains current in textbooks for sleep medicine.

In keeping with the tradition of sleep research for the Soldier, WRAIR focused on developing products to counter the effects of acute sleep loss during continuous or sustained operations. For instance, WRAIR contracted with SAIC to develop a sleep and performance model that could be built into a wrist-worn activity monitor to provide the Soldier and unit leaders immediate real-time feedback on Soldier capability. I met with the scientists at WRAIR, including Lieutenant Colonel Greg Belenky (Medical Corps) and Tom Balkin. We discussed the various lines of research and the limitations of prior modeling. It was noted that one of the implications of the AURA model was that anyone lacking the required 8 hours of sleep per day would eventually exhaust their performance resources and degrade to zero effectiveness. While no direct laboratory research was available to test that implication, it was the opinion of the sleep experts that while restricted sleep degraded performance, equilibrium was eventually reached and performance did not degrade to zero. It was known that Army units getting 4 or 5 hours of sleep were degraded but not totally incapacitated. This set of observations and deductions has turned out to be pivotal for both the eventual research agenda and the shape of fatigue modeling.

First, we realized that fatigue modeling was limited by the lack of necessary data. We had ample information on baseline performance when individuals were fully rested and extensive data on performance when individuals were totally deprived of sleep for as long as 84 hours. What we lacked were data on "sleep doses" between 8 and 0 hours per day. All modeling had to interpolate between those extremes with no guidance on the effects of partial sleep deprivation or sleep restriction. This crucial knowledge gap led to a line of research that continues today to document the effects of various patterns of sleep restriction on cognitive performance. Several crucial experiments at WRAIR and the University of Pennsylvania ultimately changed the basic mechanics of fatigue modeling (Belenky et al., 2003; Van Dongen, Maislin, Mullington, and Dinges, 2003).

Second, we realized that similar to many other physiological processes, fatigue models needed to be homeostatic. During the winter of 1996, I began to sketch out the basic mathematical structure of a homeostatic fatigue model in a simple spreadsheet calculator. That spreadsheet model evolved many times but ultimately became the basis of the WRAIR Sleep and Performance Model (SPM). It was delivered to the Army the following spring and implemented in a wrist activity monitor that could detect and score sleep, calculate performance effectiveness, and report to the Soldier on an embedded display. The concept was later the subject of several Army patents.

From Vision to Application

The initial model was then validated against data from various studies conducted by WRAIR and other labs and was found to have surprising generality and utility. At that point, we did not have the results of studies using chronic sleep restriction, but hypothetical experiments with the model suggested reasonable predictions that would be tested in planned experiments.

Due to the Army's continued interest in SPM, the Natick Research and Development Center sponsored a follow-on contract to have the SPM further extended to be included in the Integrated Unit Simulation System (IUSS). Over the next several years, SPM evolved further and became the Sleep, Activity, Fatigue, and Task Effectiveness (SAFTE) model—a three-process biomathematical simulation of the combined effects of sleep pattern, time of day, and sleep inertia. As the name implies, available estimates from wrist activity monitors continued to be a prime source of sleep information to drive the model. This work culminated in a Natick technical report that documented the equations and parameters of the model in early 1998 (Hursh, 1998). That report subsequently became the basis for a patent now held by the Army.

The initial military application of a sleep and performance model rapidly expanded into a broader vision of a model that could be used in the civilian environment for fatigue estimation associated with work schedules. In the spring of 1996, I proposed to the Department of Transportation that fatigue modeling could revolutionize the management of fatigue in all modes of transport. While the vision had merit, it was premature. What was first needed was a clear validation that such a model would be useful and predictive when applied to performance in the workplace. It would be 10 years before that vision was fulfilled, but the roadmap was clear. We needed user-friendly software to implement the model, a method to estimate sleep associated with civilian work schedules, and a large epidemiological study to verify that fatigue predictions from such a model could accurately predict changes in accident risk associated with work schedules.

The Evolution of a Product

The next step was to develop software that would take the crude developmental platform for SAFTE using a spreadsheet and transform it into a user-friendly, standalone computer program. In 1999, I visited the Air Force sleep and fatigue research program at Brooks Air Force Research Laboratory. The key leaders were research psychologists Jon French and Bill Storm. I presented a series of graphics illustrating the application of the SAFTE model to a range of typical military scenarios. By coincidence, they had been experimenting with their own form of fatigue model but quickly realized that a physiological simulation like SAFTE was a superior approach. Their support contractor, NTI (under the direction of Doug Eddy), coordinated with SAIC to develop a convenient software interface and display shell for the SAFTE model (Hursh, 2001; Hursh, Balkin, Miller, and Eddy, 2004).

Soon thereafter, the impact of Army research psychology again emerged. An old colleague at WRAIR and a former uniformed research psychologist, Tim Elsmore, had been doing programming software to read, score, and display wrist activity data (actigraphs) for Ambulatory Monitoring, Inc. Elsmore was retained by NTI to develop the software implementation of the SAFTE model, a program subsequently named the Fatigue Avoidance Scheduling Tool (FAST). That program became the first Air Force implementation of the SAFTE model in 2000 and later was further advanced by investment from the Department of Transportation to allow for modeling of fatigue in civilians.

The Military Product

The Air Force version of FAST was primarily a platform for entry of duty periods and sleep episodes along a convenient timeline. The tool took that simple timeline of events and processed them through the SAFTE model to graph predicted performance, in this case, percent change in cognitive performance in a battery of tests similar to the Automated Neuropsychological Assessment Metrics (ANAM) performance battery. The user could position the cursor at any point in the schedule and get a prediction of performance that was a percent of the best performance of a fully rested person. The model provided a useful tool for flight surgeons to estimate how a projected mission would alter the capability of pilots. It became a standard tool for advising squadron commanders and to this day is taught to all flight surgeons at the School of Aerospace Medicine.

Meanwhile, the Army continued to evolve the SPM and later adopted the SAFTE model as well. In 2002, Department of Defense scientists from the Army, Navy, and Air Force agreed on the SAFTE model as the foundation for all future fatigue modeling in DOD (Hursh, Redmond, Johnson, Thorne, Belenky, Balkin et al., 2003). True to its roots in Army sleep research, FAST can draw a graph of performance lapses like those reported in the early studies of sleep deprivation in 1959. Currently in development are enhancements to overlay pharmacological countermeasures, a Web-based fatigue assessment tool for flight operations, and a Web-based tool for aviation accident investigations.

As if to punctuate continuing Army support for SAFTE, the Army Research Laboratory has sponsored work to have the model incorporated into IMPRINT Pro, its human factors model for evaluating materiel. As of 2009, the SAFTE model was an optional plug-in to that software so Army units and materiel can be evaluated under the effects of sleep deprivation or chronic sleep restriction prior to a mission.

The Spinoff to Civilian Science and Application

With the DOD product well developed, I then responded to a request for proposals to develop fatigue management tools for the Department of Transportation. The promise of SAFTE and FAST for predicting and managing fatigue in civilian transportation applications was immediately recognized by Thomas Raslear, the head of human factors research for the Federal Railroad Administration (FRA) and a former WRAIR Army research psychologist. The FRA invested in significant additions to FAST that allowed it to be used for fatigue estimation associated with work schedules.

The leap from a military tool to a civilian product required an important additional step; it was not likely that the civilian employer would have access to either wrist activity data or records of employee sleep patterns, so we had to develop a way to estimate sleep directly from the work record. Fortunately, John Pollard had done a study of railroad engineers who maintained logs of their sleep and work. From that naturalistic data, I was able to develop a computer simulation of the decision processes that were the foundation of the sleep habits of railroad engineers.

The algorithm (AutoSleep) proved surprisingly accurate in predicting the average pattern of sleep and was sufficient to provide an aggregate assessment of fatigue associated with any work schedule (Hursh and Eddy, 2005). In recent validation experiments against wrist activity records and sleep logs, AutoSleep has proven to be about 88–92 percent accurate in predicting the timing of sleep under most shift work and on-call schedules. For greatest accuracy, especially for certain challenging schedules like night duty and split shifts, the algorithm should be trained with sleep estimates from wrist activity data and then used for prospective modeling of proposed new schedules.

The FAST was subsequently enhanced with AutoSleep and became a useful tool for assessment of fatigue associated with civilian work schedules. However, the promise of this tool for broad application in the workplace was limited by the absence of operational validation. The tool could accurately predict changes in cognitive performance using laboratory cognitive tests but it had yet to be shown that it could predict performance in the workplace. More importantly, as a tool for improved safety, could it predict an elevation of accident risk when provided with only work schedule data?

The FRA noted this limitation and sponsored an aggressive study to evaluate work schedules preceding 1,400 freight railroad accidents over 2½ years. Each accident was associated with two employees in the locomotive. We processed 30-day work histories of these 2,800 employees through a special batch processor version of FAST developed by Tim Elsmore. It implemented SAFTE and AutoSleep and could read schedule data directly from an Access database. The program could process an entire 30-day history in several seconds, recording estimates from the model for every 30-minute interval of work. I was aided by Joe Fanzone, the statistician I first encountered as an Army research psychologist and who was later an SAIC colleague. When the study was complete, we had processed over 57,000 work starts and over a million 30-minute work periods.

Of the 1,400 accidents, 400 were attributed to an error by the crew. For those accidents, the SAFTE model was able to predict a highly significant relationship between decreased performance and elevated accident risk. At the most extreme levels of predicted performance degradation (fatigue), the model accurately predicted a 65 percent elevation in risk. As expected, no such relationship existed for the 1,000 accidents attributed to nonhuman causes (Hursh, Raslear, Kaye, and Fanzone, 2006). Subsequent work has shown that the model accurately predicts an 86 percent elevation in damage risk when fatigued, and a 250 percent increase in average accident cost, relative to well-rested workers. Furthermore, the model also showed that being rested was protective, reducing damage risk 57 percent and average accident damage cost 50 percent relative to the average accident.

On a parallel track, the Army was continuing its evaluation of the effects of sleep restriction. Results of that work were published (Balkin, Thorne, Sing, Thomas, Redmond, Wesensten et al., 2000), and the SAFTE model was updated with an expression that adapts the model to the prolonged effects of sleep restriction and optimized it for prediction of the most sensitive performance test, the Psychomotor Vigilance Task (Balkin, Bliese, Belenky, Sing, Thorne, Thomas et al., 2004). This feature of the SAFTE model is key to accurate applications in the civilian world, where chronic sleep restriction, not total sleep deprivation, is the dominant sleep pattern that threatens performance in the workplace. It is currently the only fatigue model with that capability, which may account for the findings of a 2004 independent peer review published by the University of Pennsylvania that judged SAFTE to have the lowest estimation error of any fatigue model (Van Dongen, 2004; see also Hursh et al., 2004).

The Future of the Product

The SAFTE model has now been implemented in a range of software tools that can apply it to mass evaluation of work schedules. For example, it was used to evaluate fatigue under civilian flight schedules of over 2,000 pilots and was found to have an accuracy of 79 percent in predicting when a pilot would call in too fatigued to carry out an assigned trip. Another tool has been developed that makes it easy for an accident investigator to enter work and sleep data into a laptop program while in the field and get a comprehensive report of potential fatigue factors associated with the accident. The tool combines the predictions of the SAFTE model with a set of standard fatigue factors used by the National Transportation Safety Board. The FAST software is used by the Federal Aviation Administration, the Canadian Defence Forces, the Canadian workmen's compensation board, and a wide range of civilian industries in the United States and abroad.

The predictions of the model and the validation work with the model were important inputs to recent legislation mandating more stringent hours of service rules for the rail industry and recommending the use of scientific fatigue risk management tools like SAFTE for limiting fatigue in railroad workers. The trend in transportation and other industries is away from further prescriptive rules for limiting fatigue and toward the use of fatigue models like SAFTE to target the specific causes of fatigue and design mitigations that are more effective, economical, and practical. Fatigue models like SAFTE are a key component of fatigue risk management systems that provide objective metrics for fatigue, specific mitigations to reduce fatigue, and tools for auditing the results for continuous performance improvement.

In the last year, we have seen a radical reduction in the cost of measuring sleep with wrist activity monitors utilizing off-the-shelf accelerometers. With such devices now costing less than a tenth of what they cost just 3 years ago, it is possible to equip employees in safety-sensitive positions with a noninvasive monitor to track actual sleep patterns and drive the fatigue model with near laboratory precision. This improves the accuracy of real-time performance prediction and provides high-quality data to train AutoSleep for prospective schedule evaluation and design.

A Glance Backward and a Lesson for the Future

Looking back on this progression, we can now see that a parochial project aimed at modeling the military impact of Army research on sleep and performance has evolved into a useful product. This history illustrates how basic laboratory research can be translated into a tool for improving the safety and well-being of thousands of workers in a wide range of military and civilian environments as well as the millions of people these industries serve. Military psychologists led the way. They shared a common vision, a systematic and empirical approach to problemsolving, and a willingness to tackle difficult issues confronting human performance.

However, this is not a story about talented people. Rather, as Joe Brady often observed, this is a record of the "gifts of a fortuitous environment." Consistent with our appreciation for the control of behavior by its environment, Army research psychologists understand that we ourselves have been and will be shaped by the research environments created for us by the Army medical research establishment and, beyond our military careers, by the many civilian research opportunities that can aptly leverage our Army experiences. One need only have a vision, a tenacious adherence to that vision, and a willingness to exploit opportunities offered up by a "fortuitous environment" to take discoveries from science to solutions.

References

Balkin, T.J., Bliese, P.D., Belenky, G., Sing, H., Thorne, D.R., Thomas, M., Redmond, D.P., Russo, M., and Wesensten, N.J. (2004). Comparative utility of instruments for monitoring sleepiness-related performance decrements in the operational environment. *Journal of Sleep Research* 13, 219–227.

Balkin, T.J., Thorne, D., Sing, H., Thomas, M., Redmond, D., Wesensten, N.J., Williams, J., Hall, S., and Belenky, G. (2000). *Effects of sleep schedules on commercial driver performance* (Report No. DOT–MC–00–133). Washington, DC: U.S. Department of Transportation, Federal Motor Carrier Safety Administration.

Belenky, G., Wesensten, N.J., Thorne, D.R., Thomas, M.L., Sing, H.C., Redmond, D.P., Russo, M.B., and Balkin, T.J. (2003). "Patterns of performance degradation and restoration during sleep restriction and subsequent recovery: a sleep dose-response study." *Journal of Sleep Research* 12, 1–12.

Hursh, S.R. (2001). *Fatigue and alertness management using FAST*. Presentation at the 2001 Human Performance, Root Cause, and Trending Conference, Baltimore, MD.

Hursh, S.R., Balkin, T.J., Miller, J.C., and Eddy, D.R. (2004). The fatigue avoidance scheduling tool: Modeling to minimize the effects of fatigue on cognitive performance. *SAE Transactions* 113 (1), 111–119.

Hursh, S.R., and Eddy, D.R. (2005). Fatigue modeling as a tool for managing fatigue in transportation operations. *Proceedings of the 2005 International Conference on Fatigue Management in Transportation Operations, September 2005* (TP 14620E). Seattle, WA.

Hursh, S.R., and McNally, R. (1993). Modeling human performance to predict unit effectiveness. In Kamely, D., Bannister, K., and Sasmor, R. (eds.). *Army science: The new frontiers, military and civilian applications* (309–328). Saratoga, WY: Borg Biomedical Books.

Hursh, S.R., Raslear, T.G., Kaye, A.S., and Fanzone, J.F. (2006). *Validation and calibration of a fatigue assessment tool for railroad work schedules, summary report.* (Report No. DOT/FRA/ORD–06/21). Washington, DC: Federal Railroad Administration.

Hursh, S.R., Redmond, D.P., Johnson, M.L., Thorne, D.R., Belenky, G., Balkin, T.J., Miller, J.C., Eddy, D.R., and Storm, W.F. (2003). The DOD sleep, activity, fatigue, and task effectiveness model. In *Proceedings of the 12th Conference on Behavior Representation in Modeling and Simulation.* Scottsdale, AZ.

Hursh, S.R., Redmond, D.P., Johnson, M.L., Thorne, D.R., Belenky, G., Balkin, T.J., Storm, W.F., Miller, J.C., and Eddy, D.R. (2004). Fatigue models for applied research in warfighting. *Aviation, Space, and Environmental Medicine* 75 (3), 44–53.

Van Dongen, H.P.A. (2004). Comparison of mathematical model predictions to experimental data of fatigue and performance. *Aviation, Space, and Environmental Medicine* 75, A15–A36.

Van Dongen, H.P.A., Maislin, G., Mullington, J.M., and Dinges, D.F. (2003). The cumulative cost of additional wakefulness: Dose-response effects on neurobehavioral functions and sleep physiology from chronic sleep restriction and total sleep deprivation. *Sleep* 26, 117–126.

Williams, H.L., Lubin, A., and Goodnow, J.J. (1959). Impaired performance with acute sleep loss. *Psychological Monographs: General and Applied* 73, 1–26.

Chapter 3

Asleep at the Trigger: Warfighter Judgment and Decisionmaking during Prolonged Wakefulness

William D.S. Killgore

Clear judgment and effective decisionmaking are critical tools in the arsenal of the modern Soldier and leader, but these capacities are susceptible to many of the stresses inherent to combat operations. As the strategies and tactics employed by our enemies are changing rapidly, American Soldiers must acquire greater cognitive agility to adapt to these new environments. Urban combat requires that Soldiers make sound decisions in the face of pervasive uncertainty and fluctuating rules of engagement. These cognitive demands are compounded by the rapid implementation of new technologies that present the individual Soldier with an unprecedented stream of information that he or she must evaluate and attend to. Operational success under these conditions requires sound judgment, clear reasoning, and highly effective decisionmaking skills. Unfortunately, many of these critical decisionmaking capacities can be significantly degraded by the emotional, cognitive, and physical stresses associated with combat operations.

Perhaps one of the most cognitively degrading yet potentially modifiable stresses Soldiers face is lack of sleep. While it may not be possible to eliminate sleep loss during military operations, it is important for Soldiers, and particularly for their leaders, to understand the profoundly adverse effects that sleep loss can have on cognitive functioning—effects that can have life and death consequences. While the effects of sleep loss on simple alertness and vigilance are well documented and broadly understood, the more subtle and insidious impact on complex mental functions such as judgment and decisionmaking has only recently been explored. The present chapter introduces the problems associated with poor judgment and decisionmaking in combat, outlines two major theories of decisionmaking, provides a neurobiological model that describes the major brain regions involved in judgment and the more intuitive aspects of the decisionmaking processes, and describes how these brain systems may be disrupted by inadequate sleep.

Errors in Judgment and Decisionmaking during Combat

While errors in judgment and decisionmaking have always been a by-product of the murky confusion and emotionally charged context of war (Lieberman, Bathalon, Falco, Morgan, Niro, and Tharion, 2005), technological advances in modern weapons systems have greatly amplified the lethality as well as the financial and social cost of even minor mistakes (Wilson, Salas, Priest, and Andrews, 2007). A brief lapse in judgment or a poorly considered decision on the battlefield that might have easily gone unnoticed or had only minor consequence in previous conflicts is now likely to receive instantaneous worldwide dissemination and scrutiny due to modern media, satellite communication, and the World Wide Web. This became particularly apparent on April 17, 2002, when two U.S. Air Force pilots on a 20-hour patrol near Kandahar, Afghanistan, mistook a nighttime training exercise by the Canadian Light Infantry for threatening enemy fire. The U.S. F–16 fighter pilots, thinking they were under attack by enemy forces, requested permission to fire on the location of the ground activity. Despite being directed to "hold fire" by the airborne warning and control system aircraft overseeing the mission, one of the pilots decided to release a 500-pound laser-guided bomb on what was later discovered to be friendly forces, killing four Canadians and wounding eight. This grave error in judgment quickly received global attention and impacted U.S. international relations. As the conflicts in Iraq and Afghanistan continue, additional examples of lapses in judgment or failed decisionmaking are likely to emerge. While it is easy in hindsight to see critical errors in judgment and the pathways leading to poor decisionmaking, mental clarity is often difficult to sustain during intense conflict, when time is of the essence and a high-stakes decision must be made (Lieberman, Bathalon, Falco, Kramer, Morgan, and Niro, 2005; Lieberman et al., 2005).

Classical Approaches to Judgment and Decisionmaking

Early in their military education, junior Army officers are indoctrinated with classical theories of decisionmaking and problemsolving. These theories outline rational approaches to decisionmaking that follow a logical sequence: defining the problem, gathering relevant information, establishing goals, identifying potential solutions, systematically evaluating all possible solutions, selecting and executing a solution, and assessing the outcome. According to rational decisionmaking models, logical decisions can be arrived at by systematically developing multiple courses of action, objectively specifying evaluation criteria that bear on the problem, and then rating each potential course of action on each criterion. These ratings can then be multiplied by a rationally derived numerical weight that represents the importance of each criterion. This approach permits the decisionmaker to sum the weighted outcome values and determine the highest scoring and presumably the best course of action. Some of the advantages of the rational decisionmaking model include its thoroughness, reliability, and objectivity, which are particularly useful for novice or inexperienced decisionmakers or when there is ample time for such a meticulous, cover-all-the-bases approach (Klein, 1998). The objectivity of the approach makes it ideal for presentation in doctrinal publications and for didactic exposition in the classroom. Many critical combat and urban peacekeeping missions, however, do not lend themselves to methodical decisionmaking. A fascinating line of research has focused on the types of split-second judgments and decisions that are made by trained and experienced professionals when confronted with serious, often life threatening situations under conditions of severe time pressure (Klein, 1998). This line of research has led to the development of specialized models of how people make decisions in time-constrained crisis situations.

Naturalistic Decisionmaking

In contrast to the systematic comparative approach of the rational decisionmaking model, it appears that decisionmaking in naturalistic time-constrained situations involves the use of cognitive heuristics or timesaving strategies that quickly focus the decisionmaker toward a satisfactory (as opposed to ideal) solution (Orasanu, 2005). The *recognition primed decisionmaking* model attempts to account for the types of decisions that experienced professionals such as fire commanders, military leaders, and intensive care nurses make when faced with confusing and ambiguous situations, rapidly changing goals, and severe time pressures (Klein, 1998). Under these conditions, experts are able to rely on their vast experience to recognize patterns within each new situation as either consistent with a prototype, thus suggesting a typical course of action, or as a deviation from expectation and therefore necessitating additional diagnosis and clarification. Thus, rather than comparing multiple courses of action as in the rational decisionmaking model, an expert is led by experience to pigeonhole the current situation based on its similarity to many others encountered previously, thereby drastically reducing the relevant courses of action to be considered. Drawing from experience, the expert decisionmaker quickly identifies the typical course of action for that prototype and then rapidly conducts a mental simulation to determine whether it will succeed or fail. The course of action does not necessarily have to be the "best" possible solution; it just has to be sufficient to accomplish the task or solve the problem at hand. If the first course of action works, that solution is executed. If the mental simulation or actual implementation reveals that the first course of action is flawed, then the decisionmaker modifies the approach or switches to the next workable course of action.

Intuition is a term often used to describe the gut-level response that guides a person to make a decision or solve a problem despite the inability to verbally articulate the reasoning that led to the decision (Hayashi, 2001). The intuitive decisionmaker "just knows" when a course of action "feels right" or "feels wrong." Many cognitive psychologists now believe that intuition is not a random or supernatural process but may actually be based on well-developed cognitive heuristics refined through experience and years of repetitive exposure to similar circumstances so a sense of familiarity and typicality are formed (Slovic, Finucane, Peters, and MacGregor, 2004). Intuition has been described as a kind of well-developed pattern-matching ability that allows an expert to see when things fit a typical pattern, when typical patterns have been violated, or if a key element is absent altogether (Klein, 1998). Intuitive decisions based on pattern matching are reached more efficiently because the decisionmaker does not even need to consider most of the nonpertinent information. Like a folder with a brightly marked tab in an overstuffed file cabinet, the relevant course of action is highlighted and obvious to the expert, despite an overwhelming number of alternatives. Much of this pattern matching appears to occur automatically and outside of conscious awareness, which makes the concept of intuition seem mysterious and nonrational. Recent advances in neuroscience, however, have shown that the intuitive responses that affect our decisionmaking appear to have a neurobiological basis and are not as mysterious as once believed (Bierman, Destrebecqz, and Cleeremans, 2005).

Somatic Marker Hypothesis

One influential theory, known as the *somatic marker hypothesis*, explains how these intuitive decisionmaking processes are formed within the brain. This hypothesis posits that the gut-level reactions are essentially learned associations between environmental stimuli or life experiences and visceral somatic states that coincide with those experiences. A person, place, object, or other experience that is associated with a positive somatic emotional feeling will tend to reactivate a similar somatic state in the individual when similar stimuli or experiences are encountered again. This causes some stimuli and experiences to produce pleasant visceral sensations or somatic markers of "goodness" and others to produce unpleasant somatic markers indicating "badness." Consequently, in subsequent encounters these somatic markers can serve to bias an individual toward or away from specific stimuli or behaviors when they emerge. Evidence suggests that the associative process that forms somatic markers relies heavily on a brain region known as the ventromedial prefrontal cortex (Damasio, 1994), a sector of cerebral tissue located along the inferior and medial aspects of the brain a few millimeters behind the eyes. This brain region appears to form associational links between classes of situations (for example, public speaking or walking alone at night) and emotionally relevant somatic states encountered during such experiences (such as heart pounding, sweating, and muscle tension). Through these ventromedial prefrontal linkages, activation of components of specific classes of situations within other regions of the prefrontal cortex (perhaps the thought of walking to one's car late at night in an empty lot) will evoke a somatosensory emotional response via emotional centers such as the amygdala and insular cortex, which quickly indicate whether an object or situation is judged as good or bad (that is, a gut reaction; Damasio, 1994). When the associated somatic marker is strong, it provides the decisionmaker with a clear gut-level response as to whether an anticipated course of action is positive or negative. Somatic markers can also operate outside of conscious awareness, serving to bias a decisionmaker toward or away from a particular course of action without overt awareness of the process (Damasio, 1994). Intuitive decisionmaking is believed to rely heavily on the use of somatic markers to bias an individual away from courses of action that have previously led to negative outcomes and toward courses of action that have a higher probability of success (Bierman et al., 2005). This appears to be one basis of the intuitive reactions that streamline the decisionmaking process for expert decisionmakers.

Effects of Sleep Loss on Decisionmaking

Lack of sleep is a problematic but frequently unavoidable aspect of modern military operations. It is well known that insufficient sleep has a direct and profoundly negative impact on the ability to perform simple tasks involving alertness and vigilance (Dinges, Williams, Gillen, Powell, Ott, Aptowicz, and Pack, 1997; Glenville, Broughton, Wing, and Wilkinson, 1978; Horne, Anderson, and Wilkinson, 1983; Wesensten, Belenky, Thorne, Kautz, and Balkin, 2004), a fact that can be attested to by nearly any Soldier who has performed all-night sentry duty or been involved in continuous operations. A normal waking period for most humans is 16 to 18 hours. When waking is extended beyond this duration, the drive to sleep begins to overpower the will to remain awake, and decrements in alertness and vigilance become noticeable and progressively more severe (Wesensten, Killgore, and Balkin, 2005). Without adequate sleep, Soldiers are likely to have difficulty sustaining their attention on task, become more inconsistent in their responses, and take longer than usual to respond to stimulation or changing demands. Eventually, the pressure for sleep can become so strong that even the most motivated Soldiers will show significant lapses in attention and eventually fall asleep.

While the adverse impact that reduced alertness and attention may have on military operations cannot be overemphasized, the effects of sleep loss appear to have even more insidious and potentially far-reaching effects beyond simple decrements in alertness. Recent evidence suggests that even one night of sleep loss can impair certain types of higher level executive functions such as supervisory control (Nilsson, Soderstrom, Karlsson, Lekander, Akerstedt, Lindroth, and Axelsson, 2005); problemsolving, divergent thinking capacity (Horne, 1988; Linde and Bergstrom, 1992); verbal creativity, flexibility, inhibitory control (Harrison and Horne, 1998); and cognitive set shifting (Wimmer, Hoffmann, Bonato, and Moffitt, 1992). Results from a brain imaging technique known as positron emission tomography (PET) suggest that as little as 24 hours of sleep deprivation (one night of sleep loss) significantly reduces the metabolic activity within the prefrontal cortex

(Thomas, Sing, Belenky, Holcomb, Mayberg, Dannals et al., 2000). Moreover, sleep deprivation reduces the normal modulatory control exerted by the medial prefrontal cortex over the more primitive limbic regions of the brain, resulting in elevated amygdala responses to negative emotional stimuli during periods of inadequate sleep (Yoo, Gujar, Hu, Jolesz, and Walker, 2007). Indeed, some of the most prominent findings during sleep deprivation include altered mood (Scott, McNaughton, and Polman, 2006), emotional processing (Killgore, Kahn-Greene, Lipizzi, Newman, Kamimori, and Balkin, 2008), dysphoria (Kahn-Greene, Killgore, Kamimori, Balkin, and Killgore, 2007), and poor responses to frustrating situations (Kahn-Greene, Lipizzi, Conrad, Kamimori, and Killgore, 2006), possibly as a consequence of prefrontal limbic dysregulation. Thus, sleep deprivation not only reduces the ability to remain vigilant and stay on task, but it also directly impairs the prefrontal and associated limbic regions of the brain that are most critical to intuitive judgment and decisionmaking. Because of the ubiquitous nature of sleep loss in the military, it is imperative that the leadership fully understand the potential consequences on their own decisionmaking and judgment capacities and that of their Soldiers.

Insensitivity to Future Consequences

A recent study by Killgore, Balkin, and Wesensten (2006) examined the effect of sleep deprivation on the Iowa Gambling Task (IGT) by means of an experimental paradigm designed specifically to examine emotionally based decisionmaking (Bechara, Damasio, Tranel, and Damasio, 1997). During the IGT, subjects were presented with 4 decks of cards on a computer screen and were told to select cards, one at a time, from any of the decks; every card would win them some amount of money and some cards would also include a penalty that would lose some amount. The subjects were also told that some were "good decks" and that playing them would ultimately lead them to win money while others were "bad decks" that should be avoided or else they would ultimately lose money. The specific payout schedules of the decks were not made known to the subjects, although the program was written so that two of the decks yielded high immediate payoffs but were accompanied by relatively larger long-term penalties. Selecting predominantly from those decks would ultimately lead to a net loss over time. The other two decks yielded low immediate payoffs, but the occasional losses were very small. Selecting consistently from those two decks would lead to an overall net gain by the end of the game (Bechara, Damasio, Damasio, and Lee, 1999). Perhaps the most intriguing aspect of the IGT is the finding from early studies showing that when healthy subjects were queried about their choices midway through the game, most consciously reported that they were uncertain about their choices and had no explicit knowledge of the reinforcement conditions, despite obvious improvement in performance as the game progressed. Moreover, analyses of skin conductance responses showed that subjects were "unconsciously" learning the task and showed increased autonomic activity when deciding to select from riskier options. Together, those findings suggest that successful performance of the IGT involves nonconscious emotional learning, akin to a gut-level feeling or hunch (Turnbull, Evans, Bunce, Carzolio, and O'Connor, 2005). The IGT has been shown to discriminate patients with damage to the ventromedial prefrontal cortex from healthy controls and other groups of brain injured patients (Bechara, Damasio, Damasio, and Anderson, 1994; Bechara et al., 1997; Bechara, Tranel, and Damasio, 2000). Brain imaging studies also suggest that this task specifically activates the ventromedial prefrontal regions of the cerebral cortex (London, Ernst, Grant, Bonson, and Weinstein, 2000).

Because sleep deprivation significantly reduces glucose metabolism within the same ventromedial prefrontal regions as those required for successful completion of the IGT, we hypothesized that sleep-deprived volunteers would show a deficit in decisionmaking that would mimic the performance of patients with damage to these same brain regions (Killgore, Balkin et al., 2006). Thirty-four healthy adults were recruited to undergo several nights of sleep deprivation and to complete the IGT when fully rested at baseline and again following 49.5 hours without sleep. When tested on the IGT while fully rested, our volunteers performed in a manner similar to that seen in most samples of healthy normal individuals, learning easily to avoid the high-risk "bad" decks and shift their strategy toward selecting predominantly from advantageous, low-risk "good" decks as the game progressed. In other words, rested subjects were adept at using their intuitive emotional reactions to guide their decisionmaking. After two nights of sleep loss, however, volunteers showed a dramatic impairment of performance. Sleep-deprived individuals tended to select more frequently from risky "bad" decks as the game progressed, preferring short-term gains and ignoring the high probably of long-term losses. In other words, sleep deprivation led to impairment in the ability to use emotional experiences (somatic markers) to guide ongoing decisionmaking, leading to an increase in risky choices. This phenomenon has been described as "myopia for the future" (Bechara et al., 2000) and is a hallmark of patients with ventromedial prefrontal lobe damage. These data raise the possibility that sleep deprivation may produce a "functional" (and presumably reversible) brain lesion within the ventromedial prefrontal cortex.

A second important finding also emerged: the tendency to make risky decisions following sleep loss was significantly positively correlated with age (Killgore, Balkin et al., 2006). Participants in their late teens and twenties were less likely to make risky choices following sleep loss than those in their thirties. This latter finding may be particularly relevant to military decisionmaking, as most unit leaders and decisionmakers are likely to be older than those they command, with field grade officers typically being in their thirties or above and strategic leaders frequently being in their forties or fifties. The finding of greater sleep-deprived decisionmaking impairment with age contrasts with recent data regarding vigilance performance and subjective ratings of sleepiness. For tasks involving simple alertness, vigilance, and sleepiness, older individuals appear to be less affected by sleep loss than their younger counterparts (Adam, Retey, Khatami, and Landolt, 2006). The implications of these findings for the military are considerable. Older military leaders may "feel" less affected by sleep loss and may show objectively better vigilance in performance than their younger Soldiers. However, this may provide a false sense of security, as they may in fact be more susceptible to making critical errors in judgment when sleep deprived, a dangerous effect that might be masked by their objectively sustained vigilance and subjective sense of alertness. It is particularly important, therefore, for leaders to obtain adequate sleep in order to maximize their ability to make sound decisions regardless of their general sense of sleepiness or objective level of vigilance.

Effect of Duration of Sleep Deprivation

Some evidence utilizing PET imaging techniques suggests that glucose metabolism within the prefrontal cortex continues to decline up to at least 72 hours of sleep deprivation. Because it was unclear whether the previously described deficits in decisionmaking observed on the IGT would continue to decline with periods of sleep loss beyond 49 hours, Killgore, Lipizzi, Kamimori, and Balkin (2007) conducted a subsequent study that examined IGT performance at 51 and 75 hours of continuous wakefulness. Twenty-six healthy Soldiers were administered the IGT at baseline and completed alternate versions of the task after 2 and 3 nights of sleep deprivation. As in the previous study, well-rested Soldiers showed no difficulty learning to avoid the disadvantageous decks and progressively increasing their monetary winnings by selecting from the "good" decks. By 51 hours of sleep deprivation, however, the findings were strikingly similar to those reported in the previous study (Killgore, Balkin et al., 2006)—sleep-deprived Soldiers showed deficits in decisionmaking and tended to select more frequently from the "bad" decks by the end of the game relative to their performance at rested baseline. However, contrary to our expectations, we did not find any further decline in risky decisionmaking between 51 and 75 hours awake, with similar levels of impairment evident at both sessions. These findings suggest that sleep deprivation for 2 nights produces significant decrements in decisionmaking, but that these decrements are not significantly worsened with additional sleep loss.

Effects of Sleep Loss on Moral Judgment

When operating in hostile environments, Soldiers are likely to be confronted with difficult moral dilemmas. There may be times when a Soldier must make a split-second decision that may ultimately cost some lives in order to achieve an objective, accomplish the larger mission, or to save the lives of others. While realistic training can improve Soldiers' ability to make sound judgments under stressful situations, many other factors can influence the quality of such judgments. As noted, the ability to use emotions and somatic markers to guide intuitive decisionmaking appears to rely heavily on the functioning of the ventromedial prefrontal cortex (Damasio, 1994). This seems to be true of some types of moral judgments as well. In a particularly compelling study, Greene, Sommerville, Nystrom, Darley, and Cohen (2001) used functional neuroimaging to study neural activity in the brain as participants considered a series of narratives describing dilemmas of three different types: mundane nonmoral (NM) dilemmas, moral impersonal (MI) dilemmas that were low in emotional arousal and immediacy, and moral personal (MP) dilemmas that were high in emotional arousal and personal immediacy. In that neuroimaging study, MP dilemmas produced significantly greater activity in the ventromedial prefrontal cortex relative to either the MI or NM dilemmas, and participants were significantly slower to respond when deciding to approve of solutions requiring MP violations than when disapproving of them, a difference that was not observed for MI violations. Recent evidence further suggests that this medial prefrontal region of the brain is critical to a variety of social and emotional reasoning processes, many of which have recently been classified under the term "emotional intelligence" (Bar-On, Tranel, Denburg, and Bechara, 2003). Emotionally intelligent individuals are effective at monitoring their own emotions and the emotions of others, able to discriminate among subtle aspects of those emotions, and skilled at applying emotional information effectively to make decisions and guide behavior (Salovey and Mayer, 1990).

As described above, the ventromedial prefrontal cortex, which is so critical to emotionally intelligent behavior and the ability to form moral judgments, appears to be particularly vulnerable to prolonged sleep deprivation, which produces significant reductions of the metabolic activity within this region (Thomas et al., 2000). Unfortunately, sleep loss is common in many military operations, and Soldiers must frequently make morally relevant decisions with less than adequate sleep. Therefore, in a recent study (Killgore, 2007; Killgore et al., 2007), we examined the effects of sustained wakefulness on emotional intelligence, the types of moral judgments reached, and the speed with which such judgments were made. At rested baseline and again following 53.5 hours of continuous wakefulness, 26 healthy Soldiers were given an emotional intelligence test and then presented with a series of scenarios that included NM, MI, and MP dilemmas. For each moral scenario, a solution was proposed and the Soldier made a judgment as to whether that solution was "appropriate" or "inappropriate" within the context of the various contingencies present in the situation.

We found that measured emotional intelligence scores significantly declined after 2 nights of sleep deprivation. This change was most apparent in intrapersonal and interpersonal skills and functioning, as well as stress management abilities. This suggests that sleep-deprived Soldiers are likely to have difficulty understanding their own emotions as well as the emotions of their unit members and may be more prone to making errors in judgments that rely heavily on emotional processes. Moreover, there were significant changes in the moral judgments of Soldiers following sleep deprivation, but these changes were restricted exclusively to the highly emotionally charged dilemmas that have been shown in previous research to recruit the medial prefrontal cortex (Greene et al., 2001). In other words, sleep deprivation selectively

70 THE 71F ADVANTAGE

slowed the ability of Soldiers to make judgments about highly emotionally charged moral dilemmas but did not affect judgments about similar dilemmas with low or no emotional involvement (Killgore, 2007; Killgore et al., 2007). Finally, sleep loss led to a greater propensity to judge controversial solutions to moral dilemmas as "appropriate," but this effect appeared to be mitigated by premorbid levels of emotional intelligence. Specifically, Soldiers with normal levels of emotional intelligence were susceptible to an alteration in moral reasoning due to sleep loss, whereas highly emotionally intelligent Soldiers remained stable even when sleep deprived.

The present findings have implications for Soldiers who are required to function under conditions of highly restricted sleep in order to achieve operational objectives. During continuous operations or during periods when sleep is fragmented or restricted, military personnel may experience increased difficulty in reaching morally based decisions, particularly when the judgments evoke strong emotions or feelings of personal responsibility. Under such circumstances, Soldiers are likely to experience significant slowing of the thought processes required to reach a judgment or decision about the best course of action. Moreover, the evidence suggests that the likely course of action chosen during sleep deprivation appears to differ significantly from the choice that would have been made by the same Soldier in a fully rested state, with a tendency toward reduced empathy and degraded interpersonal awareness. Military leaders and decisionmakers need to consider the effects of sleep loss on the judgment and decisionmaking capacities in themselves and in their Soldiers.

Effects of Sleep Loss on the Willingness to Take Risks

Another aspect of warfighter judgment involves the willingness or propensity of Soldiers and leaders to take risks when faced with uncertainty. One of the major stresses of combat is the uncertainty that is present when facing a mortal enemy. Soldiers then must make judgments regarding the risks involved in each course of action or inaction. Some evidence suggests that sleep deprivation affects the way risk is perceived as well as the willingness to take risks. As described earlier, it has been demonstrated that when presented with a forced choice among several alternatives, subjects became more willing after sleep deprivation to select from tantalizing high-risk card decks that were likely to lead to long-term losses rather than from less exciting, low-risk decks that promised consistent long-term gains (Killgore, Balkin et al., 2006). A recent study by McKenna, Dicjinson, Orff, and Drummond (2007) showed that sleep deprivation alters the willingness to take risks but that this effect depends on whether the risk was perceived in terms of gains or losses. Specifically, they found that sleep-deprived subjects were more likely than usual to take a risk if they considered the outcome in terms of potential gains, but less likely to if the outcome was framed in terms of potential losses (McKenna et al., 2007). Further research using functional neuroimaging has shown that sleep deprivation enhances activation within some reward regions of the brain following risky decisions, suggesting that sleep loss may lead to an increased expectation of gains (Venkatraman, Chuah, Huettel, and Chee, 2007). Interestingly, when sleep-deprived subjects experienced losses, they showed reduced activity in brain regions associated with aversion and punishment (Venkatraman et al., 2007). Together, these findings suggest that sleep deprivation may alter the risk-related judgments of individuals by modifying the normal functional activity within brain networks involved in the evaluation of rewards and punishments.

To evaluate risk-taking propensity following sleep deprivation, Killgore (2007) administered a battery of risk-taking questionnaires and a behavioral test of risk-taking when subjects were well rested, then following 1 night of sleep deprivation, and finally after a 12-hour night of recovery sleep. The study found that relative to well-rested and post-recovery sleep periods, subjects who were awake for 1 night expressed a reduced willingness to engage in high-risk, sensation-seeking activities. Similarly, when the sleep-deprived subjects had to repeatedly press a response key to engage in a risky task to earn money, they tended to behave relatively conservatively, cashing in their winnings well before the task became risky. The common element among the tasks and questionnaires in that battery was the requirement for expending energy and putting forth effort to engage in the risky behavior. These findings, together with the preceding work described earlier, suggest that the judgment and decisionmaking processes involved in assessing risk appear to be impaired by sleep deprivation, along with the energy and motivation to act on those judgments. While speculative, it seems likely that the fatigue that normally accompanies sleep loss may actually provide a modest level of protection against the tendency to act on the poor judgment that occurs with inadequate sleep. This may have implications for the use of stimulants such as caffeine to reduce fatigue in sleep-deprived Soldiers. Specifically, if stimulants increase the propensity to take risks but do not equally restore the capacity to judge their appropriateness, poor decisions with grave consequences could result.

Effects of Caffeine

Caffeine is the most frequently used stimulant in the world and is routinely taken by individuals to sustain alertness and concentration during prolonged wakefulness (Fredholm, Battig, Holmen, Nehlig, and Zvartau, 1999). The alerting effects of caffeine are well documented, and its performance-enhancing effects include temporary improvements in alertness, psychomotor vigilance, subjective sleepiness, and mood (Childs and de Wit, 2008; Penetar, McCann, Thorne, Kamimori, Galinski, Sing et al., 1993; Wesensten et al., 2005). The alerting properties of caffeine have proven value under conditions of sustained military operations by enhancing Soldier marksmanship, alertness, and physical fitness performance (McLellan, Kamimori, Bell, Smith, Johnson, and Belenky, 2005; McLellan, Kamimori, Voss, Bell, Cole, and Johnson, 2005). Some research shows, however, that caffeine can enhance the speed at which a Soldier can engage a target but may not improve shooting accuracy (Gillingham, Keefe, and Tikuisis, 2004; Tikuisis, Keefe, McLellan, and Kamimori, 2004). The ability to respond quickly and to decisively engage a target with lethal firepower is important to military success. It is of even greater importance, however, to ensure that the decision to engage lethal force is the most appropriate option for the situation and that the correct target has been selected. These are matters of judgment and decisionmaking. While the ability of a sleep-deprived Soldier to remain alert and rapidly engage a target appears to be enhanced by caffeine, it remains uncertain whether caffeine also facilitates the judgment and decisionmaking processes of sleep-deprived Soldiers to a commensurate level.

Presently, there are very few available studies addressing whether caffeine affects judgment and decisionmaking capabilities in sleep-deprived individuals. One study recently evaluated the effectiveness of repeated doses of caffeine on the emotional decisionmaking of Soldiers on the IGT after 51 and 75 hours of sleep deprivation (Killgore, Lipizzi et al., 2007). In that study, Soldiers were administered the IGT at rested baseline and again after 2 and 3 nights without sleep. Half of the Soldiers were given four doses of caffeinated gum (200 milligrams [mg] every 2 hours for a total of 800 mg) between 1:00 a.m. and 7:00 a.m., while the other half received an identical placebo gum over the same schedule. The IGT was administered at 10:00 a.m. each session (3 hours after the last dose). The results of that study clearly showed that caffeine did not provide any significant advantage over placebo after 51 and 75 hours of sleep deprivation. Specifically, both groups showed similar patterns of degraded ability to make advantageous decisions. Instead, both groups tended to make riskier choices than they had when rested, similar to the findings of our previous study and suggesting that caffeine, in the dosing schedule used, was essentially ineffective at restoring emotion-based decisionmaking during sleep deprivation.

The effects of sleep deprivation have been examined on other aspects of judgment as well. In another study, we evaluated whether sleep-deprived subjects were able to accurately judge the humorous content of verbal statements and visual images (Killgore, McBride, Killgore, and Balkin, 2006). Humor appreciation, a highly developed cognitive-affective ability, was found to be significantly impaired following 2 nights without sleep and was not significantly altered by the administration of 600 mg of caffeine 5.5 hours earlier. Because psychomotor vigilance was sustained relative to placebo at this same time period, it was concluded that although caffeine sustained alertness and vigilance, it had no significant effect on the judgment of humor. Similarly, in another report, we found that 600 mg of caffeine had no effect beyond placebo on the ability to estimate unknown quantities, distances, weights, and durations following 2 nights of sleep loss (Killgore, Muckle, Grugle, Killgore, and Balkin, forthcoming). Interestingly, one aspect of judgment during sleep deprivation that does appear to be improved by stimulant medications such as caffeine is the ability to make subtle discriminations regarding complex emotional blends in facial expressions (Huck, McBride, Kendall, Grugle, and Killgore, 2008). In that study, subjects viewed series of photographs of facial expressions, some of which were clear exemplars of specific emotions and others that were complex blends of two emotions created using computerized image morphing software. Sleep deprivation had no effect on the ability to identify the emotions in the simple expressions, but produced a dramatic impairment in the ability to accurately judge the dominant emotion in the blended expressions. This impairment was reversed by a 600 mg dose of caffeine given 2.3 hours before the test. The effect was not specific to caffeine however, as other stimulants including dextroamphetamine (20 mg) and modafinil (400 mg) were equally effective at restoring that capacity.

Given the paucity of studies examining the effects of caffeine on judgment and decisionmaking during sleep deprivation, it is premature to draw firm conclusions. At present, however, the available evidence tentatively suggests that while caffeine is highly effective at restoring simple alertness and vigilance during prolonged wakefulness, it does not appear to provide commensurate restoration of a broad range of complex judgment and decisionmaking capabilities. Some aspects of judgment, such as complex discrimination of facial expressions of emotion, may be selectively enhanced, although it is not clear whether these effects might be due simply to the enhanced alertness provided by caffeine, which improved the ability of subjects to remain focused on a monotonous task, or to the neural enhancement of specific complex emotional judgment capacities. Clearly, further research is necessary to determine whether stimulants such as caffeine can be used to restore effective judgment and intuitive decisionmaking abilities in Soldiers during periods of prolonged wakefulness.

Conclusion

The social, emotional, and mental requirements of warfare are constantly changing. To optimize performance under these stresses, Soldiers must possess sound judgment and the ability to utilize intuitive decisionmaking skills under fire, capacities that often involve the integration of emotion and cognition and are mediated, to a large extent, by specific regions of the prefrontal cortex. The quality of judgments and decisions that emerge at a critical moment will depend on the neural integrity of these brain structures and the types of emotionally relevant training and experiences that have repeatedly molded their interconnections. A growing body of research suggests, however, that these critical capacities and the brain regions that mediate them are significantly degraded during periods of prolonged wakefulness. Without adequate sleep, Soldiers are prone to lose sight of long-term goals in favor of short-term benefits, are more likely to waver in their moral judgments and show indecisiveness during emotionally charged situations, and may demonstrate changes in their normal level of empathy, social awareness, and tolerance for risk and ambiguity. These decrements impair warfighter decisionmaking capabilities. Moreover, limited evidence to date suggests that these decrements in judgment and decisionmaking capacities may not be fully restored by commonly used stimulants such as caffeine. Thus, it is important for military leaders to understand the profoundly degrading effects that sleep loss can have on Soldier readiness and performance and make every effort to provide their Soldiers with adequate opportunities for sleep.

References

Adam, M., Retey, J.V., Khatami, R., and Landolt, H.P. (2006). Age-related changes in the time course of vigilant attention during 40 hours without sleep in men. *Sleep* 29 (1), 55–57.

Bar-On, R., Tranel, D., Denburg, N.L., and Bechara, A. (2003). Exploring the neurological substrate of emotional and social intelligence. *Brain* 126 (Pt. 8), 1790–1800.

Bechara, A., Damasio, A.R., Damasio, H., and Anderson, S.W. (1994). Insensitivity to future consequences following damage to human prefrontal cortex. *Cognition* 50 (1–3), 7–15.

Bechara, A., Damasio, H., Damasio, A.R., and Lee, G.P. (1999). Different contributions of the human amygdala and ventromedial prefrontal cortex to decisionmaking. *Journal of Neuroscience* 19 (13), 5473–5481.

Bechara, A., Damasio, H., Tranel, D., and Damasio, A.R. (1997). Deciding advantageously before knowing the advantageous strategy. *Science* 275 (5304), 1293–1295.

Bechara, A., Tranel, D., and Damasio, H. (2000). Characterization of the decisionmaking deficit of patients with ventromedial prefrontal cortex lesions. *Brain* 123 (pt. 11), 2189–2202.

Bierman, D.J., Destrebecqz, A., and Cleeremans, A. (2005). Intuitive decisionmaking in complex situations: Somatic markers in an artificial grammar learning task. *Cognitive Affective and Behavioral Neuroscience* 5 (3), 297–305.

Childs, E., and de Wit, H. (2008). Enhanced mood and psychomotor performance by a caffeine-containing energy capsule in fatigued individuals. *Experimental and Clinical Psychopharmacology* 16 (1), 13–21.

Damasio, A.R. (1994). *Descartes' error: Emotion, reason and the human brain*. New York: Grosset/Putnam.

Dinges, D.F., Pack, F., Williams, K., Gillen, K.A., Powell, J.W., Ott, G.E., Aptowicz, C., and Pack, A.I. (1997). Cumulative sleepiness, mood disturbance, and psychomotor vigilance performance decrements during a week of sleep restricted to 4–5 hours per night. *Sleep* 20 (4), 267–277.

Fredholm, B.B., Battig, K., Holmen, J., Nehlig, A., and Zvartau, E.E. (1999). Actions of caffeine in the brain with special reference to factors that contribute to its widespread use. *Pharmacological Reviews* 51 (1), 83–133.

Gillingham, R.L., Keefe, A.A., and Tikuisis, P. (2004). Acute caffeine intake before and after fatiguing exercise improves target shooting engagement time. *Aviation, Space, and Environmental Medicine* 75 (10), 865–871.

Glenville, M., Broughton, R., Wing, A.M., and Wilkinson, R.T. (1978). Effects of sleep deprivation on short duration performance measures compared to the Wilkinson auditory vigilance task. *Sleep* 1 (2), 169–176.

Greene, J.D., Sommerville, R.B., Nystrom, L.E., Darley, J.M., and Cohen, J.D. (2001). An FMRI investigation of emotional engagement in moral judgment. *Science* 293 (5537), 2105–2108.

Harrison, Y., and Horne, J.A. (1998). Sleep loss impairs short and novel language tasks having a prefrontal focus. *Journal of Sleep Research* 7 (2), 95–100.

Hayashi, A.M. (2001). When to trust your gut. *Harvard Business Review* 79 (2), 58–65, 155.

Horne, J.A. (1988). Sleep loss and "divergent" thinking ability. *Sleep* 11 (6), 528–536.

Horne, J.A., Anderson, N.R., and Wilkinson, R.T. (1983). Effects of sleep deprivation on signal detection measures of vigilance: Implications for sleep function. *Sleep* 6 (4), 347–358.

Huck, N.O., McBride, S.A., Kendall, A.P., Grugle, N.L., and Killgore, W.D. (2008). The effects of modafinil, caffeine, and dextroamphetamine on judgments of simple versus complex emotional expressions following sleep deprivation. *International Journal of Neuroscience* 118 (4), 487–502.

Kahn-Greene, E.T., Killgore, D.B., Kamimori, G.H., Balkin, T.J., and Killgore, W.D.S. (2007). The effects of sleep deprivation on symptoms of psychopathology in healthy adults. *Sleep Medicine* 8 (3), 215–221.

Kahn-Greene, E.T., Lipizzi, E.L., Conrad, A.K., Kamimori, G.H., and Killgore, W.D.S. (2006). Sleep deprivation adversely affects interpersonal responses to frustration. *Personality and Individual Differences* 41, 1433–1443.

Killgore, W.D.S. (2007). Effects of sleep deprivation and morningness-eveningness traits on risk-taking. *Psychological Reports* 100 (2), 613–626.

Killgore, W.D.S., Balkin, T.J., and Wesensten, N.J. (2006). Impaired decisionmaking following 49 hours of sleep deprivation. *Journal of Sleep Research* 15, 7–13.

Killgore, W.D.S., Kahn-Greene, E.T., Lipizzi, E.L., Newman, R.A., Kamimori, G.H., and Balkin, T.J. (2008). Sleep deprivation reduces perceived emotional intelligence and constructive thinking skills. *Sleep Medicine* 9 (5), 517–526.

Killgore, W.D.S., Killgore, D.B., Day, L.M., Li, C., Kamimori, G.H., and Balkin, T.J. (2007). The effects of 53 hours of sleep deprivation on moral judgment. *Sleep* 30 (3), 345–352.

Killgore, W.D.S., Lipizzi, E.L., Kamimori, G.H., and Balkin, T.J. (2007). Caffeine effects on risky decisionmaking after 75 hours of sleep deprivation. *Aviation, Space, and Environmental Medicine* 78 (10), 957–962.

Killgore, W.D.S., McBride, S.A., Killgore, D.B., and Balkin, T.J. (2006). The effects of caffeine, dextroamphetamine, and modafinil on humor appreciation during sleep deprivation. *Sleep* 29 (6), 841–847.

Killgore, W.D.S., Muckle, A.E., Grugle, N.L., Killgore, D.B., and Balkin, T.J. (forthcoming). Sex differences in cognitive estimation during sleep deprivation: Effects of stimulant countermeasures. *International Journal of Neuroscience.*

Klein, G. (1998). *Sources of power.* Cambridge, MA: MIT Press.

Lieberman, H.R., Bathalon, G.P., Falco, C.M., Kramer, F.M., Morgan, C.A., and Niro, P. (2005). Severe decrements in cognition function and mood induced by sleep loss, heat, dehydration, and undernutrition during simulated combat. *Biological Psychiatry* 57 (4), 422–429.

Lieberman, H.R., Bathalon, G.P., Falco, C.M., Morgan, C.A., Niro, P.J., and Tharion, W.J. (2005). The fog of war: Decrements in cognitive performance and mood associated with combat-like stress. *Aviation, Space, and Environmental Medicine,* 76 (7 Suppl), C7–14.

Linde, L., and Bergstrom, M. (1992). The effect of one night without sleep on problem-solving and immediate recall. *Psychological Research* 54 (2), 127–136.

London, E.D., Ernst, M., Grant, S., Bonson, K., and Weinstein, A. (2000). Orbitofrontal cortex and human drug abuse: Functional imaging. *Cerebral Cortex* 10 (3), 334–342.

McKenna, B.S., Dickinson, D.L., Orff, H.J., and Drummond, S.P. (2007). The effects of one night of sleep deprivation on known-risk and ambiguous-risk decisions. *Journal of Sleep Research* 16 (3), 245–252.

McLellan, T.M., Kamimori, G.H., Bell, D.G., Smith, I.F., Johnson, D., and Belenky, G. (2005). Caffeine maintains vigilance and marksmanship in simulated urban operations with sleep deprivation. *Aviation, Space, and Environmental Medicine* 76 (1), 39–45.

McLellan, T.M., Kamimori, G.H., Voss, D.M., Bell, D.G., Cole, K.G., and Johnson, D. (2005). Caffeine maintains vigilance and improves run times during night operations for Special Forces. *Aviation, Space, and Environmental Medicine* 76 (7), 647–654.

Nilsson, J.P., Soderstrom, M., Karlsson, A.U., Lekander, M., Akerstedt, T., Lindroth, N.E., and Axelsson, J. (2005). Less effective executive functioning after one night's sleep deprivation. *Journal of Sleep Research* 14 (1), 1–6.

Orasanu, J. (2005). Crew collaboration in space: A naturalistic decisionmaking perspective. *Aviation, Space, and Environmental Medicine* 76 (6 Suppl), B154–163.

Penetar, D., McCann, U., Thorne, D., Kamimori, G., Galinski, C., Sing, H., Thomas, M., and Belenky, G. (1993). Caffeine reversal of sleep deprivation effects on alertness and mood. *Psychopharmacology Series-Berlin* 112 (2–3), 359–365.

Salovey, P., and Mayer, J.D. (1990). Emotional intelligence. *Imagination, Cognition, and Personality* 9, 185–211.

Scott, J.P., McNaughton, L.R., and Polman, R.C. (2006). Effects of sleep deprivation and exercise on cognitive, motor performance and mood. *Physiology and Behavior* 87 (2), 396–408.

Slovic, P., Finucane, M.L., Peters, E., and MacGregor, D.G. (2004). Risk as analysis and risk as feelings: Some thoughts about affect, reason, risk, and rationality. *Risk Analysis* 24 (2), 311–322.

Thomas, M., Sing, H., Belenky, G., Holcomb, H., Mayberg, H., Dannals, R. et al. (2000). Neural basis of alertness and cognitive performance impairments during sleepiness. I. Effects of 24 h of sleep deprivation on waking human regional brain activity. *Journal of Sleep Research* 9 (4), 335–352.

Tikuisis, P., Keefe, A.A., McLellan, T.M., and Kamimori, G. (2004). Caffeine restores engagement speed but not shooting precision following 22 h of active wakefulness. *Aviation, Space, and Environmental Medicine* 75 (9), 771–776.

Turnbull, O.H., Evans, C.E., Bunce, A., Carzolio, B., and O'Connor, J. (2005). Emotion-based learning and central executive resources: An investigation of intuition and the Iowa Gambling Task. *Brain and Cognition* 57 (3), 244–247.

Venkatraman, V., Chuah, Y.M., Huettel, S.A., and Chee, M.W. (2007). Sleep deprivation elevates expectation of gains and attenuates response to losses following risky decisions. *Sleep* 30 (5), 603–609.

Wesensten, N.J., Belenky, G., Thorne, D.R., Kautz, M.A., and Balkin, T.J. (2004). Modafinil vs. caffeine: Effects on fatigue during sleep deprivation. *Aviation, Space, and Environmental Medicine* 75 (6), 520–525.

Wesensten, N.J., Killgore, W.D.S., and Balkin, T.J. (2005). Performance and alertness effects of caffeine, dextroamphetamine, and modafinil during sleep deprivation. *Journal of Sleep Research* 14, 255–266.

Wilson, K.A., Salas, E., Priest, H.A., and Andrews, D. (2007). Errors in the heat of battle: Taking a closer look at shared cognition breakdowns through teamwork. *Human Factors* 49 (2), 243–256.

Wimmer, F., Hoffmann, R.F., Bonato, R.A., and Moffitt, A.R. (1992). The effects of sleep deprivation on divergent thinking and attention processes. *Journal of Sleep Research* 1 (4), 223–230.

Yoo, S.S., Gujar, N., Hu, P., Jolesz, F.A., and Walker, M.P. (2007). The human emotional brain without sleep: A prefrontal amygdala disconnect. *Current Biology* 17 (20), R877–878.

Chapter 4

Advances in Measuring Higher Order Cognitive Performance in Warfighters under Environmental Stress

Gina E. Adam

Maintaining a high level of both physical and cognitive performance in the face of environmental and operational stressors has long been a challenge for military forces. These stressors include exposure to heat, cold, and high altitude as well as sleep deprivation and the potential for inadequate hydration and nutrition. In addition to these physical stressors, military operations impose increasing cognitive demands on warfighters. These stressors and challenges are neither new nor unknown. In fact, researchers have been studying the performance of Servicemembers for years and have often looked to the operational environment to provide a framework for research questions. This chapter provides an overview of the assessment of cognitive performance of military personnel, chronicling both the changes in the cognitive abilities required for successful operations and the research adaptations that have been made to assess the relevant cognitive concepts. While this topic is quite broad, the scope of this chapter is focused on work that has been conducted within the Army Medical Research and Materiel Command laboratories. Furthermore, the review provided here covers recent investigations that utilize computerized tests and performance metrics designed to assess the effects of environmental and operational stressors such as altitude, high workload, and caloric deprivation on multiple aspects of cognitive performance.

The cognitive performance of warfighters has been extensively studied over the years, often prompted by the observation of performance declines in conjunction with the cognitive demands of warfare. One early instance of anecdotal reports leading to scientific study was the observation of vigilance decrements in radar observers during World War II that led to Mackworth's (1948) investigation of this performance decline. By presenting brief and barely detectable signals, Mackworth demonstrated that sustained attention deteriorates over time.

Additional research has confirmed these early findings and shown that performance on vigilance tasks can be influenced by many factors, such as the modality of sensory input, signal salience, predictability or pace of the signal, the context or background within which critical cues are embedded, and the complexity of the vigilance task itself (Warm and Berch, 1985; Warm, Dember, and Hancock, 1996).

Scientists at the Army Medical Research and Materiel Command (MRMC) laboratories have continued this tradition of observing cognitive patterns of military personnel and studying cognitive performance under controlled conditions. Vigilance performance has frequently been studied with the use of a sentry duty paradigm showing that target detection times deteriorate over a 3-hour sentry duty vigil, with impairments clearly evident after 60 minutes (Johnson and Merullo, 1996; McBride, Johnson, Merullo, and Bartow, 2001). Decrements in sustained attention performance have also been studied through the use of a computerized vigilance task that requires participants to scan a computer screen for brief, intermittent stimuli (Scanning visual vigilance; Lieberman, Coffey, and Kobrick, 1998). Both methods of studying sustained attention have also demonstrated the efficacy of caffeine administration for sustaining vigilance performance as well as the exacerbation of vigilance decrements after the consumption of drugs with hypnotic properties, such as antihistamines (Johnson and McMenemy, 1989; Johnson and Merullo, 1996; Fine, Kobrick, Lieberman, Riley, Marlowe, and Tharion, 1994).

Yet warfighters are often beset by harsh environmental conditions and hydration and nutritional challenges. Recent work has examined sustained attention in conditions simulating operational stressors such as a double-blind study conducted with a Ranger battalion in garrison (Lieberman, Falco, and Slade, 2002a). In this study, Lieberman et al. observed significant impairments in vigilance performance during a brief (10-hour) simulated combat operation. Furthermore, cognitive performance in studies of operational stressors can be improved with carbohydrate supplementation (Lieberman et al., 2002a) and the administration of caffeine (Lieberman, Tharion, Shukitt-Hale, Speckman, and Tulley, 2002). Additionally, a laboratory study of the effects of cold and dehydration (Adam et al., 2008) found that performance on a sentry duty task deteriorated in the cold, but performance on a visual vigilance task was not impaired. These studies have shown that environmentally stressful conditions can have deleterious effects on the performance of tasks requiring vigilance.

Changing Cognitive Demands

While vigilance performance continues to be important to successful completion of operational duties by military personnel, the requirement to divide attention between two tasks has become increasingly prevalent. The observation of dual-task demands in military jobs led to new research questions, such as how well individuals could sustain attention to one task when an additional task was imposed. Using the sentry duty paradigm, McBride, Merullo, Johnson, Banderet, and Robinson (2007) manipulated both the rate of targets requiring a friend-foe decision and the need for participants to respond to a secondary auditory detection task. Results showed that the addition of a very simple auditory task to the primary task did not affect performance on any of the sentry duty measures. However, by moderately increasing the required work rate on the sentry duty task, performance improved as compared to when a low number of friend-foe judgments were required. Building on these results, Adam and Merullo (2006) increased the cognitive demands of this dual-task paradigm by increasing both the work rate of the sentry duty task and the cognitive demand of the auditory task. They again found that sentry duty performance declined when demands on this task were lower, especially when auditory task demands were high; but they also showed that performance on the auditory task declined when that task's cognitive demands were highest.

However, the cognitive demands of the battlefield are ever-changing, and while sentry duty experiments can assess sustained attention and dual-task demands, many aspects of cognitive performance relevant to warfighters are not well represented in this paradigm. In fact, it can be argued that military operations impose increasingly complex cognitive demands and that personnel must perform their tasks while using equipment that is often capable of providing a vast amount of information. Furthermore, the current battlefield environment may include such challenges as small unit patrols, exposure to asymmetric warfare, and changing mission priorities. This transforming environment requires individual warfighters to monitor multiple sources of information, develop and maintain battlefield situation awareness, quickly make judgments and decisions to take appropriate action, and manage simultaneous tasks. In addition to these cognitive demands, the operational and environmental demands still impose stress on military personnel.

As the cognitive demands of the battlefield and its tools have increased, so has our ability to study performance in laboratory simulations of operational and environmental stress. In one of the earliest acknowledgments of the shift in cognitive demands affecting military personnel, Killgore introduced executive function tests to sleep deprivation studies, enabling the assessment of higher order cognitive functions such as risk-taking, judgment, and decisionmaking (Killgore, Balkin, and Wesensten, 2006; Killgore, Lipizzi, Kamimori, and Balkin, 2007). These studies showed that decisionmaking as measured by a computerized gambling task was impaired after 50 and 75 hours of sleep deprivation and not improved by the administration of caffeine. By utilizing these tasks in studies at the Walter Reed Army Institute of Research, Killgore et al. encouraged the use of these kinds of metrics in other studies conducted by MRMC laboratories, such as the Army Research Institute of Environmental Medicine.

Assessing Executive Functions

Although there are many ways to categorize higher order cognitive performance, sometimes called executive function, the types of cognitive functions considered here are the ability to consider alternatives, evaluate risk, make decisions, and plan tasks. Many computerized tests of these cognitive functions exist, but the ones discussed here are limited to tests recently used in MRMC laboratories to assess cognitive performance in various environmental and operational stress conditions. Each test is briefly introduced before current studies and results are discussed.

The Iowa Gambling Task (IGT) (Bechara, Damasio, Tranel, and Damasio, 1997; Bechara, Damasio, Damasio, and Lee, 1999) has been used extensively to study various samples of neurologic patients (for example, those with prefrontal cortex damage, Bechara et al., 1997; ventromedial prefrontal cortex and/or amygdala damage, Bechara et al., 1999) and to examine brain region activation in normal volunteers (Bolla, Eldreth, Matochik, and Cadet, 2004; Ernst et al., 2002). The ability to adjust behavior to changing reward/punishment contingencies and to learn to identify risky choices is an important aspect of sound judgment and has been shown to be impaired in patients with lesions to the ventromedial prefrontal cortex (Bechara, Tranel, and Damasio, 2000). The IGT measures risk-taking and decisionmaking by having participants select from decks of cards that provide monetary rewards and punishments. Some decks are generally good (lower reward and lower punishment) and others are generally bad (higher reward but higher punishment). Evaluation of the participants' deck choices provides a measure of decisionmaking patterns.

Another measure of risk-taking and decisionmaking is the Balloon Analogue Risk Task (BART) (Lejuez et al., 2002), which asks individuals to fill a simulated balloon with air without letting it pop, thereby testing the individual's willingness to take risks versus "playing it safe." Points are given for maintaining the flow of air and keeping the volume of the balloon as full as possible: the more the balloon expands, the more points are earned. However, all points are lost if the balloon pops.

Decisionmaking is often conducted within a social context. The ability to detect and understand others' emotions is an important aspect in judgments and decisionmaking when interacting with others. Often, these judgments regarding others' emotions are made by determining emotion by facial expressions. By utilizing pictures of faces with different emotions displayed, the Ekman 60 and Emotion Hexagon Tests provide computerized tests of emotion detection (Young, Perrett, Calder, Sprengelmeyer, and Ekman, 2002) that allow researchers to assess simple and complex emotion judgment. Both include a series of faces (pictures of people) to which the participant must respond with a judgment regarding the emotion displayed (happiness, surprise, fear, sadness, disgust, or anger). The Ekman 60, the simpler of the two tasks, includes pictures showing the prototypical expression of the emotion (for example, the face displaying happiness is shown smiling). The Emotion Hexagon contains stimuli created by combining the facial expressions of two emotions (such as happiness and disgust), which results in a more complex emotion judgment (Young et al., 2002).

Finally, the ability to attend to multiple tasks and/or multiple sources of information is essential to the completion of many military jobs. The Synthetic Work Test (SynWin) (Elsmore, 1994) is a multitask performance assessment program that displays four simultaneous subtasks on one computer screen. This multitasking test provides a measure of participants' ability to divide attention while maintaining performance across multiple tasks. The four screen quadrants constitute a memory task, an arithmetic task, and continuous visual monitoring and auditory detection tasks. In the memory task, the volunteer memorizes a set of six letters, which soon disappear. Then, single letters intermittently appear in that same quadrant and the volunteer must indicate whether or not that particular letter was in the original set. The arithmetic task consists of the addition of two 3- or 4-digit numbers and is self-paced. In the visual monitoring quadrant, participants prevent a moving pointer (in the form of a fuel gauge) from reaching "empty" by clicking in the gauge and returning the pointer to the "full" position. The auditory detection task requires the participant to correctly recognize a particular sound (high tone), and then respond with a mouse click on a button on the screen. Low tones do not require a response. Multitask performance ability is reflected in the overall score, but individual subtasks can be independently scored.

Recent Experiments and Results

In addition to work by Killgore et al. (2006, 2007) with regard to decisionmaking and sleep deprivation, the tests outlined above have been included in several recent studies of environmental or operational stressors. One recent study was designed to evaluate the impact of caloric deprivation on cognitive performance. Military personnel engaged in field operations are often at risk of being unable to consume sufficient calories to offset their energy expenditure needs. While this study used a laboratory-based, low energy expenditure scenario, it also allowed for a double-blind, placebo-controlled energy intake paradigm whereby volunteers consumed hydrocolloid gel meals calibrated to provide either energy-balanced caloric needs or caloric deprivation. This study assessed a mixture of cognitive abilities over the 2-day period and showed that caloric deprivation did not negatively impact vigilance, four-choice reaction time, memory, or grammatical reasoning (Lieberman et al., 2008).

Caloric Deprivation and Executive Functions

However, the effects of caloric deprivation on tests of risk-taking and decisionmaking showed a different pattern of results. Caloric deprivation did not affect decisionmaking and risk-taking performance as measured by the BART. Over time, participants made more responses, ultimately popping more balloons and also increasing the average number of pumps on redeemed balloons; but there were no changes in performance related to the diets consumed (Adam, Szelenyi, Killgore, and Lieberman, 2007). Conversely, results from performance on the IGT showed that decisionmaking changes were related to the diets consumed. When looking at the rate of good deck choices (net score, number of good deck choices minus bad deck choices), results showed that participants made significantly more good choices when in the caloric deprivation diet condition than in the two full-calorie diet conditions. Figure 4–1 displays these results and shows that in the latter half of the task (approximately choices 50–100), responses made in the caloric deprivation

condition are significantly different than those in the other meal conditions, reflecting a better risk-taking strategy (Adam et al., 2007).

Figure 4–1. **Comparison of Iowa Gambling Task Performance when
under Caloric Deprivation Diet versus Two Full-calorie
Diets**

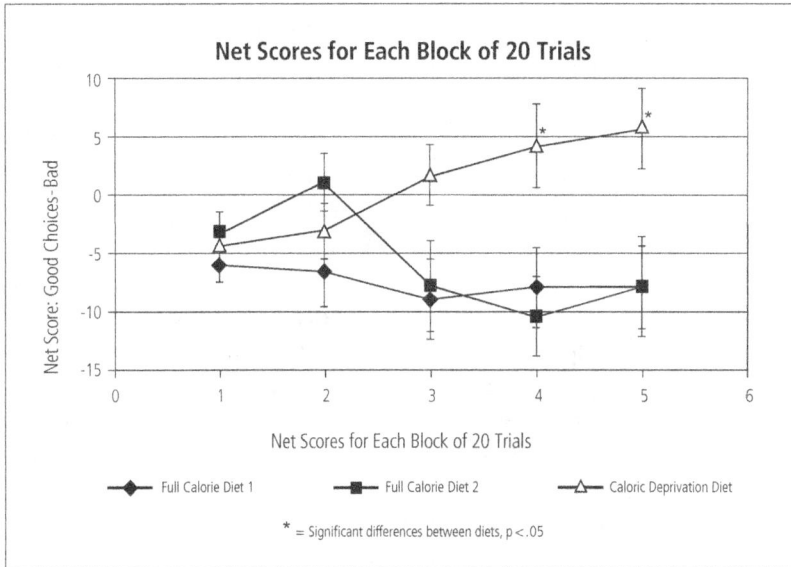

This study is one of the first to demonstrate changes in higher order cognitive performance related to a military stressor other than sleep
deprivation. It is important to note that at no time did the volunteers
know which diet they were consuming, so there is no possibility that behavior changed as a result of participant knowledge about their nutritional status. Furthermore, in the caloric deprivation conditions, interstitial glucose levels remained stable across the entire test period, never
falling below clinically acceptable levels (Lieberman et al., 2008). Thus,
we examined nutritional changes and observed a different card-choice
strategy within the calorie-deficient diet such that it appeared that
volunteers consuming a very low calorie diet adopted a more conservative strategy than those consuming full-calorie diets without knowing
their diet status. From these results, it is clear that real-world decisionmaking, judgments, and other higher order cognitive functions may be
impacted by the nutritional stress of caloric deprivation.

Executive Functions at High Altitude

Another operational stressor that is common to deployed military personnel is the need to ascend rapidly to high altitudes, which generally results in reduced oxygenation of the blood (hypoxemia) and neurological changes such as increased cerebral blood flow and hypoxia in the hippocampal and limbic systems (Virués-Ortega, Buela-Casal, Garrido, and Alcázar, 2004; Finnoff, 2008). Additionally, performance is thought to decline in complex cognitive tasks more often than simple tasks (Banderet and Shukitt-Hale, 2002). Due to the operational importance of high altitude environments, two recent projects have included assessments of higher order cognitive functions. One study evaluated multitask performance using the SynWin at sea level, moderate altitude (2,200 meters [m]), and high altitude (4,300 m, both simulated and actual). This multitask performance assessment was part of a larger study investigating the effects of staging at a moderate altitude as an intervention for the deleterious effects of high altitude, and allowed for the first known evaluation of multitask ability at high altitude. After a training period, participants engaged in a full day of tests at sea level, a day of testing at simulated high altitude (preacclimatization), a 6-day stay at moderate altitude with retraining (acclimatization period), and a final assessment period at high altitude (postacclimatization).

While this study produced a wealth of data regarding the multitask assessment and performance on the individual subtasks, initial analyses concentrated on evaluating multitask performance at sea level and high altitude, both pre- and postacclimatization. Adam, Fulco, and Muza (2008) recently presented results showing that multitask performance declined at high altitude as compared to sea level (a 7 percent decrease in performance) but that this decrement in performance was absent at high altitude after an intermediate staging period at moderate altitude (see figure 4–2). Individual task results showed that auditory score performance was higher at sea level than high altitude at preacclimatization (18 percent decline in performance), but there were no differences in the other subtasks of the SynWin at either of the high altitude time points relative to sea level performance. Additionally, multitask performance during the moderate altitude staging period was not different from either sea level or high altitude performance, indicating a stable level of performance that reflects a positive effect of acclimatization due to staging. In fact, other measures from this study provided evidence of acclimatization, such as the decreased incidence of acute mountain sickness and increased oxygenation of the blood at high altitude after acclimatization as compared to the preacclimatization high altitude assessment period.

Figure 4–2. **Overall Task Performance on SynWin**

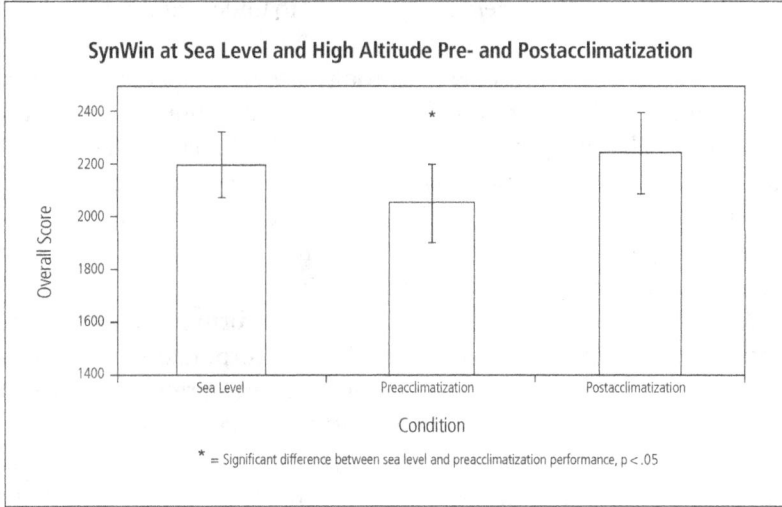

The results of this study are the first indication that multitask performance is negatively impacted by exposure to high altitude while unacclimatized. Moreover, the study was the first to demonstrate the importance of a moderate altitude staging intervention. A second study has recently furthered these results. It expanded the breadth of executive functions studied, again utilizing sea level and high altitude (4,300 m, simulated) and incorporating not only the SynWin, but also the IGT, Ekman 60, and Emotion Hexagon tests. Data from the study will provide a thorough overview of the impact of high altitude exposure on multitasking, judgments, risk-taking, and decisionmaking. The pattern of results obtained will greatly advance our understanding of complex cognitive performance in conditions of environmental stress.

Conclusion

As we have seen from recent combat operations, exposure of warfighters to environments with extremes of temperature and altitude is not only likely but commonplace. Likewise, high operational tempos, the need for sustained and continuous operations, and the challenge of providing sufficient food and water will continue to plague any fighting force. Recent research has shown that decisionmaking, risk-taking, and multitask ability are affected by sleep deprivation, caloric deprivation, and exposure to high altitude. Thus, the operational and environmental stressors on today's battlefield are important to understanding cognitive performance by military personnel.

MRMC laboratories have a strong history in the assessment of the cognitive demands of warfighters and the translation of that understanding to laboratory studies of these cognitive functions. The recent adoption of various computer-based tests of risk-taking, judgment, and decisionmaking will help these laboratories keep pace with the ever-increasing cognitive demands on today's military personnel in environmentally and operationally stressful conditions.

Future studies designed to evaluate the performance effects of operational or environmental stressors should incorporate as many tests of cognitive function as possible. From the results presented above, it is clear that more cognitively complex tasks may be more susceptible to the negative effects of stressors than simpler tasks. It is understood that cognitive demands on warriors may be multiple, complex, and dynamic, and future studies should attempt to capture this dynamism. What has not been explicitly studied is how some of these higher order cognitive functions affect each other. For example, one open question is: What happens to decisionmaking ability when an individual is already cognitively loaded by handling multiple other tasks or sources of information? It may be that an individual reduces the number of tasks attended while making the decision or that the quality of decisionmaking suffers due to attention paid to the other simultaneous tasks. Answering such questions is of paramount importance, and future work should account for these dynamic cognitive demands and assess the effects of environmental stressors on performance accordingly.

Psychologists should continue to study the cognitive performance of military members in all conditions, including environmental or operational stress. These researchers should keep an eye on the military operational environment to fully understand the ways that cognitive demands will change for military personnel. Finally, creative studies of higher order cognitive function should be designed to better understand how performance in different cognitive domains may change in the face of multiple stressors, as well as how individuals develop increased abilities in these domains with time and experience.

Acknowledgments

The research presented here is the result of several collaborative efforts between the author and other investigators at the Army Research Institute of Environmental Medicine. For these collaborations, the author is thankful to Harris N. Lieberman, Ph.D., Steve R. Muza, Ph.D., Juli Jones, Ph.D., and Chuck S. Fulco, Ph.D.

The investigators have adhered to the policies for protection of human subjects as prescribed in Army Regulation 70–25 and USAM-RMC Regulation 70–25, and the research was conducted in adherence with the provisions of 32 CFR Part 219.

References

Adam, G.E., Carter, R., Cheuvront, S., Merullo, D.J., Castellani, J.W., Lieberman, H.R., and Sawka, M.N. (2008). Hydration effects on cognitive performance during military tasks in temperate and cold environments. *Physiology and Behavior* 93, 748–756.

Adam, G.E., Fulco, C.S., and Muza, S.R. (2008). Multitask performance at sea level and high altitude. *Proceedings of the 25th Army Science Conference.*

Adam, G.E., and Merullo, D. (2006). Sentry duty performance is robust under high cognitive load. *Proceedings of the 50th Annual Meeting of the Human Factors and Ergonomics Society.*

Adam, G.E., Szelenyi, E., Killgore, W.D.S., and Lieberman, H.R. (2007). A double-blind study of two days of caloric deprivation: Effects on judgment and decisionmaking. Aerospace Medical Association 78th Annual Scientific Meeting, New Orleans, LA.

Banderet, L.E., and Shukitt-Hale, B. (2002). Cognitive performance, mood, and neurological status at high terrestrial elevation. In Lounsbury, D.E., Bellamy, R.F., and Zajtchuk, R. (eds.). *Medical aspects of harsh environments.* Washington, DC: Office of the Surgeon General, Borden Institute, 729–763.

Bechara, A., Damasio, H., Tranel, D., and Damasio, A.R. (1997). Deciding advantageously before knowing the advantageous strategy. *Science* 275, 1293–1295.

Bechara, A., Damasio, H., Damasio, A.R., and Lee, G.P. (1999). Different contributions of the human amygdala and ventromedial prefrontal cortex to decisionmaking. *The Journal of Neuroscience* 19 (13), 5473–5481.

Bechara, A., Tranel, D., and Damasio, H. (2000). Characterization of the decisionmaking deficit of patients with ventromedial prefrontal cortex lesions. *Brain* 123, 2189–2202.

Bolla, K.I., Eldreth, D.A., Matochik, J.A., and Cadet, J.L. (2004). Sex-related differences in a gambling task and its neurological correlates. *Cerebral Cortex* 14, 1226–1232.

Elsmore, T.F. (1994). SYNWORK1: A PC-based tool for assessment of performance in a simulated work environment. *Behavior Research Methods, Instruments, and Computers* 26, 421–426.

Ernst, M., Bolla, K., Mouratidis, M., Contoreggi, C., Matochik, J.A., Kurian, V., Cadet, J.L., Kimes, A.S., and London, E.D. (2002). Decisionmaking in a risk-taking task: A PET study. *Neuropsychopharmacology* 26 (5), 682–691.

Fine, B.J., Kobrick, J.L., Lieberman, H.R., Riley, R.H., Marlowe, B., and Tharion, W.J. (1994). Effects of caffeine or diphenhydramine on visual vigilance. *Psychopharmacology (Berlin)* 114, 233–238.

Finnoff, J.T. (2008). Environmental effects on brain function. *Current Sports Medicine Reports* 7, 28–32.

Johnson, R.F., and McMenemy, D.J. (1989). Antihistamines and sentry duty: Effects of terfenadine and diphenhydramine on target detection and rifle marksmanship. *Proceedings of the 1989 Medical Defense Bioscience Review*, 823–826. Aberdeen Proving Ground, MD: U.S. Army Medical Research Institute of Chemical Defense.

Johnson, R.F., and Merullo, D.J. (1996). Effects of caffeine and gender on vigilance and marksmanship. *Proceedings of the Human Factors Society 40th Annual Meeting* 40, 1217–1221.

Killgore, W.D.S., Balkin, T.J., and Wesensten, N.J. (2006). Impaired decisionmaking following 49 h of sleep deprivation. *Journal of Sleep Research* 15, 7–13.

Killgore, W.D.S., Lipizzi, E.L., Kamimori, G.H., and Balkin, T.J. (2007). Caffeine effects on risky decisionmaking after 75 hours of sleep deprivation. *Aviation, Space, and Environmental Medicine* 78 (10), 957–962.

Lejuez, C.W., Read, J.P., Kahler, C.W., Richards, J.B., Ramsey, S.E., Stuart, G.L. et al. (2002). Evaluation of a behavioral measure of risk taking: The balloon analogue risk task (BART). *Journal of Experimental Psychology: Applied* 8, 75–84.

Lieberman, H.R., Caruso, C.M., Niro, P.J., Adam, G.E., Kellogg, M.D., Nindl, B. et al. (2008). A double-blind, placebo-controlled test of 2 d of calorie deprivation: Effects on cognition, activity, sleep, and interstitial glucose concentrations. *American Journal of Clinical Nutrition* 88, 667–76.

Lieberman, H.R., Coffey, B., and Kobrick, J. (1998). A vigilance task sensitive to the effects of stimulants, hypnotics, and environmental stress: The scanning visual vigilance test. *Behavior Research Methods, Instruments and Computers* 30 (3), 416–422.

Lieberman, H.R., Falco, C.M., and Slade, S.S. (2002a). Carbohydrate administration during a day of sustained aerobic activity improves vigilance, as assessed by a novel ambulatory monitoring device, and mood. *American Journal of Clinical Nutrition* 76, 120–127.

Lieberman, H.R., Tharion, W.J., Shukitt-Hale, B., Speckman, K., and Tulley, R. (2002). Effects of caffeine, sleep loss, and stress on cognitive performance and mood during U.S. Navy SEAL training. *Psychopharmacology* 164, 250–261.

Mackworth, N.H. (1948). The breakdown of vigilance during prolonged visual search. *Quarterly Journal of Experimental Psychology* 1, 6–21.

McBride, S.A., Johnson, R.F., Merullo, D.J., and Bartow, R.E. (2001). *Effects of intermittent stimuli on marksmanship and vigilance during simulated sentry duty*. Technical Report T01-7, AD No. A390246. Natick, MA: U.S. Army Research Institute of Environmental Medicine.

McBride, S.A., Merullo, D.J., Johnson, R.F., Banderet, L.E., and Robinson, R.T. (2007). Performance during a 3-hour simulated sentry duty task under varied work rates and secondary task demands. *Military Psychology* 19 (2), 103–117.

Virués-Ortega, J., Buela-Casal, G., Garrido, E., and Alcázar, B. (2004). Neuropsychological functioning associated with high-altitude exposure. *Neuropsychology Review* 14, 197–224.

Warm, J.S., and Berch, D.B. (1985). Sustained attention in the mentally retarded: The vigilance paradigm. In Ellis, N.R., and Bray, N.W. (eds.). *International review of research in mental retardation, Volume 13*, 1–41. Orlando, FL: Academic Press.

Warm, J.S., Dember, W.N., and Hancock, P.A. (1996). Vigilance and workload in automated systems. In Parasuraman, R., and Mouloua, M. (eds.). *Automation and human performance: Theory and applications* (183–200). Hillsdale, NJ: Erlbaum.

Young, A.W., Perrett, D., Calder, A., Sprengelmeyer, R., and Ekman, P. (2002). *Facial expressions of emotion: Stimuli and test*. Bury St. Edmunds, England: Thames Valley Test Co.

Chapter 5

Rapid Assessment of Eye Injuries on the Battlefield

Michael W. Boye

Soldiers' eyes are their most important sensors on the modern battlefield. The risk of eye injuries from laser radiation and other causes has risen substantially with the increased use of several types of laser-based weapons systems. The absence of expertise and the unavailability of ophthalmic instruments in the field have rendered tools to screen or assess visual function deficits an essential and growing requirement. However, nearly all of the visual assessment methodologies currently available have been designed for use exclusively in the eye care clinic. The need for tools that provide much-needed visual information in the field of battle and other settings outside of the clinic was the impetus for developing the Aidman Vision Screener (AVS) (NSN 6549 NCM040061).

The AVS is a black-and-white vision screening measure designed specifically to assist in triaging retinal abnormalities in settings lacking advanced vision assessment equipment. It has several advantages, including portability, ease of use, and nearly immediate assessment results. However, color vision and visual function under low contrast conditions are not currently assessed by the AVS. An effort is under way to develop an upgraded version that can assess these visual abilities, thus increasing sensitivity and accuracy. This chapter discusses the impetus for this developmental effort, the methods used to create the specialized eye charts for this tool, and the initial validation research for this new metric. This tool will enhance the capabilities of medical personnel to quickly triage and treat eye injuries occurring in battlefield settings, thus increasing the operational effectiveness of the U.S. military.

Introduction

The use of directed energy systems (for example, lasers) on the modern battlefield has become prevalent. Lasers are widely used by armed forces for purposes such as range finding, target designation, weapon guidance, detection and safe detonation of landmines and bombs, detection of battlefield contaminants, slowdown of approaching vehicles, and as training devices. This has led to significant numbers of laser eye injuries (Harris, Lincoln, Amoroso, Stuck, and Sliney, 2003). Further, a new class of weapons recently has emerged in which one or more laser beams is used to damage electro-optical sensors.

Most of these sensors operate within the visible and near-infrared (IR) spectrum and are vulnerable within the same region of the electromagnetic spectrum. The retina is the most susceptible organ of the eye to this class of weapons because the cornea and lens focus light on it, thus greatly magnifying the effect of a laser on this very small area. The deployment of lasers during combat that are designed to cause permanent blindness of unenhanced vision has been prohibited by international laws (Anderberg, Bring, and Wolbarsht, 1992; Doswald-Beck, 1996). Nonetheless, it is probable that enemy forces will use these types of weapons in the future. This raises a new threat to the visual ability of military personnel and may potentially increase the number of casualties substantially. In light of the increased use of lasers on the battlefield and the heightened probability of their use by enemy forces, tools that can diagnose visual function have become an urgent need. The plethora of lasers on the modern battlefield, along with the documentation of several incidents and injuries involving lasers as tools of battle, provide the impetus for the development of an upgraded tool for assessing and triaging retinal injuries and abnormalities in field settings where no advanced ophthalmological equipment is available.

Lasers on the Modern Battlefield

The application of lasers in combat situations and settings has broadened considerably the last several years. Laser range finders and designators have been a part of military inventories for decades. More recently, laser illuminators, detection systems, and training devices have been added and are frequently used among military forces, thus increasing risks to the eyes. Many of these laser applications have been classified at dangerous levels, according to the Army Radiation Safety Program (Department of the Army, 1999), which utilizes the ANSI–Z136.1–2007 laser standard (Laser Institute of America, 2007) to measure the safety of laser equipment. This standard is also well accepted among leaders in the field of laser bioeffects.

The following is an overview of several types of laser systems being used on the modern battlefield. This review is far from exhaustive; many other laser systems are currently being used in battlefield settings. The purpose here is to provide a rudimentary understanding of the diversity of lasers and laser systems that may pose a threat to the modern Soldier and to demonstrate the need for a visual function screening tool that can be used in the absence of advanced ophthalmological equipment.

Range Finders

Laser range finders were the first military application of the laser as a tool of war (Neuenswander, 2001). A laser range finder is a laser light source that uses a laser beam to determine the distance to a reflective object. It has two basic parts, the laser and a laser receiver. After the laser is fired, the object it hits reflects the laser energy. Some of this dispersed energy is reflected back toward the laser and is detected by the receiver, which processes the speed that the laser energy took to travel to the target and return. Distance to the target is then computed based on the speed of the laser light and the duration of time it traveled. This process is completed quickly and accurately, and it substantially increases the probability that first shots will be hits for many military weapons. However, unintentional blindness can result from accidents involving laser range finders, and several retinal injuries have occurred because of these tools. Injuries involving these range finders can take place in a number of ways, including unintended use of the tools (for example, looking into the aperture where the beams exit), careless adjustment, and viewing an active range finder from a downrange position.

An example of an accident involving a laser range finder is provided by Manster, Stuck, and Brown (2004). A Soldier viewing the exit aperture of a laser range finder he was holding exposed his right eye to several powerful laser pulses in the near infrared range (that is, a 1,064-nanometer [nm] wavelength). He reported no pain but noticed an immediate decrease in vision in that eye. A professional examination showed that blood that had leaked into the vitreous humor of his eye covered two retinal holes in the right fovea. Five days after the laser exposure, the eye contained three prominent lesions. Central scarring progressed, and visual acuity 18 months after the incident was 20/400 (with his right eye, the Soldier could see at 20 feet what other people could see at 400).

Target Acquisition Designation Sight. The Target Acquisition Designation Sight (TADS) laser range finder was built specifically for the Apache AH–64A Longbow Attack Helicopter (Northrop Grumman,

2008). It enables the Longbow *Hellfire missile system* and other laser-guided munitions to precisely locate and lock onto targets. The TADS contains stabilized electro-optical sensors and a laser target designator (discussed below), along with the laser range finder.

AH–1S laser range finder. The AH–1S laser range finder is a laser system mounted on the nose of the AH–1 Cobra helicopter. It can cause eye damage at distances of 6.3 kilometers (km) when viewed with no optical equipment and 40 km when viewed through magnifying optical instruments.

The AH–1S laser system utilizes a neodymium-doped yttrium aluminum garnet laser, which operates at a wavelength of 1,064 nm. This wavelength is invisible to the human eye under most conditions, but is still within the range of wavelengths that can damage a human retina (approximately 400 to 1,400 nm). The AH–1S is a high-powered laser that can cause considerable damage, including permanent blindness, to the eyes of Soldiers.

Designators

A laser designator is a laser range finder with enhanced targeting capabilities. In addition to the normal functions of a laser range finder, laser designators can be used to guide weapons to a target (Neuenswander, 2001). Laser designators provide targeting for laser-guided bombs, missiles, and other weaponry. To use a designator, the operator shines the laser on a target during the last seconds of the laser-guided weapon's flight. The weapon is then guided by the laser energy reflected from the target.

Because laser designators are laser range finders, they pose all the same risks to personnel as laser range finders (laser exposure due to unintended use, careless adjustment, or viewing a projected laser beam from a downrange position). They also present further dangers to the eyes of Soldiers and others due to possible inaccurate illumination of the target area. When this happens, civilians and other nonparticipants in the battle may be unintentionally exposed to the laser beam of the designator.

The AN/AAS–38A laser designator system is designed to be mounted to the F/A–18 Hornet fighter jet (Federation of American Scientists Military Analysis Network, 1998). It allows U.S. Navy and Marine Corps pilots to attack ground targets day or night with a precision strike capability, and enhances the Hornet's night attack capability by providing real-time TV-like forward looking infrared (FLIR) thermal imagery on a cockpit display. The AN/AAS–38A FLIR can be fully integrated with other Hornet avionics, and data from the unit are used for the calculation of weapons release solutions. The system assists in the delivery of both laser-guided and conventional weapons.

Illuminators

Visible and infrared laser illuminators are used routinely in Operations *Iraqi Freedom* and *Enduring Freedom* to illuminate or point out targets (in much the same way a laser pointer is used in a lecture hall) or areas of interest in the area of operations. Visible illuminators have also been used effectively in an optical warning or deterrence mode to warn or discourage potential combatants from approaching fixed military assets such as the ends of convoys. Laser illuminators are mounted on vehicles and selected weapons systems, and thousands of handheld laser illuminators have been issued to Soldiers.

Both visible light laser illuminators and infrared laser illuminators are present on the battlefield. Visible light laser illuminators generally operate at either red wavelengths (632 to 670 nm) or a green wavelength (532 nm). They are typically used to temporarily blind parties who are approaching a fixed point, such as an entry control point to a restricted area. These illuminators are generally used to send a visual warning and make it difficult for the parties to move closer. Infrared laser illuminators operate in the near infrared range (wavelengths of approximately 700 to 1,400 nm), which is invisible to the human eye except at the very short wavelength end of the range. Lasers operating within this wavelength, however, can be harmful to the human retina, depending on factors such as power and pulse length. Infrared laser illuminators are designed to be detected by weapons systems and by parties wearing night vision goggles. They are usually used for precise weapon aiming and target acquisition at long ranges.

Green handheld optical surveillance and targeting system (GHOST). The GHOST laser system is a day/night green laser system with a variety of applications. It combines four green lasers in a small package that can be rail mounted. When fitted with a scope, it adds an optical augmentation capability that can detect optical systems. The laser source for the GHOST is currently doubled neodymium: yttrium aluminum garnet lasing at 532 nm. The system also has multiple modes of operation. Models vary slightly, but generally speaking the GHOST includes a day, a night, and one or more strobe or pulse modes. In the day mode, all four of the lasers fire continually. In an earlier model of GHOST, the night mode was pulsed; in later models, two of the four lasers fire continually. Various combinations of strobe modes, varying by model, are generated with varied numbers of the lasers firing in pulses.

Helios. The Helios laser system is also a day/night green laser system designed as a warning system for approaching parties. The Helios contains seven individual 532 nm lasers; each has a specified output of 70 milliwatts. The lasers are diode pumped (with 808 nm) neodymium: yttrium vanadate crystal with a potassium titanium oxide phosphate crystal, which doubles the frequency resulting in a 532 nm wavelength. The seven lasers are configured in a circle with six lasers surrounding the seventh in the center. In the day mode, all seven lasers of the Helios are on, whereas only two of the lasers are on in the night mode.

Green laser-baton illuminator. The green laser-baton illuminator is a low powered diode pumped, solid-state green light (532 nm) handheld laser illuminator designed to induce glare and possibly flashblindedness, both of which are temporary effects. The eye hazard zone of this laser tool is only 1.4 meters. It is very safe outside of this zone because of a large beam divergence and long pulse duration. However, it can cause permanent eye damage if the laser beam is stared into at close range. Many of these illuminators have been issued to Soldiers.

Effects of Laser Exposure

Exposure or even suspected exposure to a laser may have several adverse effects, including severe vision problems, skin burns, and psychological reactions. While skin is susceptible to laser damage, the greatest risk of damage is to the eye, because the human eye increases the irradiance or radiant exposure of laser light with parallel waves (that is, collimated laser light) at the retina. For visible and near-infrared laser exposure of the human eye, the radiant exposure at the retina can be 100,000 times greater than that at the cornea or the skin surrounding the eye due to focusing or imaging by the eye. Laser emission is generally well collimated; that is, the diameter of the beam increases very little with distance (low divergence). Thus, the energy contained in the beam diminishes only slightly over great distances. When taken in combination, low divergence of a laser emission and the increased radiant exposure due to ocular focusing means that low-powered lasers, such as range finders, pose little hazard to the skin at short ranges, but pose significant eye hazards at tactical ranges. Optical instruments such as binoculars or day sights increase light-collecting capabilities, thereby increasing the radiant exposure. This increases the range at which eye injuries can occur. For visible lasers, this focusing results in seeing extremely bright light at distances that exceed anticipated eye injury ranges.

Injuries result when the energy from the laser is absorbed by various anatomical structures. The most vulnerable structure is the eye, but other structures such as the skin can also be affected. The wavelength of the laser radiation determines which structure absorbs the energy, and the power density of the laser determines the damage level.

The biological effects of laser radiation on the eyes vary with the laser's wavelength, pulse duration, and intensity. The cornea and lens focus visible and near-infrared laser radiation onto the retina where the concentrated energy directly impacts the photoreceptor cell and supporting tissue. The cornea and lens absorb ultraviolet and mid- to far-infrared laser radiation. Alteration can occur in these tissues, but the retina will be spared.

The retina is the back inside of the eye where images are formed. During laser exposure of the retina, no image is formed and all energy is simply focused to a pinpoint less than 50 micrometers in diameter. A laser exposure occurring in the retinal periphery will have a minimal effect on normal vision functions, unless large portions of the retina are involved. A laser exposure in the central retina (called the macula, which includes the fovea) can cause serious visual impairment since that is the only part of the eye where high acuity and color vision occur.

Lower Level Effects

At the lowest levels of laser energy, temporary changes in the ability to see can be produced without permanent damage to the retina. Continuous wave or repetitively pulsed visible wavelength lasers can produce veiling glare while the laser is on but disappear when the laser is off. The laser simply appears so bright that it is difficult to see anything else around it. At slightly higher energy, these same lasers can saturate the photoreceptor cells in the retina. This saturation results in afterimaging that fades after the laser is turned off. Only visible lasers will produce veiling glare and afterimages; near-infrared lasers will not produce these effects even though the laser energy reaches the photoreceptor cells. Note that these effects can also be caused by other bright light sources, such as searchlights, flares, and strobes. Further increases in laser energy levels result in irreversible retinal damage.

Longer term exposure to lower level laser radiation may cause small to significant deficits in color vision. A small amount of research evidence (Robbins and Zwick, 1980; Zwick, Bedell, and Bloom, 1974) shows that repeated exposure to lower level laser radiation (for example, a one millijoule Q-switched Ruby laser pulse) brought about fairly substantial color vision deficits that did not heal quickly with time. Color vision discrepancies induced by lower level laser exposure in this research were shown to exist 30 days and 6 months after exposure.

Ultraviolet and far-infrared radiation can injure the cornea, a condition that is painful and visually handicapping. At lower powers, this injury is primarily due to a photochemical reaction. A latency period of hours may exist between the time of exposure and the development of the corneal pathology. Minimal corneal lesions, which usually produce a decrement in visual performance, heal within a few days and generally result in a full recovery. More severe corneal lesions may scar.

Higher Level Effects

When the retina is exposed to a high-energy laser, the tissue is superheated and undergoes an explosive change of state, creating shock waves that mechanically disrupt the tissue and spread the area of damage. If more energy is introduced, the injured area will become larger. The mechanical force produced can puncture a hole through the retina and choroid, resulting in hemorrhaging and possibly leading to severe visual loss. The blood can collect beneath the photoreceptor cell layer of the retina, disturbing its contact with the retinal pigment epithelium and resulting in retinal detachment. A subretinal hemorrhage can result in the death of the photoreceptor cells, and a "spot" may form in the vision that is much larger and more disruptive to the person than the thermal burn or mechanical disruption. The blood may also move into the vitreous humor through the disrupted retina, where it may obstruct the passage of light through the eye. An extensive or centrally located hemorrhage can produce a significant loss of vision. Blood in the vitreous is absorbed very slowly, but in most cases it is absorbed. The visual impairment remains as long as the blood persists, but vision may improve to normal with absorption of the blood. Persistent vitreous hemorrhages may be removed by a complicated surgical technique called vitrectomy. This procedure may also return vision to "near normal" level if the underlying damage does not involve the central portion of the retina.

Laser injury to the retina may damage the conducting fibers (axons) of the retina, producing a visual field defect peripheral to the site of injury. Laser damage to the retinal/choroidal areas may produce brief, severe pain. A major long-term effect of laser retinal injury is a scarring process that may degrade vision for weeks or even months.

High-energy, far-infrared laser radiation is absorbed mainly by the cornea, producing immediate burns at all corneal layers. An infrared laser can produce a burn that results in immediate visual incapacitation and that may lead to corneal scarring. Very high energy can perforate the cornea; this perforation may lead to loss of the eye.

Psychological Effects

Psychological reactions to lasers and their effects can also be severe (DeVour, 1993; Fitzgerald, 1993; Nahin, 1977; Warren, 1993). The hazards of current laser range finders, target designators, and laser illuminators (from flash effects to blindness) risk substantial psychological effects. It is inevitable that some Soldiers will suffer laser injuries to their eyes. The reactions of the injured Soldiers and their comrades will depend on:

- their response to the stress of a new, silent, futuristic weapon
- their training in and knowledge about laser weapons
- the treatment they receive.

Laser injuries may be an especially stressful type of wound. Vision is one of our primary means of relating to the world around us, and the fact or prospect of being deprived of vision will be a source of strong fear. In the imagination of some Soldiers whose careers, activities, and self-image depend heavily on vision, blindness may rank high as a crippling wound that makes a person helpless and an object of pity. Yet bystanders accustomed to seeing external wounds of combat may be reluctant to accept a disabling injury that produces no outward evidence.

Several aspects of the laser threat increase mental stress relative to that produced by other weapons. The fact that laser radiation travels at the speed of light along line-of-sight gives a new urgency to the saying, "If you can be seen, you can be hit." This may produce the types of psychological stress reactions and inhibitions of combat initiative that have been described in response to sniper fire. However, unlike the sniper's bullet, most lasers produce serious injury only to the extent that the target is looking at the laser source and/or through optical equipment.

Because the danger is so specific, Soldiers may be especially inhibited from performing critical surveillance, target acquisition, and aiming tasks. This is especially likely if they have just seen their fellow Soldiers suffer the effects of lasers while performing their tasks.

Laser training must provide the Soldier with the knowledge to protect himself (Department of the Army, 1990; Department of the Army, 1999; Laser Institute of America, 2007). He must be aware that protective equipment for certain laser frequencies is available and that additional protection is anticipated from ongoing research. Understanding lasers requires a certain amount of technical information. Now that lasers are widespread on the battlefield, many Soldiers are undoubtedly accustomed to their use. However, the Soldiers' fear of laser injury may increase as the instance of such injuries increases. Flashes of blinking lights are now fairly common on the battlefield and often must be looked at carefully to determine if they should be engaged as targets.

Calm, professional treatment at each echelon of medical care is mandatory following a laser injury, including reassurance that the injury is not life-threatening and that chances for some, if not total, recovery are good. The psychological effects of lasers could be enormous. It is imperative that secondary gain be minimized by prompt return to duty of those individuals with temporary flashblindness or noncritical (nonfoveal) burns of the retina without hemorrhage. If an error is to be made, it should be on the side of return to duty of Soldiers with questionable injuries. Medical management of stress reactions for patients suffering from real or imagined laser injuries is like stress management of other injuries. Care providers should repeat the reassurance that symptoms will improve with rest, nutrition, hygiene, and the expectancy of an early return to the Soldier's unit. Army Field Manual 8–51 outlines specific combat stress control procedures.

Strategies need to be developed to reduce the psychological impact on Soldiers by the increase in lasers on the modern battlefield. As with individual or collective chemical protective equipment, the use of protective eyewear is not a simple solution that guarantees 100 percent safety. The leader and the individual Soldier must be trained to accept necessary risk to accomplish the mission in accordance with mission, enemy, terrain, troops available, time, and civilian considerations, taking into account the risk of laser injury from both enemy and friendly lasers.

Current Assessment Approaches

Effects of exposure to laser energy range from concern and anxiety about possible exposure to hemorrhaging and loss of vision or severe skin burns. All of these effects are detrimental to some degree. Once a Soldier's eye has been exposed to directed energy, it is important to determine the extent to which tissue and visual function have been affected. The following section addresses many of the tools available to assess the physiology and functioning of the eyes.

Ocular Observation and Imaging

Ophthalmoscopy. Ophthalmoscopy is a method of examining the vitreous and fundus in great detail through an ophthalmoscope (Stamper, 1991). It is performed exclusively by the ophthalmologist and usually requires the pupils to be dilated with eyedrops to permit the viewing of a greater area of the retina and vitreous. Because pupillary dilation can alter the results of certain visual acuity and papillary tests, ophthalmoscopy is usually performed after these procedures.

Although the ophthalmoscope is a simple tool, the conditions that it can uncover are diverse. Ophthalmoscopy is often used with patients complaining of headaches to identify problems such as swollen discs. Raised intracranial pressure and arterial hypertension, which can be symptoms of serious conditions in the body, can be detected using the ophthalmoscopic technique. Further, ophthalmologists and other ophthalmic professionals can regularly assess the ocular conditions of diabetics to prevent diabetic retinopathy or diabetes-related vision loss. They can also see "cupped" optic discs, which are often seen in glaucoma cases.

The two main types of instruments used to perform ophthalmoscopic procedures are the direct and indirect ophthalmoscopes (Stamper, 1991). The direct ophthalmoscope is a hand-held tool with a handle, a concave mirror, a light source, and an eyepiece for the ophthalmic professional conducting the examination. The light-and-mirror system provides a narrow 15-times-magnified view of the fundus, and the light source is powered by rechargeable batteries located in the handle. The indirect ophthalmoscope is worn on the head of the ophthalmologist. The headset consists of a binocular viewing device and an adjustable lighting system wired to a transformer power source. The ophthalmologist holds one of a variety of magnifying lenses a few inches from the patient's eye, and the ophthalmoscope headset provides both the lighting and the stereoscopic vision for the examination. Unlike the upright (direct) view provided by the direct ophthalmoscope, the view of the fundus as seen by the physician is inverted but provides a wider, two to four times magnified, field of view.

Fundus camera. The fundus camera is, in effect, a large ophthalmoscope that can produce color photographs of the retina (Stamper, 1991). It consists of an optical system to view the retina, a light/flash system for illumination, and a camera in back. To allow the vision professional to obtain the best pictures of the retina, eyes are usually dilated with eyedrops.

Optometrists, ophthalmologists, and trained medical professionals often use the fundus camera to diagnose a disease and monitor its progress, or for screening programs where the pictures can be analyzed later. A high level of skill is needed to properly operate the camera and interpret and use the information.

Fluorescein and indocyanine green angiography. Angiography is a diagnostic test used by ophthalmologists to photograph structures in the back of the eye and is especially useful in finding damage to the blood vessels that nourish the retina (VisionRX, 2005). The two types of angiography are fluorescein and indocyanine green (ICG). Both techniques use the fundus camera. Fluorescein angiography is used primarily to study blood circulation in and just beneath the surface of the retina, while ICG angiography is better for photographing the deeper choroidal vessels.

The purpose of either type of angiography is to determine whether there are irregularities in the circulatory system of the retina. Several serious eye disorders, such as diabetic retinopathy, affect retinal circulation and are usually studied with the fluorescein procedure. Other problems, such as age-related macular degeneration, are caused by leakage from the deeper choroidal blood vessels. In these cases, the ICG procedure can provide additional information that may not be available through a fluorescein angiography.

In both angiography procedures, a small amount of colored dye is injected into a vein in the arm where it travels through the circulatory system and into the vessels in the eye. The dye makes these vessels visible to a special camera that is used to take photographs used by the doctor to diagnose various eye disorders.

Side effects from either procedure are minimal. The dye used in fluorescein angiography sometimes causes temporary, harmless discoloration of the skin and urine, and a few patients may feel slightly nauseated for a short time during the initial phase of the procedure. Allergic reactions to the ICG dye are rare. However, patients allergic to iodine or shellfish, or those with a history of liver disease, should advise their physician.

Optical coherence tomography. Optical coherence tomography (OCT) is a noninvasive, noncontact imaging technology that uses optical signal acquisition and processing to clearly image the structures of the retina. Using OCT, the anatomic layers within the retina can be differentiated and retinal thickness can be measured. OCT has been shown to be clinically useful for imaging selected macular diseases including macular holes, macular edema, age-related macular degeneration, central serous chorioretinopathy, epiretinal membranes, schisis cavities associated with optic disc pits, and retinal inflammatory diseases (New England Eye Center, 2008).

Scanning laser ophthalmoscopy. Scanning laser ophthalmoscopy (SLO) is a diagnostic imaging technique in which the fundus is illuminated by a narrow laser beam scanned horizontally and vertically over the surface of the retina (Nagpal and Banait, 2002). The timing of this procedure is designated to synchronize with a wide range of computer and video equipment. A detector captures the reflected light from the laser and the image of the fundus is viewed in real time on a monitor.

SLO has several major advantages over more traditional camera and imaging systems. First, pupillary dilatation is not mandatory because the technique can collect images even from a small aperture. There is less iridescence from the retinal surface, and the patient can tolerate the procedure better because it does not involve bright flashes. Further, the resolution provided by SLO is better as compared to conventional camera systems. Finally, the infrared wavelengths of the SLO can penetrate media opacities such as cataracts, corneal opacities, and hemorrhages and can thus provide useful information that some imaging systems, such as the fundus camera, cannot.

Measures of Vision Acuity

Eye charts are a typical part of routine eye examinations. Several types of charts are available but almost all of them assess one and only one capability of the eye—visual acuity under high contrast conditions. For each of the charts, the general procedure is for patients to be placed at a specified distance from the chart and asked to read aloud the line of smallest letters they can distinguish.

Snellen chart. Most people have seen the traditional Snellen eye chart in school or community vision screening programs. It is printed with 11 lines of block letters. The first line consists of one very large letter, which may be one of several letters (for example, E, A, or N). Rows below the first have increasing numbers of letters in decreasing size. The sizes of the letters are standardized so the letters in each row should be clearly legible at a designated distance to a person with normal vision.

Despite its widespread use, several criticisms of the Snellen chart have been made. For example, subjects can memorize the chart either before or between tests. Also, there are fairly large and uneven jumps in acuity between the rows, and some of the letters are harder to distinguish than others (for example, C versus G, Q versus O).

Landolt C chart. The characters on a Landolt C (or Landolt ring) eye chart are rings that have gaps, thus resembling the letter C. The gap can be at various positions—usually left, right, bottom, top, and the 45° positions in between. These rings are arranged on the chart in a manner similar to the letters of a Snellen chart, with large rings in the top row and rows below containing increasing numbers of rings that decrease in size.

A primary advantage of the Landolt C chart, as compared to the Snellen chart, is that the rings are merely the same targets with the gap in different positions. Thus, none of the characters are harder to distinguish than others. Further, the Landolt C chart may be more difficult to memorize than the Snellen chart.

Measures of Contrast Sensitivity

Even if a Soldier's visual acuity is sharp or considered normal for high contrast under high luminance conditions, his contrast sensitivity may be poor or compromised. Contrast sensitivity is the ability to visually perceive differences between an object and its background, to detect boundaries and slight changes in luminance at regions without distinct contours. If a Soldier has poor contrast sensitivity, his vision may be similar to viewing objects under twilight or foggy conditions.

A number of measures have been developed to assess contrast sensitivity. Several of them, such as the Mars Letter Contrast Sensitivity Test and the Pelli-Robson Test, present the patient with a printed chart showing letters or symbols in a faint gray print rather than the usual sharp, black-on-white characters of standard charts. Other more technical assessments involve the presentation of graded patterns or letters on an oscilloscope screen, such as the VectorVision CSV–1000E.

Measures of Glare Sensitivity

Glare occurs when light from a single bright source, such as the sun, an automobile headlight, or a light emitting diode, scatters across the entire visual field (Stamper, 1991). Such scattering often dazzles the sight and substantially reduces the quality of the image received by the retina. Laser dazzlers used on the battlefield are designed to produce this effect in greater or lesser degrees. Glare can also be quite painful. Glare sensitivity testing measures the extent to which a patient's vision is degraded in the presence of a bright light. This allows a vision professional to determine whether glare sensitivity is contributing to a patient's visual symptoms. Examples of glare sensitivity measures are the Vector-Vision CSV–1000HGT Halogen Glare Test and the Berkeley Glare Test.

Chromatic Assessment Tools

Discrepancies in color vision are not usually disabling. However, they can hinder individuals from pursuing or excelling in certain specialized careers. Further, in some situations, such as when driving or attempting to identify dangerous stimuli, the loss of color vision can be life-threatening. Several assessment tools have been developed to evaluate color vision in clinical settings.

Ishihara Color Test. The Ishihara Color Test is especially designed to detect red-green color deficiencies. Each eye is tested separately. The test consists of a number of colored plates containing a circle of dots that have been randomized in size and color. Patients with normal color vision can easily detect numbers and figures made up of, and embedded in, the multicolored dots. Patients with color vision deficits cannot see, or have great difficulty seeing, them. Combinations of colors are used to identify the nature of the color vision deficit (Stamper, 1991).

Farnsworth-Munsell 100 Hue Test. The Farnsworth-Munsell 100 Hue Test assesses hue discrimination ability often used in vocational and diagnostic applications. This test consists of 4 trays of 85 total colored removable color reference caps (incremental hue variation) spanning the visible spectrum. Color vision abnormalities and aptitude are detected by the ability of the test subject to place the color caps in order of hue. The four sequences of hues are difficult to discern for those with color deficits.

Visual Field Tests

A visual field examination tests the expanse and sensitivity of a patient's noncentral (peripheral) vision, that is, the perception of light and objects surrounding the direct line of sight (Stamper, 1991). Defects in peripheral vision are often not as apparent to the patient as are central vision deficiencies and may go unnoticed. However, peripheral vision problems are often caused by diseases to the retina, optic nerve, or structures to the visual pathway in the brain. Some of these diseases can be life-threatening as well as vision-threatening. Early detection by a visual field examination allows treatment initiation that may stop further disease progression and prevent irreversible vision loss.

Two visual field screening measures, the confrontational field test and the Amsler Grid, are addressed here first and can be used to obtain a rapid nontechnical evaluation of the patient's peripheral vision. The results of these tests may indicate the need for more advanced testing.

Confrontational field test. This test compares the boundaries of the patient's visual field to those of the examiner, who is assumed to have normal vision. The patient, seated two to three feet away facing the examiner, closes one eye while the examiner closes the opposite eye (for example, the patient's left eye and examiner's right eye are closed). The examiner then moves his fingers in from the far visual periphery to the near periphery in four separate quadrants. If the patient does not see one or more of the examiner's fingers at the same time as the examiner, the patient is assumed to have a reduced field of vision in the corresponding quadrant or quadrants and may be scheduled for further testing.

Amsler grid. The Amsler Grid assists the patient in reporting deficits in the central portion of his visual field. This measure contains a handheld card with a set of vertical and horizontal lines arranged in a grid pattern on either a black or white background with a dot in the center. Distortions in the patient's visual field are generally seen as distortions in specific areas of the grid.

Static and kinetic perimetry. Both of these tests sample the field of vision in a clinical setting, and accomplish this by increasing the intensity of a stimulus to a known area of the retina until the stimulus is seen. Static perimetry uses a stationary target that can be changed in size, brightness, and position, whereas kinetic perimetry utilizes a moving test object of a predetermined size and brightness. Types of static perimetry include suprathreshold and threshold, and the Humphreys Field Analyzer

II is a visual field assessment tool that uses static perimetry. There are many assessment tools that use kinetic perimetry, including the Tangent Screen Test, Goldmann Perimeter, Arc Perimeter, and Autoplot.

Other tools can evaluate most of the visual capabilities addressed in this chapter. However, those discussed here highlight an important truth concerning visual assessment. Except for a very few visual assessment procedures (such as the Confrontational Field Test and the Amsler Grid), these methods have been designed exclusively for clinical settings. The vast majority of visual measurement devices require special circumstances that can be provided and controlled in the clinic or laboratory. For example, the eye charts discussed above require standardized conditions of lighting and distance. The color vision assessment techniques addressed also require standardized lighting and would be difficult to complete outside of a quiet environment with few distractions. The fundus camera and varied types of equipment used for perimetry assessment are specifically designed for use in an optical clinic. Almost none of these procedures or tools can provide vital assessment information in battlefield or other settings outside of the eye clinic.

Aidman Vision Screener. Tools for providing vision information in the field of battle, particularly after a vision-related injury takes place, have been sorely lacking. This need provided the impetus for the development of the first Aidman Vision Screener (AVS) (Gunzerhauser, 1990), a portable vision assessment tool that can easily be used by an Aidman or medic supporting far-forward military operations. One of the mission goals of the U.S. Army Medical Research Detachment (USAMRD) is to improve medical triage for laser-induced ocular trauma. The development of an advanced AVS supports the mission of developing a quick and effective triage tool for making informed decisions concerning the use and evacuation of personnel who were dazzled or possibly injured by a laser or some other type of directed energy system. Effectively used, this tool will facilitate the medical evacuation of Soldiers with actual eye injuries to a higher level of care. Conversely, Soldiers who experience a bright light exposure but have no resultant eye injury can be rapidly assessed and sent back to continue their mission. Hence, the range of psychological effects resulting from the exposure can be rapidly mitigated and the Soldiers' duty effectively resumed. Clearly, the advanced AVS will also have value for assessing battlefield eye injuries from other etiologies.

The original AVS is a portable black-and-white card (approximately 5 by 8 inches) designed to triage retinal disturbances and assist in determining whether more advanced examination is needed. The AVS was developed by vision experts at USAMRD in San Antonio, Texas, and includes two near visual acuity tests, one composed of English letters and the other made up of 12 rows of Landolt rings. In both of these measures, the rows of characters are in order of descending size, and a metric is included that enables the user to make the conversion from the number and size of characters correctly identified to the typically used Snellen visual acuity measurement (for example, 20/20). In the chart made up of Landolt rings, the magnitude of the size decrease from each row to the row beneath is 0.2 log units. The AVS also includes the near visual field test discussed above, the Amsler Grid. The Amsler Grid of the AVS contains a set of white vertical and horizontal lines arranged in a grid pattern on a black background, with two diagonal white lines superimposed on the grid in an "X" pattern. Each of these diagonal lines extends from one of the corners of the grid to the opposite corner. This grid provides a measure of the extent to which any retinal injury or abnormality distorts the subject's vision. Finally, the AVS includes a table that contains criteria for triage. Based on results from the provided metric, the table will recommend that the respondent be evacuated for further evaluation, returned to duty, or reevaluated in 15 minutes.

The AVS has several features that have made it a valuable tool in the supply pack of those who would provide help for eye injuries on the battlefield. It is fairly small and exceedingly lightweight, which is important considering that the deployed Soldier often carries over 100 pounds of equipment on his person. Second, it is easy to use. Results can be obtained by following a simple set of instructions that include holding the AVS approximately two lengths of the tool in front of the Soldier and following a simple scoring scheme. Third, results can be obtained almost immediately, thus minimizing the time between the occurrence and the treatment.

The original AVS has been found by many to be helpful in the assessment of retinal disturbances. However, it was developed using the highest level of contrast possible (100 percent contrast using black characters on a white background). Thus, it is unable to detect vision deficits at low levels of contrast, a type of contrast sensitivity. This visual attribute is extremely important for Soldier tasks such as driving, enemy detection, and target discrimination. Lives often depend on these capabilities. Further, because the AVS is achromatic (black and white), it is unable to detect abnormalities in color vision. Color vision is also important to the aforementioned tasks and thus to the well being of Soldiers. A fairly large and diverse body of research (Jackson, Ong, and Ripley, 2004; King-Smith, Lubow, and Benes, 1984; Krill, 1972; Page and Crognale, 2005; Williams and Leaver, 1980; Zwick, Bedell, and Bloom, 1974) indicates that chromatic measures (measures with color) can obtain information that is helpful in detecting retinal abnormalities and injuries that achromatic measures cannot provide. These studies, considered together, point toward the need to add chromatic assessment tools to those that are achromatic (black-on-white measures of visual acuity) when attempting to determine the presence of decrements to visual function.

A research program has been launched to add lower contrast and chromatic measures into the Aidman Vision Screener, which is currently achromatic. This screener is being developed to assist medical providers in making informed decisions concerning the use and evacuation of military personnel who were dazzled or possibly injured by a laser or some other type of directed energy system. As part of the larger project of fielding an advanced AVS, validation research will be conducted to explore the functionality of the experimental eye charts under various conditions of exposure to bright light. Subjects will read the charts in three experimental conditions, which will be created by the presentation of three separate light emitting devices. The eye-chart development work described below is presented as a critical first stage of this work.

Development of the Advanced AVS

To provide more sensitive measures of visual function than are currently available in settings without advanced ophthalmological equipment, an effort is being made to upgrade the AVS. High contrast, lower contrast, and chromatic metrics are included in the advanced AVS. Gunzenhauser (1990) describes the work that was done to develop the original version of this tool. Because of the similarities between the first AVS and the advanced version, this work is also important to the advanced AVS. Two of the visual function tests contained in the original AVS, the Amsler Grid and the 100 percent contrast near visual acuity test made up of Landolt rings, are also included in the experimental version of the advanced AVS.

Because detecting objects under low luminance and contrast conditions is so important to many tasks conducted on the battlefield (such as tracking, driving, and enemy detection), a set of eye charts was developed to assess near visual acuity under lower levels of contrast. These charts, taken together, provide information on both how large an object must be and how much contrast it must have in comparison to its background to be seen by the subject. These charts each consist of 12 rows of Landolt rings, arranged in the same pattern of decreasing size (that is, a size decrease of 0.2 log units from each row to the next beneath) as they are arranged on the 100 percent contrast eye chart contained in the original AVS. These eye charts were developed using contrast levels that have been shown to differentiate foveal from parafoveal functioning and adequately represent the range of contrast sensitivity in previous research (Ness, Zwick, and Molchany, J., 1996; Zwick, Ness, Molchany, Stuck, and Loveday, 1998). These levels of contrast are 70, 50, 30, 10, and 2 percent. This set of charts is designed to assess contrast sensitivity for the smallest gap in the Landolt rings. Only the 100, 30, 10, and 2 percent charts are being used for the initial research, to simplify the design and shorten the time needed to complete it. To assess the functioning of the individual long, medium, and short wavelength cone systems of the retina, three charts consisting of colored Landolt rings on a white background were also developed. The theory providing the foundation for this development effort is relatively straightforward. When individual cone systems are presented with a very bright stimulus at the approximate wavelength for which they are the most sensitive, their functionality will temporarily decrease and they should perform more poorly in identifying targets of approximately the same wavelength. This developmental work parallels in some respects the work of Rabin (2004), who was the first to develop a chromatic eye chart that assesses the functionality of the individual cone systems for clinical use, thus applying a growing body of research highlighting the importance of assessing color vision in addition to achromatic visual acuity. This work also parallels the work of Rabin in that it employs logarithmic decreases in the size of the targets on the charts from each row to the next beneath. As with the black-and-white charts described above, the colored charts to be used in this research each consist of 12 rows of rings decreasing 0.2 log units in size from each row to the next beneath. The test targets are yellow in one chart, green in another, and blue in a third, corresponding to the peak sensitivities of the long, medium, and short wavelength cone systems, respectively. The targets of these charts are placed on a white background. To minimize radiance as a factor in their evaluation, the eye charts were designed to be as close to equal in radiances as possible. Table 5–1 displays the radiance measures of the final pigments used, as measured with an incandescent 2,803 Kelvin light source mounted at a 45-degree angle in relation to the charts. As shown in table 5–1, all radiance measurements are within one order of magnitude.

Table 5–1. **Radiance of Pigments in Colored Target on White Background Near Visual Acuity Charts**

Color	Radiance (Watts/m²sr)*
Yellow	.1081
Green	.0354
Blue	.0275
White	.1197

* = Watts per steradian per square meter

A third set of eye charts was developed to assess the neural processing of the retina that is used to create color opponency (Hurvich and Jameson, 1957). The identical 12-row Landolt ring pattern was used in the construction of this third set of charts. The chromaticities of pairs of pigments (for example, pairs of blue and yellow in various shades, pairs of magenta and green in various shades) were assessed in CIE x-y space (Wyszecki and Stiles, 2000). Color pairs were chosen for the charts largely by whether the color plots indicated that they were complementary. In a color pair, color A is complementary to color B if it produces white when mixed in an appropriate portion with color B. In x-y color space, colors are shown to be complementary if two points representing the plotted colors A and B are on opposite sides of the achromatic center point (representing "white") in a linear configuration of "color point, center point, color point" order. X-y space was used for this color plotting due to the ease with which complementary colors can be evaluated in this type of color space. Figures 5–3 through 5–5 provide examples illustrating how these color plots were evaluated. Figure 5–1 displays a shade of blue and a shade of green that are not complementary, plotted in x-y color space. As can be seen in the figure, their color coordinates place them at nearly right angles to each other in the color space. These two color shades would be evaluated as noncomplementary due to the extent of their deviation from a linear pattern.

Figure 5–1. **Blue and Green Plotted in x-y Color Space**

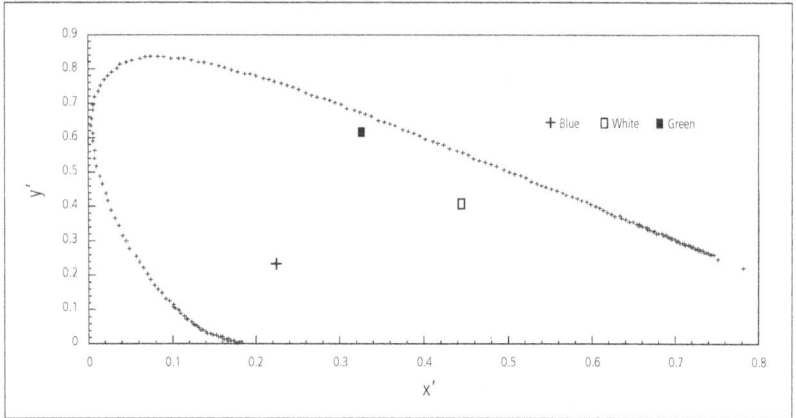

Figure 5–2 displays the blue-yellow pairs used in three of the charts in this set, and figure 5–3 displays the magenta-green pairs used in the other four charts. Examining figures 5–2 and 5–3, it can be seen for each pair that the blues are directly across from the yellows in x-y space, and that the magentas are almost directly across from the greens. This indicates that these pairs are complementary (or very nearly so, in the case of the pairs of magenta and green).

Figure 5–2. **Yellows and Blues in Advanced Aidman Vision Screener Plotted in x-y Color Space**

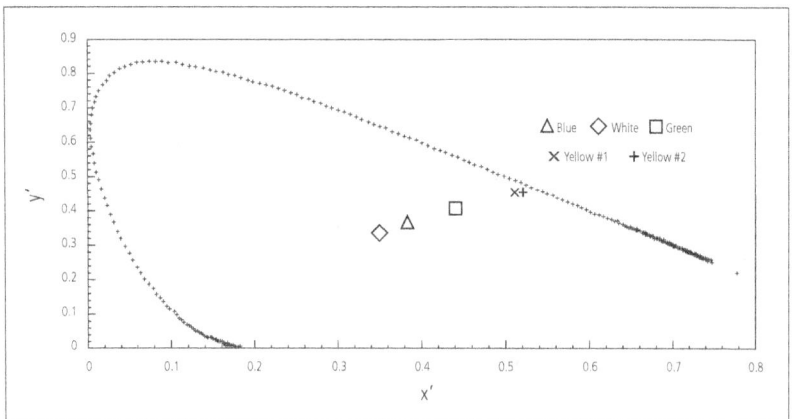

Figure 5–3. **Magentas and Greens in Advanced Aidman Vision Screener Plotted in x-y Color Space**

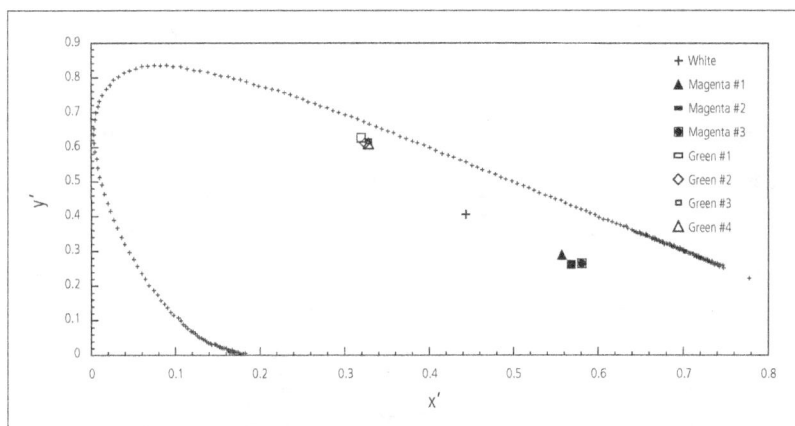

Reflected energy was also used as a criterion for color pair selection, to control its potential confounding effect on the experimental results. A color pair was not selected unless the level of energy reflected from one pigment in the dyad was on the same order of magnitude as the energy level reflected by the other. These charts each consist of Landolt rings of one color against a background of a complementary (opponent) color of similar radiance, using the same layout of decreasing row sizes as the charts described above. Table 5–2 displays the radiance measurements of the color pairs in each of the seven opponent-colored experimental eye charts. As can be seen in table 5–2, no one chart has differentials in radiance that exceed an 11:5 ratio. Further, all the differentials are small, with none exceeding a difference of .0384 watts per steradian per square meter. Thus, seven experimental charts were created to assess neural processing in the inner retina. Three are composed of shades of blue and yellow; the other four are made up of shades of magenta and green. To simplify the initial research and shorten the time needed to complete it, only two blue and yellow charts and two magenta and green charts are being used. The research program launched to explore the effectiveness of these eye charts is being conducted both to ascertain that they work and to guide any modification needed to improve the effectiveness of the advanced AVS. Subjects in this research will be asked to respond to the eye charts under varied lighting conditions. In

Table 5-2. **Radiance of Pigments in Each Opponent-colored Near Visual Acuity Chart**

Chart	Color	Radiance (Watts/m²sr)*
1	Light blue	.0623
1	Yellow	.0778
2	Light blue	.0744
2	Yellow	.0657
3	Light blue	.0614
3	Yellow	.0645
4	Green	.0337
4	Magenta	.0394
5	Green	.0349
5	Magenta	.0643
6	Green	.0368
6	Magenta	.0720
7	Green	.0349
7	Magenta	.0733

* = Watts per steradian per square meter

the experimental conditions of this research, subjects will look into light emitting devices and then respond to the charts while experiencing a large temporary afterimage in their visual field.

Directions for Future Research

The primary purpose for developing an advanced Aidman Vision Screener is to provide a more sensitive and accurate tool for visual function assessment in battlefield settings than the original AVS. An assumption driving the development of the advanced AVS is that accuracy will be increased by adding capabilities for the assessment of color vision and lower contrast visual acuity. This increase in accuracy is probable due to the increased ability to detect visual function deficits among those not previously identified as experiencing vision problems. However, the accuracy of this tool in identifying retinal abnormalities has yet to be determined. It is possible that the advanced AVS may also make more "false positives" than the original version, unnecessarily classifying more Soldiers as needing to be evacuated due to vision deficits. Because of difficulties in recruiting sufficient subjects with possible retinal abnormalities for a laboratory study, such a study may not be completed for some time. However, careful gathering of data from military operations may allow the accuracy of the advanced AVS to be estimated. Additionally, further research using methodologies similar to that briefly mentioned above (inducing temporary visual deficits) may assist in the further validation of this upgraded metric.

An issue often discussed among military vision scientists concerns the extent to which visual physiology and/or functioning must be compromised before evacuation from a military operation is deemed appropriate. A Soldier may be experiencing significant visual deficits and have one or more physical alterations of the retina (for example, retinal lesions), yet be deemed capable of performing his duties because the injuries cannot be seen without advanced ophthalmological equipment and are not substantially altering his behavior or job performance. Many retinal injuries rapidly become worse if they are not detected and treated. It would be of interest to determine the extent to which the advanced AVS can detect problems of intermediate severity, and whether it can detect most retinal alterations quickly after onset.

Given the extent of laser use on the modern battlefield, it is probable that some Soldiers have acquired or are currently developing subtle pathologies of vision due to lower level laser effects that have not yet had significant effects on the performance of their military duties. It may be that the physiology of these Soldiers' eyes has been slightly degraded and is growing steadily worse, and/or that visual function has been compromised to a slight to moderate extent but has yet to be detected due to its gradual onset. It would be important for the military to be able to detect these types of pathologies in battlefield settings. Early intervention in such cases could very possibly change the course of many of these types of retinal disorders. It would be of interest to know the extent to which the advanced AVS is able to detect subtle changes in visual pathology or functioning, and how severe a vision pathology has to be before is it detectable by this new vision assessment metric.

Conclusion

The proliferation of lasers on the modern battlefield has substantially increased the level of risk to Soldiers. Laser exposure can bring about severe injuries to the eyes and skin. Soldiers' exposure to lasers in combat settings may also result in fear and anxiety relating to performing many warrior tasks due to fear of incurring laser-related injuries.

While the skin is susceptible to laser damage, the risk of damage is much higher for the eye because the eye increases the irradiance or radiant exposure of collimated laser light at the retina. A variety of tools and measures have been developed to assess the many aspects of vision, including measures of eye physiology, visual acuity, contrast sensitivity, glare sensitivity, color vision, and the visual field. A major drawback of the vast majority of vision screening tools, however, is that they are designed exclusively for use in clinical settings. They require special conditions for their use that need to be provided and controlled in the clinic or laboratory. Because of this drawback, tools for providing vision information in the field of battle have been sorely needed.

The need for vision screening tools that can be used in combat settings provided the impetus for the development of the Aidman Vision Screener, an achromatic visual function assessment device that can be used to evaluate both high-contrast visual acuity and the visual field of the Soldier. The AVS has several features that make it a valuable tool in the supply pack of military medical personnel and others who would provide help for eye injuries on the battlefield. These features include portability, ease of use, and nearly immediate assessment results. However, several aspects of vision are not currently assessed by the AVS. These include color vision and visual function under low levels of contrast. An effort is under way to develop an upgraded version of the AVS that will have the means to assess these visual abilities. This chapter presented the methodology used to develop the eye charts being used for the advanced AVS, along with a brief description of research being conducted to collect validation evidence supporting this new metric.

References

Anderberg, B., Bring, O.E., and Wolbersht, M.L. (1992). Blinding laser weapons and international humanitarian law. *Journal of Peace Research* 29 (3), 287–297.

Department of the Army (1990). Prevention and medical management of laser injuries. (Field Manual 8–50). Washington, DC: Headquarters Department of the Army.

———. (1999). The Army radiation safety program (Army Regulation 11–9). Washington, DC: Headquarters Department of the Army.

DeVour, R. (1993). Possible psychological and societal effects of sudden permanent blindness of military personnel caused by battlefield use of laser weapons. In L. Doswald-Beck (ed.). *Blinding weapons: Reports of the meetings of experts convened by the International Committee of the Red Cross on battlefield laser weapons, 1989–1991* (46–51). Geneva: International Committee of the Red Cross.

Doswald-Beck, L. (1996). New protocol on blinding laser weapons. *International Review of the Red Cross* 312, 272–299.

Federation of American Scientists Military Analysis Network (1998). *AAS–38A/B Nite Hawk*. Retrieved July 20, 2009, from <www.fas.org/man/dod-101/sys/smart/aas-38.htm>.

Fitzgerald, R. (1993). Psychiatric, psychosomatic and socio-economic status reactions to blindness. In L. Doswald-Beck (ed.). *Blinding weapons: Reports of the meetings of experts convened by the International Committee of the Red Cross on battlefield laser weapons, 1989–1991* (258–268). Geneva: International Committee of the Red Cross.

Gunzenhauser, J.D. (1990). *Issues in the development of the Aidman vision screener* (Laboratory Note No. 90–81). San Antonio, TX: United States Army Medical Research Detachment.

Harris, M.D., Lincoln, A.E., Amoroso, P.J., Stuck, B., and Sliney, D. (2003). Laser eye injuries in military occupations. *Aviation, Space, and Environmental Medicine* 74, 947–952.

Hurvich, L.M., and Jameson, D. (1957). An opponent-process theory of color vision. *Psychological Review* 64, 384–404.

Jackson, T.L., Ong, G.L., and Ripley, L.G. (2004). Orientational contrast sensitivity and chromatic contrast thresholds in multiple sclerosis. *American Journal of Ophthalmology* 137 (2), 283–286.

King-Smith, P.E., Lubow, M., and Benes, S.C. (1984). Selective damage to chromatic mechanisms in neuro-ophthalmic diseases I. Review of published evidence. *Documenta Ophthalmologica* 58 (3), 241–250.

Krill, A.E. (1972). *Hereditary retinal and choroidal diseases* (vol. 1, 337). Hagerstown, MD: Harper and Row.

Laser Institute of America. (2007). *American national standard for the safe use of lasers*. Orlando, FL: Laser Institute of America.

Mainster, M.A., Stuck, B.E., and Brown, J. (2004). Assessment of alleged retinal laser injuries. *Archives of Ophthalmology* 122, 1210–1217.

Nagpal, M., and Banait, M. (2002). Scanning laser ophthalmoscopy in macular degeneration. *Bombay Hospital Journal* 44 (3), 333–339.

Nahin, P.J. (1977). The laser BMD and other radiant energy weapons: Some thoughts. *IEEE Transactions on Aerospace and Electronic Systems* 13 (2), 96–107.

Ness, J., Zwick, H., and Molchany, J. (1996). Laser-inflicted eye injuries: Epidemiology, prevention, and treatment. *Proceedings of the International Society for Optical Engineering (SPIE)* 2674, 131–135.

Neuenswander, D. (2001). Joint laser interoperability, tomorrow's answer to precision engagement. *Air and Space Power Journal—Chronicles Online Journal, 15*(2).

New England Eye Center (2008). *Optical coherence tomography (OCT)*. Retrieved July 20, 2009, from <www.neec.com/ Pages/Services/Glaucoma/OCT.php>.

Northrop Grumman (2008). *Target acquisition designation sight (TADS) laser rangefinder/ designator (LRF/D)*. Retrieved July 20, 2009, from <www.es.northropgrumman.com/ solutions/tads/>.

Page, J.W., and Crognale, M.A. (2005). Differential aging of chromatic and achromatic visual pathways: Behavior and electrophysiology. *Vision research* 45 (11), 1481–1489.

Rabin, J. (2004). Quantification of color vision with cone contrast sensitivity. *Visual Neuroscience* 21, 483–485.

Robbins, D.O., and Zwick, H. (1980). Changes in spectral acuity following brief laser (647 nm) exposure. In Verriest, G. (ed.). *Colour vision deficiencies* (106–110). London: Adam Hilger Ltd.

Stamper, R.L. (1991). *Ophthalmic medical assisting: An independent study course.* San Francisco, CA: American Academy of Ophthalmology.

VisionRX (2005). Angiography. In *Eye care library encyclopedia.* Retrieved July 20, 2009, from <www.visionrx.com/library/enc/enc_angiography.asp>.

Warren, D. (1993). Psychological effects of total permanent blindness occurring in early adulthood. In Doswald-Beck, L. (ed.). *Blinding weapons: Reports of the meetings of experts convened by the International Committee of the Red Cross on battlefield laser weapons, 1989–1991* (52–68). Geneva: International Committee of the Red Cross.

Williams, C.M., and Leaver, P.K. (1980). Visual acuity and colour vision in central serous retinopathy. In G. Verriest (ed.). *Colour vision deficiencies* (306–309). London: Adam Hilger Ltd.

Wyszecki, G., and Stiles, W.S. (2000). *Color science: Concepts and methods, quantitative data and formulae* (2^d ed.). New York: John Wiley and Sons.

Zwick, H., Bedell, R.B., and Bloom, K. (1974). Spectral and visual deficits associated with laser irradiation. In G. Verriest (ed.). *Modern Problems in Ophthalmology* (299–306). Basel, Switzerland: S. Karger, AG.

Zwick, H., Ness, J.W., Molchany, J.M., Stuck, B.E., and Loveday, D. (1998). Neural motor ocular strategies associated with the development of a pseudofovea following laser induced macular damage and artificial macular occlusion: Is the fovea replaceable? *Journal of Laser Applications* 10 (3), 144–147.

Part III

Psychology of Leadership

Chapter 6

New Wine in Old Bottles: Leadership and Personality in the Military Organization

Paul T. Bartone

Leadership is a topic of abiding interest to the military, as well as to social scientists who study military organizations and people. In seeking to understand the personal qualities associated with effective leadership, psychologists have largely focused on intelligence and other mental abilities, to the neglect of factors such as personality traits. This chapter summarizes the major theoretical perspectives on leadership that have dominated the field for the last 50 years, and shows how normal personality has reemerged as a legitimate topic in the study of leadership. Examples are provided of recent studies that document the influence of five main personality factors on leadership, and also the influence of personality "hardiness." In addition to empirical studies showing that hardiness is associated with healthy response to stress and effective leader performance, the hardiness construct provides a theoretical framework for understanding the leader influence process under high-stress conditions. Especially as regards stressful and ambiguous events, military leaders can apply hardiness qualities to facilitate generalized positive interpretations or "sense-making" among unit members. Leaders do this through actions, policies, and personal example, and in this way they increase stress resilience throughout the organization.

Introduction

Military organizations have long emphasized the importance of leadership and sought to train or develop highly effective leaders in a multitude of ways. For example, in the epic classical stories of *The Iliad* and *The Odyssey*, Homer portrays a number of personal leader qualities including courage, strength in adversity, loyalty, persistence, and mental agility (Homer, 1975). More recently, Jonathan Shay has provided an interesting analysis of Homer's Odysseus, drawing lessons for the impact of leadership and policy on the mental health of American troops today (Shay, 2002). Leadership was also an important topic for the Greek philosophers; for example, in Plato's *Republic*, the ideal leader is described as a "philosopher-king," one who is intelligent, highly moral, and selfless (Plato, 1991).

As roles and missions have expanded in the postmodern military, demands on military personnel have increased substantially (Moskos, Williams, and Segal, 2000; Nagl, 2005). Our forces today are called upon to serve in a broader range of missions that are complex and changeable. The challenges for leaders are especially severe in this new environment (Shamir and Ben-Ari, 2000).

In this context, understanding the range of factors that contribute to effective leadership is more important than ever. What are the personal qualities that contribute to effective leadership? There is considerable speculation on this question, and answers too often rely on anecdotes and hearsay. Historically, social scientists who have tried to address the question empirically have tended to focus on intelligence and various mental abilities, while personality factors have drawn considerably less attention. In recent years, however, more studies have examined personality variables as predictors of leader performance. This trend toward increased scientific study of the role of personality promises to bring new understanding for enhancing the quality of leadership in military organizations.

The chapter begins with a summary of the more important theoretical perspectives on leadership from recent decades. Next, it shows how these theoretical ideas, together with a growing number of empirical studies, converge on an understanding of leadership as primarily a "sense-making" activity in which both personal and contextual factors have important influence. Especially in organizations like the military, a key aspect of leadership concerns the ways in which events get interpreted by members of the unit. Empirical studies provide some evidence that highly effective military leaders possess certain personality attributes that allow them to adapt quickly and make constructive sense of rapidly changing and ambiguous situations. Perhaps more importantly, these qualities are also used to inspire subordinates and peers to construct similar positive interpretations of experience. The influence of leaders on subordinates in this regard may be even greater under highly stressful external conditions.

What Is Military Leadership?

Before going further, it is useful to define *military leadership*. One option is simply that it is leadership within the military organization or context, across the range of missions and activities in which military organizations are engaged. But this definition is unsatisfying, mainly because it leaves the key term unspecified. The real problem in defining military leadership lies in specifying what is meant by *leadership*, and then applying it to the military case.

In a detailed review of the leadership literature, Stogdill (1974) observed, "There are almost as many definitions of leadership as there are persons who have attempted to define the concept." The confusion goes beyond definitions. The literature is vast, extending from rigorous scientific studies including experiments and correlational and case studies, all the way to folk wisdom, popular how-to guides, metaphors of every kind, magic formulas, and rules for success. While some of the confusion can be overcome by focusing only on careful theoretical formulations, case studies, and empirical investigations, it is not always easy to separate science from superstition when it comes to leadership. It is also fair to say there is little agreement even among the more scientific approaches as to just what leadership is. According to a recent special report of the National Research Council (Druckman, Singer, and Van Cott, 1997), "the leadership literature is full of ambiguous theory and contradictory research findings."

How, then, should leadership be defined? This brief report will not attempt to cover all the definitions. For this, the interested reader is referred to comprehensive summaries by Stogdill (1974) and Bass (1990). Yukl (1998) also provides an excellent overview of leadership definitions, theories, and research. Similar to Yukl, I will here define leadership in broad terms as both a social influence process as well as a particular, specialized role of "the leader." Leaders provide leadership by influencing the choice of objectives and strategies, organizing tasks and activities, developing skills, committing and motivating people in an organization or group, and, most importantly, by influencing how events get interpreted.

Leadership Models and Theories

As with definitions of leadership, there is likewise a multiplicity of models, frameworks, and metaphors for understanding the leadership process. As an example, the U.S. Army's doctrinal manual on leadership, Field Manual 22–100, *Army Leadership*, puts it in terms of the "Be–Know–Do" framework of leadership, which is said to include character, values, and other personal attributes (the "Be" dimension), knowledge and skills (the "Know" dimension), and behaviors or actions (the "Do" dimension). Similar to many other models of leadership, the "Be–Know–Do" model is both descriptive and prescriptive, defining both what leaders are and what they should be, all in somewhat global terms. Military organizations like the U.S. Army also favor geometric and spatial metaphors (including *pyramids, pillars,* and *building blocks*) to describe how leadership is composed, and to suggest how leader development might occur. As another example, Hughes, Ginnett, and Curphy (1996) rely on Venn diagrams to indicate overlapping influences between leaders, followers, and situations. While such models and metaphors can stimulate thinking and communication about leadership, they generally do not tell us anything about the nature of leadership or its underlying processes. For example, to assert that "leadership" occurs at the overlap between followers, leaders, and situations sounds sensible, but actually explains little. A danger with such leadership models is that they can create the illusion of understanding, leading to complacency and failure to seek more substantial explanations of the complex phenomenon of leadership.

Another common model or framework for understanding leadership attempts to partition it into component pieces, or levels-of-analysis. This is the approach in the required course on military leadership at the U.S. Military Academy, West Point (United States Military Academy, 2000). The syllabus divides the course into four main content areas: Individual System, Group System, Leadership System (focusing on leader-follower interactions as well as situational factors), and Organizational System (macro-level factors including culture). This approach has the virtue of being comprehensive, and it exposes students to multiple theoretical perspectives but does not attempt to compare these perspectives critically or determine their relative importance to effective leader performance. Many other models have been proposed as aids or heuristics for understanding leadership.

One of the most important and influential theories of leadership is known as the contingency model of leader effectiveness, developed by Fiedler (1964, 1967). Prior to Fiedler, most research focused on two styles of leader, authoritarian versus nonauthoritarian, also known as directive versus nondirective, or task-oriented versus human relations–oriented (Katz and Kahn, 1952; Shaw, 1955; Shaw and Blum, 1966). The Ohio State University leadership studies conducted in the 1950s and 1960s established a strong research tradition of evaluating performance correlates of rated leader behaviors. These studies led to the identification of two leadership styles, "consideration" (of subordinates) and "initiating structure," or defining structures and tasks aimed at achieving the goals of the organization (Fleishman and Harris, 1962). About the same time, researchers at the University of Michigan described three distinctive types of leader behavior: task-oriented, relations-oriented, and participative (Likert, 1967).

Fiedler in 1964 developed a specialized instrument, the least preferred coworker (LPC), to assess the extent to which a leader was more task-oriented (low LPC scores) or relations-oriented (high LPC scores) (Fiedler, 1964). In this framework, high scores reflect more forgiving or lenient ratings of one's "least preferred" coworker, while low scores show more harsh and unfavorable ratings. Those who are more accepting toward their least preferred coworker are presumed to be more relations-oriented and supportive of coworkers in general. Over hundreds of studies, Fiedler and colleagues found essentially that leader effectiveness is a function of the interaction between leader style (task- or relations-oriented) and the favorable or unfavorable nature of the situation. The model basically indicates that low LPC leaders (task-oriented) are more effective in extremely unfavorable or favorable situations, whereas high LPC leaders (relationship-oriented) are more effective in intermediate situations. The performance effects of leadership style are thus contingent on aspects of the situation or work context, so this is known as a "contingency theory" of leadership.

A special value of Fiedler's contingency model (and others like it) is that it directs attention to the importance of contextual factors in understanding the effects of various leader styles and approaches. Research findings are generally supportive of the model, but among other criticisms, it has been faulted for not paying enough attention to medium LPC leaders (more balanced in task and relational tendencies), who likely outnumber the extreme types (Yukl, 1998). Nevertheless, until fairly recently Fiedler's contingency model was viewed by many as "probably the best theory of leadership proposed to date" (Shaw and Costanzo, 1982).

Another influential contingency theory is known as the "path-goal theory" of leadership (House, 1971, 1996). This approach emphasizes the influence leaders can have on the motivation of subordinates by identifying goals, establishing environments in which people are rewarded for attaining them, and clearly communicating the pathways for achieving them (thus the term path-goal theory). Early versions of the path-goal theory identified two main approaches, supportive leadership and directive leadership (similar to the relations-oriented and task-oriented styles discussed earlier). Interestingly, a later and less familiar version of path-goal theory included two additional behavior orientations: participative leadership and achievement leadership (House and Mitchell, 1974; House, 1988). As with Fiedler's model, the contingency aspect of path-goal theory has to do with which type of leader approach is considered most effective under various external conditions. When situations are tedious, dangerous, or stressful, supportive leadership is posited to be more effective. When situations are ambiguous and complex and subordinates are inexperienced, directive leadership is presumed to be preferable. Similar to other contingency theories, research on path-goal theory has yielded mixed results regarding these hypothesized interaction effects. However, it is clear that various situational as well as leader characteristics have measurable direct effects on performance, and that at least in some cases situation and leader variables interact (Wofford and Liska, 1993; Podsakoff, MacKenzie, Ahearne, and Bommer, 1995).

A later modification of Fiedler's contingency theory emphasizes the importance of the leader's "cognitive resources" (Fiedler, 1986). According to this model, intelligence and experience are two different types of cognitive resources that have more or less relative value depending on the nature of the situation and whether the leader tends to be directive or not. Leader intelligence presumably leads to better performance across a broad range of circumstances, especially when the leader is also directive and the task is complex. However, under highly stressful conditions, intelligence fades in importance (and may even be a detractor), and performance is more a function of the leader's practical experience. Interestingly then, this later version of Fiedler's contingency theory of leadership incorporates person-variables that go beyond intelligence or traditional cognitive abilities. Research has provided some support for this model (Blythe, 1987; Potter and Fiedler, 1981).

Robert Sternberg presented a somewhat related theory focusing directly on "practical intelligence" or experience-based knowledge— "tacit knowledge"—as important for leader success, and applied this model specifically to the military (Sternberg, Forsythe, Hedlund, Horvath, Wagner, Williams, Snook, and Grigorenko, 2000; Hedlund, Forsythe, Horvath, Williams, Snook, and Sternberg, 2003). Previously, Stephen Zaccaro and colleagues presented a similar model that emphasized the importance of experience-based knowledge for leader flexibility and effectiveness across a range of situations (Zaccaro, Gilbert, Thor, and Mumford, 1991). In general, this tacit knowledge perspective on leadership is universalistic, not contingency-oriented. That is, according to this approach, increased levels of practical intelligence and tacit knowledge will be of value to leaders across the entire spectrum of situations they may encounter (although it may be of even greater value in novel and ambiguous circumstances). Thus far, not much research has been done on this new concept in military leadership, but early results are promising (Hedlund et al., 2003).

Other significant leadership research in recent years has emphasized cognitive abilities and problem-solving skills that may distinguish effective from ineffective leaders (Jacobs and Jaques, 1986; Mumford, Zaccaro, Harding, Fleishman, and Reiter-Palmon, 1993; Phillips and Hunt, 1992; Mumford, Zaccaro, Harding, Jacobs, and Fleishman, 2000). In part this emphasis on cognitive variables comes as a result of claims that personality traits do not appear to predict leadership very well, at least independently of other variables (Hollander and Julian, 1969; Stodgill, 1948; House, 1988). But interest has persisted in personal traits other than cognitive abilities that might distinguish good from bad leaders (Bass, 1998; Burns, 1978; House and Howell, 1992). This may be in part because cognitive abilities, while important, leave much unexplained about variance in effective leadership. In a review of research and theory on military leadership, Alan Lau (1998) suggests that some of the early lack of support for a trait or personality approach is due to measurement limitations. When better measures are utilized, results generally support the importance of trait variables (Lau, 1998; Bass, 1990). Robert Lord and colleagues (Lord, DeVader, and Alliger, 1986), in a meta-analysis of 19 samples first presented by R.D. Mann in 1959, also observed that methodological factors influenced early results and conclusions. Lord points out that Mann and Stogdill in their reviews did not examine personality traits in relation to actual leader performance, but rather in relation to leader emergence or "attained leadership status." This leads Lord to the conclusion that the early work on leadership and personality traits has been misinterpreted, and that personality variables merit renewed attention.

Probably the most influential theoretical perspective on leadership in recent years is referred to as transformational leadership, developed first by Burns and elaborated by Bass and Avolio (1989). Judge and Bono (2000) found that of all articles dealing with leadership published since

1990, over half dealt with transformational leadership in some significant way. Transformational leadership is closely related to the earlier concept of charismatic leadership. One of the first scholars to discuss charismatic leadership was Max Weber (1947), who described it as a form of social authority based on perceived exceptional and inspirational qualities of the leader, as opposed to rational-legal authority or authority based on tradition. Several leadership theories emphasize the charismatic influence of leaders: House (1977) focuses on the traits and behaviors of charismatic leaders; Conger (1989) puts additional emphasis on the attributions made by followers about their charismatic and visionary leaders; and Shamir, House, and Arthur (1993) argue that charismatic leaders are influential mainly through the need followers have to build up their self-concepts through social identification with powerful leaders. Several authors have also discussed the "dark side" of charisma, or ways in which charismatic leaders can have destructive effects on organizations and people (Yukl, 1989; Conger, 1989; Hogan, Raskin, and Fazzini, 1990).

Bernard Bass developed his model of transformational leadership following on Burns's (1978) ideas about transforming leadership as something that inspires followers to higher levels of motivation and moral behavior (Bass, 1985). The Bass model posits basically two kinds of leadership styles: transformational and transactional. The transactional leader operates mainly on the basis of exchange, or transactions, rewarding subordinates for desired behaviors and sometimes punishing them for undesired ones. In contrast, the transformational leader seeks to empower and inspire subordinates, providing an exciting vision of the future while also communicating personal concern and support. Bass and colleagues developed a simple instrument, the Multifactor Leadership Questionnaire, for assessing transformational and transactional behaviors (Bass and Avolio, 1989; Avolio, 1999). Despite some limitations, this instrument has been widely used in studies showing significant associations between transformational leadership style and performance. But even Bass admits that little is known about how leaders develop to be transformational versus transactional (Bass, 1998).

Personality and Leadership

The reemergence of personality variables is apparent in several domains related to leadership (Sashkin, 1992). For example, stratified systems theory, which puts heavy emphasis on cognitive abilities that are important for leaders at the senior executive organizational levels, has recently incorporated variables that go beyond cognitive abilities. Self-esteem and "social perspective-taking" are now included in a broadened concept of "cognitive complexity," something Lewis and Jacobs (1992) consider essential for effective strategic leadership. Growing interest in the concept of "emotional intelligence" in both popular and professional literature also suggests a tendency to incorporate personality and other noncognitive ability variables into an expanded definition of intelligence or mental capacities (Salovey and Mayer, 1990; Goleman, 1995). David Ryback (1998) is one of several authors who have discussed potential applications of emotional intelligence to the leadership domain.

One of the major developments in personality psychology in recent years is a growing consensus on the so-called five-factor model (FFM), or "Big Five," of normal personality. These factors are neuroticism, extraversion, openness to experience, agreeableness, and conscientiousness (Digman, 1990; McCrae, 1992). Progress in conceptualizing and measuring normal personality has also contributed to the renewed interest in applying personality to leadership research and theorizing (Barrick and Mount, 1999; Judge and Bono, 2000; Rubenzer, Faschingbauer, and Ones, 2000). Conceptually, personality traits are generalized consistencies in styles of thinking, feeling, and acting, and thus can be expected to affect many aspects of behavior, including leadership. While the content of a leader's goals, concerns, and plans is influenced by organizational factors and situational contexts, basic personality traits of the individual may influence how goals and plans are pursued, how efficiently they are formulated, and how conscientiously they are followed. In a recent study examining cognitive and personality factors in West Point cadets, the FFM personality dimensions of conscientiousness and agreeableness emerged as significant predictors of cadet leader performance over a 3- to 4-year time span (Bartone, Snook, and Tremble, 2002). In a similar study with Australian Army officers, leader achievement was predicted by high conscientiousness and high openness, and low extraversion (introversion) (McCormack and Mellor, 2002).

Recent studies have also begun to investigate the influence of personality factors on transformational leadership style. For example, Timothy Judge and Joyce Bono conducted a meta-analysis of 14 samples, exploring the relations of the five-factor model to transformational leadership (Judge and Bono, 2000). They found that agreeableness and extraversion predicted transformational leadership, as rated by subordinates.

Their findings, while provocative, left much of the variance in transformational leadership unexplained. In a followup to this work, Bono and Judge (2004) again used meta-analysis to look at the five-factor model personality dimensions as predictors of both transformational and transactional leadership. They found the strongest effects for extraversion and (-)neuroticism, but concluded that other personality dimensions besides the Big Five should be explored in future research.

While the five-factor model certainly represents an important scientific advance in the field of personality, it need not be viewed as describing the sum total of normal personality. As argued by Jack Block (1995) and others, there may be traits not fully represented in the Big Five paradigm that have considerable influence on behavior in various domains including leadership (Widiger and Trull, 1997; Dawda, 1997; Hough, 1992). Transformational leadership style and its component elements, as described by Bass and Avolio, may thus incorporate or be influenced by personality traits or tendencies not completely covered by the Big Five model and that are more theoretically relevant to leadership styles (Bono and Judge, 2004).

An especially promising personality dimension for predicting military leader performance and also transformational leadership is described as psychological hardiness. Conceptually rooted in existential psychology, *hardiness* involves a high sense of commitment to life and work, a strong belief in one's ability to control events and influence outcomes, and greater openness to changes and challenges in life (Kobasa, 1979; Maddi and Kobasa, 1984; Maddi, 1999). Persons high in hardiness are more resilient when exposed to a range of environmental stressors and tend to remain healthy and perform well despite high stress levels (Bartone, 1989; 1998a; 1999a). The underlying mechanisms whereby hardiness confers resiliency, though still poorly understood, most likely involve the interpretation or meaning individuals impute to events around them, and how they see themselves in regard to the world of experiences (Maddi, 1967). Highly hardy people typically interpret experience as interesting and worthwhile overall, something they can exert control over, and challenging, presenting opportunities to learn and grow. Many studies have now documented positive health and performance effects for hardiness in a variety of groups, including the military (Bartone, Ursano, Wright, and Ingraham, 1989; Bartone, 2000; Westman, 1990).

In military units, this "meaning-making" process is likely something leaders can influence. The units are highly interdependent, group-oriented structures. Most tasks require the coordinated actions of individuals. Also, units are organized hierarchically, with those at the top having the greatest influence over policies, procedures, and goals. Through their policies, advice, and examples, leaders may help shape the way their subordinates make sense of events and experiences (Weick, 1995). Many sociologists and social psychologists have advanced the idea that how meaning gets constructed at the individual level is to some degree a reflection of social processes (Berger and Luckmann, 1966; Janis, 1972; Allport, 1985). It thus seems likely that leaders, as well as peers and the entire organizational culture, can have an impact on how individuals interpret their experiences.

Several recent research studies support the idea of the influence of personality hardiness on leadership style and performance. For example, a longitudinal study of West Point cadets identified both hardiness and transformational leadership style as significant independent predictors of leader performance, as measured by military development grades (Bartone, 1999b). Examined alongside several other personality and cognitive abilities measures, personality hardiness was found to be the strongest predictor of leader performance. These results show that people who are high in hardiness—a characteristic sense of commitment, control, and challenge—are more effective in positions of leadership in a military organization. Additional analyses found that the hardiness facet of commitment was moderately correlated with transformational leadership style, although total hardiness scores were not. This suggests that those high in psychological hardiness are more apt to develop a transformational leadership style, but only under certain external conditions. In another study of West Point cadets, hardiness was a stronger predictor of leader performance for female cadets (Bartone and Snook, 2000).

A similar study of Norwegian Navy officer cadets found that in the context of a highly stressful training exercise, hardiness and leadership interacted to increase small unit cohesion levels (Bartone, Johnsen, Eid, Brun, and Laberg, 2002). This provides further evidence that psychological hardiness can affect leadership, and that there are important consequences for relevant outcomes such as unit cohesion. An important question for future research thus concerns to what degree hardiness can be trained or developed in military leaders. There is preliminary evidence that hardiness levels can be increased through specialized training (Judkins, Reid, and Furlow, 2006; Maddi, 1987; Maddi, Khoshaba, Jensen, Carter, Lu, and Harvey, 2002), but more work is needed.

Also arguing for the importance of noncognitive variables that can influence leader effectiveness, Gal (1987) suggests that more demanding military operations call for leaders who can increase the commitment of subordinates. Gal posits that this is the critical contribution of transformational leaders. The research summarized above on hardiness and leader performance lends credence to this view. A critical aspect of this commitment-building process likely concerns how experiences get interpreted individually and as groups.

A number of other personality dimensions relevant to military leadership are also being actively investigated. For example, a series of studies with cadets and Army officers indicate that a person's stage of psychosocial development, as articulated by Harvard University psychologist Robert Kegan, can have broad influence on leader style and performance (Kegan, 1982; Forsythe, Snook, Lewis, and Bartone, 2002; Bartone, Snook, Forsythe, Lewis, and Bullis, 2007). Additional research has identified possible links between Kegan's constructive-developmental framework and transformational leadership style in Army officers, and also found significant correlations between constructive-developmental stage level and "conceptual work capacity," a measure of strategic leader conceptual ability (Kuhnert and Lewis, 1987; Lewis and Jacobs, 1992). Significant associations have also been found between military leadership (variously assessed) and personality measures that do not seem fully encompassed by the five-factor model, including the Hogan Personality Inventory, the California Personality Inventory, and the Myers-Briggs Type Indicator (Lall, Holmes, Brinkmyer, Johnson, and Yatko, 1999; Thomas, Dickson, and Bliese, 2001; Gough, 1990; McCaulley, 1990). At this point, then, studies of personality and leader performance should probably not be unduly restricted to the Big Five model, but should consider other dimensions as well, such as hardiness and Kegan's constructive-developmental level. The same point is made by Bono and Judge in their report on a comprehensive meta-analytic study, in which they examine the effects of the Big Five personality dimensions on transformational and transactional leadership styles (Bono and Judge, 2004).

Gender Differences in Personality and Leadership

As the proportion of women military leaders continues to rise in the U.S. all-volunteer force (Segal, 1995), it is important to understand gender differences in factors influencing leader styles and performance.

Bass summarizes evidence showing that women tend to be higher in transformational leadership style than men; studies also suggest that more "feminine" (concerned leadership, individualized attention) approaches may lead to better performance in modern organizational cultures that tend to emphasize cooperation and collaboration within relatively flat and flexible structures, as compared to "male" task-oriented approaches which are characterized by instrumental strategies in more hierarchical and rigid structures (Bass, 1998; Eagley, Karau, and Makhijani, 1995).

In their meta-analytic study, Judge and Bono (2000) identified agreeableness and extraversion as significant predictors of transformational leadership style. While these authors do not address gender differences directly, it happens that women made up 57 percent of their sample, a factor that might have influenced their results. Costa and McCrae likewise report a general finding across many samples that women tend to score higher in agreeableness (Costa and McCrae, 1992). Female cadets at West Point also fit this pattern, scoring higher in agreeableness than their male counterparts (Bartone, Snook, and Tremble, 2002). This higher agreeableness may incline female cadets toward a more transformational style, which in turn is more effective and valued in modern organizations. Interestingly, it has also been reported that female cadets are higher in hardiness than male cadets at West Point, and that hardiness appears to be a stronger moderator of psychosocial stress for female cadets (Bartone and Priest, 2001). This suggests that under stressful conditions, hardiness may be a more important influence on leader performance for women than for men. Future research is needed to explore this intriguing possibility.

The U.S. Army is currently placing greater emphasis on efforts to increase self-awareness, adaptability, flexibility, and mentoring of subordinates, as evidenced, for example, in the formation in 2001 of a special panel to examine leader development programs and future needs for Army officers (*Army Training and Leader Development Panel Officer Study: Report to the Army*, 2001). This shift is also apparent at the Army's primary leader development institution, the U.S. Military Academy (United States Military Academy, 2002). But these programmatic efforts need to be guided by sound research into the underlying processes involved, and clarifying what works and what does not. Important areas for future research include examining the possible direct and interacting effects of gender, personality, transformational leadership style, job, life experiences, and organizational factors on leader adjustment and performance.

Leader Development

If certain individual characteristics, attributes, or skills are associated with effective leader performance under various conditions, a vital practical question for organizations such as the military concerns how to develop or cultivate these characteristics. While a detailed review of the leader development literature is beyond the scope of this chapter (see Day, 2001, for a comprehensive treatment), some trends of special relevance to the military can be identified. One is a growing recognition that effective development of leaders must involve not only skills acquisition, but also personal growth and increasing individual maturity. Simple training, no matter how elaborate or intensive, is not likely to produce the kind of sophistication and breadth of perspective that the military needs in its leaders (Forsythe et al., 2002; Bartone et al., 2007). Reflecting this trend, West Point has recently incorporated an explicitly developmental philosophy into its overall leader development program (United States Military Academy, 2002; Bartone, 2001). Another trend is the increased attention to personality and cognitive factors that may be related to effective leader performance (Bartone, Snook, and Tremble, 2002; McCormack and Mellor, 2002; Hedlund et al., 2003). Both of these developments reflect a recognition of and appreciation for abiding individual differences and at the same time an understanding that growth and change can occur across the lifespan. Military academies are paying more attention to the assessment of personality factors in cadets, while also providing personalized feedback in a manner designed to facilitate improved self-understanding and growth in those characteristics associated with positive leader performance (Costa, Bartone, Herbst, Brazil, Kelly, Friedman, and McCrae, 2002).

Various personality measures are also being used increasingly in the developmental programs for leaders at U.S. military senior Service colleges. For example, both the U.S. Army War College and the Industrial College of the Armed Forces at National Defense University have implemented comprehensive leader development assessment programs that include measures of the Big Five personality factors and involve detailed feedback sessions for students. Another development worth noting is increased cross-national comparative research into factors that may influence military leader development. For example, a five-nation cross-national study applied a "grounded theory" approach to identify common career and job experiences that appear to spur leader development, possibly in interaction with personality variables (Larsson et al., 2006). The South African military is also using more personality measures in selection and developmental applications (van Dyke, 2008; van Wijk and Waters, 2000).

Conclusion

This chapter has summarized several major theoretical perspectives on leadership and reviewed recent conceptual and empirical work in personality, leadership, and performance in the military. While a profusion of models and approaches continues to mark the leadership field, advances in personality theory and assessment have led to a growing consensus on the importance of individual or personality variables as well as situational ones. It is now clear that personality variables merit serious consideration in high-risk occupations like the military. With a better understanding of how personality factors can influence leader and Soldier performance under rapidly shifting, high-stress conditions, military organizations will be better able to select and develop the best leaders and troops for the future.

References

Allport, G.W. (1985). The historical background of social psychology. In G. Lindzey and E. Aronson (eds.). *Handbook of social psychology* (3ᵈ ed., vol. 1) (1–46). New York: Random House.

Army training and leader development panel officer study: Report to the Army (May 2001). Retrieved January 15, 2009, from <www.army.mil/features/ATLD/report.pdf>.

Avolio, B.J. (1999). *Full leadership development.* Thousand Oaks, CA: Sage.

Barrick, M.R., and Mount, M.K. (1999). The big five personality dimensions and job performance: a meta-analysis. *Personnel Psychology* 44, 1–26.

Bartone, P.T. (August 1988). *Stress, hardiness and symptoms in Bosnia deployed soldiers.* Paper presented at the American Psychological Association Convention, San Francisco, CA.

———. (1989). Predictors of stress-related illness in city bus drivers. *Journal of Occupational Medicine* 31, 657–663.

———. (1999a). Hardiness protects against war-related stress in Army Reserve forces. *Consulting Psychology Journal* 51, 72–82.

———. (1999b). *Personality hardiness as a predictor of officer cadet leadership performance.* Paper presented at the International Military Testing Association Meeting, Monterey, CA.

———. (2000). Hardiness as a resiliency factor for United States forces in the Gulf War. In J.M. Violanti, D. Paton, and C. Dunning (eds.). *Posttraumatic stress intervention: Challenges, issues, and perspectives* (115–133). Springfield, IL: Charles Thomas.

———. (2001). *Leader development programs and research at the U.S. Military Academy, West Point.* Paper presented at the American Psychological Association annual convention, Symposium on Leader Development at the Service Academies: Current and Future Directions, San Francisco, CA.

Bartone, P.T., Johnsen, B.H., Eid, J., Brun, W., and Laberg, J.C. (2002). Factors influencing small-unit cohesion in Norwegian Navy officer cadets. *Military Psychology* 14, 1–22.

Bartone, P.T., and Priest, R.F. (2001). *Sex differences in hardiness and health among West Point cadets.* Paper presented at the American Psychological Society Meeting, Toronto, Canada.

Bartone, P.T., and Snook, S.A. (2000). *Gender differences in predictors of leader performance over time.* Paper presented at the American Psychological Society 12ᵗʰ Annual Convention, Miami Beach, FL.

Bartone, P.T., Snook, S.A., Forsythe, G.B., Lewis, P., and Bullis, R.C. (2007). Psychosocial development and leader performance of military officer cadets. *Leadership Quarterly* 18, 490–504.

Bartone, P.T., Snook, S.A., and Tremble, T. (2002). Cognitive and personality predictors of leader performance in West Point cadets. *Military Psychology* 14, 321–328.

Bartone, P.T., Ursano, R.J., Wright, K.W., and Ingraham, L.H. (1989). The impact of a military air disaster on the health of assistance workers: A prospective study. *Journal of Nervous and Mental Disease* 177, 317–328.

Bass, B.M. (1985). *Leadership and performance beyond expectations.* New York: Free Press.

———. (1990). *Bass and Stogdill's handbook of leadership.* New York: Free Press.

———. (1998). *Transformational leadership.* Mahweh, NJ: Lawrence Erlbaum.

Bass, B.M., and Avolio, B.J. (1989). *Manual: Multifactor leadership questionnaire.* Palo Alto, CA: Consulting Psychologists Press.

Berger, P.L., and Luckmann, T. (1966). *The social construction of reality.* Garden City, NJ: Doubleday.

Block, J. (1995). A contrarian view of the five-factor approach to personality description. *Psychological Bulletin* 117, 187–215.

Blyth, D.E. (1987). *Leader and subordinate expertise as moderators of the relationship between directive leader behavior and performance.* Doctoral dissertation, University of Washington, Seattle, WA.

Bono, J.E., and Judge, T.A. (2004). Personality and transformational and transactional leadership: A meta-analysis. *Journal of Applied Psychology* 89, 901–910.

Burns, J.M. (1978). *Leadership.* New York: Harper and Row.

Conger, J.A. (1989). *The charismatic leader: Behind the mystique of exceptional leadership.* San Francisco, CA: Jossey-Bass.

Costa, P.T., Bartone, P.T., Herbst, J.H., Brazil, D., Kelly, D., Friedman, S.B., and McCrae, R.R. *Five-factor model of personality and leader performance among United States Army cadets: Prediction equations and feedback strategies.* Unpublished manuscript.

Costa, P.T., and McCrae, R.R. (1992). *The NEO PI-R professional manual.* Odessa, FL: Psychological Assessment Resources.

Dawda, D. (Summer 1997). Personality or factor-analytically developed, lay person, self-reported, single-word, adjectival descriptors of global characteristics of personality structure: The NEO five-factor model and skimming the surface of the wetlands of personality. *Psybernetika.* Retrieved January 15, 2009, from <http://psybernetika.ca/issues/1997/summer/dawda.htm>.

Day, D. (2001). Leadership development: A review in context. *Leadership Quarterly* 11, 581–613.

Department of the Army (1990). Field Manual 22-100, *Army Leadership.* Washington, DC: U.S. Government Printing Office.

Digman, J.M. (1990). Personality structure: Emergence of the five-factor model. *Annual Review of Psychology* 41, 417–440.

Druckman, D., Singer, J.E., and Van Cott, H. (eds.). (1997). *Enhancing organizational performance.* Washington, DC: National Academy Press, 98.

Druskat, V.U. (1994). Gender and leadership style: Transformational and transactional leadership in the Roman Catholic Church. *Leadership Quarterly* 5, 99–119.

Eagley, A.H., Karau, S.J., and Makhijani, M.G. (1995). Gender and the effectiveness of leaders: A meta-analysis. *Psychological Bulletin* 117, 125–145.

Fiedler, F.E. (1964). A contingency model of leader effectiveness. In Leonard Berkowitz (ed.). *Advances in experimental social psychology,* vol. I (149–190). New York: Academic Press.

———. (1967). *A theory of leadership effectiveness.* New York: McGraw-Hill.

———. (1986). The contribution of cognitive resources to leadership performance. *Journal of Applied Social Psychology* 16, 532–548.

Fleishman, E.A., and Harris, E.F. (1962). Patterns of leadership behavior related to employee grievances and turnover. *Personnel Psychology* 15, 43–56.

Forsythe, G.B., Snook, S.A., Lewis, P., and Bartone, P.T. (2002). Making sense of officership: Developing a professional identity for 21st century Army officers. In D.M. Snider, G.L. Watkins, and L.J. Matthews (eds.). *The future of the Army profession* (357–378). New York: McGraw Hill.

Gal, R. (1987). Military leadership for the 1990s: Commitment-derived leadership. In L. Atwater and R. Penn (eds.). *Military leadership: Traditions and future trends.* Annapolis, MD: U.S. Naval Academy.

Goleman, D. (1995). *Emotional Intelligence.* New York: Bantam Books.

Gough, H. (1990). Testing for leadership with the California psychological inventory. In K.E. Clark and M.B. Clark (eds.). *Measures of leadership* (355–379). West Orange, NJ: Leadership Library of America, Inc.

Hedlund, J., Forsythe, G.B., Horvath, J.A., Williams, W.M., Snook, S.A., and Sternberg, R.J. (2003). Identifying and assessing tacit knowledge: Understanding the practical intelligence of military leaders. *Leadership Quarterly* 14, 117–140.

Hogan, R., Raskin, R., and Fazzini, D. (1990). The dark side of charisma. In K.E. Clark and M.B. Clark (eds.). *Measures of leadership* (343–354). West Orange, NJ: Leadership Library of America, Inc.

Hollander, E.P., and Julian, J.W. (1969). Contemporary trends in the analysis of leadership processes. *Psychological Bulletin* 71, 387–397.

Homer (1975). *The odyssey of Homer* (R. Lattimore, trans.). New York: Harper and Row.

Hough, L. (1992). The "big five" personality variables—construct confusion: Description versus prediction. *Human Performance* 5, 139–155.

House, R.J. (1971). A path-goal theory of leader effectiveness. *Administrative Science Quarterly* 16, 321–339.

———. (1977). A 1976 theory of charismatic leadership. In J.G. Hunt and L.L. Larson (eds.). *Leadership: The cutting edge*. Carbondale, IL: Southern Illinois University Press.

———. (1988). Leadership research: Some forgotten, ignored, or overlooked findings. In J.G. Hunt, B.R. Baliga, H.P. Dachler, and C.A. Schriesheim (eds.). *Emerging leadership vistas*. Lexington, MA: Lexington Books.

———. (1988). Power and personality in complex organizations. In L.L. Cummings and B.M. Straw (eds.). *Research in Organizational Behavior*. Greenwich, CT: JAI Press.

———. (1996). Path-goal theory of leadership: Lessons, legacy, and a reformulated theory. *Leadership Quarterly* 7, 323–352.

House, R.J., and Howell, J.M. (1992). Personality and charismatic leadership. *Leadership Quarterly* 3, 81–108.

House, R.J., and Mitchell, T.R. (Fall 1974). Path-goal theory of leadership. *Contemporary Business* 3, 81–98.

Hughes, R.L., Ginnett, R.C., and Curphy, G.K. (1996). *Leadership: Enhancing the lessons of experience*. Chicago: Irwin.

Jacobs, T.O., and Jaques, E. (1986). Leadership in complex systems. In J. Zeidner (ed.). *Human productivity enhancement: Volume 2, Organizations, personnel and decision-making* (7–65). New York: Praeger.

Janis, I. (1972). *Victims of groupthink*. Boston, MA: Houghton-Mifflin.

Judge, T.A., and Bono, J.E. (2000). Five-factor model of personality and transformational leadership. *Journal of Applied Psychology* 85, 751–765.

Katz, D., and Kahn, R.L. (1952). Some recent findings in human relations research in industry. In G.E. Swanson, T.M. Newcomb, and E.L. Hartley (eds.). *Readings in social psychology*. New York: Holt Rinehart and Winston.

Kegan, R. (1982). *The evolving self: Problem and process in human development*. Cambridge, MA: Harvard University Press.

Kobasa, S.C. (1979). Stressful life events, personality, and health: An inquiry into hardiness. *Journal of Personality and Social Psychology* 37, 1–11.

Kuhnert, K.W., and Lewis, P. (1987). Transactional and transformational leadership: A constructive/developmental analysis. *Academy of Management Review* 12, 648–657.

Lall, R., Holmes, E.K., Brinkmyer, K.R., Johnson, W.B., and Yatko, B.R. (1999). Personality characteristics of future military leaders. *Military Medicine* 164, 906–910.

Larsson, G., Bartone, P.T., Bos-Bakx, M., Danielsson, E., Eid, J., Jelusie, L., Johansson, E., Moelker, R., Sjöberg, M., Vrbanjac, A., Bartone, J., Forsythe, G.B., Pruefert, A., and Wachowicz, M. (2006). Leader development in natural context: A grounded theory approach to discovering how military leaders grow. *Military Psychology* 18S, 69–81.

Lau, A. (1990). Military leadership. In C. Cronin (ed.). *Military psychology: An introduction* (49–69). Needham, MA: Simon and Schuster.

Lewis, P. and Jacobs, T.O. (1992). Individual differences in strategic leadership capacity: A constructive/developmental view. In R.L. Phillips and J.G. Hunt (eds.), *Strategic leadership: A multiorganizational-level perspective* (121–137). Westport, CT: Quorum Books.

Likert, R. (1967). *The human organization: Its management and value*. New York: McGraw-Hill.

Lord, R.G., DeVader, C.L., and Alliger, G.M. (1986). A meta-analysis of the relation between personality traits and leadership perceptions: An application of validity generalization procedures. *Journal of Applied Psychology* 71, 402–410.

Maddi, S.R. (1967). The existential neurosis. *Journal of Abnormal Psychology* 72, 311–325.

———. (1987). Hardiness training at Illinois Bell Telephone. In J.P. Opatz (ed.). *Health promotion evaluation.* Stevens Point, WI: National Wellness Institute.

———. (1999). Comments on trends in hardiness research and theorizing. *Consulting Psychology Journal* 51, 67–71.

Maddi, S.R., Khoshaba, D.M., Jensen, K., Carter, E., Lu, J.L., and Harvey, R.H. (2002). Hardiness training for high risk undergraduates. *NACADA Journal* 22, 45–55.

Maddi, S.R., and Kobasa, S.C. (1984). *The hardy executive: Health under stress.* Homewood, IL: Dow Jones-Irwin.

Mann, R.D. (1959). A review of the relationship between personality and performance in small groups. *Psychological Bulletin* 56, 241–270.

McCaulley, M.H. (1990). The Myers-Briggs type indicator and leadership. In K.E. Clark and M.B. Clark (eds.). *Measures of leadership* (381–418). West Orange, NJ: Leadership Library of America, Inc.

McCormack, L. and Mellor, D. (2002). The role of personality in leadership: An application of the five-factor model in the Australian military. *Military Psychology* 14, 179–197.

McCrae, R.R. (ed.). (1992). The five-factor model: Issues and applications. [Special Issue]. *Journal of Personality* 60 (2).

Moskos, C.C., Williams, J.A., and Segal, D.R. (2000). *The postmodern military: Armed forces after the Cold War.* New York: Oxford University Press.

Mumford, M.D., Zaccaro, S.J., Harding, F.D., Fleishman, E.A., and Reiter-Palmon, R. (1993). *Cognitive and temperament predictors of executive ability: Principles for developing leadership capacity* (ARI Technical Report 977, AD A267 589). Alexandria, VA: U.S. Army Research Institute for the Behavioral and Social Sciences.

Mumford, M.D., Zaccaro, S.J., Harding, F.D., Jacobs, T.O., and Fleishman, E.A. (2000). Leadership skills for a changing world. *Leadership Quarterly* 11, 11–35.

Nagl, J.A. (2005). *Learning to eat soup with a knife: Counterinsurgency lessons from Malaya and Vietnam* (2^d ed.). Chicago, IL: University of Chicago Press.

Phillips R.L., and Hunt, J.G. (eds.). (1992). *Strategic leadership: A multiorganizational level perspective.* Westport, CT: Quorum Books.

Plato (1991). *The republic* (Allan Bloom, trans.). New York: Basic Books.

Podsakoff, P.M., MacKenzie, S.B., Ahearne, M., and Bommer, W.H. (1995). Searching for a needle in a haystack: Trying to identify the illusive moderators of leadership behaviors. *Journal of Management* 21, 423–470.

Potter, E.H., and Fiedler, F.E. (1981). The utilization of staff member intelligence and feedback under high and low stress. *Academy of Management Journal* 24, 361–376.

Rubenzer, S.J., Faschingbauer, T.R., and Ones, D.S. (2000). Assessing the U.S. Presidents using the Revised NEO Personality Inventory. *Assessment* 7, 403–420.

Ryback, D. (1998). *Putting emotional intelligence to work: Successful leadership is more than IQ.* Boston, MA: Butterworth-Heinemann.

Salovey, P. and Mayer, J.D. (1990). Emotional intelligence. *Imagination, Cognition and Personality* 9, 185–211.

Sashkin, M. (1992). Strategic leadership competencies. In R.L. Phillips and J.G. Hunt (eds.). *Strategic leadership: A multiorganizational-level perspective* (139–160). Westport, CT: Quorum Books.

Segal, M.S. (1995). Women's military roles cross-nationally: Past, present and future. *Gender and Society* 9, 757–775.

Shamir, B., and Ben-Ari, E. (2000). Challenges of military leadership in changing armies. *Journal of Political and Military Sociology* 28, 43–59.

Shamir, B., House, R.J., and Arthur, M.B. (1993). The motivational effects of charismatic leadership: A self-concept based theory. *Organizational Science* 4, 1–17.

Shaw, M.E. (1955). A comparison of two types of leadership in various communication nets. *Journal of Abnormal and Social Psychology* 50, 127–134.

Shaw, M.E., and Blum, J.M. (1966). Effects of leadership style upon group performance as a function of task structure. *Journal of Personality and Social Psychology*, 238–242.

Shaw, M.E., and Costanzo, P.R. (1982). *Theories of social psychology.* New York: McGraw-Hill, 388.

Shay, J. (2002). *Odysseus in America: Combat trauma and the trials of homecoming.* New York: Scribner.

Sternberg, R.J., Forsythe, G.B., Hedlund, J., Horvath, J.A., Wagner, R.K., Williams, W.M. et al. (2000). *Practical intelligence in everyday life.* New York: Cambridge University Press.

Stodgill, R.M. (1948). Personal factors associated with leadership: A survey of the literature. *Journal of Psychology* 25, 35–71.

———. (1974). *Handbook of leadership.* New York: Free Press, 259.

Thomas, J.L., Dickson, M.W., and Bliese, P.D. (2001). Values predicting leader performance in the U.S. Army Reserve Officer Training Corps Assessment Center: Evidence for a personality-mediated model. *Leadership Quarterly* 12, 181–196.

United States Military Academy (2000). *PL300 military leadership course guide.* West Point, NY: Department of Behavioral Sciences and Leadership, U.S. Military Academy.

United States Military Academy (June 2002). *Cadet leader development system* (USMA Circular 1–101). West Point, NY: U.S. Military Academy, Office of the Superintendent.

van Dyk, G. (September 2008). *A profile analysis of students at the South Africa Military Academy.* Paper presented at First International Conference on Strategic Challenges for African Armed Forces for the Next Decade, Saldanha, South Africa.

van Wijk, C., and Waters, A.H. (2000). Personality characteristics of South African Navy submarine personnel. *Military Medicine* 165, 656–658.

Weber, M. (1947). *The theory of social and economic organizations* (T. Parsons, trans.). New York: Free Press.

Weik, K.E. (1995). *Sensemaking in organizations.* Thousand Oaks, CA: Sage.

Westman, M. (1990). The relationship between stress and performance: The moderating effect of hardiness. *Human Performance* 3, 141–155.

Widiger, T.A., and Trull, T.J. (1997). Assessment of the five-factor model of personality. *Journal of Personality Assessment* 68, 228–250.

Wofford, J.C., and Liska, L.Z. (1993). Path-goal theories of leadership: A meta-analysis. *Journal of Management* 19, 858–876.

Yukl, G. (1998). *Leadership in organizations.* Upper Saddle River, NJ: Prentice Hall.

Zaccaro, S.J., Gilbert, J., Thor, K.K., and Mumford, M.D. (1991). Leadership and social intelligence: Linking social perceptiveness and behavioral flexibility to leader effectiveness. *Leadership Quarterly* 2, 317–331.

Chapter 7

Development and Fielding of the U.S. Army Leader Self-development Portfolio

James Ness, Thomas A. Kolditz, Philip Lewis, and David M. Lam

Leader Self-development in the U.S. Army

In a 2002 review of Army leader training and development, the Army Training and Leader Development Panel (2003) determined that leader adaptability and self-awareness are critical for the leadership of our future force. The panel was chartered by then–Chief of Staff of the Army (CSA), General Peter Schoomaker, to examine the state of training, education, and leader development programs and the extent to which these programs meet the Army's transformation objectives. In October 2003, as a result of the panel's recommendations, General Schoomaker announced "The Bench" as one of the Army's 15 focus areas to address the leader self-development needs identified by the panel (Institute of Land Warfare, 2004).

The broad purpose of The Bench initiative was to develop leader adaptability and self-awareness within the Army. The Army War College (AWC), under the leadership of Major General David Huntoon, accepted strategic oversight of this initiative, which included policy change initiatives and the implementation of an Army-wide virtual leader self-development program. On July 2, 2004, the AWC received authorization from the office of the CSA to execute a strategy to accomplish Bench initiatives. This authorization included guidance that the Department of Behavioral Sciences and Leadership (BS&L) at the U.S. Military Academy design and implement a virtual leader self-development program that included an online portfolio providing leaders a view over time of the development of their leadership perspective and decisionmaking processes.

In this chapter, we will describe the development and implementation of the online Leader Development Portfolio (LDP) and its transition to the U.S. Army Training and Doctrine Command (TRADOC) in October 2007. We will begin with a discussion of the process of turning the CSA's intent into reality and of how constraints were identified and managed through leveraging resources extant on Army Knowledge On-line/Defense Knowledge Online (AKO/DKO).[1] We will then describe how "self-awareness" was defined. This was critical in developing methods to generate information to provide feedback to the leader. We will then turn to the structure of the LDP and discuss the goal of developing a software system that was scalable, adaptable, and usable, and the means by which the goal was accomplished. Finally, we will provide data showing the effectiveness of the system and how the metadata were used to modify the structure of the system to better diffuse leadership lessons and best practices throughout the force.

Building the Leader Development Portfolio

Constraints and Their Resolution

The U.S. Military Academy (USMA) received $500,000 from the Army War College shortly after being tasked to build the LDP. These resources were gladly accepted, but the challenge was how to build the system specified by the CSA for $500,000. Further, the money had to be obligated by the end of the fiscal year. That is, a contract had to be initiated before the end of October 2004, about 2 months from the time we received the funding.[2] The constraints of limited dollars and time to obligate the dollars were significant in our planning process.

Some initial decisions about what was and was not possible had to be made.[3] First, it was recognized that necessary information assurances required for the system could not be instituted in a timely manner (HQDA, 2007). Second, there was not enough lead time for Defense Contracting to establish a new contract. Third, there was not enough money to build a unique standalone system, purchase the necessary hardware, and hire personnel to program and maintain the system. Asking for more funding and time was considered, but it was felt that with some creative thinking, a functional system could be built with the available resources. One idea was to house the system on USMA servers, but information assurance was still an issue and there were no programmers organic to USMA available to work on the system. The other idea was to work with AKO and leverage its resources.

After a few cold calls, we found our way to the AKO program manager, Lieutenant Colonel Kenneth Blakely. He agreed in principle to assist, and we arranged a meeting at USMA to work out an agreement. The result was that AKO would accept our $500,000 and amend its contract to support our effort. AKO was interested in both bolstering the functionality of its portal and hearing our ideas, especially since it had ambitions to explore similar LDP functionality, such as private blogs. With AKO came a solution to our information assurance, contracting, storage, programming, and access issues. Our next step was to specifically define the requirements for an operational prototype.

LDP Concept of Leader Self-development

Before constructing an online leader self-development system, we had to resolve what leader self-development meant in the context of CSA intent. In defining the issues, we consulted our colleagues within BS&L, the AWC, the U.S. Army Research Institute, the Center for Army Leadership, and the Army's clinical mental health fields.[4] Without a well-defined conceptual framework, the specific functional requirements for the software engineers to build an online system could not be written. The conceptual framework required a theoretical underpinning supported in the professional scientific literature. It needed to be congruent with doctrinal concepts of U.S. Army leader development, to meet the CSA's leader development intent and specified functionality, and have broad acceptance from the community of intended users.

Doctrinal and CSA Guidance

In the year leading up to the announcement of The Bench initiative at the October 2003 Association of the U.S. Army (AUSA) meeting, the Army Training and Leader Development Panel (ATLDP) released its findings in a report to the Army (ATLDP, 2003). The overall study included Officer, Noncommissioned Officer (NCO), Warrant Officer, and Civilian panels (Army Leader Development Panel Study Reports [n.d.]; AUSA, 2008). The major finding was that to meet transformation objectives, training and leader development programs should produce self-aware and adaptive leaders. The panels found that formal educational and operational experiences cannot provide all the knowledge needed to be self-aware and adaptive leaders. Further, the panels commented that self-development is essential to lifelong learning and provides the training and education that formal educational and operational experiences cannot supply.

From this recommendation, we chose as our goal for the LDP to improve self-awareness for the purpose of improving leader adaptability. To begin the requirements process for the LDP, we adopted from the report those leader development and training principles related to self-development and self-awareness:

- *Know Yourself.* Self-aware and adaptive leaders are the basis for success in full-spectrum operations. The relationship between self-awareness and adaptability is symbiotic. The greater the self-awareness gained by assessment against measurable standards, the more adaptive the leader. Through a commitment to lifelong learning enabled by self-development, leaders can narrow the knowledge gaps not filled or addressed through educational and operational experiences.

- *Lifelong Learning.* Part of the Army's culture should be the commitment to lifelong learning by its leaders. Learning organizations support self-awareness and adaptability. Lifelong learning requires standards, with related tools for assessment, feedback, and self-development.

- *Mentorship.* Mentoring is not a formal program, but part of the stock and trade of the profession. It focuses on the art of leadership.

The "Know Yourself" aphorism is at the core of the LDP's leader developmental goal. The LDP must help leaders know how they tend to think, feel, and react, and who they tend to surround themselves with in the various contexts of decisionmaking to be able to lead with full awareness of the nature and effects of their decisions. The "Lifelong Learning" principle addresses the need for the LDP to be accepted and compelling as part of the culture of the Army. Further, the principle specifically states that the LDP requires standards, tools, and feedback. The "Mentorship" principle was adopted because it is part of the profession and fundamental to accelerating life experience through the experiences and wisdom of others. Working through mentors adds perspective to issues confronting a leader, puts feedback in context, guides resolution, and mitigates repeating one's own and others' past errors.

Drawing from the ATLDP study and the CSA's tasks, we began to sketch out the requirements for the LDP. We knew from the CSA specified task that we needed a leader self-development system that was online and on-demand, maintained user anonymity, and had a two-store security system. From the ATLDP, we knew that the system needed a toolbox of standardized methods for promoting self-development relevant to current and future leader positions, a mechanism for feedback, and a means for mentoring. Further, this system had to be accepted by the force.

Although the Army wanted self-aware and adaptive leaders, there was no developmental pedagogy offered to achieve this endstate. Identifying the leader development need was the charge of the panel. Development of mechanisms to achieve the endstate was the panel's tasking to the leader development community. Thus, we had to develop a working definition of self-awareness within a self-development context so we could measure it and provide structured developmental feedback that would improve leader adaptability.

Self-awareness Defined

Engaging the USMA library staff, we performed an exhaustive review of the literature to find a theory of self and self-awareness that was parsimonious, easily applied, and effective within the context of U.S. Army leader self-development. Duval and Wicklund's (1972) concept of self-awareness was decided on. Most appealing was the fact that this theory only had two states, the subjective and the objective. All other theories had many more divisions of the self and simply lacked the parsimony to be practically applied. Second, there were specific findings that showed that indeed becoming more self-aware, under their rubric, had positive effects to include those that could be categorized as adaptable.

However, self-awareness was also shown to have significant negative effects, illustrating that one must be careful with how one facilitates self-awareness and provides feedback in a self-development context. As an analogy, a system that facilitates self-awareness is like a mirror that directs attention to reflections from that mirror. A mirror that provides enough resolution to comb one's hair helps improve one's appearance. However, a mirror that provides too much detail about one's "flaws" can bring about unhealthy obsessive or compulsive body-focused behaviors.

For Duval and Wicklund, the self is indivisible; however, consciousness can act in two ways in relation to the self. It can be focused on the task in which the self is engaged, a state referred to as the "Subjective Self." Consciousness can also be focused on the self, a state referred to as the "Objective Self." While in the Subjective Self state, one is focused on external events, where one is typically absorbed in the task, with the purpose of achieving goals or otherwise influencing task-related outcomes. While in the Objective Self state, one is focused on the self as the object of study. In this state, attention is on, for example, a state of arousal, likes and dislikes, and personal history. It is in this state that the individual self-monitors and self-reflects. Both objective and subjective states are susceptible to conscious activity with attention as the limiting resource. Under this rubric, our goal was to promote a healthy amount of self-monitoring in situations demanding a focus on external events to check stereotypic or ritualized behavior often employed to solve problems. In this way, the evaluation of such behaviors could promote more creative and critical thinking about a problem, and potentially improve perspective-taking when engaging others.

A number of studies showed that self-awareness promotes constructive outcomes in the areas of perspective-taking, self-control, creative achievement, and self-esteem (Silvia and Duval, 2001; Wicklund and Gollwitzer, 1987). However, a unique set of feedback structures and tools is required. In addition, a context of privacy, anonymity, and security are needed to promote self-reflection in order to engender self-awareness and diminish the threat of self-consciousness. If the system used to promote self-reflection did not take care to incorporate psychological safeguards, then instead of healthy outcomes, the system could yield self-consciousness, an exacerbation of the negative, a limited focus, and a decrease in intrinsic motivation (Fenigstein, Scheier, and Buss, 1975). Thus, when building the LDP, care was taken in structuring the tools, context, and feedback to promote self-awareness without inducing self-consciousness or control by a "generalized other" (Mead, 1934). In other words, we did not want people to strive toward a mean or to view a norm as a "minimum standard," but to come to know their strengths and weaknesses within the context of self (versus in comparison to the generalized other). Further, the goal was that the knowledge gained would be useful to better perform duties and achieve personal goals. For this reason, the tools employed did not make comparisons to populations or norms. Comparisons were made across natural groupings within the Army's organizational and social structure (for example, peers, subordinates, seniors, NCOs, officers, acquaintances, and friends). Developmentally, we followed the Army's doctrinal progression from junior to senior leader for both uniformed and civilian leaders, and focused on aspects particular to each period of progression.

LDP Functional Components

The LDP system was programmed such that it was scalable, adaptable, and usable.[5] By *scalable*, we mean the ability of the system to adapt to increased demand. By *adaptable*, we mean the ability to easily incorporate functionality written for the AKO portal and the capacity of the system to meet the self-development needs of the individual leader and leader development community. By *usable*, we refer to a system that is easy to learn and use, efficient, well laid out, effective, and compelling. To ensure adaptability, the LDP was written as an integrated system of major functional components on which other components can call when that particular functionality is needed. The major functional components of the LDP are the User Interface, the Management Console, the Developmental Box, and the Administrator Console. Subcomponents of each of these major components were phased in to meet timelines and budget constraints. Each major component is discussed in detail in the subsequent sections.

User Interface

Figure 7–1 shows the major components of the User Interface: the leader development portfolio, the database, the results engine, and the two-store security system where the LDP context and database are linked by way of the Web services bridge. Major components of the portfolio include the leader development toolbox, where one can initiate a self-assessment instrument, engage a mentor, solicit feedback through the respondent chooser, view assessment feedback, reach the Help Desk, and access the Benchworks supplemental leader development Web site.

Figure 7–1. **Leader Development Portfolio Major Components**

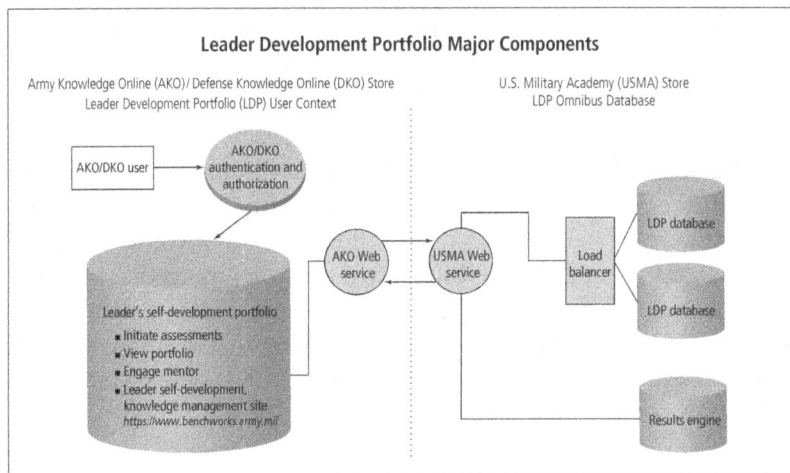

Leader Development Portfolio Major Components

Army Knowledge Online (AKO)/ Defense Knowledge Online (DKO) Store
Leader Development Portfolio (LDP) User Context

U.S. Military Academy (USMA) Store
LDP Omnibus Database

AKO/DKO user

AKO/DKO authentication and authorization

AKO Web service

USMA Web service

Load balancer

LDP database

LDP database

Leader's self-development portfolio
- Initiate assessments
- View portfolio
- Engage mentor
- Leader self-development, knowledge management site
  https://www.benchworks.army.mil

Results engine

Two-store security system. The two-store security system posed a bit of a problem. The original concept as presented to us was to physically remove the data from the network, load it onto a "standalone" storage device that was not connected to the net, and, when requested, physically move the data back to the server from the storage device. This was deemed impractical, as it would significantly hinder the ability to have an online on-demand system if we were continually taking individual leader development portfolios off- and online. However, the CSA's specified tasks remained: no one was to be able to access this information except the user, and the security system would be a two-tiered system. Our solution was to separate the data from the context and to only allow access to the data through the AKO portal authentication process. The data reside on servers outside of AKO's physical plant, and the data backups are separated from the active data servers. The program itself, or context, resides on the AKO portal. The only way to access the data is through the portal, which activates a Web services encryption bridge that allows access only to that user's information. We took great pains to set up this procedure for two reasons: the CSA directed such a system; and pedagogically, in order to promote self-reflection, the user has to feel a sense of privacy and security. Thus, there is no way to access another's data, even at the administrative level. The program is designed to be anonymous and used exclusively for individual leader self-developmental purposes. We specifically put in safeguards so LDP data could not be incorporated into the formal evaluation system or used to assess command climate.

Leader Development Toolbox. The Leader Development Toolbox consists of tools designed to solicit others' feedback on those competencies identified by the Army for leaders to perform well. The toolbox tailors available content based on the user's position within the Army. The individual leader's developmental needs within the Army are defined in the LDP by the individual's rank and position. The principal tool is the 360 (Toegel and Conger, 2003). This tool consists of 12 to 15 questions about doctrinally identified competencies required to perform well in one's rank and position. The 360 programs were developed separately for different groups of Army leaders: NCOs, company grade officers, field grade officers, civilian personnel, and some specialty-specific groups, such as chaplains.

To execute a 360, the leader first answers the questions for himself and then solicits peers, subordinates, and seniors to anonymously respond to the same questions. The responses are not presented to the initiator until a minimum number within a category have responded. The feedback is presented as a box plot of the pooled data within a category. Minimum numbers of respondents in each category (6 subordinates, 6 peers, 3 seniors) were an optimization of respondent anonymity, timeliness of feedback, and available pool of individuals within each respondent category. Usability tests showed that on average, a leader could expect a completed 360 assessment within 30 days. There were provisions within the system to remind respondents of outstanding feedback, to add respondents to an ongoing assessment, or to close out an assessment and only be allowed to view feedback from those respondent categories that met the minimums.

The 360 was designed to provide insights to prepare the leader for the next level of leadership. Thus, the 360 assessments were constructed from research identifying leader core competencies from doctrine and from interviews and focus-group testing. The Company Grade 360 was developed by members of the USMA BS&L through work with company grade officers returning from Iraq and Afghanistan. The Field Grade 360 was derived from the Leadership Lessons at Division Command Level study with permission of the lead author (Ulmer et al., 2004). The Civilian Leadership 360 was developed by BS&L through work with Department of Defense Senior Executive Service personnel. The NCO 360 was developed in collaboration with the U.S. Army Sergeants Major Academy. Other 360s such as that for the Chaplain Corps were provided directly by the leader developers within the respective corps.

Also within the toolbox is a coach's journal where leaders can privately share results of their portfolio with a mentor and create private blogs to discuss leadership issues within their selected groups. The purpose of this system is to support the stock and trade of the profession, facilitating contact between leaders and their mentors, and to provide leader forums regardless of distance. Finally, prior to transition to TRADOC, BS&L developed and added a personality tool to the toolbox, which will be discussed later.

Respondent Chooser. The program itself is adaptable in that it calls a number of AKO functions, such as the White Pages function. The White Pages call within the Respondent Chooser allows the initiator of an assessment to quickly and efficiently choose colleagues from whom input is desired. Within the Respondent Chooser, the initiator can search for respondents by all or part of a respondent's first name, last name, service, rank, and branch. With this functionality, the initiator can find individuals with a registered AKO/DKO account and choose them to respond to an assessment. The system automatically emails the potential respondents with a link to respond or decline. The initiator is provided feedback on how many individuals within each category have responded, have responses pending, or have declined, but the system does not track which individuals choose or decline to participate. The initiator can choose to send out reminder emails or add new respondents if the trends indicate that a particular category may not reach the minimum number of respondents. The initiator can also choose to close out an assessment at any time after 30 days in process. However, only the feedback from those respondent categories that reach the required minimum number of responses will be displayed.

The dynamic that we had to optimize in defining an assessment time window so feedback would keep up with developmental events (promotion, position change, critical incidents, and so forth) was the balance between the initiator's need for timely feedback and the respondent's need for anonymity and a reasonable interval to respond. We began with a system that reminded respondents weekly of their requirement to respond or decline. That strategy produced many angry emails and telephone calls to the Help Desk. We needed a different tactic.

For those who may foray into a software development project, a rule of thumb is that for usability issues, the best solution is typically to give users sufficient information that they can take control. In this case, we provided a table that was updated in real time as to the number of respondents registered by category, as well as the number who had responded, declined, or were pending. Then we gave users the ability to send out email reminders as well as information about the last time they sent out reminders and the ability to add more respondents.

Further, after 30 days, we gave users three options. Option 1 provided the ability to leave an assessment open, which meant they could continually add new respondents and send out reminders, but would have to wait to view their results until all category minimums were met and they could not initiate a new assessment of that current type. Option 2 provided the ability to close an assessment, which meant users could view the results from the categories of respondents in which the minimums were reached and that the results would continue to update as respondents provided feedback. Under this option, reminder emails could be sent, but no further respondents could be added and they could not initiate a new assessment of that current type. Option 3 provided the ability to complete an assessment, which meant the results would be posted for the respondent categories that met the minimums. No further reminders could be sent, no additional respondents could be added, and no further updates to the results were made. Under this condition, the initiator could begin a new assessment.

We chose this course of action because in our tests of the system, the majority of assessments were completed within 30 days. Therefore, we wanted to give the respondents that period to answer and to prevent the initiator from prematurely closing an assessment. However, the initiator had the power to send reminders, monitor the numbers of individuals responding, declining, or pending, and add more respondents to more quickly meet minimums for timely feedback. After 30 days, we turned control over to the initiator to continue with an open assessment or to close or complete an assessment. We maintained a distinction between *close* and *complete* so the portfolio would maintain a continuous timeline of assessments. Thus, when leaders looked back over their assessment portfolio, clear chronological demarcations and naming conventions (rank, position, and unit are suggested in the program) would be evident. This structure was put in place to facilitate recall when reflecting on leader developmental issues over the course of a career.

Help Desk. Over the first year of the program, the Help Desk was the first author's AKO email address. This involved good news and bad news. We went with the first author's address because we did not have the resources to institute a Help Desk, and at the time AKO was a bit stingy with corporate accounts. The bad news about the Help Desk was that whatever can go wrong with new software does go wrong, and it was always the first author's fault. In terms of responding to Help Desk issues, we discovered that people react with different levels of intensity despite the similarity of the problem. Notwithstanding that, all are grateful that there is a real person at the other end of the line. However, this is not enough. For a system to remain usable, the Help Desk must respond to queries as soon as possible, provide updates on progress, and deliver resolution that restores user satisfaction with and confidence in the system and gives users a sense that they constructively contribute to system improvement.

Even though we ran usability tests that revealed gross human factor failures with the user interface and brainstormed new functionality in our weekly meetings, the Help Desk was our best source of information to refine system usability. Although time-consuming, it was a means for the first author to understand user concerns and habits to ensure that programming requirements were refined in a manner that produced a more usable system. Knowing the system from the user's perspective also improved our brainstorming process through shaping our big ideas into good ideas that rendered a more scalable, adaptable, and usable system.

AKO eventually granted us a corporate account, and we were also able to hire a professional staff person who managed the Help Desk and tabulated issues for consideration at our weekly meetings to determine how to approach our fixes. The approaches included instructional fixes such as pop-up boxes, additional to-the-point instructions, improved buffering to prevent aberrant mouse clicks, redesigns of user interfaces, and the addition of user functionality. From the logging of Help Desk issues, we were able to improve the LDP. One such improvement was the design of our Management Console, which will be discussed below. Its design included the ability to change and publish instructions without having to wait for the AKO build cycle.[6]

BenchWorks Web site. A Benchworks Web site was developed and deployed as a response to many Help Desk calls as to what leaders could do next to support their own leader development.[7] The Web site contained resources to further research those issues that emerged in the self-reflection process.[8] It also offered other leader developmental tools, which were tested for incorporation into the LDP toolbox. The approach was twofold: there was plenty of leader development information in the literature and simply synthesizing that information could provide useful applications that could be tested on the Web site; and with a nominal investment we could buy a limited number of site licenses of leader self-development "serious games" and offer them as resources to the Army through the Web site.

From the first approach, we developed a critical thinking tool. The tool was based on the research of the U.S. Army Research Institute on the Army Green Program (Hinkle, Karrasch, and Burke, 2006). The Army Green Program recognized that "platoon leaders (PL) are often required to resolve non-tactical problems on which they have received little to no formal training. Many of these problems are atypical, undocumented, and require judgmental resolution by PLs with limited experience as they take command of their first platoon." The Army Green Program researched common issues on which junior leaders often make errors in judgment and compiled reflections from senior leaders as to what newer leaders should do. However, these scenarios were didactic and did not provide a means for junior leaders to explore the extent of the implications of decisions. For this purpose, Vroom's theory of leadership style (Vroom, 2000) was overlaid on the scenarios to generate a series of plausible decisions from which the junior leader could select and then explore the consequences. The scenarios were introduced to remind leaders that decisions are made to optimize a future outcome. The outcome itself cannot be known until the decision is made and its consequences are manifest. In this sense, a decision uses what is known in the present to affect the future. Although outcomes of decisions are not perfectly predictable, they are far from random. From Vroom's theory, critical to an optimal outcome are:

- an understanding of one's habits in assessing the quality of present information
- the leader's and the team's expertise to make the decision
- the competence and motivation of the team to execute the decision.

Appendix A presents the Army Physical Fitness Test (APFT)[9] scenario, one of 10 scenarios available on the Benchworks Web site. The senior leader commentary is based on actual reported mentorship from battalion and brigade commanders.

Concerning the second approach, and working with the Army Green findings, the question arose as to how to accelerate the development of military leadership skills outside the context of one's own real-world decisions. Knowledge gained from real-world experience can take years. Moreover, stakes are high and the breadth of possible decisionmaking practices cannot be systematically explored. Within a simulation, the leader can gain insight into strengths and weaknesses of courses of action and explore the breadth of possible decisions to gain the full awareness of their nature and effects.

Our view of leader development and the tools needed to facilitate the skills and habits of good leaders was that if a tool or program was successful in venues outside the military, it was highly likely it could be scaled and made relevant to the particulars of leader development within the military. We reviewed a number of serious games and practicewares that we could deploy on our Benchworks Web site. We decided on and negotiated a limited site license with Simulearn for Army leaders to access the VLeader program (Simulearn, 2007). The VLeader program has as its theoretical underpinning that leaders must balance tension, ideas, and power to accomplish work (Aldrich, 2004). The VLeader program met with success within the LDP and was transitioned as a commercial-off-the-shelf (COTS) program with the transition of the Leader Development Portfolio to TRADOC. The VLeader program has since been scaled to meet additional Army Leadership requirements and developed into the Army Influence Trainer (PRWeb, 2008).

Management and Administrator Consoles
and the Developmental Box

The Management Console has multiple functionalities. The console's primary function is to allow designated administrators to add tools to the LDP toolbox and thus build an instrument with the ability to apply a variety of scales (for example, Likert, Semantic Differential, Bogardus-like distance scales [Parrillo and Donoghue, 2005], and others), write customized instructions, target content to specific AKO/DKO users and groups, apply a variety of feedback displays via an integrated display generator, and designate respondent groups (such as seniors, peers, subordinates, friends, acquaintances). The ability to create new content allowed us to be immediately responsive to leader developer communities within the Army (like the Chaplain Center and School) through tailoring tools to subject matter experts' specifications and posting content to their communities of interest.

Besides instrument development functionality, the Management Console also has a user log to monitor usage and support Help Desk functions. It provides information on the state of individual assessments to support "bug tracking," a functionality allowing administrators to be immediately responsive to Help Desk calls and to triage problems that need to be forwarded to the software developers. This functionality greatly enhanced user satisfaction through expedient resolution of issues.

The function of the Administrator Console is to change the state of LDP functionality from read/write, to read only, to offline. This console is used for system maintenance. The administrator can post a warning of temporary outage and status of the LDP to users, and also perform local builds to apply security patches, upgrade hardware, and apply software upgrades of applications independent of the AKO portal. Upgrades that involve the AKO portal are performed by AKO and have to meet build-cycle timelines and software quality control. These upgrades cannot be fixed, undone, or otherwise modified until the next scheduled monthly build. For this reason, we built a Developmental Box, which is separate from and mirrors the functionality and systems of the "live" system. The Developmental Box was used to test new code and functionality using test accounts outside of the live system. The testing ensured the code would behave as expected when released on the live system. (As an aside, AKO personnel liked the Developmental Box idea so much that they implemented their own.)

Two further Management Console developments that were planned but not instituted were passed on to TRADOC as recommendations: the ability of an initiator to voluntarily send his or her anonymized data to a Test Psychometrics Repository, and creation of levels of administrator control within the Management Console. The repository would entail the development of a database to store anonymized data and would include procedures for obtaining informed consent to use the data for research purposes to improve current tools and develop new ones.

The creation of levels of administrator control was to give leader development principals the ability to control leader self-development content for their communities of practice. The idea was to stratify control based on the existing organizational structure of the U.S. Army Combined Arms Center subordinate schools, with the Center for Army Leadership and USMA as omnibus administrators.

The Army Personality Tool

One of the charges to the leader development team from the Army War College was that the LDP should not be just a 360 tool. At the time of its development, the U.S. Military Academy was using the Revised NEO Personality Inventory (NEO–PI–R) (Costa and McCrae, 2003) instrument as a developmental feedback tool with cadets taking the test in their freshman and again in their junior year. Informed consent was also collected so longitudinal data existed. However, what was lacking was a focus for applying the data and a theoretical framework for interpreting the practical significance of personality on military performance.

Personality is a construct long debated in the field of psychology and a construct that the authors struggled with in terms of the "so what" of knowing that you are an introvert or extrovert. Our vexation centered on the question of stability of the personality phenotype. We readdressed this debate because it was critical to determining how the individual leader could use the information developmentally and to advantage.

There is little doubt that any given individual tends to behave in a relatively stable manner (Siegel and Bergman, 2002). However, there is considerable debate as to the source of the stability (Mischel, 1977). Arguments generally emphasize either personal or situational variables, or some combination of them. Although analogous to nature/nurture dichotomies, the debate over the stability of personality has its roots in a learning theory debate over the specificity versus the generality of learning experience. The catalyst for the debate was a series of transfer of training studies led by Hartshorne (1928, 1929, 1930). The studies showed significant inconsistencies across situations in children's moral behavior. Notwithstanding that, consistent but low intercorrelations across situations in behavior were also revealed. Herein lies the crux of the debate: how to explain the source of low but consistent intercorrelations. Hartshorne attributed the consistencies to identical elements in the stimulus conditions across the test situations. In other words, Hartshorne argued that the stability for the psychological construct of personality exists in the situational variables.

The notion that situational variables afford consistent behavior originated with connectionist-oriented learning theories. In 1903, Thorndike articulated the specialist viewpoint, arguing the theory of common elements. The theory states that a change in one function alters any other only insofar as the two functions have common elements. Thus, to the extent stimulus characteristics spur equivalent responses, habits will form and behavior will generalize across these situations. This consistency is only understood in relation to the experiences of the individual and not as a result of any dispositional characteristics. This suggests that ontologically, what develops in terms of the personality is an ability to see consistencies in the stimulative environment and behave in a like manner in response to perceptions of consistent stimulus events.

Despite the appeal of causal stimulus-response connections to explain learning, connectionist theorists were unable to uniquely define the stimulus-response connections across contexts. The inability to define "identical elements" a priori across conditions rendered the notion of "identical elements" arbitrary to the post hoc justifications of the researcher. This weakness led those who maintained the potency of situational variables to shift their focus from respondent behavior, a class of responses defined in terms of stimuli that reliably produce them, to operant behavior, a class of behavior modifiable by the consequences of the behavior. Skinner (1977) makes the case for operant behavior, suggesting that what guides behavior is an ability to respond to contingencies in the environment. What develops is the ability to perceive relevant contingencies in the environment acquired through a reinforcement history. Thus, situational variables account for the stability in behavior, but the relevant stimulus conditions are made salient through a reinforcement history initiated by the organism emitting a behavior. The set of relevant stimulus conditions is thus constructed by the organism. This, of course, spurred a further debate as to whether there are mediating cognitions, which Skinner (1977) argued against.

Even though the environment is predictable and reliable, it is at the same time a source of instability. The problem in determining what matters behaviorally for environmental consistency and inconsistency is that an environment cannot be defined devoid of an organism (Lickliter and Ness, 1990). Organisms certainly have biological structures that render certain stimuli relevant or salient and other stimuli irrelevant. Further, the extent of behavior an organism can emit, or the conditions an organism finds reinforcing, is certainly shaped by biology. The human visual system, for example, renders perception of certain spatial and temporal characteristics of stimuli more easily than others under specific lighting conditions. Certainly, experience is essential (Held and Hein, 1963), but the underlying sensitivities for which perceptions are built begin with the limits of the resolving power of the eye. In this sense, there are dispositional characteristics that impose organization on the stimulus environment and thus impose stability on behavior. Examples of such perceptual tendencies are well articulated by gestalt psychologists (Ellis, 1997).

On the generalist side of the generalist/specialist learning theory debate are arguments for such dispositional characteristics. In 1937, Gordon Allport hypothesized the existence of personal dispositions, which are more or less stable determinants of behavior. These dispositions were hypothesized to have the capacity to render many stimuli functionally equivalent, and to initiate and guide consistent or equivalent forms of behavior. Allport's discussion of dispositional variables is careful not to posit a reduction to ever finer fundamental elements. The suggestion was that the organism itself is a source of stability, given its structural characteristics. However, for personality, these dispositional characteristics or traits are only knowable through inference. Statistically, they are constructs defined using correlations and co-variation (Eysenck, 1970; McPhee, 1983; Winkelspecht, Lewis, and Thomas, 2006).

Given reasonable evidence supporting dispositional and situational variables, a natural compromise to the stability question was to suggest that some portion of the stability of personality exists in the situation and some in the person. However, since situations are never devoid of a person, and personality is always measured in a context or through questions of preference for a context, person and context are inextricably confounded and possibly indivisible for constructs such as personality. Thus, we propose that the observed stability is constructed and develops over one's lifespan.

Although not espousing a theory of mind here, we do believe that situational and dispositional variables meet and are integrated to construct personality in the brain. In this perspective, personality is viewed as something one produces, rather than as something one has. Antecedents are an active personal construction and the resulting personality is not uniquely understandable by the structural preconditions of person or environment, or both. Rather, personality is understood through the functions or goals of the individual or group. In this way, it is induced through the social group and ongoing activity, and is not predictable by material entity, whether stimuli, trait, or a combination thereof.

A means to test whether personality is predictable in terms of material entity or function is to longitudinally evaluate personality under supernormal stimulus conditions (Hinde, 1970; Tinbergen, 1951). Some environments, such as the academic and social environment at West Point, are "strong" and are designed to instill a set of organizational values and beliefs. West Point appears to be an environment that pervasively and repeatedly prompts and reinforces particular personality attributes, and discourages others with their associated behaviors.

If personality is predictable in terms of material entity, then either situational variables or dispositional variables should dominate in the strong environment. If dispositional variables are at play, then the presence of a strong environment should result in a greater than expected attrition of those cadets not possessing ideal characteristics. Moreover, for those who remain, the further cadets are from an ideal, the poorer their overall performance will be by standard Academy metrics. If situational variables are at play, then attrition and overall performance should not be predicted by distance from ideal personality characteristics. Cadets who remain at the Academy should show strong uniform changes in their personality profiles toward the ideal that are not attributable to maturation. Academic performance should not correlate with personality measures given a regression of personality characteristics to an Academy ideal. If personality is a result of an interaction between dispositional and situational variables, then personality facets should show some uniform intermediary movement to the ideal with entry personality measures moderately predictive of attrition and Academy performance metrics.

If personality is not predictable in terms of material entity, then neither situational variables nor dispositional variables should dominate. Personality facets should adapt to the pressures of the organization; however, the organizational ideal should also show shifts toward the reality of the cadets. Further, only those facets relevant to the organization should show this compromise. Personality facets should not correlate with attrition. However, since entering students come with a developmental history, they should self-select into the environment and show differences from the population on ideal characteristics. Personality facets also should not be predictive of academic performance, but should be predictive of more subjective, socially based Academy metrics. That is, evaluators will grade like individuals on qualitative measures (for example, leader potential) more favorably than those who deviate from the ideal.

Much contemporary research on personality traits has been conducted using the "five-factor model" of personality (Goldberg, 1981). A central feature of the model is the proposition that personality dispositions can be subsumed under five broad traits. The most popular measure of these five broad traits (extraversion, agreeableness, conscientiousness, emotional stability [neuroticism], and openness to experience) is the NEO–PI–R (Costa and McCrae, 1992). Longitudinal administration of the inventory at West Point permitted an exploration of some assumptions about the nature and malleability of personality traits.

Members of USMA Class of 2006 (N = 945; 142 females, 803 males) were given the NEO–PI–R during their freshman year. The remaining members of that class (N = 766; 113 females, 653 males) were again given the NEO–PI–R their junior year. Randomly selected faculty (N = 18), staff (N = 14), and seniors (N = 25) responded to the inventory with the "ideal cadet" in mind.

The results showed that freshmen scores did not differ reliably between those who remained at the Academy and those who resigned, nor did they predict academic performance (Bartone, Snook, and Tremble, 2002). NEO–PI–R scores for the total cohort differed from population norms on Neuroticism and Conscientiousness for males and on Neuroticism for females. Cadets tended to be higher on Conscientiousness and lower on Neuroticism. Scores on the NEO–PI–R did not predict attrition. For those who remained at the Academy, Neuroticism and Conscientiousness scores migrated reliably from freshman to junior year toward the "ideal"—that is, lower and higher, respectively. These factor scores were strongly agreed upon by the faculty and staff, and differed greatly from actual cadet scores, constituting a "strong environment" (Ness et al., 2008). Figure 7–2 shows a slight but statistically significant uniform movement in the Neuroticism scores from freshman to junior year toward the ideal. There was no difference between the scores of those freshmen who remained at and those who resigned from the Academy. As with Neuroticism, there was strong consistent agreement as to the ideal for Conscientiousness. The Conscientiousness factor showed a similarly small but consistent uniform migration of scores from freshman to junior year toward the ideal.

The data support a developmental influence on the personality phenotype and the idea that personality is understandable in terms of the

Figure 7–2. **Change in Distribution of Neuroticism Scores by Class Year and Compared to "Ideal"**

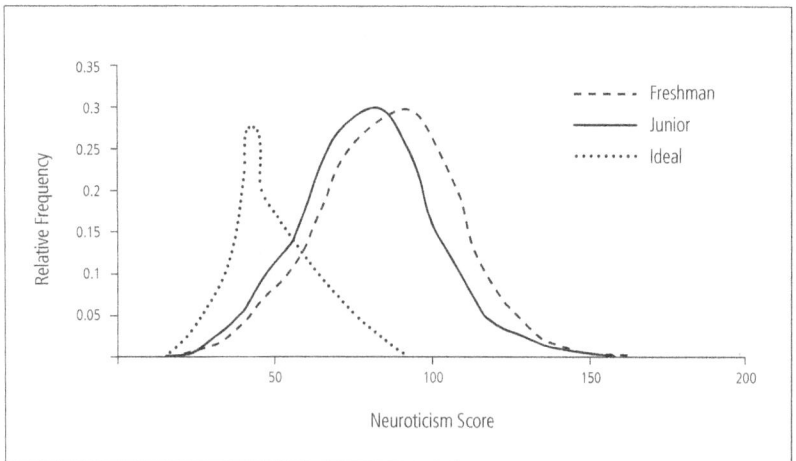

functions or goals of the individual or group. In this framework, personality is induced by the present organization and the ongoing activity, and is not predictable in terms of material entity, whether stimuli, trait, or combination thereof. Here we emphasize the epigenetic nature of personality and the nature of its stability. Stability of the personality is as much a function of individuals seeking certain niches as of adapting to those niches. Evidence for self-selection is demonstrated in the fact that those students entering the U.S. Military Academy were only different from their peers on Conscientiousness and Neuroticism. However, there was no evidence of an effect of lack of fitness for the environment since no personality factor or facet predicted attrition, whether transformed to reflect distance from the ideal or as the untransformed raw score. Further, this lack of predictability was not due to a lack of variability, as shown in the distribution. There was evidence of adaptation to the influences of the strong environment. Evident from the graph is the reliable shift toward the ideal from freshman to junior year. Experience served to shape and channel the expression of the personality phenotype. Thus, experiences become foundational but not determinant of individual expression.

The responses on the NEO–PI–R do not appear to be biased. All cadets reported that the test was completed honestly and assumed adherence to the cadet honor code, anonymity of the feedback for personal development, and course preparation,[10] supporting self-development arguments against the likelihood of response bias. Developmentally, these shifts in behavior become foundational for more permanent shifts maintained through habit strength, social conformity, engendered values, and reinforcing instruction and evaluations. The hope of the Academy is that these foundational shifts in desired behavior will lead to a relatively permanent change in behavior given the cadet's future responsibility with lethal force.

In the field of Army psychology, it is not enough to know; one must also apply. The purpose of the research was to develop a self-development tool specific to developing one's positive personal presence and bearing on the job. From the study above, we abstracted four qualities that present an archetypical Soldier based on those facets that showed reliable concordance among the staff and faculty, and that showed a reliable change from freshman to junior year: Gets the Job Done, Trusts Others, Friendly, and Resilient. These four qualities were then measured by 36 questions selected from the International Personality Item Pool[11] based on psychometric correspondence with the like items on the NEO–PI–R identified in the study. The resulting instrument was the Army Personality Inventory.

The Army Personality Inventory was deployed to the Army in a manner similar to that described by Bem and Allen (1974). The instructions introduced the concept of the Army as a "strong" environment, which places heavy demands on the leader. Adapting to these demands can be exhilarating and rewarding, as well as exhausting. Presenting oneself consistently and well in these varied situations is important to getting one's message out. The Army Personality Inventory was designed to provide some insights into how leaders present themselves to others by gauging those qualities important to their message being heard. In executing the tool, the leader was asked to choose a group of individuals (friends) who have observed him in various contexts and know him relatively well. This group should have insights into the leader's character and should be best able to reflect his consistent qualities. The leader was also asked to choose a group of individuals with whom he has had occasion to work in a particular context (acquaintances). This group should have opinions of what others perceive the leader to be like within a particular context while not being influenced by prior and varied experiences with him. The leader could also choose a group of acquaintances of homogeneous character, such as foreign officers or ones of a particular ethnic/gender background, and thereby use the inventory as a cultural awareness tool.

Upon receipt of feedback, the leader was encouraged to engage in a thoughtful and personal reflective process. All the tools of the LDP were designed to foster healthy self-awareness through honestly evaluating the degree of concordance between how leaders believe they are presenting themselves and how they are perceived. The leaders were advised to use the coach's journal to compile thoughts and positions resulting from the reflection process. They were also encouraged to sort through scenarios considering alternate courses of action they might have to take and work through them with a mentor.

The Army Personality Inventory was designed to provide the leader with feedback on the subjective, socially based metrics of the Army. It was also developed to broaden leadership feedback by soliciting it from colleagues beyond the 360 structure and thus further diffusing leadership lessons and best practices throughout the force. The tool was well received, and over the 3-month period from deploying the tool to the Army until transition of the LDP to TRADOC, 260 assessments were initiated with a total of 4,469 respondents providing leader feedback.

Use of Metadata to Structure Diffusion of Knowledge

Help Desk calls and usability studies are great sources of information to improve systems, as are big ideas, provided there is honest deliberation to evaluate them. However, these were not the only sources for improving the system. Another source was the information within the metadata. Metadata are an often forgotten source that gives information about communication within the system and between its users. The data depicted in figure 7–3 show one of our last investigative efforts prior to transitioning the LDP to TRADOC.[12] The purpose of this effort was to build and maintain a distributive knowledge management system for leader self-development and self-study. The figure shows the networks among users in the soliciting and distribution of leader-related information. This map was rendered after the personality tool was released to the Army. The Coach's Journal was still under development, and the few nodes seen are from the software developer's test of the module. The other represented instruments are the various 360s, with which feedback from peers, subordinates, and seniors is solicited.

With the 360 tools, the figure shows that warrant officers only seek information from within their ranks and appear relatively isolated from the other ranks. Future steps were planned to restructure tools to better integrate warrant officers into the process but were never executed

Figure 7–3. **Leadership Development Profile Network Clouds Showing Connections among User Types and Functions**

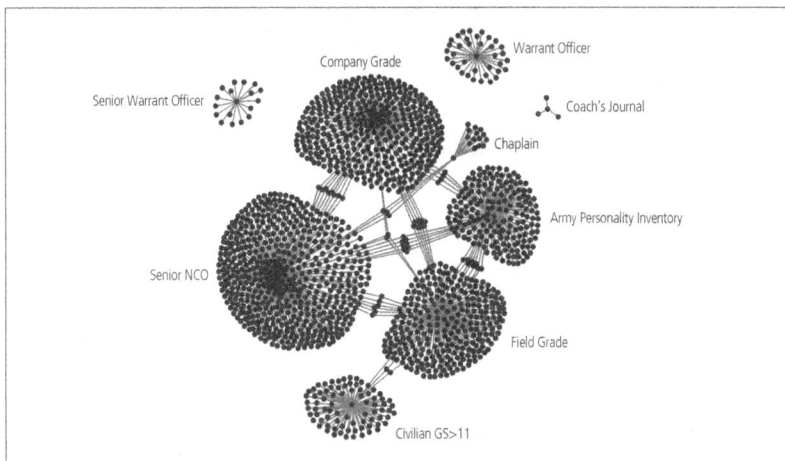

due to the transition to TRADOC. The Chaplain's tool was developed by the Chaplain Center and School to solicit feedback from members of a chaplain's religious community for feedback on his ministry. It was expected that a broad representation of rank would be a result of the open structure. However, the metadata indicate that the chaplains nearly exclusively solicited feedback from NCOs. It is quite possible that they only solicited feedback from their assistants and not from those who participated in their religious community. The information was passed to the Chaplain Center and School, and an instructional fix was planned. Senior civilians appeared to solicit information from themselves and field grade officers. This outcome was not expected and indicates an isolation of civilians from other ranks, which was most likely a result of the forcing function of the structure of the 360 to their seniors, peers, and subordinates. NCOs and company and field grade officers reliably solicited feedback from each other. NCOs and company grade officers rarely solicited feedback from civilians, again possibly because of the nature of the structure of the 360.

The data concerning the networks rendered by the various leader self-development tools were known prior to the development and implementation of the Army Personality Inventory. This latter tool was designed to solicit a broader range of respondents for feedback to the leader. Our hypothesis was that the isolation, particularly for civilians with NCOs and company grade officers, was due to the structure imposed by the 360. To yield feedback from across the spectrum of leaders, the Army Personality Inventory was structured in a manner that would allow for feedback from across organizations and ranks. The cluster labeled Army Personality Inventory in figure 7–3 shows direct connections between company grade officers, field grade officers, and senior NCOs with civilians. The links from the Civilians to the Army Personality Inventory run under the Field Grade cloud and are not visible in this rotation. This distributional change in the flow of information supports the hypothesis that the structure of the tool constrains feedback. The authors highly encourage the use of this analysis to gain an understanding about the likelihood that the leadership message is being distributed to ensure the breadth and scope of the diffusion of lessons and best practices. The pedagogical motivation here was to provide a rich base of content knowledge to the leader (Chi, 1981). In theory, the richer the content knowledge, the greater the formation of networks of associations that link different entities to each other and thus improve discernment of features that indicate novel aspects and distinctive approaches to leadership challenges. This remains an empirical question and was presented to the Center for Army Leadership to explore with the transition of the LDP to TRADOC.

Conclusion

The Leader Development Portfolio tools were designed to provide feedback showing the differences and similarities between how leaders believe they are presenting themselves to organizationally (peers, subordinates, seniors) or socially (friends, acquaintances, males, females) defined subgroups. The feedback indicates to leaders how they are viewed along organizationally critical aspects of their personality and leader competencies. This self-knowledge is meant to motivate them to act on the information through discerned self-awareness. This feedback, along with the coach's input and self-reflection journal, provides the structure of a powerful online, on-demand, secure, and anonymous leader self-development system.

The system itself is flexible and seamlessly integrated into AKO/DKO such that without any additional programming, one can add new tools to the assessment toolbox, tailor the portfolio to target audiences, tailor feedback results, and establish online mentorship of leaders. In addition, the system has a support Web site for additional information and support tools that can be used individually or as part of unit officer and NCO professional development programs. The integrated Help Desk tracks user logs allowing administrators to resolve most support issues within 24 hours.

In October 2007, the concepts, processes, and content of the LDP were successfully transitioned to TRADOC under their Multi-Source Assessment and Feedback program.[13] In the year between the launch of the LDP to the Army and its transition to TRADOC, the system boasted over 10,000 assessments. During this time, the Company Grade and Field Grade 360s were deployed for 1 year, the NCO 360 for 9 months, the Civilian and Warrant Officer 360s for 6 months, and the Army Personality Inventory for 3 months. At the transition, there were over 230,000 respondents and 50 to 100 new assessments initiated weekly. For an unpublicized, voluntary, self-development tool, that is arguably a high volume of usage.

The question remains whether the LDP is effective in promoting self-awareness and motivating positive leadership behaviors. We will close with an anecdote from the many comments posted to the Help Desk. A female company commander from 1ˢᵗ Cavalry Division initiated a 360. She had the personal opinion that she was fair, open, and honest. When the results of her 360 feedback came back, her subordinates gave her particularly low ratings on fairness, and in the free text comments they accused her of favoritism toward the females in the unit. The captain was devastated. After a period of rumination about what she thought of herself and what was reflected back to her, she was prepared to reengage the system (self-control). The instructions for feedback reminded her to work through these dissimilarities and to put them in context.

The first thing the captain noticed was that her peers and seniors saw her as she saw herself. She then reflected on the discrepancy between subordinates and self, as well as peers and seniors. It was then she realized that members of her company might have felt as they did because she was granting special privileges to a female in her company (perspective-taking). This particular female specialist was early in what her physicians deemed a high-risk pregnancy. The specialist could work, but only on desk duty for limited hours. The specialist asked the captain to keep the pregnancy a secret in the event things did not go well, so she would not have to relive an experience of a lost child through answering the questions of acquaintances. The request was granted. With this self-awareness, the captain rearranged work schedules (creative achievement) to give less of an appearance of favoritism, but was comfortable in the decision she made given her full awareness of the nature and effects of her decision (self-esteem).

Appendix

Vignette: Failing the Army Physical Fitness Test

Lieutenant (LT) Davis had just taken over as platoon leader when orders were received for deployment to a combat theater of operations in 2 months. A review of physical training (PT) records for the platoon revealed that Private First Class (PFC) Freeman failed the last two Army Physical Fitness Tests (APFTs). However, there is no record that the private was flagged, counseled, or given any special PT program. PFC Freeman is given the test and once again fails due to an inability to run without periodically stopping to walk. The private is an otherwise good Soldier with a strong desire to serve.

Courses of Action

1. Protect unit deployment readiness by giving a passing APFT with the understanding that PFC Freeman will not be promoted until he passes.
2. This is an indiscipline problem; put PFC Freeman on a remedial PT program.
3. Work with platoon sergeant (PSG), squad leader, and PFC Freeman to remediate him.
4. Too busy with deployment issues; delegate the issue to the PSG and focus on preparing for the deployment.

Outcome Assessment 1

This is a directive approach to the decision that takes into account mission, a belief in the Soldier, and confidence in one's own ability to motivate. The directive approach is most appropriate in instances when the quality of the information relevant to the decision is high, the decisionmaker has the expertise to make the decision, subordinates can and will comply, and time is of the essence.

Quality of information. There is a lack of historical data as to why PFC Freeman does not have a profile or a flag. The leader assumed that Freeman's chronic problem was motivation.

Expertise. The platoon leader is new to the unit and failed to consult the PSG concerning the private. He also shows lack of expertise in the belief that a platoon leader can stop the promotion of a subordinate after promotion requirements are documented.

Compliance. There is no evidence that PFC Freeman is physically capable of passing the APFT. He may have a medical condition that impacts running performance and may or may not be aware of it. The private was not asked, and this was not ruled out. Further, the NCO chain was not brought in on the decision or to monitor compliance.

Time. Two months presents a time constraint, given a significant increase in unit tempo in preparing for deployment.

Outcome Assessment 2

This is a directive approach to the decision that takes into account PFC Freeman's shortfall and seeks to direct the Soldier to pass the APFT.

Quality of information. There was a lack of historical data as to why PFC Freeman did not have a profile or a flag. The platoon leader assumed that PFC Freeman's chronic problem was a result of indiscipline. The lack of a profile, flag, or other documentation could be a result of poor recordkeeping or a misplaced file.

Expertise. The platoon leader is new to the unit and failed to consult the PSG concerning PFC Freeman or the personnel administration center (PAC) NCO about the records.

Compliance. There is no evidence that PFC Friedman is physically capable of passing the APFT. PFC Freeman may have a medical condition that impacts running performance and may or may not be aware of it. PFC Freeman was not asked, and this was not ruled out. The NCO chain was not brought in on the decision or to monitor compliance.

Time. Two months presents a time constraint, given a significant increase in unit tempo in preparing for deployment.

Outcome Assessment 3

This is a participative approach to the decision that takes into account the ambiguity of the information concerning why the Soldier was not flagged or put on profile. The participative approach is most appropriate when the quality of the information relevant to the decision is poor, the decisionmaker has limited expertise to make the decision, subordinates may be unwilling to comply, and time is moderately critical.

Quality of information. There was a lack of historical data as to why PFC Freeman did not have a profile or a flag. Improving the quality of the information is critical to an optimal outcome.

Expertise. The platoon leader is new to the unit. Consulting the NCO chain and those in recordkeeping will leverage unit expertise to augment leader expertise.

Compliance. There is no evidence that PFC Freeman is physically capable of passing the APFT. There is a reason why he periodically has to stop running (for example, smoker, asthma, allergy, fitness). Informally counseling Freeman may bring out the underlying cause and invest him in the solution. Consulting the NCO chain invests them in the course of action to ensure followthrough.

Time. Two months presents a time constraint, given a significant increase in unit tempo in preparing for deployment.

Outcome Assessment 4

This is a delegate approach to the decision that takes into account the Soldier's shortfall and seeks to establish a trusting relationship with the PSG. The delegate approach is most appropriate in those instances when you wish to develop subordinate leaders, their commitment is high, and the subordinates are competent to make the decision. This approach can be used purely as developmental in those cases where the significance of the decision is low.

Developing. Developing Soldiers is something every leader must do. However, developing an NCO requires some history or knowledge of the NCO. The platoon leader was new to the unit and has limited or no such knowledge.

Importance of decision. This decision will present a first impression to the platoon leader's rater and senior rater. This makes the outcome critical to setting expectations and a power base.

Competence. The platoon leader has reason to suspect some hindrance to competence given that there is no profile or flag in PFC Freeman's file. Given the ambiguity of the information on this point, it is hard to judge competence. The platoon leader should discuss this issue with the PSG.

Commitment. There is evidence of some barrier to PFC Freeman passing the APFT. PFC Freeman may have a medical issue or be an indiscipline problem. The NCO chain of responsibility may have some hindrance keeping them from managing the issue, or there may be a problem with filing or recordkeeping in the PAC.

Time. Implicit to a delegate approach is the time and space to develop. Two months to deployment presents a time constraint, given a significant increase in unit tempo in preparing for deployment.

Senior Leader Coaching

Themes

- Be a role model
- Be a good communicator
- Be consistent
- Be a learner
- When in charge, take charge
- Know your Soldiers.

In the real situations on which this vignette is based, the lieutenant believed the private could be motivated to pass the APFT. The unit deployed with PFC Freeman. LT Davis put the Soldier on a remedial PT program for running and went to great lengths to motivate him to run. Six months later, Freeman failed again. Everyone in the platoon was aware of the situation. PFC Freeman appeared on the promotion list, along with those who actually passed the APFT. LT Davis was powerless to stop the private from being promoted.

The Soldier had a physical ailment preventing completion of the 2-mile run. Had this been simply a motivational or indiscipline issue, the decisions based on these assumptions would have produced a good outcome. Since the quality of the data in this case could not disambiguate these assumptions, the leader in this case should seek more information as to the cause of the chronic PT failure and the lack of a record of unit remediation.

In general, a leader needs to balance Army standards with relationship development and safety. In the reference for this case, there was a clinical reason the Soldier could not run. The private's safety and those of the unit can be at risk if this is not addressed before deployment. A full medical checkup was warranted. Moreover, the leader should understand that once the Soldier passed the APFT, "forged" or legitimate, this promotion requirement is considered fulfilled.

In terms of developing/motivating the PFC, barring a medical condition causing the poor performance, the lieutenant did not know the Soldier well enough to be able to motivate him. Especially being new to the unit, the lieutenant should at least consult the NCOs individually. The lieutenant must provide fair and consistent treatment for each Soldier or risk negatively affecting unit morale. Falsifying a report will damage credibility with the unit and up the chain of command (if someone will falsify a report, what else will that person do?).

To recover, the lieutenant must see that PFC Freeman gets medically evaluated, counsel the Soldier that future APFTs will not be falsified, and work with the PSG on a program that will enhance the private's chances of passing. He should admit the mistake to the PSG and ask for help with damage control. The lieutenant must admit the mistake to the company commander. The lesson learned is to never compromise standards and commit to upholding standards in the future.

References

Aldrich, C. (2004). *Simulations and the Future of Learning: An innovative (and perhaps revolutionary) approach to e-Learning.* San Francisco, CA: John Wiley and Sons, Inc.

Allport, G. (1937). The functional autonomy of motives. *American Journal of Psychology* 50, 141–156.

Army Training and Leader Development Panel (2003). *The Army Training and Leader Development Panel Officer Study Report to the Army.* Washington, DC: Defense Technical Information Center, Accession No. AD415810. Retrieved from <http://handle.dtic.mil/100.2/ADA415810>.

Army Leader Development Panel Study: Officer, NCO, and Warrant Officer Reports. Retrieved from <www.army.mil/features/ATLD/ATLD.htm>.

Association of the United States Army (October 1, 2008). *Army to enhance civilian career programs.* Retrieved from <www.ausa.org/publications/ausanews/specialreports/09_08/Pages/Armytoenhanceciviliancareerprograms.aspx>.

Bartone, P., Snook, S., and Tremble, T. (2002). Cognitive and personality predictors of leader performance in West Point cadets. *Military Psychology* 14 (4), 321–338.

Bem, D., and Allen, A. (1974). On predicting some of the people some of the time: The search for cross-situational consistencies in behavior. *Psychological Review* 81 (6), 506–520.

Chi, M. (1981). Knowledge development and memory performance. In J.P. Pas and N. O'Conner (eds.). *Intelligence and learning.* New York: Plenum Press.

Costa, P.T., and McCrae, R.R. (1992). Professional manual for the NEO Personality Inventory. Odessa, FL: Psychological Assessment Resources.

———. (2003). *Revised NEO Personality Inventory (NEO-PI-R).* Lutz, FL: Psychological Assessment Resources, Inc.

Duval, T.S. and Wicklund, R.A. (1972). *A theory of objective self-awareness.* New York: Academic.

Ellis, W. (1997). *A source book of Gestalt psychology.* Berlin, Germany: Gestalt Journal Press.

Eysenck, H. (1970). *The structure of human personality* (3^d ed.). London: Methuen.

Fenigstein, A., Scheier, M., and Buss, A. (1975). Public and private self-consciousness: Assessment and theory. *Journal of Consulting and Clinical Psychology* 43 (4), 522–527.

Goldberg, L. (1981). Language and individual differences: The search for universals in personality lexicons. In L. Wheeler (ed.), *Review of personality and social psychology*, vol. 2 (141–165). Beverly Hills, CA: Sage.

Hartshorne, H., and May, M. (1928). *Studies in the nature of character: Volume 1, Studies in deceit.* New York: McMillan.

———. (1929). *Studies in the nature of character: Volume 2, Studies in service and self control.* New York: McMillan.

Hartshorne, H., May, M., and Shuttleworth, F. (1930). *Studies in the nature of character: Volume 3, Studies in the organization of character.* New York: McMillan.

Headquarters Department of the Army (October 24, 2007). *Army Regulation 25-5: Information Assurance.* Retrieved from <www.fas.org/irp/doddir/army/ar25-2.pdf>.

Held, R., and Hein, A. (1963). Movement-produced stimulation in the development of visually guided behavior. *Journal of Comparative and Physiological Psychology* 56, 872–876.

Hinde, R.A. (1970). *Animal behaviour* (2^d ed.). Tokyo: McGraw-Hill Kogakusha, Ltd.

Hinkle, R., Karrasch, A., and Burke, T. (July 1, 2006). *Army Green: Training non-tactical problem solving by platoon leaders.* U.S. Army Research Institute for the Behavioral and Social Sciences, Research Report 1855. Accession No. ADA454772. Retrieved from <http://handle.dtic.mil/100.2/ADA454772>.

Institute of Land Warfare (February 2004). The U.S. Army and beyond: Strategically agile and adaptive. *Association of the United States Army, Torchbearer.* Retrieved from <www.ausa.org/programs/torchbearer/nsr/National%20Security%20Reports/TB_StratBook.pdf>.

Lickliter, B., and Ness, J. (1990). Domestication and comparative psychology: Status and strategy. *Journal of Comparative Psychology* 104, 211–218.

MacPhee, D. (April 1983). *What do ratings of infant temperament really measure?* Poster presented at the Biennial Meeting of the Society for Research in Child Development, Detroit, MI. (ERIC Document Reproduction Service No. ED233817). Retrieved from <http://eric.ed.gov/ERICDocs/data/ericdocs2sql/content_storage_01/0000019b/80/30/1a/61.pdf>.

Mead, G.H. (1934). *Mind, self, and society.* Chicago: University of Chicago Press.

Mischel, W. (1977). On the future of personality measurement. *American Psychologist* 32, 246–254.

Ness, J., Dieper, V., Lewis, P., and Brazil, D. (August 2008). Personality development in strong environments. *Proceedings XXIX International Congress of Psychology.* Berlin, Germany.

Parrillo, V., and Donoghue, C. (2005). Updating the Bogardus social distance studies: A new national survey. *The Social Sciences Journal* 42, 257–271.

PRWeb (May 28, 2008). *U.S. Army partners with Simulearn to train soldiers in the art of influence.* Retrieved from <www.prweb.com/releases/leadership_development/simulation_training/prweb976254.htm>.

Siegal, M., and Bergman, A. (2002). Waddington's canalization revisited: Developmental stability and evolution. *Proceedings of the National Academy of Sciences* 99, 10528–10532. Retrieved from <www.pnas.org/cgi/content/full/99/16/10528>.

Silvia, P., and Duval, T. (2001). Objective self-awareness theory: Recent progress and enduring problems. *Personality and Social Psychology Review* 5 (3), 230–241.

Simulearn (2007). *VLeader.* Simulearn Inc. Norwalk, CT. Retrieved from <www.simulearn.net/leadership_training/leadership_training.html>.

Skinner, B.F. (1977). Why I am not a cognitive psychologist. *Behaviorism* 5, 1–11.

Thorndike, E. (1903). *Educational psychology.* New York: Lemcke and Buechner.

Tinbergen, N. (1951). *The study of instinct.* New York and London: Oxford University Press.

Toegel, G. and Conger, J. (2003). 360-degree assessment: Time for reinvention. *Academy of Management Learning and Education* 2, 297–311.

Ulmer, W., Shaler, M., Bullis, R., DiClemente, D., Jacobs, T., and Shambach, S. (November 5, 2004). *Leadership Lessons at Division Command Level.* Army War College, Carlisle Barracks, PA. Accession No. ADA435928. Retrieved from <http://handle.dtic.mil/100.2/ASA435928>.

Vroom, V. (2000). Leadership and the decision making process. *Organizational Dynamics* 28, 82–94.

Wicklund, R., and Gollwitzer, P. (1987). The fallacy of the private-public self-focus distinction. *Journal of Personality* 55 (3), 491–523.

Winkelspecht, C., Lewis, P., and Thomas, A. (2006). Potential effects of faking on the NEO–PI–R: Willingness and ability to fake changes who gets hired in simulated selection decisions. *Journal of Business and Psychology* 21, 243–259.

Notes

[1] Army Knowledge Online (AKO) is the U.S. Army's corporate intranet, which was expanded to include the entire Department of Defense in 2007. AKO/DKO is accessed through <http://www.us.army.mil>.

[2] Even though fiscal years run from October to October, the closeout of the Leadership Development Portfolio account was extended to November 1 to give us time to transfer funds to AKO and for AKO to accept and obligate the funds.

[3] We began our planning process using a Strengths, Weaknesses, Opportunities, and Threats Analysis. The results of this analysis were used to evaluate and direct our initial courses of action.

[4] Of special acknowledgment in helping to formulate the concept of self-awareness and how to reflect information back to the leader are Dr. Don Campbell, MAJ Silas Martinez, and COL Donna Brazil from BS&L, Dr. Lenny Wong and Dr. Craig Bullis from the Army War College, Dr. Angela Karrasch from the Center for Army Leadership, and COL Sally Harvey from Clinical Mental Health, Fort Bragg, NC, not to mention the occasional "skull sessions" with LTC Todd Henshaw of BS&L.

[5] Michael Stewart and Adrian Hartline from Appian Corporation, who held the contract for AKO/DKO software development, were the lead software engineers on the LDP project (see <www.appian.com>).

[6] AKO updates all portal code on a regular basis. At these times, typically once a month, updates are published to the portal to fix bugs or to add new functionality. There is an involved process for getting code into the build cycle and minimizing dependency on the build cycle greatly improves agility to meet user needs.

[7] The Benchworks site was designed and maintained by Marianne Carlson, MAJ Dennis O'Neil, and John VanVynck.

[8] With the transition to TRADOC, the benchworks.army.mil site was shut down and the information reconstructed at <https://msaf.army.mil/ResourcesAndTraining.aspx>.

[9] The Army Physical Fitness Test is a test of minimum fitness to perform operational duties. The test consists of 3 timed events, tested every 6 months, with different standards of performance by age grouping for men and women. For a summary of purpose, events, standards, and relevant links, see <http://en.wikipedia.org/wiki/Army_Physical_Fitness_Test_(APFT)>.

[10] The NEO–PI–R was used as a self-development tool in the required freshman and junior level leader development courses. The administration of the NEO–PI–R followed a block of instruction on self-development and self-awareness.

[11] See <http://ipip.ori.org>.

[12] The graph was rendered based on discussions with LTC John Graham of BS&L, whose area of research is in social networks.

[13] See <https://msaf.army.mil/Default.aspx>.

Chapter 8

Understanding and Measuring Creative Thinking in Leaders

Mark A. Vaitkus

To address complex problems related to changing global markets and security concerns, corporations and government agencies are increasingly interested in diagnosing and enhancing the potential of their leaders to generate flexible, multifaceted, and innovative solutions. In the research reported here, government, military, and industry leaders attending a senior Service college were studied to discover the extent and degree to which leaders could be classified as having a propensity toward creativity, originality, and innovation based on the so-called five-factor model of personality. Two measures of "openness" were included. While both measures performed well on standard tests of internal reliability and construct validity, they yielded significantly different estimates of high openness in the study sample (30 percent high versus 4 percent high). This chapter considers possible reasons for these disparate results, including variable question wording, scale metrics, and subfactor conceptualization. The conclusions offer some important guidelines and caveats regarding possible approaches for identifying and developing creative thinking potential in senior leaders.

Introduction

In an era of global markets and rapid technological change, few question the value of creativity and the innovations it generates within organizations that wish to remain viable and competitive. To the extent that leaders are viewed as critical agents of organizational change (Kotter, 1990), it would seem imperative that they at least be open to the products of creativity and imagination, if not key to their genesis. Indeed, Puccio, Murdock, and Mance (2007) view creativity as a "core leadership competence" that can facilitate effective change through questioning the status quo, considering new ideas, experimenting and risk-taking, and developing novel approaches for meeting challenges.

This chapter examines and compares two conventional psychological measures of creative disposition within a senior leader population. Both measures view such a disposition under the rubric of "openness" within the so-called big five or five-factor model of personality. The analysis will attempt to answer the following questions:

- Is each measure of openness reliable and valid in its ability to differentiate this factor from other personality factors?
- How different are the measures in terms of describing this population with respect to creative proclivities, and what accounts for these differences?
- What are the caveats that leader developers should keep in mind when using these and other "standard" measures of personality to screen for creative qualities, and how should such caveats inform feedback and training programs?

Personality Conceptualization and the Five-factor Model

Since the early 1990s, there has been considerable convergence in the field of personality assessment toward a five-factor or "Big Five" measurement model (John and Srivastava, 1999). Movement toward a five-factor taxonomic model of personality can be traced back to Fiske (1949), Tupes and Christal (1961/1992), and Norman (1963), who built on, and partially replicated, Cattell's (1945) factor analytic work, concluding that five factors adequately described the analytical space of extant personality trait measures (John and Srivastava, 1999). In their review of published studies to that time, Digman and Takemoto-Chock (1981) and Goldberg (1981, 1993) reaffirmed that the Big Five did indeed capture the most generalizable categorization of personality characteristics. Following Digman (1990), we may describe these as follows:

- *Surgency or extraversion.* Those who score higher on this factor tend to enjoy the stimulation that comes from interacting with large numbers of people in whose company they are generally talkative and assertive. They are not shy and have a high degree of positive energy that leads them to actively engage in a wide variety of activities.
- *Agreeableness.* Individuals who display higher agreeableness tend to be kind and value cooperation and getting along with

others. They are more sensitive to, and concerned about, the feelings of others, more likely to give people the benefit of the doubt, and will often compromise to avoid conflict and the prospect that others will not like them.

- *Conscientiousness.* People reporting higher levels of conscientiousness desire orderliness, organization, and efficiency in their lives. They take their perceived responsibilities and duties very seriously, and have considerable investment in viewing themselves as dependable, thorough, and achievement-oriented.

- *Emotional stability.* Those exhibiting greater emotional stability report relative freedom from experiencing "negative feelings" such as anger, anxiety, and depression. They are also less likely to be moody, mercurial, and temperamental. In general, they will display calmness and less nervousness than those who score low on this factor. Some researchers have preferred to concentrate on its opposite pole instead (that is, the negative feelings themselves), and thus label it "neuroticism."

- *Openness.* Variably called culture, intellect, and openness to experience, this factor attempts to measure dispositions that skew toward higher curiosity, imagination, and creativity. Those on the higher end may think of themselves as being more open-minded and appreciative of the arts, while also valuing originality, new experiences, and deep or intellectual reflection.

The dimension labeled *openness* clearly comes closest to describing the personality factor linked to a creative disposition and its contingent leanings or preferences, and therefore the one on which I will focus in this chapter.

Openness and Organizational Leader Effectiveness

Openness is probably the least understood of the five personality factors. Part of the problem lies squarely with poor conceptualization and attempts to variably describe the factor as a disposition toward critical thinking and inquisitiveness (intellect), a need for variety and new experiences, a tendency to be creative and imaginative, an interest in aesthetic pursuits and the arts, an antiauthoritarian orientation, and an empathy with or understanding of diverse others (McCrae and Costa, 1997). Such subdimensions do not always correlate well with one another in analyses, certainly when compared to the other four factors. Yet such effects of openness on the endorsement of liberal political and social values (McCrae, 1996; Van Hiel and Mervielde, 2004), as well as liberal candidates for political office (Caprara and Zimbardo, 2004), are fairly well documented. Of the Big Five, openness also comes closest to predicting personal exploration of spiritual phenomena (versus holding conventional beliefs) and the profession of virtues such as forgiveness (MacDonald, 2000). To the extent that we are talking about an active curiosity about the unknown, a desire to experience the unfamiliar, taking an original and innovative approach to problems, and not shying away from the complex or abstract, such qualities of the "open" personality may indeed be functional for the executive leader operating in an ambiguous and culturally diverse strategic environment.

Psychometric measures of basic intelligence are correlated with openness (DeYoung, Peterson, and Higgins, 2005; McCrae and Costa, 1997). There is also little doubt that openness is an important ingredient in what Thomas (2006; cf. Earley, 2002) calls "cultural intelligence" (or "cultural quotient" [CQ]) (Thomas, 2007, and Earley, 2002) or "the ability to interact effectively with people who are culturally different." Coupled with a desire for new cultural experiences, another aspect of openness, CQ is further enhanced. However, while personality factors such as openness are related to the raw capacity for CQ development, the latter remains distinct from its *actual* development or exhibition in a particular circumstance, which depends on interactional experiences as much as the individual's motivation (Thomas).

In terms of studies that have looked at Big Five models of personality with respect to leadership criteria variables, Judge, Ilies, Bono, and Gerhardt (2002) find the following correlational averages with effective leadership outcomes across 73 samples: extraversion (.31), conscientiousness (.28), openness to experience (.24), neuroticism (-.24), and agreeableness (.08). While just 23 percent of the variance in positive leadership outcomes across samples was explained by personality differences, this is a significant number and demonstrates that personality, including openness, should indeed be considered part of the variable solution set for modeling leadership.

In his most comprehensive attempt to lay out an integrated theory of executive leadership based on a meta-analysis of dozens of Army Research Institute and other studies, Zaccaro (2001) identifies such personality characteristics as openness, curiosity, flexibility, adaptability, and risk propensity as all being requisite for executive effectiveness beyond expertise, intelligence, and other cognitive capacities. Streufert and Swezey (1986) also provide evidence in the civilian sector that executives who possess greater "multidimensional thinking" make better decisions and, therefore, more valuable contributions to their organizations.

When it comes to the special case of transformational leadership, the results for Big Five linkages are somewhat different than for leadership generally. In their review of study results from 14 samples of leaders representing over 200 organizations, Judge and Bono (2000) report that transformational leader behaviors had moderate zero-order associations with higher extraversion, agreeableness, and openness. Surprisingly, there were no significant effects for neuroticism and conscientiousness. In a more sophisticated analysis incorporating a military sample, Ployhart, Lim, and Chan (2001) employed structural equation modeling that differentiated "typical" versus "maximum" performance criteria related to transformational leadership to show that extraversion was strongly predictive of both types of criteria, but openness had predictive power only for the maximum criteria, and low neuroticism was exclusively related to the typical performance criteria.

Along with a highly developed conceptual capacity, deeply held ethical values, and a mature and flexible perspective, Jacobs (2006) believes that openness is a critical attribute of transformational leaders because it enables them to consider new ideas and solutions without dismissing them based on preconceived notions. Indeed, Jacobs states flatly that openness, in terms of a disposition toward intellectual curiosity, imagination, and creative problem-solving, may well be the most important Big Five dimension for strategic leadership. This is partially because strategic leaders must not only be able to apply a systems perspective to gauging their organization's environment in time and space, but they also must take that necessarily complex understanding, fashion a vision for an organizational path forward consistent with that understanding, and communicate that vision in an effectively parsimonious, meaningful, clever, and ideally original and inspiring manner.

It should be noted that openness is not a panacea, especially when taken to an extreme. Jacobs (2006) makes it clear that "[t]op-level executives must understand the necessity to fix it even 'when it ain't broke'" and that openness helps a strategic leader be functionally proactive in this regard. At the same time, an executive with an extremely high drive for openness may initiate change for its own sake to fulfill a thirst for stimulation and excitement, thus squandering resources and putting the organization at unnecessary risk (Jacobs). Openness is ultimately no different, therefore, than the other dimensions of personality in terms of the need to achieve a reasonable balance. However, there is little doubt that its elements are closely tied to an affinity for the kind of creative problem-solving and complex thinking skills associated with effective leadership in volatile and changing environments (Puccio, Murdock, and Mance, 2007; Yukl, 2002), and therefore with the leader behaviors associated with transformational, charismatic, and visionary leadership (House and Shamir, 1993).

Method

Participants

The subjects in this study were students attending a U.S. Government senior Service college during the 2006–2007 academic year. The present study was reviewed and approved by the Institutional Review Board at the National Defense University, and completion of any one or more of the instruments was voluntary with the guarantee that all individual results would remain confidential.

All 292 U.S. students in the class completed either or both of the two instruments under examination here. Four international fellows also did so, but these subjects were removed from the analysis. Demographic data were available for 266 of the U.S. students. Of these, 75.6 percent (201) were male, and 24.4 percent (65) female. The youngest participant was 32, and the oldest 59, with a mean age of 44.2 (standard deviation = 4.5) and both a median and modal age of 43. A majority of the participants were in the military (70.1 percent) with 29.9 percent holding civilian positions. Of the 261 who reported their Service or component, 53 (20.3 percent) said Army, 64 (24.5 percent) Air Force, 43 (16.5 percent) Navy, 19 (7.3 percent) Marine Corps, 4 (1.5 percent) Coast Guard, 53 (20.3 percent) Department of Defense civilian, 19 (7.3 percent) other Government agency civilian, and 6 (2.3 percent) reported that they were civilian industry fellows. Finally, regarding ethnicity, of the 250 responding to this question, 187 (74.8 percent) stated they were non-Hispanic white, 31 (12.4 percent) African-American,

9 (3.6 percent) Hispanic, 7 (2.8 percent) Asian and Pacific Islander, and 16 (6.4 percent) chose "other."

Measures

Mini-markers. Gerard Saucier (1994) developed the mini-markers (MM) instrument as a shortened version of Lewis Goldberg's (1992) 100-item set of single adjective, unipolar markers for measuring the Big Five personality dimensions. It is representative of the lexical approach. While the 100-item set had respectable loadings on five factors as predicted in factor analyses (Goldberg, 1992), they produced scales that were intercorrelated at moderately high levels (over .40) and included some "user-unfriendly" terms such as "imperturbable" and "uncharitable" (Saucier). By choosing "homogeneous" items that loaded highest on their respective predicted factors and less than half as highly on any other factor across a dozen samples, Saucier successfully whittled down the adjective list to 40, with 8 items representing each of the Big Five dimensions. Following Goldberg's (1992) original designations, he labeled these extraversion or surgency, agreeableness, conscientiousness, emotional stability, and intellect or openness. The resulting scales had better interitem correlations and orthogonal properties than those reported by Goldberg, although reliability scores (coefficient alphas) were predictably, but marginally, sacrificed and remained above .73 (Saucier). Dwight, Cummings, and Glenar (1998) reported that the validity of the mini-markers compared favorably with that of Goldberg's full set of markers.

The 40 items of the original mini-markers used in this study are found in table 8–1. Eight single adjective items are included to measure each of the Big Five factors. Participants were asked to rate how accurately each trait presented describes them. Responses are then scored on a 9-point scale, from 1 being extremely inaccurate to 9 being extremely accurate. MM scale scores were constructed using Saucier's (1994) original method. The appropriate 20 adjective responses addressing the negative pole of the relevant factor were reverse-coded (from 9 to 1) before being added together with the response scores of their positive-pole counterparts to produce the respective Big Five scale score. Each of the five scale scores was then divided by 8 to rescale the sum to the original metric. There were no missing data by individual item response, which rendered a total of 289 cases with complete MM data for analysis.

Table 8–1. Mini-markers Factor Analysis: 40 Item Factor Loadings for Five-factor Solution (N = 289)

	Component*				
	1	2	3	4	5
Talkative	**.747**	.208	.176	.024	.131
Extroverted	**.841**	.138	.074	-.067	.077
Bold	**.540**	-.167	-.018	-.006	.275
Energetic	**.452**	.172	-.128	-.197	.243
Shy	**-.797**	-.059	.131	.059	-.047
Quiet	**-.838**	.015	-.003	-.042	.016
Bashful	**-.766**	-.010	.101	.135	-.062
Withdrawn	**-.676**	-.295	.178	.146	.063
Sympathetic	.019	**.761**	.055	-.072	.133
Warm	.272	**.684**	-.078	-.116	.055
Kind	.117	**.735**	-.044	-.139	.118
Cooperative	.030	**.579**	-.005	-.139	.038
Cold	-.185	**-.691**	.069	.097	.017
Unsympathetic	-.018	**-.727**	.073	.093	-.154
Rude	.039	**-.564**	.448	.140	.119
Harsh	.144	**-.482**	.470	-.021	.008
Organized	.108	.084	.012	**-.789**	-.095
Efficient	.126	.060	-.026	**-.645**	.158
Systematic	.051	.028	.027	**-.602**	.131
Practical	-.010	.152	-.063	**-.454**	.075
Disorganized	-.011	-.052	.021	**.751**	.084
Sloppy	-.064	-.099	.094	**.656**	.082
Inefficient	-.110	-.089	.119	**.746**	-.065
Careless	.092	-.183	.215	**.502**	-.127
Unenvious	.094	-.102	**-.531**	-.041	.122
Relaxed	.011	.200	**-.441**	.244	.008
Moody	-.138	-.144	**.657**	.098	.126
Jealous	-.094	-.096	**.728**	.103	-.071
Temperamental	.093	-.130	**.713**	.099	.117
Envious	.020	-.044	**.714**	.065	-.189
Touchy	.097	.018	**.592**	.134	.095

Fretful	-.198	.051	**.621**	.037	-.068
Creative	.248	.208	-.093	.096	**.731**
Imaginative	.228	.235	-.075	.074	**.724**
Philosophical	-.075	.073	.046	.042	**.646**
Intellectual	.099	-.080	-.041	-.300	**.589**
Complex	-.029	-.147	.138	-.025	**.615**
Deep	-.042	.065	.082	-.065	**.686**
Uncreative	-.256	-.193	.172	.002	**-.679**
Unintellectual	-.133	-.035	.129	.273	**-.550**

*Rotated Component Matrix (Varimax Rotation with Kaiser Normalization)

International Personality Item Pool. The second instrument examined in this study is based on items drawn from the International Personality Item Pool (IPIP). Goldberg and colleagues (1999; Goldberg, Johnson, Eber, Hogan, Ashton, Cloninger, and Gough, 2006) constructed the IPIP in 1996 as a free and open source, online repository of personality measurement items collected from multiple public-domain sources. Their goal was to accelerate advancement in the field of personality research by alleviating copyright and other restrictions related to commercial test publishing and, thereby, increasing the prospect for collaborative research networks, data-sharing, and study replication. They have since referred to their project as a "collaboratory," defined as "a computer-supported system that allows scientists to work with each other, facilities, and data bases without regard to geographical location" (Finholt and Olson, 1997). The IPIP currently has over 2,000 items on its Web site, which is maintained by the Oregon Research Institute.

Johnson's short form IPIP-based representation of the NEO–PI–R. The NEO–PI–R (Neuroticism, Extraversion, Openness Personality Inventory–Revised; Costa and McCrae, 1992), a licensed commercial product, is perhaps the most widely recognized and accepted standardized tool for measuring the five-factor domains and the facets that are hypothesized to underlie them (see Johnson, 2000a). The NEO–PI–R itself contains 240 items representing 30 facets, or 6 facets for each of the overarching 5 factor domains: neuroticism, extraversion, openness, agreeableness, and conscientiousness (sometimes rearranged to form the acronym OCEAN). The "middle-level" personality categories reflected by the 30 facets of the NEO–PI–R are perhaps the most frequently measured in empirical studies (John and Srivastava, 1999). In short, the comprehensiveness of the NEO–PI–R and its high concurrent validity with a host of other personality instruments have often rendered it the preferred instrument of choice for many researchers and clinicians.

Johnson (2001) took Goldberg's 300-item proxy for the NEO–PI–R found in the IPIP and uploaded it to his own Web site on September 1, 1998. In 2000, using standard correlational, factor, reliability, and content analytic techniques, Johnson produced a new, abbreviated instrument with somewhat lower scale and facet alpha reliability scores. This short form, which he added to his Web site as an alternative to the 300-item inventory, consists of 120 items with 30 four-item sets intended to measure the NEO–PI–R facets, for a total of 24 items (6 facets with 4 items per facet) representing each of the Big Five factor domains. It is a collection of brief, descriptive phrases and represents the more deductive, theoretical approach in contemporary personality psychology as opposed to the lexical approach. Following Johnson (2001), this NEO–PI–R proxy instrument is labeled the IPIP–NEO. Johnson labels his five factors in the same manner as Costa and McCrae for the NEO–PI–R, although he has somewhat varying names for the 30 facets. Johnson's facet labels for the openness factor, with Costa and McCrae's (1992) labels given in parentheses, are: O1–Imagination (Fantasy), O2–Artistic Interests (Aesthetics), O3–Emotionality (Feelings), O4–Adventurousness (Actions), O5–Intellect (Ideas), and O6–Liberalism (Values). The 24 items comprising the openness factor can be found in appendix A.

Participants were asked to rate how accurately each of the 120 phrases presented describes them on a 5-point scale, with 1 being very inaccurate to 5 being very accurate. The IPIP–NEO Big Five domain scales and the 30 facet scales were constructed using Johnson's (2001) methodology. The appropriate 55 short-phrase responses addressing the negative pole of the relevant factor or facet were reverse-coded (from 5 to 1) before being added together with the response scores of their positive-pole counterparts to produce their respective Big Five domain scale or facet score. Each of the 5 domain scale scores was then divided by 24, and each of the 30 facet scores by 4, to rescale the sums to the original metric. There were 156 instances of missing item data across the 120 item variables, leaving 174 cases with complete IPIP–NEO data. A missing data analysis showed that items with missing data were apparently randomly distributed across facets and scales.

Results

Factor Analyses

The issue of construct validity for the included items on the two different openness scales is examined through the use of factor analysis. For the mini-markers, the initial principal components analysis of the 40 items comprising the instrument revealed a 10-factor solution with the minimum eigenvalue = 1.0 criterion. However, an examination of the associated scree plot for the analysis showed a decided drop in initial eigenvalues from 2.74 to 1.50 between the fifth and sixth factors, giving some support for a five-factor solution.

Saucier's (1994) published factor analysis of the mini-markers (N = 636) was replicated by forcing a five-factor solution using varimax rotation with Kaiser normalization. This rotation converged in six iterations and the five factors accounted for a total of 51 percent of sums of squared loadings (variance). The results are shown in table 8–1 and closely match those of Saucier. Although the factor judged to be emotional stability comes up third instead of fourth in the rotated component order, and conscientiousness fourth rather than third, the factor loadings of the individual items are very clear in their relatively high loadings (over .40 absolute value) on the factors predicted, and simultaneously low loadings (.30 or less) on the other four factors. For the items purporting to measure openness in particular (see the eight highlighted factor loadings under the fifth component), there are no cases where predicted items fail to load highly on the presumed factor.

Factor analyses conducted with the IPIP–NEO proved to be more difficult to interpret. In terms of the initial principal components solution with all 120 items and the minimum eigenvalue = 1.0 criterion, 32 factors were extracted. The scree plot was of little help in determining a "natural break" in the number of factors based on initial eigenvalues, which do not drop below 2.0 until after the 13th factor. If we take an IPIP–NEO facet point of view, there is also no such break between the 30th and subsequent factors. Forcing a 30-factor solution does not produce the expected loading patterns for the 30-facet item groupings. Forcing the five-factor solution using the varimax rotation method with all 120 items accounts for just 35 percent of the sums of squared loadings or variance. Although there are clear clusterings of items on expected factors, there are dozens of exceptions to our standard of having items with (absolute value) loadings over

.40 on the apparent predicted factor and .30 or less on any of the other four factors.

As a method for coping with this complexity, Johnson (2000b; 2001; cf. Buchanan, Johnson, and Goldberg, 2005) suggests using the 30 facet scores instead of the 120 individual items for conducting factor analysis. In his findings based on a forced five-factor solution using varimax rotation and a sample of 20,993 cases gathered from the Internet, Johnson found that the solution converged in 11 iterations. Results based on an attempt to replicate Johnson's Internet findings are found in table 8-2. The factor solution converged in 13 iterations and the five-factor components produced accounted for 54 percent of the variance (sums of squared loadings). However, while the factor loadings led in the right direction, they were not as tightly arrayed as those for the mini-markers and fell somewhat short of those reported by Johnson as well. For the openness factor in particular (see the six highlighted factor loadings under the fifth component), the following facets did not load highly (over .40) on the predicted factor: emotionality (.34) and liberalism (.398). Furthermore, there were three instances where openness facets loaded above .30 (absolute value) on one or more of the other four factors, with a high loading (.47) for O3–Emotionality on the presumed neuroticism factor (component 3).

Table 8–2. **IPIP–NEO Factor Analysis: 30 Facet Factor Loadings for Five-factor Solution (N = 174)**

	Component*				
	1	**2**	**3**	**4**	**5**
A1 Trust	.381	-.014	-.099	**.519**	-.036
A2 Morality	.023	.514	-.158	**.444**	-.017
A3 Altruism	.452	.239	.111	**.618**	.295
A4 Cooperative	-.172	.299	-.312	**.592**	.125
A5 Modesty	-.129	-.023	.031	**.661**	-.076
A6 Sympathy	.110	.147	.062	**.576**	.426
C1 Self-efficacy	.226	**.619**	-.122	-.292	.272
C2 Orderliness	.081	**.489**	-.015	.079	-.292
C3 Dutiful	.031	**.635**	.054	.177	-.120

C4 Achievement	.238	**.557**	.038	.062	.320
C5 Self-discipline	.239	**.589**	-.332	-.129	.074
C6 Cautious	-.324	**.592**	-.228	.129	.025
E1 Friendliness	**.849**	.117	-.100	.182	.053
E2 Gregariousness	**.825**	-.039	-.021	-.067	.095
E3 Assertiveness	**.494**	.422	.063	-.474	.058
E4 Activity	**.352**	.159	-.114	-.053	.077
E5 Excitement	**.569**	-.426	.087	-.158	.309
E6 Cheerful	**.650**	-.028	-.464	.193	.125
N1 Anxiety	-.262	-.107	**.782**	.008	-.094
N2 Anger	.028	-.029	**.679**	-.405	.001
N3 Depression	-.413	-.187	**.664**	-.001	.034
N4 Self-conscious	-.780	-.090	**.148**	.022	-.078
N5 Immoderation	.064	-.421	**.374**	-.040	.223
N6 Vulnerability	-.195	-.419	**.611**	.044	-.270
O1 Imagination	.151	-.307	.165	-.079	**.554**
O2 Artistic	.160	.087	-.083	.230	**.619**
O3 Emotionality	.372	.135	.466	.266	**.341**
O4 Adventurous	.253	-.027	-.256	.005	**.646**
O5 Intellect	.030	.096	.090	-.105	**.685**
O6 Liberalism	-.107	-.116	-.139	.242	**.398**

*Rotated Component Matrix (Varimax Rotation with Kaiser Normalization)

These factor analyses lead to the conclusion that the mini-markers items show better construct validity than either the IPIP–NEO individual items or facet scores as far as the presumed Big Five factor domains, and openness specifically, are concerned. However, both measures are certainly reasonable in their ability to differentiate openness items from those hypothesized to be related to other personality dimensions.

Correlational Analyses

Concurrent validity. The question of concurrent validity across the two instruments on their respective Big Five scales is addressed by examining the correlation matrix relating each instrument's scales with those of the other (see table 8–3). These results generally support the assertion that each instrument's scales are roughly measuring the same Big Five dimension. The critical correlations in terms of concurrent validity are found along the diagonal, where they range from an absolute value of .57 to .77, with that between the two openness measures being .60. At the same time, one can be encouraged that in terms of the off-diagonal correlations, there are none for openness that are above .25. While even lower off-diagonal correlations would be desirable, the mutually discriminating power of the scales across instruments is fairly good.

Table 8–3. **Mini-markers and IPIP–NEO Big Five Scale Intercorrelation Matrix**

IPIP–NEO	Mini-markers				
	Surgency	**Agree-ableness**	**Conscien-tiousness**	**Emotional Stability**	**Open-ness**
Extraversion	**.77***	.21*	.08	.17*	.18*
Agreeableness	-.06	**.57***	.19*	.30*	.01
Conscientious-ness	.11	.31*	**.64***	.17*	.17*
Neuroticism	-.32*	-.29*	-.22*	**-.65***	-.10
Openness	.25*	.19*	-.12	-.04	**.60***

Note: N = 235–242; all (*) correlations are significant at p < .05.

Within-instrument scale orthogonality. Within each instrument, one should see interscale correlations as low as possible to justify assertions that each is measuring an orthogonal or conceptually distinct Big Five personality dimension. This assumption, of course, also underlies the varimax rotation method in the factor extraction analyses. Table 8–4 presents the scale intercorrelations for both the MM and IPIP–NEO instruments on the openness measures. While openness has statistically significant correlations with extraversion/surgency and agreeableness for both instruments, these are moderate to low, which renders some confidence these respective instrument scales are at least partially orthogonal to one other.

Table 8–4. **Big Five Openness Scale Intercorrelations by Instrument**

Mini-markers	Openness	IPIP–NEO	Openness
Surgency	.26*	Extraversion	.36*
Agreeableness	.18*	Agreeableness	.21*
Conscientiousness	.13*	Conscientiousness	.02
Emotional Stability	.07	Neuroticism	-.11
N = 289		N = 241–223	

Note: All (*) correlations are significant at p < .05.

Reliability analyses. A comparison of instrument openness scales was also conducted to assess scale utility in terms of relative reliability and internal consistency. The internal consistency (coefficient alpha) results were based on 8 items for the MM openness scale and 24 items for the IPIP–NEO openness scale. As a result of the higher number of included items, the IPIP–NEO scale was expected to show a higher coefficient alpha (Schultz and Whitney, 2005; cf. Saucier, 1994). Results are presented in table 8–5.

Table 8–5. **Comparative Reliability Analyses of Mini-markers and IPIP–NEO (Item Level) for Openness**

	Range	Alpha	Inter-item r Range	Item-total r
MM Openness	Items = 8, N = 289	.83	.18 to .81 (Mean = .38)	.44 to .67
IPIP–NEO Openness	Items = 24, N = 234	.81	-.12 to .85 (Mean = .15)	.11 to .51

There is little difference in analysis results regarding the two openness scales. The MM openness scale actually has a slightly higher coefficient alpha (.83) than the IPIP–NEO one (.81), despite having fewer scale items, though both alphas are still quite strong. Saucier (1994) reports an alpha of .78 for his openness scale, and Johnson (2006) reports an alpha of .82 for his. The average inter-item correlation for the IPIP–NEO scale (.15), however, is less than half that for the MM scale (.38), with correspondingly lower item-total correlations as well. The lowest interitem correlation on the MM scale is .18, but there are 21 such correlations that are negative for the IPIP–NEO openness scale and another 85 positive ones that are less than .10. There were no weak items (net negative contribution to coefficient alpha) on the MM scale, but four on the IPIP–NEO: "experience my emotions," "believe that there is no absolute right or wrong," "rarely notice my emotional reactions" (inverted), and "believe that we should be tough on crime" (inverted).

It must be kept in mind here that the IPIP–NEO openness scale purports to be measuring six distinct facets within the factor, which would conceptually tend to weaken a certain number of interitem correlations across included facets. To address the issue of "noise" introduced by individual items tied to IPIP–NEO facets, a reliability analysis for the openness scale was repeated using just the six facet scales instead of the individual items. The results of the reanalysis are shown in table 8–6. Coefficient alpha drops from .81 to .63, but this was not surprising given the fact that the factor is now being analyzed in terms of 6 rather than 24 component variables. Importantly, the mean interfacet correlation (.22) is markedly better than the mean interitem correlation found in the previous analysis (.15), and item-total correlations were also improved, although O3–Emotionality and O6–Liberalism remain under .30.

Cross-facet weaknesses for the IPIP–NEO openness scale may be seen clearly in table 8–7 and summarized as follows: O3–Emotionality

Table 8–6. **IPIP–NEO Facet Level Reliability Analysis for Openness (N = 234)**

IPIP–NEO Openness	Zero-Order r	Item-Total r (Corrected)
Facet		
O1 Imagination	.57	.34
O2 Artistic Interests	.69	.46
O3 Emotionality	.50	.28
O4 Adventurousness	.67	.46
O5 Intellect	.64	.42
O6 Liberalism	.47	.20

Cronbach's coefficient alpha = .63
Interfacet correlations range from .04 to .39 (Mean = .22)
Note: All zero-order and item-total correlations are significant at p < .05.

is correlated at less than .20 with both O4–Adventurousness and O5–Intellect; O6–Liberalism is correlated at less than .20 (and as low as .04) with O1–Imagination, O2–Artistic Interests, O3–Emotionality, and O5–Intellect, and its deletion from the openness scale would raise coefficient alpha. While the six facets are meant to be distinct, these correlations are lower than one would expect within a common factor domain. Few to no such problems exist for the individual items comprising the MM openness scale, whose intercorrelations are shown in table 8–8. There is certainly correlational variability here, with the likelihood of facet patterns especially along "creative-imaginative" and "complex-deep" axes, but nothing of great concern.

Table 8–7. **IPIP–NEO Openness Facet Intercorrelations**

Facet	Artistic Interests	Emotionality	Adventurousness	Intellect	Liberalism
Imagination	.23	.23	.24	.25	.09 (n.s.)
Artistic Interests		.22	.33	.39	.19
Emotionality			.17	.17	.09 (n.s.)
Adventurousness				.39	.23
Intellect					.04 (n.s.)

Note: N = 234; all correlations are significant at p < .05 except (n.s.).

Table 8–8. **Mini-markers Openness Item Intercorrelations**

Item	2	3	4	5	6	7	8
1. Complex	.30	.54	.27	.30	.36	-.18	-.20
2. Creative		.36	.81	.30	.35	-.78	-.27
3. Deep			.32	.37	.50	-.30	-.33
4. Imaginative				.29	.38	-.75	-.28
5. Intellectual					.27	-.31	-.61
6. Philosophical						-.33	-.23
7. Uncreative							-.39
8. Unintellectual							

Note: N = 289; all correlations are significant at p < .05.

Descriptive Measures and Norms

Descriptive statistics for the openness scales as measured separately by the mini-markers and IPIP–NEO are presented in table 8–9, with histograms for the two scales shown respectively in figures 8–1 and 8–2. Recall that for both the MM and the IPIP–NEO, the raw additive scores were rescaled to the original item metric (that is, divided by the number of included items) for ease of interpretation. It is also useful to keep in mind that the midpoint of the MM scales is 5.0 on a 1.0 to 9.0 scale, with scores less than 5.0 indicating relatively increasing degrees of reported inaccuracy (that is, slightly, moderately, very, and extremely), and those above 5.0 indicating the same degrees of relative accuracy for the openness adjective set. Regarding the IPIP–NEO scales, the midpoint is 3.0 on a 1.0 to 5.0 scale, with scores below the midpoint representing relative inaccuracy (moderately and very), and those above 3.0 the same degrees of relative accuracy with respect to the openness set of phrases.

Table 8–9. **Descriptive Statistics for Openness Measures**

Mini-markers		IPIP–NEO	
	Openness		Openness
Valid N	289	Valid N	234
Missing	3	Missing	58
Mean	6.37976	Mean	3.1051
Median	6.50000	Median	3.0833
Standard Deviation	1.120520	Standard Deviation	.44287
Skewness	-.409	Skewness	.106
Kurtosis	-.152	Kurtosis	.422

The two openness scales give us very disparate pictures of the population distributions across instruments. While both mean scores are above their scale midpoints, for the IPIP–NEO scale this is just barely the case (mean = 3.11), while for the mini-marker, this is significantly so (mean = 6.38). The MM scale yields 85.8 percent above the midpoint, compared to just 56.8 percent on the IPIP–NEO. On the high end of these scales (above 7.0 and 4.0), the percentages are 29.8 percent and 3.8 percent, respectively. The MM openness scale is far more negatively

Figure 8–1. **Histogram for Mini-markers: Openness**

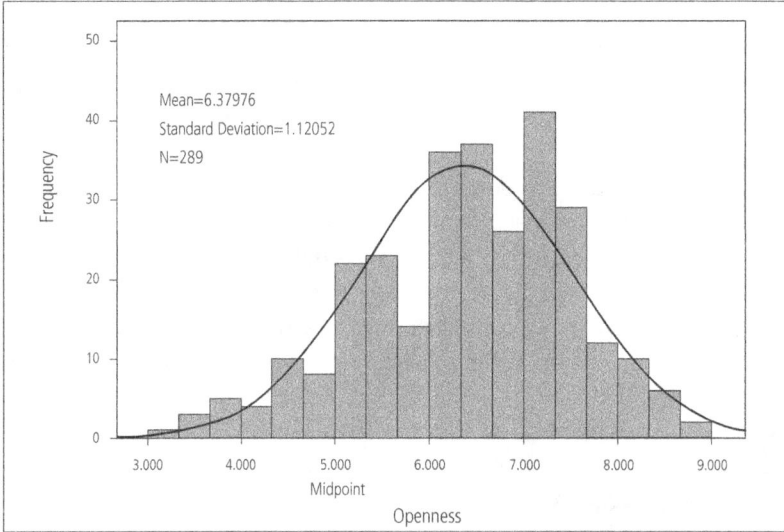

Figure 8–2. **Histogram for IPIP–NEO: Openness**

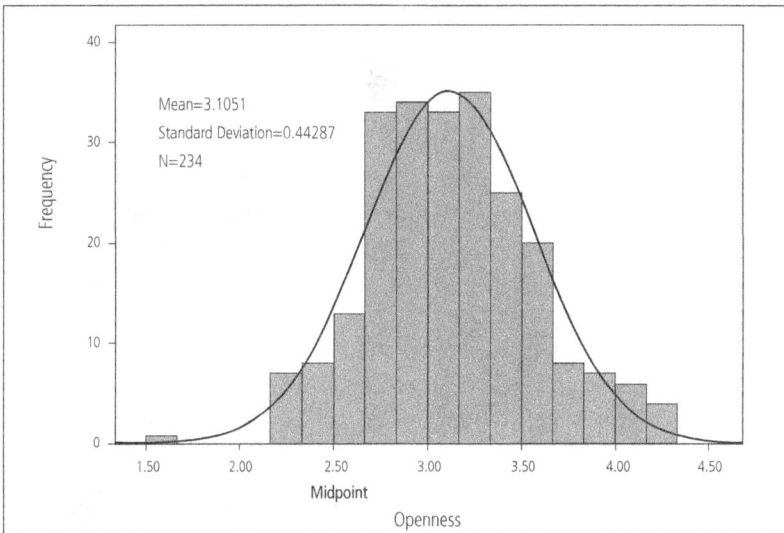

skewed (-.41 versus .11), but also considerably less leptokurtic (-.15 versus .42), meaning that the preponderance of MM openness scores are far more positively leaning than those for the IPIP–NEO, as well as having a flatter or wider distribution. These results raise the question whether the two scales are measuring the same phenomena with respect to our population, at least in terms of how they differentially distribute our subjects.

It is instructive to compare our sample's openness means on the two instruments to those from Sauciers's (1994) MM sample of 1,125 community residents of Eugene-Springfield, Oregon, and Johnson's (2006) IPIP–NEO sample of 20,993 Web users. For the mini-markers, Saucier reports an openness scale mean of 6.6, or nearly the same as ours (6.4). Johnson, however, finds an openness mean that is decidedly higher than ours (3.6 versus 3.1). This comparison alone, however, does not make clear whether the differences are due to the instruments themselves, or rather the variable nature of the sample populations.

The descriptive statistics for the individual openness MM items and IPIP–NEO facets are shown in table 8–10. These statistics help clarify the sources of variability on the respective measures. While there are statistically significant mean differences among the MM items—for example, between "intellectual" and "philosophical" ($t = 7.8, p < .05$)—differences are more pronounced on the IPIP–NEO. To cite two examples, means on both O1–Imagination and especially O6–Liberalism are below the scale midpoint (neutral) and are significantly lower than the O5–Intellect mean ($t = 11.1, p < .05$ and $t = 21.0, p < .05$, respectively). (Statistics for the individual IPIP–NEO items are found in appendix B.)

Multivariate Analysis

A multivariate regression analysis was conducted to help answer the critical question of how well the MM openness scale represents the relevant facets on the IPIP–NEO openness scale. Standardized betas are presented to show the relative predictive power of each IPIP–NEO facet vis-à-vis MM scale scores. The results are presented in table 8–11.

The conclusion from the descriptive analyses that the IPIP–NEO openness scale might well be measuring at least somewhat different phenomena than the MM openness scale is considerably reinforced by these results. The MM openness scale scores are most clearly predicted by O5–Intellect with an extremely strong standardized beta or beta weight of .62. There is only one other definitely significant effect for an associated IPIP–NEO facet, O4–Adventurousness, though its relative

Table 8–10. Descriptive Statistics for Openness Items

Raw Descriptive Statistics for Mini-markers Openness Items

Item		N	Minimum	Maximum	Mean	Standard Deviation
5	Complex	289	1.00	9.00	5.8270	1.91970
7	Creative	289	2.00	9.00	6.5121	1.65434
8	Deep	289	1.00	9.00	5.9827	1.84364
16	Imaginative	289	2.00	9.00	6.4810	1.52996
18	Intellectual	289	1.00	9.00	6.6886	1.41908
23	Philosophical	289	1.00	9.00	5.8166	1.71723
35	Uncreative	289	1.00	8.00	3.3010	1.70050
37	Unintellectual	289	1.00	8.00	2.9689	1.47045

Descriptive Statistics for IPIP–NEO Openness Facets

		N	Mean	Standard Deviation	Skewness	Kurtosis
O1	Imagination	259	2.9035	.74276	.322	.053
O2	Artistic Interests	260	3.3923	.83014	-.264	-.568
O3	Emotionality	262	3.4218	.62788	-.105	-.019
O4	Adventurousness	259	3.2191	.73004	.066	-.812
O5	Intellect	263	3.5618	.76806	-.489	.247
O6	Liberalism	259	2.1844	.74776	.222	-.649

effect (.23) is only about one-third the size as that for Intellect. O1–Imagination and O3–Emotionality display very weak to insignificant effects, and O2–Artistic Interests and O6–Liberalism do not independently predict MM openness scores, net of (or controlling for) the other facets. The zero-order correlation between O6–Liberalism and MM openness (.12) is the lowest of any of the six facets. The squared multiple correlation or model r^2 of .49 means that only about half of the variance in MM openness scores is explained by the IPIP–NEO facets.

Discussion

This analysis raises several issues for practical consideration when it comes to the goal of enhancing creative thinking among senior leaders

Table 8–11. **Multivariate Regression Analysis of Mini-markers Openness on Related IPIP–NEO Domain Facets (N = 172; d.f. = [6, 165])**

MM Openness	Zero-Order r	Partial r	Standardized Beta
IPI–NEO Facet			
O1 Imagination	.34*	.15	.12 (p<.06)
O2 Artistic Interests	.30*	-.07	-.06 (n.s.)
O3 Emotionality	.27*	.15	.12 (p<.06)
O4 Adventurousness	.46*	.27*	.23*
O5 Intellect	.64*	.52*	.62*
O6 Liberalism	.12 (n.s.)	.03	.02 (n.s.)

Model R^2 = .49
Note: All (*) correlations and betas are significant at $p < .05$ except (n.s.).

and the likelihood they will produce innovative solutions within their organizations. The first involves the value of psychological testing as a method for screening personnel for either creative predilections or abilities. If it is agreed that such screening has merit, then we must ask what instruments and feedback reporting systems are most useful for this purpose. Second, with or without such screening data in hand, the question arises of what training techniques and developmental programs are most valuable for fostering growth in this area. Finally, the roles of organizational climate and the wider culture for facilitating or dampening creative effort should be addressed.

The openness scales examined herein are just two of the over 200 instruments available for measuring creative tendencies or potential (Puccio and Murdock, 1999). Some instruments, like the popular Torrance Tests of Creative Thinking, attempt to measure global creative abilities across a number of subdimensions (Torrance, 1998). Others focus more clearly on domain-specific measures of creativity (Baer, 1998; Kaufman and Baer, 2004; cf. Chen, Himsel, Kasof, Greenberger, and Dmitrieva, 2006). Still others, such as the Creative Personality Scale, involve checklists of demonstrated creative behavior (Plucker, 1999). The examples of the five-factor personality measures used in this paper or other psychological tests that include creativity factors, such as "Innovative" (Gordon, 1993) or "New Ideas Generator" (Belbin, 2003) in their pool of traits, are often attractive because they include creativity as part of a comprehensive personality assessment, feedback, and development program.

Whatever method one uses to tap creative potential or openness to innovation, the current analysis makes it abundantly evident that different measures, though designed to tap the same underlying phenomena, may produce very different results for characterizing individuals and populations. In the case of the MM and the IPIP–NEO, while different measurement methods (single adjectives versus phrases) and metrics (five versus nine response choices) may account for some variation in the data, outcome disparities are most likely due to operationalization and content issues. It appears that a high tendency toward "openness" is much more likely to be found in our population by using the MM, because the IPIP–NEO includes Liberalism as one of its facets and its Imagination facet includes such items as "enjoy[s] wild flights of fantasy," which are underrepresented by the mini-marker that emphasizes Intellect and simply stated creative and imaginative thought. It is indeed possible on a facet such as Liberalism, which includes such items as "Tend to vote for liberal political candidates," where a total of just 11.4 percent of students found it at all accurate for themselves, and "Being tough on crime" (classified as "anti-liberal"), where less than 1 percent said it was not an accurate statement for them, that these items and their associated facet response variability are not generalizable from the larger civilian population to our government-military one.

One may argue that Liberalism, while important conceptually as an indicator of tendencies away from authoritarianism and dogmatism (McCrae and Costa, 1997) is not well constructed in the IPIP–NEO, with three of its four items referencing voting behavior ("liberal" versus "conservative" candidates) and being "tough on crime." It may also be unclear, for example, just how critical the IPIP–NEO Artistic Interests facet truly is from a strategic leader development perspective. In any event, the takeaway lesson is that it is incumbent upon the leader development and screening team to be as concrete and detailed as possible about the creative or innovative qualities desired and proceed on that basis to choose the appropriate instrument or instruments to measure these, rather than adopt some standard screening tool in wholesale fashion based on global descriptors alone. All included individual items should also be carefully evaluated for relevance, appropriateness, and bias within one's organizational context.

Assuming the leader developer has the right instrument for capturing creative abilities or attitudes deemed positive for the organization, it is then critical to know to what degree the measured characteristics are desirable and how to deliver the proper feedback and, if applicable and possible, the recommended training program for creativity enhancement. Although the goal may be to raise scores that are "too low" on creativity to foster innovative potential, scores that are "too high" (either globally or on some particular facet such as in the IPIP–NEO) may also be problematic or dysfunctional for the organization's needs. In addition, feedback and training should be tailored to the particular leader based on role. For example, it may not be necessary for certain administrators to be especially creative themselves as long as they can facilitate innovation and are open to new ideas from peers and subordinates. A counseling tool like the prescriptive Career Architect/Leadership Architect (Lombardo and Eichinger, 2004) is useful in this regard since it has separate sets of recommendations for developing creativity, managing innovation, broadening perspective, dealing with ambiguity, and becoming strategically agile.

In sum, tailored developmental programs that include feedback, coaching, role-playing, and the like can make most people better leaders, regardless of any natural advantage they may or may not have due to personality or prior experiences (Zenger and Folkman, 2002). Because there are as many kinds of creativity training as there are screening instruments, leader developers must be very careful in selecting or designing the best courses for their participants if they are to add value to their organizations given resource and time expenditure (see Birdi, 2007). For leaders, one fruitful form of training certainly involves learning how to successfully identify and evaluate *useful* original ideas (Blair and Mumford, 2007; Licuanan, Dailey, and Mumford, 2007). Increasing awareness of the variables, processes, and conditions leading to positive consideration of original ideas can provide the foundation for skill development under both simulated and real conditions. Specifically, the effects of such factors as job challenge and autonomy, social norms, group or team cohesion and diversity, and organizational identity must be understood if leaders are to effectively respond to and manipulate them to foster innovative growth (Adarves-Yorno, Postmes, and Haslam, 2007; Carmeli, Cohen-Meitar, and Elizur, 2007; Mirowsky and Ross, 2007; Nemeth and Ormiston, 2007). The best facilitators of creative thinking thoroughly understand both their own and others' thinking processes (Catling, 2007).

Ferris, Perrewe, Anthony, and Gilmore (2000), believe that personality assessment and the self-awareness and understanding that follow are critical first steps to developing the political and social skills that *do* lead to organizational innovation and success. This stance has clearly been taken to heart at such corporations as Intel, which in 1998 launched its Leadership Development Forum as an effort to develop the leadership necessary to meet the challenges of technological change and burgeoning demand (Carter, Ulrich, and Goldsmith, 2005). Its premise is that "leadership is just as much about *who we are* as it is about *what we do*" (Carter, Ulrich, and Goldsmith). Thus, a key part of the program involves a "self-discovery process" through which participants come to understand how they think and make decisions, and are encouraged to step out of their "comfort zones" and question their assumptions and paradigms. They learn that leadership necessitates continuous learning and moving beyond "known" management skills and techniques to incorporating their personal values into an innovative, passionate, even "audacious" vision or specific project for the organization that will help set corporate direction, align constituents, and inspire others for the future (Carter, Ulrich, and Goldsmith). Participants practice what they have learned through production of a leadership autobiography that embodies their self-reflection, a leader action plan, and special project. The program to date continues to enjoy concrete results in the form of organizationally beneficial project outcomes, as well as the enthusiastic support of senior and junior executives alike (Carter, Ulrich, and Goldsmith).

Ultimately, of course, leader developers can do all the personality screening and training it is possible to do and still have little impact on organizational change and implementation of innovation if managerial support and organizational climate and culture are inconsistent with creative work, and thus diminish motivation to perform the latter (Birdi, 2007; McLean, 2007; Senge, 1990). In fact, Ekvall (1999) specifically links a transformational leadership style to fostering a creative climate. The broader society must also be taken into account for possible cultural hindrances to creative expression, especially when working in international settings (Averill, Chon, and Hahn, 2001; Kim, 2007). Finally, it must be kept in mind that some but certainly not all psychological instruments are "culturally transferable" in terms of the meaningfulness of their referents.

This study has a number of limitations that affect drawing generalizations and conclusions. First, it must be remembered that the IPIP–NEO instrument used was only a proxy representation of the

NEO–PI–R itself and although that proxy may have acceptable corre-lations with the original (Goldberg, 1999), ultimately it is not the same and may be criticized on that basis. In addition, the particular IPIP–NEO proxy that was used was *the short form* of the full 300-item proxy, which further limits confidence in the instrument's power in this regard. Although there appeared to be no noteworthy patterns of response bias, the fair number of cases missing complete data on the IPIP–NEO de-tracts somewhat from a strong faith in comparisons to the mini-mark-ers. The most important limitation of the current study is that it falls short of comparing the two instruments head-to-head with respect to some external criterion, and thus we can make no claims regarding the relative external or predictive validity of the instruments with respect to outcomes of interest. Such a head-to-head comparison would indeed have provided recommendations with far heavier weight regarding fur-ther utilization of either or both of these instruments (cf. Saucier and Goldberg, 2003).

Conclusion

Openness, as the "Big Five" dimension of personality most close-ly tied to a proclivity for creative thinking and intellectual or innovative pursuits, has importance for successful executive leadership at the stra-tegic level. To enhance this disposition, leader developers have many options for what screening tools to use for personnel assessment, as well as the appropriate program to execute for feedback and development. Though each had some problems, the two personality instruments ex-amined here to measure openness had reasonable correlational proper-ties, to include both validity and internal consistency. Despite these psy-chometric properties, each produced a fairly different distribution and thus description of the subject population due to the nature of included subfacets and question wording. "Overlap" or shared variance between the instruments was only about 50 percent.

The content of this chapter serves as a caveat for those who would consider using standard psychological instruments for measuring cre-ative proclivities without carefully scrutinizing such instruments in terms of precisely what creative qualities or tendencies are the desired targets of testing. Relevant external criteria of importance to one's orga-nization must be identified for definitive decisions regarding the value of specific instruments versus others. A discerning evaluation must then be made to determine how the results of feedback and follow-on development programs will be structured to render the best or most balanced creative force for producing the level of workplace innovation desired. Efforts that focus solely on individual development in this area, however, without addressing a conducive organizational climate and culture for innovative behavior stand little chance of making a positive difference to the bottom line.

References

Adarves-Yorno, I., Postmes, T., and Haslam, S.A. (2007). Creative innovation or crazy irrelevance? The contribution of group norms and social identity to creative behavior. *Journal of Experimental Social Psychology* 43 (3), 410–416.

Averill, J.R., Chon, K.K., and Hahn, D.W. (2001). Emotions and creativity, East and West. *Asian Journal of Social Psychology* 4 (3), 165–183.

Baer, J. (1998). The case for domain specificity of creativity. *Creativity Research Journal* 11, 173–177.

Belbin, M.R. (2003). *Management teams: Why they succeed or fail* (2ᵈ ed.). Oxford, United Kingdom: Butterworth-Heinemann.

Birdi, K. (2007). A lighthouse in the desert? Evaluating the effects of creativity training on employee innovation. *Journal of Creative Behavior* 41 (4), 249–270.

Blair, C.S., and Mumford, M.D. (2007). Errors in idea evaluation: Preference for the unoriginal? *Journal of Creative Behavior* 41 (3), 197–222.

Buchanan, T., Johnson, J.A., and Goldberg, L.R. (2005). Implementing a five-factor personality inventory for use on the Internet. *European Journal of Psychological Assessment* 21 (2), 116–128.

Caprara, G.V., and Zimbardo, P.G. (2004). Personalizing politics: A congruency model of political preference. *American Psychologist* 59, 581–594.

Carmeli, A., Cohen-Meitar, R., and Elizur, D. (2007). The role of job challenge and organizational identification in enhancing creative behavior among employees in the workplace. *Journal of Creative Behavior* 41 (2), 75–90.

Carter, L., Ulrich, D., and Goldsmith, M. (eds.) (2005). *Best practices in leadership development and organization change: How the best companies ensure meaningful change and sustainable leadership* (213–231). San Francisco: Pfeiffer.

Catling, T. (2007). Creative thinking: A truly renewable energy source. *The British Journal of Administrative Management*, October/November, 22–23.

Cattell, R.B. (1945). The description of personality: Principles and findings in a factor analysis. *American Journal of Psychology* 58, 69–90.

Chen, C., Himsel, A., Kasof, J., Greenberger, E., and Dmitrieva, J. (2006). Boundless creativity: Evidence for the domain generality of individual differences in creativity. *Journal of Creative Behavior* 40 (3), 179–199.

Costa, P.T., and McCrae, R.R. (1992). *Revised NEO Personality Inventory (NEO PI-R) and NEO Five-Factor Inventory (NEO-FFI). Professional manual.* Odessa, FL: Psychological Assessment Resources.

DeYoung, C.G., Peterson, J.B., and Higgins, D.M. (2005). Sources of openness/intellect: cognitive and neuropsychological correlates of the fifth factor of personality. *Journal of Personality* 73, 825–858.

Digman, J.M. (1990). Personality structure: Emergence of the five-factor model. *Annual Review of Psychology* 41, 417–440.

Digman, J.M., and Takemoto-Chock, N.K. (1981). Factors in the natural language of personality: Reanalysis, comparison, and interpretation of six major studies. *Multivariate Behavioral Research* 16, 149–170.

Dwight, S.A., Cummings, K.M., and Glenar, J.L. (1998). Comparison of criterion-related validity coefficients for the Mini-Markers and for Goldberg's markers of the Big Five personality factors. *Journal of Personality Assessment* 70, 541–550.

Earley, P.C. (2002). Redefining interactions across cultures and organizations: Moving forward with cultural intelligence. *Research in Organizational Behavior* 24, 271–299.

Ekvall, G. (1999). Creative climate. In M.A. Runco and S.R. Pritzker (eds.). *Encyclopedia of creativity* (vol. I, 403–412). San Diego: Academic Press.

Ferris, G.R., Perrewe, P.L., Anthony, W.P., and Gilmore, D.C. (2000). Political skill at work. *Organizational Dynamics* 28 (4), 25–37.

Finholt, T.A., and Olson, G.M. (1997). From laboratories to collaboratories: A new organizational form for scientific collaboration. *Psychological Science* 8 (1), 28–36.

Fiske, D.W. (1949). Consistency of the factorial structures of personality ratings from different sources. *Journal of Abnormal and Social Psychology* 44, 329–344.

Goldberg, L.R. (1981). Language and individual differences: The search for universals in personality lexicons. *Review of Personality and Social Psychology* 2, 141–166.

———. (1992). The development of markers for the Big-Five factor structure. *Psychological Assessment* 4, 26–42.

———. (1993). The structure of phenotypic personality traits. *American Psychologist* 48, 26–34.

———. (1999). A broad-bandwidth, public domain, personality inventory measuring the lower-level facets of several five-factor models. In I. Mervielde, I. Deary, F. De Fruyt, and F. Ostendorf (eds.). *Personality psychology in Europe* (vol. 7, 7–28). Tilburg, The Netherlands: Tilburg University Press.

Goldberg, L.R., Johnson, J.A., Eber, H.W., Hogan, R., Ashton, M.C., Cloninger, C.R., and Gough, H.G. (2006). The international personality item pool and the future of public-domain personality measures. *Journal of Research in Personality* 40, 84–96.

Gordon, L.V. (1993). *Gordon Personal Profile-Inventory: Manual.* San Antonio, TX: Harcourt Assessment, Inc.

House, R.J., and Shamir, B. (1993). Toward the integration of transformational, charismatic, and visionary theories. In M.M. Chemers and R. Ayman (eds.). *Leadership theory and research: Perspectives and directions* (81–108). San Diego, CA: Academic Press.

Jacobs, T.O. (2006). *Strategic leadership: The competitive edge* (Revised ed.). Washington, DC: National Defense University.

John, O.P., and Srivastava, S. (1999). The Big-Five trait taxonomy: History, measurement, and theoretical perspectives. In L.A. Pervin and O.P. John (eds.). *Handbook of personality theory and research* (2^d ed., 102–138). New York: Guilford Press.

Johnson, J.A. (2000a). Predicting observers' ratings of the Big Five from the CPI, HPI, and NEO-PI-R: A comparative validity study. *European Journal of Personality* 14, 1–19.

———. (2000b, March). *Web-based personality assessment.* Paper presented at the 71st Annual Meeting of the Eastern Psychological Association, Baltimore, MD.

———. (2001, May 9). *Screening massively large data sets for non-responsiveness in web-based personality inventories.* Invited talk to the joint Bielefeld-Groningen Personality Research Group, University of Groningen, The Netherlands. Retrieved November 12, 2006, from <www.personal.psu.edu/~j5j/papers/screening.html>.

———. (2006). *IPIP20993iteminfo.xls.* Unpublished analysis available from Dr. Johnson at j5j@psu.edu.

Judge, T.A., and Bono, J.E. (2000). Five factor model of personality and transformational leadership. *Journal of Applied Psychology* 85, 751–765.

Judge, T.A., Ilies, R., Bono, J.E., and Gerhardt, M.W. (2002). Personality and leadership: A qualitative and quantitative review. *Journal of Applied Psychology* 87 (4), 765–780.

Kaufman, J.C., and Baer, J. (2004). Sure I'm creative—but not in mathematics! Self-reported creativity in diverse domains. *Empirical Studies of the Arts* 22, 143–155.

Kim, K.H. (2007). Exploring the interactions between Asian culture (Confucianism) and creativity. *Journal of Creative Behavior* 41 (1), 28–53.

Kotter, J.P. (1990). *A force for change: How leadership differs from management.* Boston: Harvard Business School Press.

Langer, E.J, and Moldoveanu, M. (2000). The construct of mindfulness. *Journal of Social Issues* 56, 1–9.

Licuanan, B.F., Dailey, L.R., and Mumford, M.D. (2007). Idea evaluation: Error in evaluating highly original ideas. *Journal of Creative Behavior* 41 (1), 1–27.

Lombardo, M.M., and Eichinger, R.W. (2004). *FYI For Your Improvement: A guide for development and coaching* (4th ed.). Minneapolis, MN: Lominger International.

MacDonald, D.A. (2000). Spirituality: Description, measurement, and relation to the Five Factor model of personality. *Journal of Personality* 68, 153–197.

McCrae, R.R. (1996). Social consequences of experiential openness. *Psychological Bulletin* 120, 323–337.

McCrae, R.R., and Costa, P.T. (1997). Conceptions and correlates of openness to Experience. In R. Hogan, J. Johnson, and S. Briggs (eds.). *Handbook of personality psychology* (825–847). San Diego: Academic Press.

McLean, J. (2007). The art of thinking outside the box. *The British Journal of Administrative Management*, October/November, 16.

Mirowsky, J., and Ross, C.E. (2007). Creative work and health. *Journal of Health and Social Behavior* 48 (4), 385–403.

Nemeth, C.J., and Ormiston, M. (2007). Creative idea generation: Harmony versus stimulation. *European Journal of Social Psychology* 37 (3), 524.

Norman, W.T. (1963). Toward an adequate taxonomy of personality attributes: Replicated factor structure in peer nomination personality ratings. *Journal of Abnormal and Social Psychology* 66, 574–583.

Ployhart, R.E., Lim, B.-C., and Chan, K.-Y. (2001). Exploring relations between typical and maximum performance ratings and the Five Factor Model of personality. *Personnel Psychology* 54 (4), 809–843.

Plucker, J.A. (1999). Reanalyses of student responses to creativity checklists: Evidence of content generality. *Journal of Creative Behavior* 33 (2), 126–137.

Puccio, G.J., and Murdock, M.C. (eds.) (1999). *Creativity assessment: Readings and resources.* Buffalo, NY: Creative Education Foundation Press.

Puccio, G.J., Murdock, M.C., and Mance, M. (2007). *Creative leadership: Skills that drive change.* Thousand Oaks, CA: Sage Publications.

Saucier, G. (1994). Mini-Markers: A brief version of Goldberg's unipolar Big-Five Markers. *Journal of Personality Assessment* 63 (3), 506–516.

Saucier, G., and Goldberg, L.R. (2003). The structure of personality attributes. In M.R. Barrick and A.M. Ryan (eds.). *Personality and work: Reconsidering the role of personality in organizations.* San Francisco: Jossey-Bass.

Schultz, K.S., and Whitney, D.J. (2005). *Measurement theory in action: Case studies and exercises.* Thousand Oaks, CA: Sage Publications.

Senge, P.M. (1990). *The fifth discipline: The art and practice of the learning organization.* New York: Doubleday.

Streufert, S., and Swezey, R.W. (1986). *Complexity, managers, and organizations.* Orlando, FL: Academic Press.

Thomas, D.C. (2006). Domain and development of cultural intelligence: The importance of mindfulness. *Group and Organization Management* 31 (1), 78–99.

Torrance, E.P. (1998). *Norms-technical manual: Torrance Tests of Creative Thinking.* Bensenville, IL: Scholastic Testing Service.

Tupes, E.C., and Christal, R.C. (1992). Recurrent personality factors based on trait ratings (1961). *Journal of Personality* 60, 225–251.

Van Hiel, A., and Mervielde, I. (2004). Openness to experience and boundaries in the mind: Relationships with cultural and economic conservative beliefs. *Journal of Personality* 72, 659–686.

Yukl, G. (2002). *Leadership in organizations* (5th ed.). Upper Saddle River, NJ: Prentice-Hall.

Zaccaro, S.J. (2001). *The nature of executive leadership: A conceptual and empirical analysis of success.* Washington, DC: American Psychological Association.

Zenger, J.H., and Folkman, J. (2002). *The extraordinary leader: Turning good managers into great leaders.* New York: McGraw-Hill.

Appendix A

Items Comprising the IPIP–NEO Openness Facets

Facet: Imagination
>Q3. Have a vivid imagination
>Q33. Enjoy wild flights of fantasy
>Q63. Love to daydream
>Q93. Like to get lost in thought

Facet: Artistic Interests
>Q8. Believe in the importance of art
>Q38. See beauty in things that others might not notice
>Q68. Do not like poetry
>Q98. Do not enjoy going to art museums

Facet: Emotionality
>Q13. Experience my emotions intensely
>Q43. Feel others' emotions
>Q73. Rarely notice my emotional reactions
>Q103. Don't understand people who get emotional

Facet: Adventurousness
>Q18. Prefer variety to routine
>Q48. Prefer to stick with things that I know
>Q78. Dislike changes
>Q108. Am attached to conventional ways

Facet: Intellect
>Q23. Love to read challenging material
>Q53. Avoid philosophical discussions
>Q83. Have difficulty understanding abstract ideas
>Q113. Am not interested in theoretical discussions

Facet: Liberalism
>Q28. Tend to vote for liberal political candidates
>Q58. Believe that there is no absolute right or wrong
>Q88. Tend to vote for conservative political candidates
>Q118. Believe that we should be tough on crime

Appendix B

Raw Descriptive Statistics for IPIP–NEO Openness Items

Item	N	Minimum	Maximum	Mean	Standard Deviation
Q301	265	1.00	5.00	3.5358	0.94530
Q802	263	1.00	5.00	3.5513	0.95920
Q1303	266	1.00	5.00	2.9887	1.04785
Q1804	265	1.00	5.00	3.5434	0.98040
Q2305	265	1.00	5.00	3.3623	1.02474
Q2806	265	1.00	5.00	2.1019	1.12857
Q3301	265	1.00	5.00	2.3849	1.03854
Q3802	266	1.00	5.00	3.4511	0.88566
Q4303	262	1.00	5.00	3.5840	0.85243
Q4804	265	1.00	5.00	3.1396	0.97669
Q5305	265	1.00	5.00	2.4340	1.00960
Q5806	263	1.00	5.00	2.8327	1.23040
Q6301	263	1.00	5.00	2.8365	1.04839
Q6802	264	1.00	5.00	3.0076	1.21459
Q7303	266	1.00	4.00	2.3083	0.93309
Q7804	264	1.00	5.00	2.5038	0.95128
Q8305	265	1.00	5.00	2.3132	1.00943
Q8806	265	1.00	5.00	3.7170	1.10057
Q9301	264	1.00	5.00	2.8258	1.04278
Q9802	265	1.00	5.00	2.4302	1.21381
Q10303	266	1.00	5.00	2.5865	0.98001
Q10804	263	1.00	5.00	3.0228	0.94881
Q11305	266	1.00	5.00	2.3872	1.10080
Q11806	264	2.00	5.00	4.4432	0.66740

Part IV

Soldier Psychological
Health

Chapter 9

U.S. Army Mental Health Advisory Teams

Sharon A. McBride, Jeffrey L. Thomas, Dennis McGurk, Michael D. Wood, and Paul D. Bliese

Combat and operational stress has repeatedly been documented as a factor affecting the overall effectiveness and readiness of military units (see Department of the Army Field Manual 4–02.51, *Combat and Operational Stress Control*). Soldier mental health and well-being are important facets of such overall effectiveness and readiness. As such, military leaders and policymakers have emphasized the importance of understanding the causes and prevalence of mental health problems, ensuring access to mental health care, and determining potential interventions that can mitigate the effects of mental health problems for deployed Soldiers and veterans.

A key cornerstone of the Army's strategy to understand combat and operational stress effects among Soldiers deployed to Iraq and Afghanistan is the Mental Health Advisory Team (MHAT) theater evaluation and report. The MHAT was established in July 2003 through a charter from the U.S. Army Surgeon General in order to provide recommendations to commanders and medical personnel deployed to Iraq. Since 2003, an MHAT has been deployed each year to Iraq (2003–2007) and twice to Afghanistan (2005 and 2007). In total, nearly 8,500 Soldiers have been surveyed across MHATs. The focus of this chapter will be on the specific findings from MHAT V (2007). Each of the MHATs has sought to determine the following:

- *Behavioral health of Soldiers.* What is the behavioral health status of Soldiers? How does it compare to Soldiers who deployed previously to Operation *Iraqi Freedom* (OIF) or Operation *Enduring Freedom* (OEF)? Are there other at-risk groups (for example, junior enlisted, married, or deployment-extended personnel)?
- *Behavioral health care system effectiveness.* What is the status of the behavioral health care system in Iraq and Afghanistan?

What are the systemic factors that need to be addressed to ensure that Soldiers are receiving the best care?

- *Future needs.* What are the planning and resource factors that must be considered to care for Soldiers engaged in future deployments to Iraq and Afghanistan? As both wars evolve, what can the Army do to prepare to meet behavioral health needs?

MHAT V (2007): Assessment of U.S. Soldiers in Iraq and Afghanistan

An in-theater assessment of Soldier mental health was conducted from September to November 2007. This was the fifth iteration of an MHAT deployment to Iraq and the second to Afghanistan. The team produced two separate reports—one for OIF, which includes a section on Soldiers in Kuwait, and one for OEF. The OIF and OEF reports were independent and designed to be standalone documents. At the same time, there was close coordination between the two teams.

MHAT V was staffed primarily by U.S. Army research psychologists from the Walter Reed Army Institute of Research and its subordinate unit, the U.S. Army Medical Research Unit–Europe. Both the OIF and OEF research teams used nearly identical assessment tools as well as similar analytic strategies and collaborated in writing the reports. Both research teams collected Soldier well-being data using anonymous, voluntary paper and pencil surveys. While in theater, the research teams processed and analyzed survey data, examined secondary data sources, and conducted focus group interviews with Soldiers. The MHAT V team report and recommendations were based on these data sources. Behavioral health, unit ministry team, and medical personnel were also surveyed and interviewed by the MHAT V researchers; however, this chapter focuses primarily on the Soldier survey. Detailed information about support unit findings can be found in the complete MHAT V report (see below).

The MHAT V survey instrument contains the same core measures that were used in all previous MHAT assessments. The survey focuses on four main areas that target Soldier mental health and well-being illustrated in the conceptual model below. These are risk factors, such as combat and deployment experiences; protective factors, such as training and willingness to seek care; behavioral health status and performance indices, such as individual and unit morale, depression, anxiety, and acute stress symptoms; and unethical behaviors.

Figure 9–1. **Soldier Combat and Well-being Model**

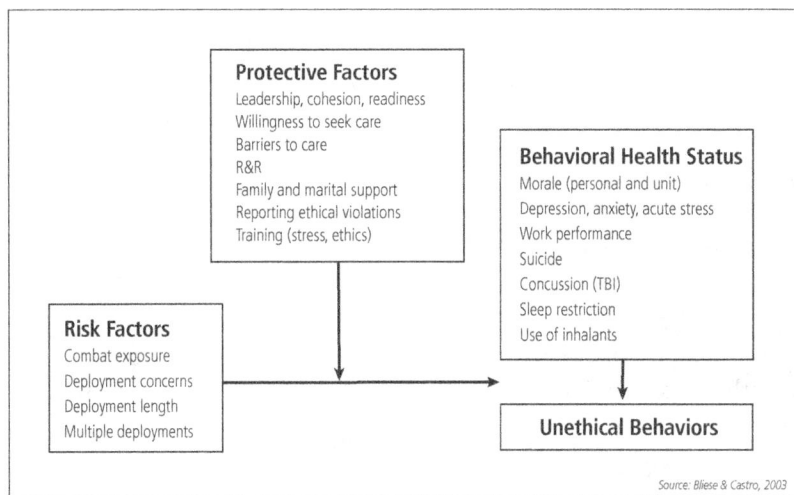

Protective Factors
Leadership, cohesion, readiness
Willingness to seek care
Barriers to care
R&R
Family and marital support
Reporting ethical violations
Training (stress, ethics)

Behavioral Health Status
Morale (personal and unit)
Depression, anxiety, acute stress
Work performance
Suicide
Concussion (TBI)
Sleep restriction
Use of inhalants

Risk Factors
Combat exposure
Deployment concerns
Deployment length
Multiple deployments

Unethical Behaviors

Source: Bliese & Castro, 2003

The findings that follow are presented by theater and represent a summary of the two independent MHAT V reports. The complete OIF and OEF MHAT report is available on the Army Medical Department Web site at <www.armymedicine.army.mil/reports/mhat/mhat_v/mhat-v.cfm>. The OIF section mainly presents Soldier responses to the 2007 survey in comparison to responses from the 2006 OIF survey. However, in some cases, MHAT V indices are interpreted within the context of data from all previous MHAT missions. The OEF section presents data from the 2007 survey in comparison to the 2005 OEF survey, as well as the 2007 OIF survey. These latter comparisons with OIF data were made because the previous OEF data collection was conducted almost 3 years earlier, and many of the responses on the surveys needed to be interpreted in a broader, more recent context.

Readers may occasionally note small discrepancies in the OIF 2007 values reported in the sections below. These differences reflect the fact that it was often necessary to adjust values for demographic and other sample differences to clearly delineate findings. For each section below, data are presented in the broad categories of Soldier Behavioral Health and Performance, Soldier Risk Factors, and Soldier Protective Factors. This is followed by key recommendations that were made to commands following examination of all collected data.

MHAT V Operation *Iraqi Freedom* Findings

Data Collection Procedures and Demographics

Anonymous surveys were completed primarily by Brigade Combat Team Soldiers in OIF. All regions within the Iraqi theater of operations with significant numbers of U.S. Army Soldiers were surveyed. Data from 2,195 surveys were compared to findings from the MHAT IV (2006) data collection, which had a sample size of 1,368. Demographic variables of gender, rank, and months in theater are statistically controlled to ensure that observed differences between both samples were not an artifact of demographic differences. For instance, when comparing combat experiences across samples, it is important to normalize the length of time Soldiers have been deployed in order to determine whether there has been either a decline or an escalation in combat intensity. Also, adjusted values are typically provided based on the relative numbers of junior enlisted male Soldiers in theater for 9 months due to this being the most prevalent group. A comparison of selected demographic variables for both samples can be found in the full online report cited above.

Soldier Behavioral Health and Performance

Soldier behavioral health and performance are viewed as outcomes that are determined by risk factors and protective factors. This section examines behavioral health and performance outcomes, and uses subsequent sections on risk factors and protective factors to interpret the behavioral health and performance results. Specific measures of behavioral health and performance that will be presented here include individual and unit morale, acute stress, depression, anxiety, stress and work performance, suicidal ideation, alcohol and substance abuse, and unethical behaviors.

Morale: individual and unit. The percentage of Soldiers who rated their individual morale as high or very high did not change significantly between 2006 and 2007 (15.3 percent and 19.1 percent, respectively). However, ratings of unit morale increased significantly over the same time period (6.4 percent and 12.7 percent).

Acute stress, depression, and anxiety. Included in the MHAT V survey was a standardized, validated scale for assessing acute stress symptoms, the Post-Traumatic Stress Disorder Checklist (Weathers, Litz, Herman, Huska, and Keane, 1993), and validated scales for assessing depression and generalized anxiety from the Patient Health Questionnaire (Spitzer, Kroenke, and Williams, 1999). These scales have been used in previous MHAT surveys and have formed the basis of several peer-reviewed publications from the Walter Reed Army Institute of Research (for example, Bliese et al., 2007; Hoge et al., 2004; Hoge et al., 2007).

The OIF rates for acute stress, depression, anxiety, or any of these mental health problems tended to be lower in 2007 than in 2006; however, using a conventional criterion of p < .05, none of the differences were statistically significant (see figure 9–2).

Figure 9–2. **Operation *Iraqi Freedom* Brigade Combat Team Soldiers Reporting Mental Health Problems (adjusted percent)**

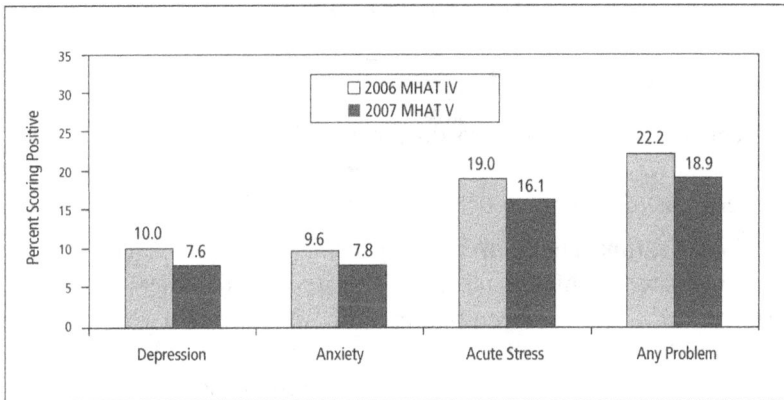

Figure 9–3 presents the 5-year trend for Soldiers scoring positive for any mental health problem. The 5-year trend for these rates has remained relatively steady, with the exception of 2004. The 2007 rate differs significantly only from that reported in 2004. These rates remain consistently higher than those reported in the general population.

Stress and work performance. In the Soldier Well-being Survey, work performance is assessed with three items. Soldiers are asked to indicate whether stress or emotional problems in the last 4 weeks have "limited your ability to do your job"; "caused you to do work less carefully than usual"; or "caused your supervisor to be concerned about your performance." After adjusting for sample differences, Soldiers in 2007 were significantly less likely than Soldiers in 2006 to report that stress or emotional problems had limited their ability to do their job (15.1 versus 18.8

Figure 9–3. **Operation** *Iraqi Freedom* **Soldiers Reporting Any Mental Health Problem (adjusted percent)**

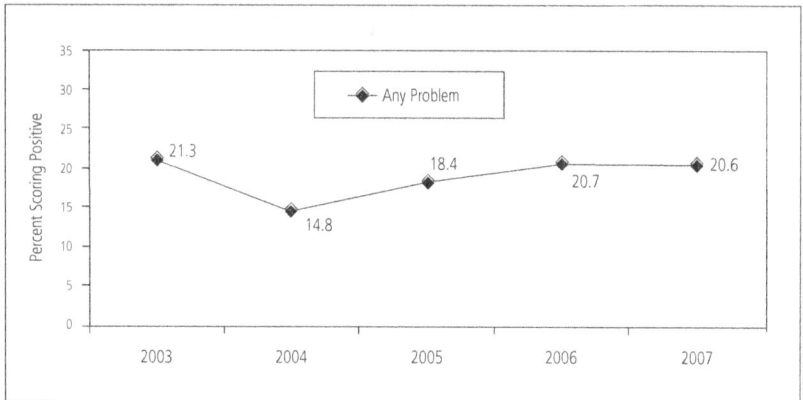

percent) and caused them to do their work less carefully than usual (23.8 versus 27.7 percent). The difference for the item about supervisor concern was not significant (15.0 versus 15.7 percent).

Suicidal ideation. The MHAT survey includes one question that is used as an indicator of suicidal ideation. This item asks Soldiers if, over the last 4 weeks, they have been bothered by "thoughts that you would be better off dead or of hurting yourself in some way." Any response other than "Not at all" was considered a positive response. In 2006, the adjusted positive response rate was 17.7 percent. In contrast, in 2007, the adjusted rate was 15.2 percent. Raw rates were 14.5 percent and 13.1 percent for 2006 and 2007, respectively. These rates did not change significantly between 2006 and 2007 and did not mirror the significant increase in rates for completed suicides found between 2006 and 2007.

Divorce intent. Another possible indication of behavioral health problems is the percentage of Soldiers who report they are considering divorce. In 2007, the in-theater adjusted percent for junior enlisted was 17.0 percent; for noncommissioned officers (NCOs), 12.3 percent; and for officers, 3.5 percent. Raw rates were 20.8 percent, 15.1 percent, and 4.3 percent for E1–E4, NCOs, and officers, respectively. Values significantly differed across ranks but did not differ from 2006 to 2007.

Alcohol and substance abuse. Although alcohol use is not permitted while deployed, 8 percent of Soldiers surveyed in 2007 reported using alcohol in Iraq. Although slightly higher, this rate was not significantly different than the reported rate of 6.8 percent in 2006. The reported rate for illegal drug use in 2007 was 1.4 percent and also did not differ significantly from the 2006 rate of 1.6 percent.

Inhalants. Inhalants have been described as a frequently overlooked substance that are readily available, low cost, and undetectable by randomized drug screens (Lacy and Ditzler, 2007). A recent U.S. Department of Health and Human Services study found that a little less than 1 percent of the population reported using inhalants in the past year (HHS SAMHSA National Survey on Drug Use and Health 2007 Report). The 2007 MHAT survey included an item about inhalant use, the first year these items were included. Overall, 3.8 percent of Soldiers in OIF reported that they "huffed" a substance. The most frequently inhaled substance was compressed air (3.1 percent), followed by paint (1.1 percent), fuels (.10 percent), aerosols (.07 percent), and nitrous oxide (.04 percent). It appears that the overall rate for Soldiers in theater is elevated when compared to rates reported for the overall U.S. population.

Unethical behavior. The 2006 and 2007 MHAT survey included ethical issues. The questions specifically addressed battlefield ethics and the adequacy of ethics training. When responses were compared, there was only one significant difference between 2006 and 2007. Percentages for 2006 and 2007, respectively, follow the items in parentheses:

- Insulted and/or cursed noncombatants in their presence (34.6 percent versus 33.0 percent)
- Damaged and/or destroyed private property when it was not necessary (10.9 percent versus 13.6 percent)
- Hit/kicked a noncombatant when it was not necessary (5.3 percent versus 6.1 percent)
- Modified the rules of engagement in order to accomplish the mission (10.0 percent versus 7.4 percent)
- Ignored the rules of engagement in order to accomplish the mission (5.7 percent versus 4.3 percent).

Unethical behavior and behavioral health status. One of the central findings from MHAT IV (2006) was that Soldiers were more likely to report they had engaged in unethical behavior if they had also screened positive for behavioral health problems. The relationship between unethical behaviors and behavioral health status was reevaluated in 2007 and the findings were replicated. Specifically, Soldiers who screened positive for mental health problems of depression, anxiety, or acute stress were significantly more likely to report engaging in unethical behaviors than those who did not. These behaviors included insulting or cursing at noncombatants (48.4 percent versus 26.5 percent), unnecessarily damaging or destroying private property (19.1 percent versus 9.9 percent), and hitting or kicking a noncombatant (10.2 percent versus 3.8 percent).

Summary of behavioral health and performance indices in OIF. The examination of OIF Soldiers' behavioral health in 2007 compared to other MHAT data indicates positive trends on several measures. Soldiers' ratings of unit morale, although still low, showed a large increase relative to 2006, and their reports of stress-related performance problems significantly declined relative to 2006. In terms of unethical behaviors, Soldiers reported a significant decline in the degree to which their units modify the rules of engagement. Other indices remained stable from 2006 to 2007. These included reported rates for individual morale, suicidal ideation, in-theater alcohol usage, and four of the five unethical behavior measures. Additionally, the positive relationship between unethical behaviors and mental health problems was consistent with findings from the 2006 MHAT.

Soldier Risk Factors

The examination of risk factors provides a theoretical basis from which to explain changes in Soldier behavioral health and reported performance indices. Two risk factors potentially unique to the 2007 deployment cycle are the length of the deployment and the potential cumulative impact of deploying multiple times. These factors and their correlates are examined in detail below. Evaluating the effect of deployment length and multiple deployments is unique to MHAT V because Soldiers deploying in 2007 were among the first to go for 15-month tours, and a fairly large number had deployed three or four times. An additional reason to examine risk factors is to specifically focus on those known factors that can be directly influenced by command and/or mental health providers. To this end, this section also focuses on the relationship between sleep deprivation and behavioral and performance-related problems.

Combat experiences. Exposure to potentially traumatic experiences is one of the principal risk factors for behavioral health problems in combat settings (Fontana and Rosenheck, 1998). In the Soldier Well-being Survey, combat experiences are measured with 33 items assessing experiences such as "Knowing someone seriously injured or killed" and "Being wounded/injured." A combat experience score (ranging from 0 to 33) was created by summing the number of reported experiences.

MHAT V OIF data reflect a decline in combat intensity from 2006 to 2007. Scores for 11 of the 33 items on the combat experience scale declined significantly and none increased substantially relative to 2006. This decline is also evident by examining results based on the number of months deployed and the data collection year. For example, at month 2, only 21.5 percent of Soldiers reported being attacked or ambushed in 2007, whereas the value had more than doubled by month 2 in 2006 (50.8 percent).

Deployment concerns. Combat experiences are intense events that put Soldiers at risk for mental health problems. From a behavioral health perspective, however, less dramatic chronic concerns have also been shown to negatively relate to health (Gilbert, Lieberman, Morewedge, and Wilson, 2004). In the MHAT surveys, less dramatic chronic events are captured with a series of 11 questions addressing deployment concerns rated on a scale from 1 (very low trouble or concern) to 5 (very high trouble or concern). Long deployment length, separation from family, boring and repetitive work, and lack of privacy or personal space are consistently rated as the top deployment concerns in MHAT surveys.

- *Deployment length.* The MHAT V OIF sample was well suited for examining the effects of deployment length on outcomes such as mental health because the sampling distribution and the large sample size of 2,195 provided a wide range of data across months deployed. This wide range provided the opportunity to model months deployed as a continuous variable.

- *Deployment length and morale.* The number of months deployed was related to both individual and unit morale. Ratings of morale were initially high and fell to their lowest levels at months 8, 9, and 10 before gradually increasing (see figure 9–4). Subsequent analyses were conducted to determine whether the changes in morale were related to cumulative combat experiences. These analyses indicated that combat experiences were unrelated to ratings of either individual or unit morale.

Figure 9–4. **Predicted Levels of Individual and Unit Morale by Months in Operation *Iraqi Freedom* Theater**

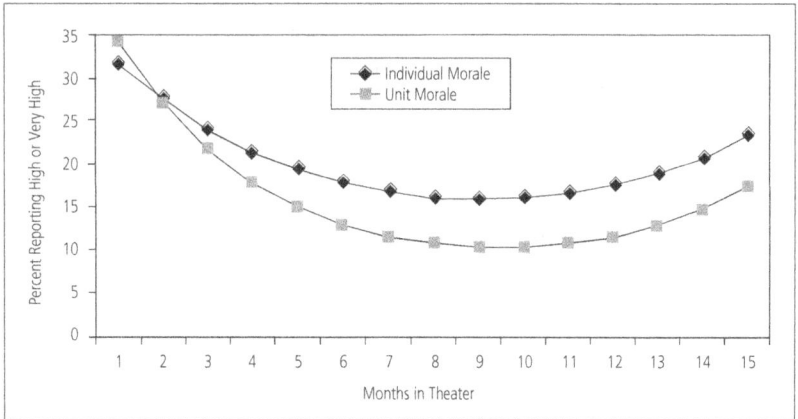

■ *Deployment length and behavioral health.* Figure 9–5 illustrates the relationship between months deployed and screening positive for depression, anxiety, or acute stress (that is, having any mental health problem). The figure shows a linear increase, with some degree of leveling off or decrease for the latter months. It is unclear why this decrease occurs. It could be due to the optimism of being able to return home, or by theater psychiatric evacuations in the earlier months of the deployment. As with morale, the highest risk times are at 8, 9, and 10 months. In figure 9–5, it is important to point out that the model predicts a threefold increase in the number of junior enlisted male Soldiers that will test positive for mental health problems by the end of a 15-month deployment. It is also important to consider that with shorter deployments, the shape of the curve might be the same as that shown in figure 9–5; however, with shorter deployments the apex of the curve might not reach the same high point as it does in the figure.

These results suggest that the post–6-month deployment period is an especially heightened risk time for mental health problems (a finding also noted in MHAT IV), and that reports of mental health problems level off in the months immediately before redeployment. As figures 9–4 and 9–5 demonstrate, the last months of deployment show increases in morale and decreases in mental health problems, which could

Figure 9–5. **Predicted Levels of Mental Health Problems by Months in Operation *Iraqi Freedom* Theater**

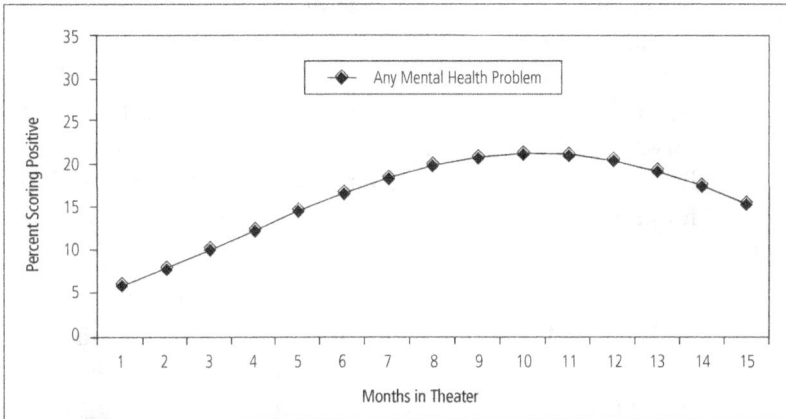

be labeled as redeployment optimism. However, even with redeployment optimism, by the end of a 15-month deployment, levels did not return to those seen at 1 month. Further redeployment research strongly suggests that rates for Soldiers reporting mental health problems will rise when Soldiers return home (Bliese et al., 2007).

- *Deployment length and suicidal ideation.* The MHAT V survey asked Soldiers, "Over the last four weeks, how often have you been bothered by thoughts that you would be better off dead or hurting yourself in some way?" Junior enlisted Soldiers were twice as likely to endorse this question during the mid-deployment window. As seen with the relationship between deployment length and behavioral health status, this mid-deployment peak decreases as redeployment nears.

- *Deployment length, stress, and work performance.* Rates for Soldiers reporting that stress or emotional problems had limited their ability to do their jobs, caused them to work less carefully than usual, and caused their supervisors to be concerned about their performance generally showed a linear increase as deployment time lengthened. The only variable to show a quadratic trend was working "less carefully," which leveled off and slightly declined after month 12. As with mental health problems, the cumulative effect of months deployed has a pronounced impact on reports of work performance by month 15.

■ *Deployment length and divorce intent.* Rates for Soldiers reporting that they plan on getting a divorce or separation increased significantly as a function of deployment length (see figure 9–6). Note that there are substantial rank differences for this measure. Although junior enlisted Soldiers are the most likely to endorse this question at all time points, the slopes for junior enlisted Soldiers and NCOs are significantly steeper than for officers.

Figure 9–6. **Predicted Divorce Intent for Officers, Noncommissioned Officers, and Junior Enlisted by Months in Operation *Iraqi Freedom* Theater**

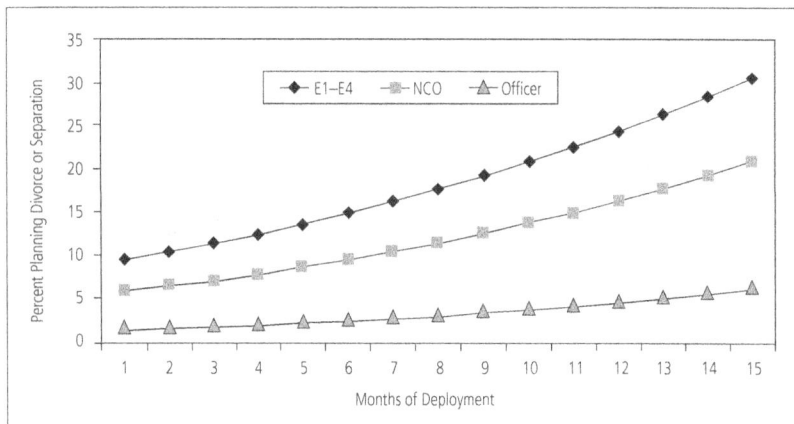

■ *Deployment length and substance abuse.* Deployment length was a significant predictor of Soldiers' reports of inhalant and alcohol use. Reported use of both alcohol and inhalants showed significant quadratic effects: reported alcohol abuse tapered off after month 11, and reported use of inhalants peaked between 8 and 9 months.

■ *Deployment length and unethical behaviors.* Deployment length was also significantly related to the probability that a Soldier would report having engaged in the unethical behaviors of insulting/cursing at noncombatants, damaging property unnecessarily, or hitting or kicking a noncombatant. The form of the relationship for all three of these variables was a rise in the first 10 months of a deployment, followed by a decline after month 9.

Multiple deployments. Both the MHAT III report in 2005 and the MHAT IV report in 2006 identified multiple deployments as a risk factor for mental health problems. In previous years, analyses have examined the effects of multiple deployments by comparing first-time deployers with those who had deployed at least once before. In both 2005 and 2006, however, the multiple deployment group was almost entirely comprised of Soldiers on their second deployment. By contrast, the 2007 sample contains a sufficiently large number of individuals on their third or fourth deployment, making it possible to create three deployment groups: first-time deployers (n = 1,496), second-time deployers (n = 538), and third-/fourth-time deployers (n = 129), with 32 Soldiers not reporting number of deployments.

In presenting the findings related to multiple deployments, results are provided for NCOs rather than for junior enlisted because Soldiers in the multiple deployer group are predominantly NCOs. Specifically, in the first-time deployer group, NCOs constitute 19 percent of the sample, whereas they constitute 60.8 percent of the second-time deployer sample and 74.4 percent of the third-/fourth-time deployer sample.

- *Multiple deployments and morale.* Figure 9–7 shows adjusted rates of morale for male NCOs deployed for 9 months. NCOs on their second or third/fourth deployments have significantly lower morale than NCOs on their first deployment (for third-/fourth-time deployers compared to first-time deployers for unit morale, p < .10). In the figure, the difference between those on a second deployment and those on their third/fourth for individual and unit morale is not statistically significant.

- *Multiple deployments and behavioral health status.* Both the MHAT III and MHAT IV reports found that the behavioral health status of Soldiers on their second deployment was significantly lower than on their first. This finding was replicated in the 2007 sample and was able to be extended to Soldiers on

Figure 9–7. **Operation *Iraqi Freedom* Noncommissioned Officer Brigade Combat Team Soldiers Reporting High Morale (adjusted percent)**

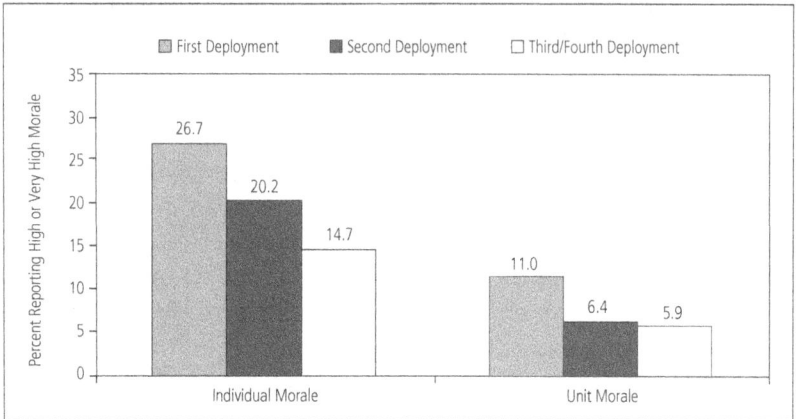

their third or fourth deployment. The form of the relationship is illustrated in figure 9–8 for the combined behavioral health measure of being positive for depression, anxiety, and/or acute stress. An NCO on his second or third/fourth deployment reported significantly more mental health problems than did an NCO on his first deployment. Furthermore, the value for NCOs on their third/fourth deployment (unadjusted percent of 29.2 percent) was significantly different from NCOs on their second deployment (unadjusted percent of 19.0 percent).

- *Multiple deployments, stress, and work performance.* Three different statistical models were run to examine whether there was a multiple deployment effect if NCOs reported stress or emotional problems while deployed. Results indicated that Soldiers in their third/fourth deployment were significantly more likely than first-time deployers to report that stress or emotional problems in the last 4 weeks limited their ability to do their job (16.6 percent versus 9.7 percent) and caused their supervisor to be concerned (14.1 percent versus 7.9 percent). Soldiers on their second deployment did not differ from either first-time deployers or third-/fourth-time deployers.

Figure 9–8. **Operation** *Iraqi Freedom* **Noncommissioned Officer Brigade Combat Team Soldiers with Mental Health Problem (unadjusted percent)**

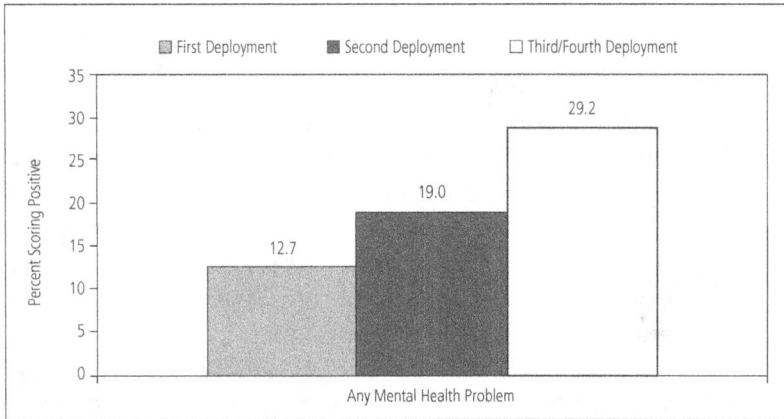

Several measures were not found to be associated with multiple deployments. NCOs' reports of suicidal ideation, divorce or separation intent, unethical behaviors, or use of inhalants or illegal drugs/substances were not found to be related to the number of deployments.

Sleep deprivation. Sleep deprivation is a risk factor for behavioral health and performance problems. Importantly from a prevention perspective, sleep problems and sleep deprivation represent risk factors that may be manageable with good leadership. In 2007, the MHAT V survey included a number of questions about Soldiers' sleep patterns while deployed.

Soldiers in OIF reported needing 6.4 hours of sleep per day to feel well rested, while they reported receiving 5.6 hours. Both of these values are less than the 8 hours a night shown to be necessary to maintain optimal cognitive functioning (Belenky et al., 2003). The difference between what Soldiers report needing and what they report receiving represents a sleep deprivation value of 0.8 hours per day. There was, however, considerable variability across individuals. In all, 44.6 percent of the Soldiers reported no sleep deprivation, 52.1 percent reported some degree of sleep deprivation, and 3.2 percent had missing data.

■ *Sleep and behavioral health.* Soldiers who reported that they were sleep deprived screened positive for a mental health problem significantly more often than those who did not report being sleep deprived. Only 11.7 percent of junior enlisted male Soldiers in theater 9 months who reported no sleep deprivation were positive for depression, anxiety, or acute stress. In contrast, 23.1 percent of the Soldiers who reported 2 hours of sleep deprivation screened positive for any mental health problem.

- *Sleep and work performance.* Sleep deprivation has a known negative link to performance. Even relatively small amounts of sleep deprivation can result in a cumulative performance decline over time (Belenky et al., 2003; Bliese et al., 2006; Van Dongen et al., 2003). In MHAT V, the relationship between sleep loss and work performance was examined. Analyses revealed that even 1 hour of reported sleep deprivation was significantly associated with increased work-related problems and accidents.

Summary of risk factors. Compared to 2006, the intensity of combat in OIF appears to have declined significantly, with the drop particularly evident among those who have been in theater for only a few months. As a whole, however, the Soldiers deployed to OIF in 2006 through 2008 have clearly witnessed a high degree of intense combat events while deployed. In particular, a high percent of the sample reported knowing someone seriously injured or killed. As with combat experiences, on a normalized basis, many deployment concerns are lower than in 2006. On an unadjusted basis, concerns about deployment length and being separated from family are high among the 2007 OIF sample.

The MHAT V OIF sample allowed for a detailed analysis of the relationship between deployment length and a variety of mental health outcomes. In some cases, such as with reports of getting a divorce or separation or of stress and emotional problems impacting work, the relationship was linear. In these cases, as deployment months increased so did the probability that a Soldier would screen positive on the outcome measure. In a number of other cases, such as with morale and reports of any mental health problems, the relationship was curvilinear so that toward the end of the deployment, the probability of problems decreased. Even with the curvilinear patterns, however, a much higher percent of Soldiers reported problems at the end of the deployment than at the beginning.

One of the most dramatic findings centered on the effects of multiple deployments. As a group, those Soldiers who were on their second or third/fourth deployment were at increased risk for low morale, mental health problems, and degraded performance due to stress or emotional problems.

Soldier Protective Factors

In the conceptual model used to guide the MHAT reports, protective factors represent the area most amenable to intervention. This section examines unit social climate (leadership, readiness, and cohesion), willingness to seek behavioral health care, reducing barriers to care, family and marital support, willingness to report ethical violations, and training as protective factors.

Leadership, readiness, and cohesion. Social factors within platoons and companies play a critical role in how well unit members respond to combat experiences. In the current report, we examine the combined variables of cohesion, readiness, and perceptions of NCO and officer leadership. These variables were measured on five-point scales, with 3 being a generally neutral response. To facilitate the presentation of results in the tables, the combined climate measure was considered positive if the mean score was above 3. There was a significant increase of 5.6 percentage points between 2006 (54.5 percent) and 2007 (60.1 percent) in ratings of positive climate for junior enlisted male Soldiers in theater for 9 months.

Figure 9–9 illustrates the importance of social climate as a protective factor in terms of Soldiers' responses to the risk factor of combat exposure. Soldiers who rate social climate positively have lower levels of acute stress than those who rate the social climate negatively across all levels of combat exposure. Just as importantly, however, figure 9–9 also shows that there is less of an increase (flatter slope) for combat exposure on acute stress for those who rate the social climate positively. This is a typical buffering effect in the social science literature (Cohen and Wills, 1985).

Behavioral health care seeking. A Soldier's willingness to seek behavioral health care may also serve as a protective factor. A key impediment to seeking care involves overcoming the associated stigma. A major challenge to providing mental health care is the fact that sensitivity to stigma appears strongest among individuals who screen positive for mental health problems (Hoge et al., 2004). Therefore, when looking at stigma rates across years, it is informative to examine those individuals.

Table 9–1 provides the adjusted percents for Soldiers who also screen positive for depression, anxiety, or acute stress. As the table

Figure 9-9. **Predicted Acute Stress Scores for Operation *Iraqi Freedom* Soldiers**

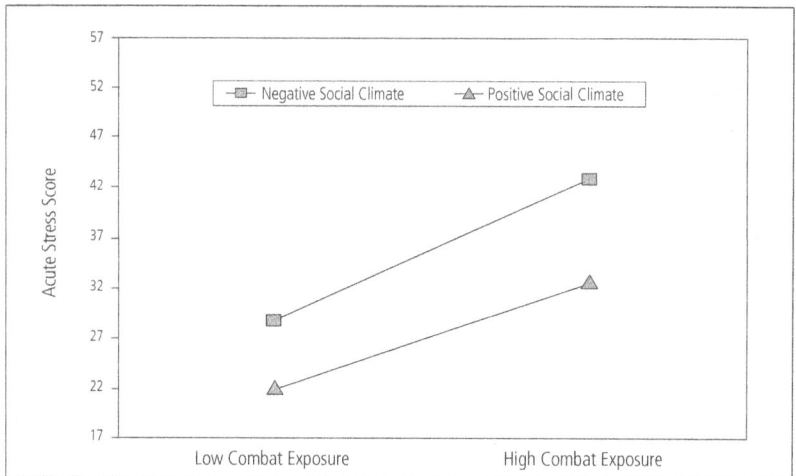

Table 9-1. **Reported Stigma-related Rates for Operation *Iraqi Freedom* Soldiers Who Screen Positive for Mental Health Problem (adjusted percent)**

Factors that Affect Your Decision to Receive Mental Health Services	Agree or Strongly Agree		p-value
	MHAT IV 2006	MHAT V 2007	
It would be too embarrassing	36.6	32.0	0.04
It would harm my career	33.9	29.1	0.02
Members of my unit might have less confidence in me	51.1	44.8	0.00
My unit membership might treat me differently	57.8	52.1	0.00
My leaders would blame me for the problem	43.0	38.5	NS
I would be seen as weak	53.2	49.8	NS

Table 9–2. **Reported Barriers to Care for Operation** *Iraqi Freedom*
Soldiers who Screen Positive for Mental Health Problem
(adjusted percent)

Factors that Affect Your Decision to Receive Mental Health Services	Agree or Strongly Agree		p-value
	MHAT IV 2006	MHAT V 2007	
Mental health services aren't available	6.8	9.6	0.053
I don't know where to get help	13.3	13.4	0.920
It is difficult to get an appointment	13.6	20.3	0.002
There would be difficulty getting time off work for treatment	41.0	40.6	0.845
It's too difficult to get to the location where the mental health specialist is	8.7	17.9	0.000
My leaders discourage the use of mental health services	14.6	21.2	0.006

indicates, rates of stigma for four of the six items were significantly lower in 2007 than in 2006. This is a positive finding and may reflect an increased emphasis by leadership to destigmatize seeking treatment.

Reducing barriers to care. Perceived barriers to care also vary depending on whether a Soldier screens positive. As with stigma issues, Soldiers who screen positive for a mental health problem typically report higher barriers to care. Table 9–2 presents perceived barriers to care and indicates that a number of them increased from 2006 to 2007.

The increases in perceived barriers to care almost certainly reflect changes in the distribution of Soldiers on the battlefield in Iraq. As a result of surge operations, more Soldiers were located on outposts in 2007 than in 2006. Survey findings indicated that those on outposts reported higher barriers to care. For instance, while 17.9 percent of all Soldiers in table 9–2 reported difficulty getting to mental health specialists, the value increased to 29.3 percent for those who reported being on outposts versus 12.9 percent for those who were not.

Marital functioning and rear detachment support. The Well-being Survey assessed Soldiers' perceptions of the quality of their marital relationships and their satisfaction with family support. Analyses revealed that there were no significant changes from 2006 to 2007. In general, a majority of respondents indicated that they have a good marriage and a positive relationship with their spouse. However, an even larger majority indicated that they were not satisfied with the rear detachment support for their family.

Reporting ethical violations. Soldiers' willingness to report unit members for unethical behaviors almost certainly runs counter to the strong sense of bonding that occurs among unit members during deployment. However, the degree to which Soldiers believe their unethical behaviors will be reported by other unit members may serve as a powerful deterrent. Therefore, given that unit morale is significantly higher in 2007, it is not surprising that Soldiers continue to be reluctant to report ethical violations by other unit members. Table 9–3 provides adjusted responses

Table 9–3. **Reported Rates of Ethical Violations for Operation** *Iraqi Freedom* **Soldiers (adjusted percent)**

Reporting Ethical Violations	Agree or Strongly Agree		p-value
	MHAT IV 2006	MHAT V 2007	
I would report a unit member for the mistreatment of a noncombatant	37.1	34.3	0.11
I would report a unit member for injuring or killing an innocent noncombatant	45.5	41.2	0.02
I would report a unit member for unnecessarily destroying private property	32.8	30.7	0.22
I would report a unit member for stealing from a noncombatant	38.9	34.8	0.02
I would report a unit member for violating the Rules of Engagement	37.1	35.9	0.52
I would report a unit member for not following General Orders	36.9	35.5	0.43

from both 2006 and 2007. Response rates to four of the six items remained statistically unchanged from 2006 to 2007. However, in 2007 Soldiers indicated less willingness to report a unit member for injuring or killing an innocent noncombatant and stealing from a noncombatant.

Training adequacy for deployment stress and suicide. Soldiers were asked whether the training they received for deployment stressors and suicide was adequate. From 2006 to 2007, there were significant increases in perceptions of training adequacy for three of the four items (see table 9–4). The only item that did not change significantly addressed Soldiers' confidence about helping fellow Servicemembers get mental health assistance.

Battlemind training and training adequacy. One of the initiatives recommended in MHAT IV was to implement Battlemind training (Castro, 2004, 2005; Castro, Hoge, and Cox, 2006). Battlemind is a mental health resiliency–building training system with different modules for pre- and post-deployment. Several modules have been shown to reduce deployment-related behavioral health problems in large-scale group randomized trials (for example, Adler, Bliese, Hoge, McGurk, and Castro, in press).

In the 2007 OIF sample, a number of deploying units reported that they received predeployment Battlemind training. Those Soldiers

Table 9–4. **Adequacy of Suicide and Stress Training for Operation** *Iraqi Freedom* **Soldiers (adjusted percent)**

Adequacy of Suicide and Stress Training	Agree or Strongly Agree MHAT IV 2006	MHAT V 2007	p-value
I am confident in my ability to help Servicemembers get mental health assistance	54.6	56.5	0.34
The training in managing the stress of deployment and/or combat was adequate	40.0	45.4	0.00
I am confident in my ability to identify Servicemembers at risk for suicide	50.9	54.8	0.04
The training for identifying Servicemembers at risk for suicide was sufficient	47.6	55.3	0.00

who reported that they had received the training were significantly more likely to agree that the training in managing the stress of deployment was adequate ($p < .01$), and the training to identify Servicemembers at risk for suicide was sufficient ($p < .01$).

Predeployment Battlemind training efficacy. Soldiers who reported that they had attended predeployment Battlemind training were significantly less likely to screen positive for a mental health problem (12 percent adjusted) than those who did not attend (20.5 percent adjusted). While these results are impressive, it should be noted that certain units implemented predeployment Battlemind training in conjunction with a number of other best practice behavioral health interventions. Therefore, the differences cannot be attributed solely to predeployment Battlemind training and may represent an increased emphasis on overall behavioral health training in certain units.

Ethics training. Soldiers were asked whether they recalled having ethics training and whether the training was adequate. In MHAT IV (2006), 80 percent of Soldiers reported that they had received ethics training; this was similar to the percent reported in the 2007 MHAT V (81 percent). Adequacy was evaluated both by directly asking if it was adequate, and also by asking if the Soldier had encountered situations that were ethically difficult despite the training, such as following the rules of engagement. In 2007, there was a significant increase ($p < .05$) in the percent of Soldiers who reported that "the training they had received in the proper (ethical) treatment of noncombatants was adequate." In 2006, 77 percent of respondents reported the training was adequate, whereas in 2007 the figure was 80 percent.

Summary of protective factors. Relative to the 2006 MHAT, the 2007 sample as a whole had significantly better perceptions of leadership, cohesion, and readiness as indexed by a unit climate variable. Second, although the sample from 2007 had significantly lower worries about stigma in seeking care, Soldiers who were symptomatic for mental health problems were somewhat less able to perceive being able to obtain care in 2007 than in 2006. Finally, Soldiers reported being better trained for the stresses of combat and part of their preparation may be attributed to receiving Battlemind training. Other factors, such as marital support, remained unchanged from 2006. Finally, in terms of ethical training, more Soldiers reported that ethical training was adequate, yet fewer Soldiers reported that they would report their unit member for the unethical behavior of stealing from a noncombatant or injuring or killing an innocent noncombatant.

MHAT V Operation *Enduring Freedom* Findings

Data Collection Procedures and Demographics

The OEF 2007 MHAT assessment of Soldiers focused on companies from brigade combat teams (BCTs) and supporting task forces located primarily in the Regional Command East section of Afghanistan, the zone that is under U.S. command and has the highest density of American ground forces. In this section of the chapter, Soldier responses to the OEF 2007 survey (n = 699) are compared to responses to the OEF 2005 survey (n = 610) and the OIF 2007 survey (n = 2,195). Comparisons across the sampled populations were evaluated using both adjusted and unadjusted percents. Adjusted values are typically provided for junior enlisted males in theater for 9 months. Key demographic variables across the three sample populations can be found in the full report, available at <www.armymedicine.army.mil/reports/mhat/mhat_v/mhat-v.cfm>.

Soldier Behavioral Health and Performance

In general, health and performance indices for OEF 2007 are examined relative to data from OEF 2005 as well as OIF 2007. However, in OEF 2007, surveys were completed by BCT and supporting task force Soldiers, whereas in OIF 2007 only BCT Soldiers completed surveys. There are differences between BCT Soldiers and supporting task force Soldiers, both demographically and in the missions they complete. Therefore, additional analyses were conducted in some cases comparing data from BCT Soldiers in OEF 2007 with BCT Soldiers in OIF 2007.

Morale: individual and unit. Soldiers' ratings of individual morale were significantly lower in OEF 2007 than OEF 2005 but similar to ratings in OIF 2007. However, ratings of unit morale did not differ significantly for the three samples. The percentages of Soldiers reporting high or very high individual and unit morale are presented in figure 9–10. When these percentages are adjusted to control for gender, rank, and months in theater, unit morale in OEF 2007 (9 percent) was significantly lower ($p < .05$) than unit morale in OIF 2007 (11.9 percent).

Acute stress, depression, and anxiety. Rates for depression, anxiety, and acute stress either singly or in combination were significantly higher in

Figure 9–10. **Individual and Unit Morale Rates for Operations** *Enduring Freedom* **and** *Iraqi Freedom* **Soldiers**

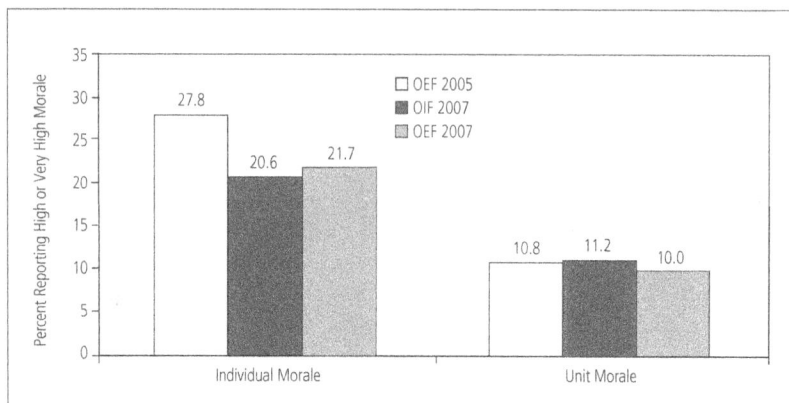

OEF 2007 (p < .001) than those reported in OEF 2005. There was a tendency for Soldiers in OEF 2007 to report higher depression and anxiety values than Soldiers in OIF 2007; however, using a conventional criterion of p < .05, these differences were not statistically significant. If these percentages are adjusted to control for gender, rank, and months in theater, then rates on all scales for OEF 2007 remain significantly greater than OEF 2005, and additionally the rate of depression in OEF 2007 was significantly higher than OIF 2007 (11.4 percent versus 7.6 percent; p < .01).

As previously stated, in OEF 2007, surveys were completed by Soldiers supporting task forces as well as BCTs. However, in OIF 2007, surveys were completed by Soldiers in BCTs only. Therefore, additional analyses were run to compare OEF 2007 BCT Soldiers with OIF 2007 BCT Soldiers. The unadjusted analyses are presented in figure 9–11. A significantly higher percentage of OEF 2007 BCT Soldiers screened positive for depression compared to OIF 2007 Soldiers using both unadjusted (p < .01) and adjusted (p < .001) rates. Although unadjusted rates for anxiety and any mental health problem in OEF 2007 BCT Soldiers tended to be higher than OIF 2007 BCT Soldiers, these differences were not statistically significant. However, when adjusted, the OEF 2007 BCT Soldiers were more likely to screen positive for depression (p < .001), anxiety (p < .01), and any mental health problem (p < .05).

Stress and work performance. Just over 14 percent of OEF 2007 Soldiers reported that stress or an emotional problem limited their ability to do

Figure 9–11. **Mental Health Rates for Operations** *Enduring Freedom* **and** *Iraqi Freedom* **Brigade Combat Team Soldiers (unadjusted percent)**

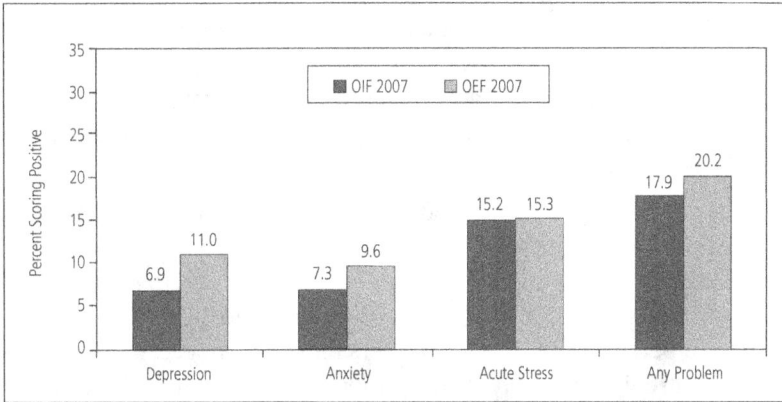

their job; 21.0 percent of OEF Soldiers reported that stress or an emotional problem had caused them to work less carefully; and 10.3 percent reported their supervisor was concerned. No significant differences were found among the three samples on these measures of stress and work performance.

Suicidal ideation. A significantly higher percentage of OEF 2007 Soldiers (p < .001) indicated suicidal ideation compared to OEF 2005 (15 percent versus 8 percent), whereas suicidal ideation was similar for both OEF 2007 and OIF 2007 (15 percent versus 13 percent, respectively). When comparing these populations using adjusted values, the same pattern of significance was found. Furthermore, while 87 percent of OEF 2007 Soldiers reported receiving suicide prevention training, only 51 percent reported the training to be sufficient, indicating the need to ensure that Soldiers receive suicide prevention training that is applicable to a combat environment.

Divorce intent. Significantly more Soldiers were planning to get divorced (p < .01) in OEF 2007 (19 percent) compared with OEF 2005 (13 percent). Soldiers' reports of their intent to divorce did not differ significantly when comparing OEF 2007 and OIF 2007 (19 percent). This pattern was the same using adjusted values.

Alcohol and substance abuse. The reported use of alcohol in OEF 2007 was significantly lower compared to OEF 2005 (7 percent versus 12 percent; p < .01). However, significant differences were not found when the values were adjusted for gender, rank, and time in theater. Reported alcohol usage in OIF 2007 (8 percent) was the same as in OEF 2007.

Reported use of illegal drugs or substances was significantly higher (p < .05) in OEF 2007 (2.6 percent) than OIF 2007 (1.4 percent). These statistical differences were also found when adjusting for gender, rank, and time in theater. Rates of illegal drug use were similar in OEF 2007 (2.6 percent) and OEF 2005 (2.3 percent) using both adjusted and unadjusted values.

Unethical behaviors. As outlined in the OIF section of this chapter, the 2007 MHAT survey included questions pertaining to ethical behaviors. These items were not included in the OEF 2005 survey; therefore, comparisons for OEF 2007 are only presented relative to OIF 2007. Approximately 10 percent of OEF 2007 Soldiers reported unnecessarily damaging or destroying property, while almost 4 percent reported that they hit or kicked noncombatants when it was not necessary. Using the conventional p-value of p < .05, the analyses reveal that for most questions, responses did not differ between the two theaters. The only significant difference (p < .001) was found for the item, "Insulted and/or cursed noncombatants in their presence," in which 36.6 percent of OEF 2007 Soldiers reported compared to 29.6 percent in OIF 2007. This relationship was also significant for adjusted values (p < .001).

As previously reported (MHAT IV and MHAT V OIF), Soldiers who screened positive for mental health problems of depression, anxiety, or acute stress were significantly more likely to report engaging in unethical behaviors. This relationship was also found in OEF 2007. Specifically, Soldiers who screened positive for any mental health problem were more than twice as likely to report engaging in unethical behaviors as those who did not. This included insulting or cursing at noncombatants (60.7 percent versus 31.7 percent), unnecessarily damaging or destroying private property (22.2 percent versus 7.2 percent), and hitting or kicking noncombatants (11.1 percent versus 2.5 percent).

This pattern was also found when evaluating reports of unethical behavior as a function of high anger levels. Reports of unethical behavior were significantly higher for Soldiers who had high rather than low levels of anger. This included insulting or cursing at noncombatants (53.4 percent versus 21.5 percent), unnecessarily damaging or destroying private property (15.1 percent versus 5.0 percent), and hitting or kicking a noncombatant (7.1 percent versus 1.1 percent). Anger therefore may serve as an additional key factor for an increased propensity to engage in unethical or inappropriate behaviors.

Summary of behavioral health and performance indices in OEF. Overall, Soldiers' behavioral health status in OEF 2007 was significantly lower than in OEF 2005. Ratings of individual morale in OEF 2007 were substantially lower than in OEF 2005. Considerably more OEF 2007 Soldiers reported planning to get a divorce compared to OEF 2005 Soldiers. Further, ratings of depression, generalized anxiety, and acute stress were significantly higher in OEF 2007 than OEF 2005.

Ratings of individual and unit morale and behavioral health were similar for OEF 2007 and OIF 2007. Other comparisons revealed that Soldiers in OEF 2007 BCTs reported significantly more overall mental health problems than OIF 2007 Soldiers in BCTs when using adjusted values. Self-reports of drug use were also higher in OEF 2007 than in OIF 2007, and more OEF Soldiers reported insulting or cursing at noncombatants. As with OIF 2007, there was also a significant relationship between reported unethical treatment of noncombatants and screening positive for a mental health problem, as well as high levels of anger in OEF 2007.

Soldier Risk Factors

This section examines behavioral health indices in relation to risk factors while deployed to OEF.

Combat experiences. Given the importance of combat experiences in terms of behavioral health, a detailed examination of differences between OEF 2007 compared to OEF 2005 and OIF 2007 follows.

- *Combat experiences for OEF 2007 compared to OEF 2005.* The average months in theater for OEF 2007 were 7.66 compared to 9.56 in OEF 2005 and 9.40 for OIF 2007. Comparisons using adjusted values based on a junior enlisted Soldier deployed for 9 months are presented here.

  Comparison across years (see table 9–5) indicates significantly higher combat intensity in OEF 2007 compared to OEF 2005. However, some combat experiences have declined. The pattern of combat experiences reported by Soldiers reflects the changing nature of the war from operations in 2005 to counterinsurgency in 2007. Overall, there is evidence that Soldiers exposure to potentially traumatic combat experiences has increased in OEF.

Table 9–5. **Combat Experiences in Operation** *Enduring Freedom* **2005 and 2007**

Combat Experiences	Percent (adjusted)		
	OEF 2005	OEF 2007	p-value
Significantly Higher			
Being attacked or ambushed	49.6	61.6	0.00
Seeing dead bodies or human remains	50.7	59.2	0.01
Seeing dead or seriously injured Americans	44.7	55.2	0.00
Knowing someone who was seriously injured or killed	65.9	73.7	0.01
Being in a threatening situation where you were unable to respond because of the rules of engagement	34.6	44.2	0.00
Being wounded/injured	5.5	13.7	0.00
Receiving incoming artillery, rocket, or mortar fire	71.3	81.5	0.00
Being directly responsible for the death of an enemy combatant	13.3	21.0	0.01
Had a close call: was shot or hit but protective gear saved you	3.0	8.0	0.01
Significantly Lower			
Seeing destroyed homes and villages	63.3	50.1	0.00
Working in areas that were mined or had improvised explosive devices	72.6	64.3	0.00
Disarming civilians	42.7	28.7	0.00
Clearing/searching homes or buildings	53.1	32.3	0.00
Clearing/searching caves or bunkers	45.3	31.2	0.00
Seeing ill/wounded women and children whom you were unable to help	46.9	33.3	0.00

- *Combat experiences for OEF 2007 BCT Soldiers compared to OIF 2007.* Although both samples reported high combat exposure, OEF 2007 BCT Soldiers reported more combat exposure than OIF 2007 Soldiers across many of the combat experience items (see table 9–6). These findings are quite striking, showing that the 2007 reported levels of combat exposure in OEF BCTs (the units most involved in direct combat) were significantly higher on 24 indicators than in OIF BCTs.
- *Combat experiences and acute stress.* The relationship between combat experiences and acute stress in OEF 2007 was examined. Soldiers were divided into low, medium, and high combat experience groups based on the number of combat events they reported experiencing during the deployment. Soldiers with high levels of combat exposure were significantly more likely to screen positive for acute stress or any mental health problem.

Deployment concerns. OEF 2007 Soldiers indicated a higher level of concern with the chronic, less traumatic aspects of being on a combat deployment compared to OEF 2005 Soldiers. When compared to OIF 2007 Soldiers, OEF 2007 Soldiers reported similar levels of chronic deployment concerns.

Effect of multiple deployments. As discussed in the OIF section, multiple deployments are a risk factor for behavioral health problems. Also, results are presented for NCOs rather than for junior enlisted because Soldiers in the multiple deployer group are predominantly NCOs. For NCOs in OEF 2007, 9.8 percent of first-time deployers screened positive for any mental health problem, whereas 14.2 percent of NCOs who had previously deployed screened positive. This difference was significant (one tailed, $p < .05$). This is consistent with the findings from other MHATs.

Sleep deprivation. Overall, 31 percent of OEF 2007 Soldiers reported high or very high concern that they were not getting enough sleep. Nearly one-quarter of OEF 2007 Soldiers reported falling asleep during convoys. Additionally, 16 percent of OEF 2007 Soldiers reported taking mental health medications, approximately half of which were sleep medications. The relationship between sleep and performance was assessed by examining Soldiers' responses to the item, "During this deployment, have you had an accident or made a mistake that affected the

Table 9–6. **Combat Experiences for Brigade Combat Team Soldiers in Operations** *Enduring Freedom* **and** *Iraqi Freedom*

Combat Experience	Percent (adjusted)		p-value
	OIF 2007	OEF 2007 BCTs	
Being attacked or ambushed	52.5	75.1	0.00
Receiving small arms fire	59.7	70.3	0.00
Seeing dead bodies or human remains	60.8	74.4	0.00
Handling or uncovering human remains	29.7	44.8	0.00
Witnessing an accident that results in serious injury or death	37.0	47.7	0.00
Witnessing violence within the local population or between ethnic groups	37.8	46.2	0.01
Seeing dead or seriously injured Americans	46.3	63.7	0.00
Knowing someone who was seriously injured or killed	72.3	87.4	0.00
Participating in demining operations	22.2	37.8	0.00
Having hostile reactions from civilians	45.6	58.8	0.00
Being in a threatening situation where you were unable to respond because of the rules of engagement	41.8	54.3	0.00
Shooting or directing fire at the enemy	38.5	62.7	0.00
Calling in fire on the enemy	11.9	31.0	0.00
Clearing/searching caves or bunkers	16.4	51.2	0.00
Being wounded/injured	11.9	24.4	0.00
Receiving incoming artillery, rocket, or mortar fire	80.7	91.6	0.00

Being directly responsible for the death of an enemy combatant	13.7	32.8	0.00
Observing abuse of the Law of War/Geneva Convention	6.2	11.2	0.01
Having a member of your unit become a casualty	54.5	76.5	0.00
Had a close call: dud landed near you	25.0	38.0	0.00
Had a close call: equipment shot off your body	4.6	15.2	0.00
Had a close call: was shot or hit but protective gear saved you	6.4	12.9	0.00
Had a buddy shot or hit who was near you	16.6	24.6	0.01
Informed unit members/friends of a Servicemember's death	10.5	22.2	0.00

mission because of sleepiness?" with 6 percent of OEF 2007 Soldiers reporting in the affirmative.

Summary of risk factors. The intensity of combat in OEF 2007 was significantly higher than in OEF 2005. As a whole, Soldiers deployed to OEF in 2007 witnessed a higher degree of intense combat and experienced more significant levels of combat activity. Additionally, the rates for OEF 2007 BCT Soldiers are significantly higher than those of OIF 2007 BCT Soldiers on 24 of the 33 combat experience scale items.

There was also a significantly higher rate of noncombat, deployment-related concerns raised by Soldiers in OEF 2007 compared to OEF 2005. Interestingly, comparisons between OIF 2007 and OEF 2007 indicate a high degree of similarity between the two theaters on noncombat deployment concerns. Finally, there was a significant relationship between mental health problems and multiple deployments in the current OEF sample. Consistent with other MHAT findings, NCOs who had deployed more than once were at increased risk for a mental health problem compared to those on their first deployment.

Soldier Protective Factors

Leadership, readiness, and cohesion. As discussed in the OIF section, we examined social climate as the combined variables of cohesion, readiness, and perceptions of NCO and officer leadership. There was a decrease of 6 percentage points between OEF 2005 (64 percent) and OEF 2007 (58 percent) in adjusted percentage reporting a positive social climate. While small in absolute terms, this value is statistically significant (p < .05). No significant differences were found between OEF 2007 (58 percent) and OIF 2007 (61 percent). As mentioned in the combat experiences section of this report, exposure to high levels of combat significantly increases the risk of reporting a mental health problem. Previous MHATs have found that good NCO leadership can, to some extent, limit the degree to which Soldiers screen positive for any mental health problem (see MHAT IV report at <www.armymedicine.army.mil/reports/mhat/mhat_iv/mhat-iv.cfm>).

Behavioral health care seeking. Soldier ratings of stigma related to seeking mental health care were not statistically different from ratings of OEF 2005 or OIF 2007. The fact that perceptions have not changed significantly from 2005 suggests that more attention should be given to outreach and education programs that emphasize reducing stigma. Some 35 percent of OEF 2007 Soldiers reported that seeking mental health care would be too embarrassing, 31 percent believe it would harm their career, 45 percent felt that members of their unit might have less confidence in them, 56 percent reported that unit members might treat them differently, 44 percent worried that leaders might blame them for the problem, and 57 percent worried about being seen as weak.

Reducing barriers to behavioral health care. When comparing barriers across years and theaters, a number of perceived barriers are higher in the OEF 2007 sample compared to both OEF 2005 and OIF 2007. Table 9–7 provides the results using adjusted values. As the table indicates, perceived barriers to care have mostly increased since 2005 and, in general, are higher in the present OEF theater than in OIF. The OEF theater has considerable transportation challenges that may contribute significantly to some of these findings. This limits the ability of behavioral health personnel to get to outlying posts as well as the ability of Soldiers to get back to behavioral health personnel at the larger forward operating bases.

Marital satisfaction. Overall reports of marital satisfaction were significantly lower in OEF 2007 than they were in OEF 2005 (see table 9–8). Significantly fewer OEF 2007 Soldiers reported that they have "a good marriage," that "my relationship with my spouse makes me happy," and that "I really feel like a part of a team with my spouse" compared to Soldiers

Table 9-7. Reported Barriers to Behavioral Health Care for Operations *Enduring Freedom* and *Iraqi Freedom* Soldiers who Screen Positive for Mental Health Problem (adjusted percent)

Factors that Affect Your Decision to Receive Mental Health Services	Agree or Strongly Agree		
	OEF 2005	OIF 2007	OEF 2007
Mental health services aren't available	21.4	11*	19.9
I don't know where to get help	17.2	14.3	15.1
It is difficult to get an appointment	17.4*	21.3	26.8
There would be difficulty getting time off for treatment	43*	43.4*	56.3
It's too difficult to get to the location where the mental health specialist is	24.2*	17.7*	32.7
My leaders discourage the use of mental health services	19.5*	21.8*	33.0

*Indicates statistically significant difference from OEF 2007 (p < .05).

in OEF 2005. On these same questions, rates for OEF 2007 Soldiers were similar to OIF 2007 Soldiers.

Reporting ethical violations. In terms or reporting ethical violations, OEF 2007 Soldiers were not significantly different than OIF 2007 Soldiers. Not surprisingly, Soldiers are reluctant to report the ethical violations of unit members, and this reluctance is consistent across theaters. Unadjusted rates were consistent with adjusted values. These questions were not included in the OEF 2005 survey.

Training. Slightly fewer OEF 2007 Soldiers (87.5 percent) and OEF 2005 (87.5 percent) Soldiers reported receiving suicide prevention training compared to OIF 2007 (93.3 percent). Similarly, fewer OEF 2007 Soldiers (80.7 percent) reported receiving training in managing the stress of deployment and/or combat prior to deployment compared to OIF 2007 (86.8 percent). Slightly fewer OEF 2007 Soldiers (63.9 percent) reported receiving predeployment Battlemind training compared to OIF 2007

Table 9–8. **Marital Satisfaction Ratings for Operations** *Enduring Freedom* **and** *Iraqi Freedom* **Soldiers (adjusted percent)**

Marital and Family Support	Agree or Strongly Agree		
	OEF 2005	OIF 2007	OEF 2007
I have a good marriage	73.5	66.8	65.6
My relationship with my spouse is very stable	70.4	63.5	62.7
My relationship with my spouse makes me happy	75.8	69.2	67.7
I really feel like a part of a team with my spouse	73.3	63.9	63.6

(67.6 percent). These last two questions were not included in the OEF 2005 survey.

With respect to ethics training, significantly fewer Soldiers in OEF 2007 reported having received such training or that the training was adequate. Additionally, fewer Soldiers reported that the training made it clear how they should behave toward noncombatants.

Summary of protective factors. On most measures, reports of barriers to behavioral health care were higher in OEF 2007 compared to OEF 2005 and OIF 2007. This may largely be due to transportation difficulties in Afghanistan. Additionally, fewer OEF 2007 Soldiers reported that the training preparing them for the stress of deployment, identifying Soldiers at risk for suicide, and ethically treating noncombatants was adequate compared to OIF 2007 Soldiers.

Recommendations

Recommendations based on the findings from the OIF and OEF MHAT V data were included in the full report and also presented at senior leadership briefings. Some applied to both theaters, and others were theater specific. A complete list of recommendations appears in the full MHAT V report. The key recommendations pertaining to both theaters are summarized below:

- Increase in-theater behavioral health assets.
- Develop a mechanism to allow Government or contracted psychiatrists, psychologists, and social workers to fill selected

behavioral health positions in theater to augment military personnel.

- Create and fill Behavioral Health Officer and NCO positions in aviation brigades.
- Mandate that all combat medics receive Battlemind Warrior Resiliency (formerly Battlemind First Aid) Training before deploying to OEF or OIF to augment behavioral health personnel.
- Change the Modified Table of Organization and Equipment to maximize the impact of organic behavioral health assets:
  - Move Division Psychiatrist position from Sustainment Brigade to Division Surgeon Cell
  - Move Brigade Behavioral Health Officer and NCO positions from Brigade Support Battalions to Brigade Surgeon cell.
- Mitigate the effects of multiple deployment:
  - Provide Soldiers who have deployed multiple times priority for temporary duty assignments
  - Ensure adequate dwell time between deployments.
- Adopt strategies to reduce suicide risk:
  - Amend TRICARE rules to cover marital and family counseling as a medical benefit
  - Tailor suicide prevention training packages to focus on phase of deployment and aim at building psychological resiliency.
- Continue and enhance training:
  - Continue emphasis on Battlemind training for Soldiers and families
  - Enhance training for NCOs at the Warrior Leader Course, Basic Noncommissioned Officer Course, and Advanced Noncommissioned Officer Course on their role in maintaining Soldier resiliency through counseling and mentorship
  - Develop and implement senior leader Battlemind training
  - Continue emphasis on ethics training.

Conclusion

This chapter has outlined the findings from the fifth Mental Health Advisory Team Report. The MHAT V report continued the practice of surveying the behavioral health and well-being of a deployed force. This report identified deployment length, combat exposure, and multiple deployments as major risk factors for mental health problems and found that reports of barriers to behavioral health care in theater had increased from previous years. This was the first MHAT to collectively evaluate and present findings from both the OEF and OIF theaters of operation. These findings indicated that combat experiences and mental health problem rates were higher in Afghanistan than in Iraq in 2007. Finally, the findings from OIF, taken with findings from previous MHAT reports, embody a collection of 5 consecutive years of in-theater mental health data. This ongoing effort represents the first cumulative record of Soldier mental health over the course of a prolonged conflict. A sixth MHAT was to be conducted in both theaters in 2009, thus continuing to build on this invaluable repository of mental health data.

References

Adler, A.B., Bliese, P.D., Hoge, C.W., McGurk, D., and Castro, C.A. (in press). Early interventions during reintegration with soldiers returning from Iraq: Randomization by platoon. *Journal of Consulting and Clinical Psychology.*

Belenky, G., Wesensten, N.J., Thorne, D.R., Thomas, M.L., Sing, H.C., Redmond, D.P., Russo, M.B., and Balkin, T.J. (2003). Patterns of performance degradation and restoration during sleep restriction and subsequent recovery: a sleep dose-response study. *Journal of Sleep Research* 12, 1–12.

Bliese, P.D. (2006). Social climates: Drivers of soldier well-being and resilience. In A.B. Adler, C.A. Castro, and T.W. Britt (eds.). *Military life: The psychology of serving in peace and combat: Vol. 2. Operational Stress* (213–234). Westport, CT: Praeger Security International.

Bliese, P.D., and Castro, C.A. (2003). The soldier adaptation model (SAM): Applications to peacekeeping research. In T.W. Britt and A.B. Adler (eds.). *The psychology of the peacekeeper.* Westport, CT: Praeger Security International.

Bliese, P.D., Wesensten, N., and Balkin, T.J. (2006). Age and individual variability in performance during sleep restriction. *Journal of Sleep Research* 15, 376–385.

Bliese, P.D., Wright, K.M., Adler, A.B., Thomas, J.L., and Hoge, C.W. (2007). Timing of postcombat mental health assessments. *Psychological Services* 4, 141–148.

Castro, C.A. (2004). How to build Battlemind. *NCO Journal*, April, 23–24.

———. (2005). Building Battlemind. *Countermeasures* 26, 6–8.

Castro, C.A., Hoge, C.W., and Cox, A.L. (2006). Battlemind training: Building soldier resiliency. In *Human Dimensions in Military Operations: Military Leaders' Strategies for Addressing Stress and Psychological Support*, 42-1 to 42-6. Proceedings of RTO-MP-HFM-134, Paper 42. Neuilly-sur-Seine, France: RTO. Retrieved from <www.rto.nato.int/abstracts.asp>.

Cohen, S., and Wills, T.A. (1985). Stress, social support, and the buffering effect. *Psychological Bulletin* 98, 310–357.

Fontana, A., and Rosenheck, R. (1998). Psychological benefits and liabilities of traumatic exposure in the war zone. *Journal of Traumatic Stress* 3, 485–503.

Gilbert, D.T., Lieberman, M.D., Morewedge, C.K., and Wilson, T.D. (2004). The peculiar longevity of things not so bad. *Psychological Science* 15, 14–19.

Hoge, C.W., Castro, C.A., Messer, S.C., McGurk, D., Cotting, D., and Koffman, R.L. (2004). Combat duty in Iraq and Afghanistan, mental health problems, and barriers to care. *New England Journal of Medicine* 351, 13–22.

Hoge, C.W., Terhakopian A., Castro, C.A., Messer, S.C., and Engel, C.C. (2007). Association of posttraumatic stress disorder with somatic symptoms, health care visits, and absenteeism among Iraq War veterans. *American Journal of Psychiatry* 164, 150–153.

Lacy, B.W., and Ditzler, T.F. (2007). Inhalant abuse in the military: An unrecognized threat. *Military Medicine* 172, 388–392.

Spitzer, R.I., Kroenke, K., and Williams, J.B. (1999). Validation and utility of a self-report version of PRIME–MD: The PHQ primary care study. Primary Care Evaluation of Mental Disorders. Patient Health Questionnaire. *Journal of the American Medical Association* 282, 1737–1744.

Van Dongen, H.P.A., Maislin, G., Mullington, J.M., and Dinges, D.F. (2003). The cumulative cost of additional wakefulness: Dose-response effects on neurobehavioral functions and sleep physiology from chronic sleep restriction and total sleep deprivation. *Sleep* 26, 117–126.

Weathers, F.W., Litz, B., Herman, D., Juska, J., and Keane, T. (1993). *The PTSD Checklist (PCL): Reliability, validity, and diagnostic utility.* Poster session presented at the annual meeting of the International Society of Traumatic Stress Studies, October 1993, San Antonio, TX.

Chapter 10

Military Benefit Finding: Turning Adversity to Advantage

Michael D. Wood and Thomas W. Britt

The war environment is treacherous and burdens Soldiers with many unique stressors, including combat exposure, separation from friends and family, barriers in communicating with home, harsh living conditions, lack of privacy, environmental stressors, and an unpredictable combination of boredom, uncertainty, and threat (Bartone, Adler, and Vaitkus, 1998; Ritzer, Campbell, and Valentine, 1999). Despite these challenges, many Soldiers emerge from war reporting that they benefited from their experiences (Schok, Kleber, Elands, and Weerts, 2007). This tendency to report benefits from traumatic or highly stressful events—including rape, car accidents, and life-threatening diseases —has also been reported in the civilian literature (Helgeson, Reynolds, and Tomich, 2006). By researching the benefits individuals derive from trauma, scientists have been able to take a salutogenic perspective in expanding treatment concepts to support human health and well-being, while addressing causative disease factors (Helgeson, Reynolds, and Tomich, 2006).

In this chapter, we describe the concept of benefit finding (BF) or "post-traumatic growth" (PTG) (Tedeschi and Calhoun, 1996) and discuss how the concept has been applied to members of the military. Most of the current body of work on the antecedents and consequences of BF has been conducted using civilian samples; the little research that has been carried out in military settings is often based on the recollections of veterans who served in combat over 20 years ago. In this respect, the work conducted by psychologists at the Walter Reed Army Institute of Research (WRAIR) is unique; the research is based on BF that occurs during and closely following military operations.

This chapter will review theoretical approaches to benefit finding, then address research on the benefits of trauma among military personnel. Finally, areas needing future research in order to better understand how the concept of benefit-finding can be used to enhance the health and well-being of military personnel are highlighted.

Theory and Research on Benefit Finding

The concepts of benefit finding, post-traumatic growth, and stress-related growth (SRG) have been considered by prior researchers for some time (see Taylor, 1983). However, Tedeschi and Calhoun (1996) and Park, Cohen, and Murch (1996) were some of the first researchers to systematically categorize and measure benefits resulting from exposure to stressful or traumatic events. These authors synthesized the concepts and developed subscales assessing New Possibilities, Relating to Others, Personal Strength, Spiritual Change, and Appreciation of Life. They found that individuals who rated their traumatic events as more stressful were more likely to report PTG than those who rated the events as less stressful. The study also showed that the results of PTG were related to the personality traits of optimism and extraversion.

Park et al. (1996) developed their SRG scale based on qualitative analyses of participants' descriptions of how stressful events had affected them. The authors ultimately decided on a 50-item scale that assessed a single general factor of growth. They also found that SRG was related to the stressfulness of the events as well as the amount of social support they reported.

Since the publication of these seminal articles, researchers have elaborated on the concept of benefit finding following stressful events and devoted more attention to the theoretical mechanisms by which exposure to stressful events results in BF or growth (Helgeson, Reynolds, and Tomich, 1996; Affleck and Tennen, 1996). Furthermore, Affleck and Tennen (1996) provided evidence for the important distinction between BF and benefit reminding. The latter term reflects the intentional use of benefits as a coping mechanism to help the individual deal with the demands imposed by stressful events. In this way, benefit reminding becomes a problem-focused coping strategy (see Lazarus and Folkman, 1984) so that individuals can "nourish the conviction that we are in some way better off than we were before" (Affleck and Tennen, 1996). The authors emphasized that it was important to identify differences between beliefs about benefits from adversity (BF) and the use of that knowledge as a strategy for managing problems (benefit reminding).

In their review, Tedeschi and Calhoun (2004) summarized the work that had been conducted on benefit finding or PTG and emphasized that positive transformative changes can result from traumatic events. For true growth to occur, the authors claim that events need to be traumatic enough to engender positive changes. The authors emphasized that they were not denying the negative consequences of stressors, but rather arguing that researchers should also study the positive changes resulting from transformative events. In fact, the authors indicated that psychological distress and the perception of benefits can occur simultaneously as the individual processes the full impact of the traumatic event. This latter point will become particularly important in the review of research on BF among military personnel.

In an important meta-analytic review, Helgeson et al. (2006) examined the cross-sectional relationship between BF and measures of health and well-being across 77 studies. The authors found that perceived benefits were related to lower depression, lower anxiety, higher positive effect, and higher levels of intrusive thoughts. The results of the meta-analysis also highlighted the relative independence of perceived benefits or growth and distress following highly stressful events.

The few longitudinal studies conducted on benefit finding in civilian literature suggest that finding benefits is prospectively related to higher levels of well-being and reduced illness symptoms (Bower, Kemeny, Taylor, and Fahey, 1998; Carver and Antoni, 2004; McMillen, Smith, and Fisher, 1997; but see Tomich and Helgeson, 2004). However, Park and Helgeson (2006) noted that more longitudinal research was needed to clarify the effects of such variables as the length of time between the traumatic event and the assessment of benefits.

Benefit Finding Among Military Personnel: World War II, Korea, and Vietnam

A significant percentage of veterans perceive positive changes in personal growth and report benefits following their war experience (Lewis, 2006; Schok, Kleber, Elands, and Weerts, 2007). Stouffer, Suchman, Devinney, Star, and Williams (1949) examined post–World War II Soldiers' adjustment following Army institutional life and war. The veterans were honorably discharged, with approximately 50 percent reporting combat experience. Soldiers were asked, "Do you feel that Army life changed you?" to which 22 percent reported desirable changes only, 20 percent reported desirable and undesirable changes, and 37 percent reported undesirable changes only. The remaining 21 percent reported no change.

Examining the positive or negative long-term consequences of wartime imprisonment, Sledge, Boydstun, and Rabe (1980) found that 221 Vietnam War prisoners of war (POWs) suffered higher levels of physical and psychological problems when compared to 415 non-POW controls. However, the study revealed higher levels of favorable change among POWs in their "mental character" than unfavorable change. POWs who reported favorable benefits also identified a greater number of injuries during war and captivity and more extreme levels of physical and psychological suffering during captivity than other POWs. The Soldiers' sense of having benefited from war imprisonment was positively correlated with the harshness of their experience.

To understand the means by which one may perceive benefits following combat, Elder and Clipp (1989) examined a longitudinal sample of 149 World War II and Korean War veterans at three points in time, including pre- and postwar, for emotional or psychological impairment and resilience. Subjects were categorized by level of combat experience: noncombatants (40 percent), light combatants (30 percent), and heavy combatants (30 percent), based on an itemized combat scale. The scale featured a selection of combat-related experiences, the frequency of which identified the combat level of intensity. All Soldiers reported both positive and negative outcomes. Sixty to 70 percent of the men selected "learned to cope with adversity" and "exercised self-discipline" and "greater independence," while 40 to 50 percent reported undesirable outcomes such as a "disrupted life" and "separation from loved ones." Less than 20 percent cited "combat anxieties." Soldiers with heavy combat experience were more likely to report both benefits and stressful symptoms later in life than noncombat veterans. The study emphasized that both benefits and detriments can occur in the same postwar individual.

The authors explained that higher levels of combat experience created valuable coping skills such as self-discipline and an increased appreciation for life, while it simultaneously led to experiencing traumatic memories, personal loss, anxiety, and nightmares. Additionally, the authors identified a potential vulnerability marker by comparing pre- and postwar psychological functioning on ego resilience and helplessness. They found that low levels of prewar ego resilience were also associated with higher levels of post-traumatic stress.

Perceptions of desirable and undesirable outcomes of combat were further investigated by Aldwin, Levenson, and Spiro (1994), who examined positive and negative military experiences as a mediator and moderator of the relationship between combat stress and post-traumatic stress disorder (PTSD). Subjects included 1,287 male veterans selected from a database that tracked Servicemembers from World Wars I and II, Korea, and Vietnam, with the majority of the sample having served in World War II and Korea. Similar to the findings of Sledge, Boydstun, and Rabe (1980), more men reported desirable than undesirable effects of military service and were generally perceived to have had "maturational and positive effects." Results showed that combat exposure significantly predicted PTSD symptoms, and that this relationship was partially mediated in the expected directions by both the desirable and undesirable effects of military service. Additionally, a linear relationship was found between benefits and combat exposure, but the authors failed to find the expected quadratic relationship (with benefits decreasing again at high levels of combat exposure). The authors explained that positive effects of combat, such as developing coping resources, were negatively related to PTSD. Thus, the lack of BF may enhance war stress, and subsequently the intensity of PTSD.

Fontana and Rosenheck (1998) similarly viewed BF as a means of coping with or managing the aftermath of war (cf. Aldwin et al., 1994). Nonetheless, the seminal contribution of their study was to identify the type of perceived benefits (self-improvement, solidarity, affirmation of patriotic beliefs) and liabilities (disillusionment of patriotic beliefs, self-impoverishment, alienation from others) as a function of the amount and type of combat exposure (fighting, killing, threat, death, atrocities). The authors used 1,198 subjects from the National Vietnam Veterans Readjustment Study (NVVRS), and measured BF, liabilities, and combat exposure using a method similar to that of Aldwin et al. (1994). Bivariate relationships revealed that the benefit category "self-improvement" and the overall benefit score were related to the following types of combat exposure: fighting, killing, perceived threat, and death of others, but not to atrocities. Liabilities appreciably correlated with fight disillusionment, fighting, perceived threat, and atrocities. Linear relationships between the type of benefit and liability and the type and level of combat exposure revealed significant quadratic relationships in the form of an inverted U. More specifically, solidarity was closely associated with perceived threat, indicating that the highest level of solidarity emerges from moderate levels of perceived threat. Additionally, self-improvement and the overall measure of BF were significantly associated with death of others, but not with a true quadratic relationship. Rather, moderately high combat exposure revealed the highest feelings of self-improvement and overall BF score. The authors described this effect as a leveling process as opposed to a true quadratic function. Conclusions yielded that some level of suffering or stress (combat exposure) may encourage coping in the form of benefits perceived (for example, solidarity or self-improvement). Additional reports have focused on how BF following combat is related to such factors as wisdom and alienation later in life. Jennings, Aldwin, Levenson, Spiro, and Mroczek (2006) surveyed 1,990 World War II and Korean War veterans nested in the Normative Aging Study (cf. Aldwin et al., 1994) to assess prior military service, degree of combat exposure, and positive and negative aspects of military service. Subjects were resurveyed 10 years later to complete measures of coping (taking positive or negative action to solve problems), wisdom (less dependent on extrinsic sources for self-fulfillment), and alienation (the absence of meaning in life and feeling isolated). The authors found that combat exposure was related to both positive and negative outcomes associated with military service, but the relationship was much stronger for negative outcomes. Positive outcomes of military service were positively related to wisdom later in life, while negative outcomes were related to alienation later in life. Additional investigation revealed that low combat exposure was related with the later development of wisdom, while high combat exposure was associated with lower levels of wisdom. The authors concluded that how one appraised and coped with problems was more important in the prediction of positive adaptation than simply experiencing stress.

The influences of BF later in life were also investigated by Dohrenwend et al. (2004), who sought to establish if perceived benefits (appraisals) represent a positive adaptation to war experience, defensive denial, or maladaptive outcomes. The authors used 1,183 subjects from the NVVRS (cf. Fontana and Rosenheck, 1998) and clustered them according to how important their war experience was (salient) and if the experience was positive or negative (valence). The findings revealed that 71 percent of male U.S. veterans appraised the effect of their war experience on their present lives as positive, and 42 percent indicated that these experiences were highly salient. Results revealed that those reporting positive salience were less likely to have PTSD, had higher rates of occupational attainment, reflected higher levels of education, and were more likely to maintain marital relationships than those reporting negative salience. The authors concluded that the perceived benefits (appraisals) of war experience are affirmations of successful wartime and postwar adaptation as opposed to defensive denials and maladaptive outcomes.

Benefit Finding Among Military Personnel: Bosnia, Kosovo, and the Iraq War

Although prior research on BF following military operations and service is informative, it is important to note that these studies involved a time lag as long as 20 to 40 years between the combat experience and the report of benefits (Aldwin, Levenson, and Spiro, 1994; Elder and Clipp, 1989; Fontana and Rozenheck, 1998; Jennings, Aldwin, Levenson, Spiro, and Mroczek, 2006). Therefore, the benefit reports likely reflected a great deal of postcombat accommodation, rationalization, and bias. A full understanding of the issue requires the examination of benefits closer to real-time military operations. Research psychologists working for WRAIR have been involved in this, and the remainder of the chapter highlights the findings and advantages of this work.

In order to better examine the multiple ways peacekeeping operations might impact Servicemembers, Britt (1998) administered a questionnaire to 800 military personnel who had just participated in Operation *Joint Endeavor* in Bosnia and were in a staging area awaiting return to their home stations. Participation was voluntary. One of the open-ended questions was, "How has this deployment impacted you?" Although a number of individuals indicated negative effects of the deployment, 33 percent of the Servicemembers indicated some personal benefit as a result of the operation ("I am a stronger person," "I can deal with stress better now"), 8 percent indicated a positive effect on work ("I got a chance to do my MOS [military occupational specialty]"), and 5 percent even indicated a positive effect of the deployment on their family ("I realized the importance of family").

Little research has examined the predictors of deriving benefits from combat or peacekeeping operations. Britt, Adler, and Bartone (2001) conducted a longitudinal study examining engagement in meaningful work (assessed as a latent variable through self-report measures of possessing a peacekeeper identity, doing important work, and being personally engaged in work) during the Bosnia operation as a possible predictor of derived benefits from the deployments 3 months after it was over. Soldiers who reported higher engagement in meaningful work during the deployment also reported benefits ("I deal with stress better because of this deployment," "This deployment made me more aware of problems in this world") from the deployment 3 months after it was over, even after controlling for the personality variable of hardiness. In addition, Britt et al. (2001) found that deployed Soldiers who reported seeing the destruction caused by the war and who interacted with the local civilians were more likely to report benefits than Soldiers who lacked this exposure.

Britt, Dickinson, Moore, Castro, and Adler (2007) replicated key aspects of these results in another longitudinal study of U.S. Soldiers on a peacekeeping mission to Kosovo. Engagement in meaningful work during the Kosovo peacekeeping deployment was a strong predictor of derived benefits from the operation 6 months after it was over. Furthermore, engagement in meaningful work was a stronger predictor of benefits than the negative experiences encountered during the deployment. Therefore, researchers working at WRAIR have identified meaningful aspects (morale, engagement in meaningful work, confidence in the unit functioning, perceptions of leadership) of deployment that predict the likelihood of finding benefits after the deployment is over.

In addition to examining benefit finding following peacekeeping operations, WRAIR researchers have investigated it following combat deployments to Iraq. One important factor that has received little attention with respect to military benefit finding is the influence of stressor duration or deployment length on BF perception. Wood, Thomas, Klocko, Bliese, and Hoge (2008) surveyed Soldiers at various times during a 15-month deployment and found that the number of months of deployment was negatively related to BF. However, this effect was qualified by a significant quadratic function in the shape of a U, indicating a mid-deployment slump in reported benefit finding. Thus, BF appeared to undergo a transition by phase throughout the 15-month deployment. The early deployment phase (approximately 0–6 months) showed a decline and may reflect a diminishing hope or mood brought about by reflecting on the remaining months of deployment. Mid-deployment (approximately 6–10 months) revealed the lowest levels of BF.

Anecdotal reports collected from focus groups indicated a mid-deployment foreboding influenced by the long separation from home (U.S. Army Surgeon General, 2008, February 14) which may have caused the slump in mid-deployment BF. In late deployment (approximately 10–15 months), researchers again found the positive relationship between combat and BF (Aldwin et al., 1994), perhaps mimicking a runner's high based on "redeployment optimism" (U.S. Army Surgeon General, 2008, February 14).

Wood, Britt, Thomas, Klocko, Bliese, and Hoge (unpublished) tested the effects of two moderators, deployment length and BF, on the relationship between combat exposure and PTSD. Findings revealed a significant three-way interaction between the predictor (combat) and two moderating variables (deployment length and BF). Figure 10–1 illustrates that high BF buffers Soldiers from the negative effects of combat exposure on PTSD regardless of deployment length. Nonetheless, subjects who report low benefits appear more vulnerable to high combat-derived PTSD symptoms early in deployment (approximately 0–7 months) than those late in deployment (approximately 8–15 months). These data indicate that high BF shows a protective influence on mental health, as in previous research (Aldwin, Levenson, and Spiro, 1994; Fontana and Rosenheck, 1998), but also clarifies that low benefit finders do develop ways of coping with PTSD symptoms relative to the increase in combat exposure and time in the war environment.

Social variables such as leadership climate and unit cohesion are also related to the perception of benefits. Wood, Kim, Bliese, Castro, and

Figure 10–1. **Interaction of Benefit Finding (± 1 SD) and Deployment Length by Combat Level on Post-traumatic Stress Disorder Symptoms (± 1 SD)**

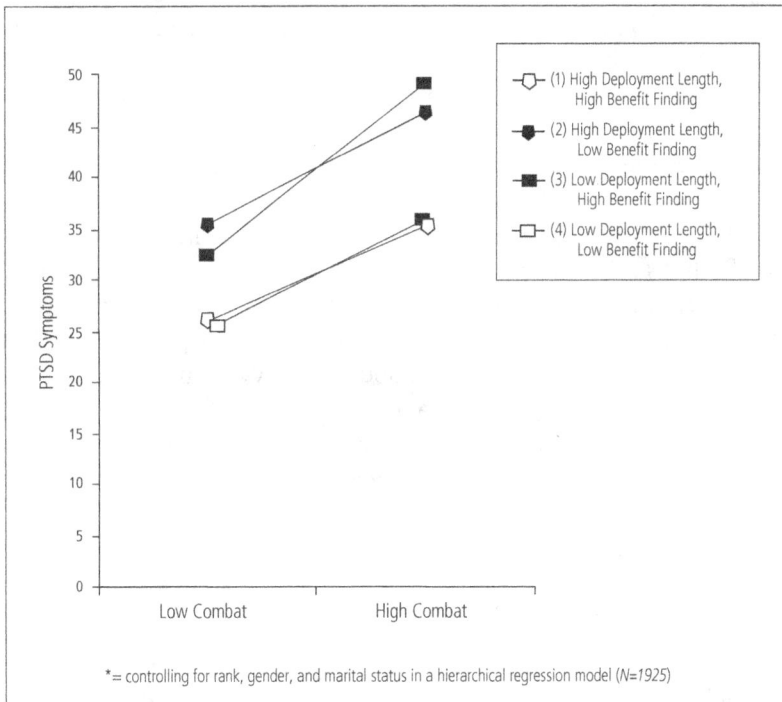

*= controlling for rank, gender, and marital status in a hierarchical regression model (N=1925)

Hoge (2008) sought to identify the buffering capability of social variables (unit cohesion and noncommissioned officer [NCO]/officer leadership) on the combat exposure–benefit finding relationship in Operation *Iraqi Freedom* veterans. However, in contrast to previous designs, the researchers introduced a two-points-in-time measure in which social variables and combat at Time 1 (3 months after redeployment) predicted BF at Time 2 (6 months after redeployment). Both positive NCO leadership and unit cohesion predicted BF. However, only the cohesion by combat interaction emerged as significant. These data revealed that Soldiers who reported higher unit cohesion showed a steeper rate of increase in reported BF. Thus, unit cohesion among Soldiers and NCOs significantly enhanced combat-derived BF.

Britt, Castro, Hoge, McGurk, Herleman, and Odle-Dusseau (2008) examined the consequences of benefit finding for PTSD, morale, combat readiness, and overall health using a longitudinal sample of Iraq War veterans. The veterans completed measures of all constructs 3 months after their redeployment from *Iraqi Freedom* and then 3 months later (6 months after redeployment). Using structural equation modeling, Britt et al. found that benefit finding at Time 1 predicted reduced levels of PTSD at Time 2, and higher levels of morale, combat readiness, and overall health at Time 2. The study found more statistical support for benefit finding as a predictor of these outcomes than for outcomes as predictors of benefit finding (see Zapf, Dormann, and Frese, 1996).

These results suggest that BF following combat may be related to psychological problems resulting from combat, but also to positive psychological states that contribute to personal thriving and resiliency. The findings echo recent calls from researchers to study not only what leads to dysfunction and psychological symptoms, but also what leads to thriving under stress and achieving a meaningful purpose in life (Britt et al., 2008; Helgeson et al., 2006; Southwick, Vythilingam, and Charney, 2005; Tugade, Fredrickson, and Barret, 2004). We do not mean to suggest that combat is desirable because it leads to particular types of benefits for Servicemembers. Nonetheless, it is worth the investment to design programs and interventions to maximize the benefits found and their concomitant positive consequences.

Future Directions and Conclusion

BF research within military populations has a long history, and psychologists' research for the Army has contributed to this literature.

However, much of the research on military benefit finding has been plagued by long time lags between the experience of stressors during deployment and the assessment of the benefits following operations. Additionally, recent research conducted by uniformed research psychologists has not used standardized measures of benefit finding that are available in the literature (see Tedeschi and Calhoun, 1996). In their meta-analytic review, Helgeson et al. (2006) noted that research on BF often includes a variety of scales that may not relate to each other. Current measures used by WRAIR were originally constructed from Soldiers' feedback on open-ended questions concerning how deployment to a peacekeeping environment had affected them (Britt et al., 2001). Although these measures show similar relationships to positive outcomes, as well as interactions with the level of exposure to stress and trauma on mental health outcomes, as those in other reports (Aldwin, Levenson, and Spiro, 1994; Elder and Clipp, 1989; Helgeson, Reynolds, and Tomich, 2006; Schok et al., 2007), future research ought to focus on restructuring the current scale and pattern questions after those found in validated measures of BF.

Helgeson et al. (2006) noted a low representation of studies measuring BF over time and emphasized the need for longitudinal research. The recent research conducted by uniformed 71Fs has used longitudinal designs that have been recommended by civilian psychologists. However, additional research is needed using multiple time lags to distinguish between benefits that reflect true personal growth following exposure to trauma versus rationalized benefits from Servicemembers who want to find something positive from horrors they have witnessed. To improve the understanding of benefit finding, WRAIR is currently involved in longitudinal research to follow veterans over multiple time points during the postdeployment period and to track the development of conscious benefits and their consequences. This represents an important area for continued and future research.

Along with additional research on the predictors and consequences of BF in combat, there is a need to begin exploring interventions that increase the likelihood of Servicemembers accruing benefits from participation in military operations. We know from longitudinal research conducted during and following peacekeeping operations that engagement in meaningful work during a deployment is related to derived benefits from the deployment months after it is over (Britt et al., 2001; Britt et al., 2007). Therefore, interventions that enhance the personal relevance and meaning associated with Servicemembers' work during a military operation may increase the benefits Soldiers derive from their participation in deployments. For example, Britt (2003) discussed aspects of predeployment training for Operation *Joint Endeavor* to Bosnia that highlighted the purpose of the mission and how Servicemembers were contributing to a lasting peace in the region. Thus, military planners and leaders should be vigilant in finding opportunities to communicate the meaning of a given military operation to Servicemembers.

In addition to understanding the importance and relevance of the mission itself, Servicemembers can be encouraged to realize the positive consequences of their mission experience as training and for stress resiliency. Many Servicemembers come through a difficult mission recognizing a strength and resiliency they were not previously aware that they possessed. As discussed above, the finding of benefits can occur simultaneously with symptoms related to stressors encountered during military operations. In light of this fact, the approach of focusing on strengths learned during deployment is actively used in the form of Battlemind training to help Soldiers cope with combat and its aftermath (Thomas et al., 2007). Adler et al. (2006) found that Battlemind training given soon after Soldiers' return from deployment significantly reduced mental health symptoms 4 months postdeployment. Thus, highlighting the strengths Servicemembers demonstrated during combat might capitalize on the salutary effects of benefits that occurred in the face of high levels of adversity.

Although BF is not the cure-all for military trauma and stress, the current literature identifies it as a valuable cognitive resource with a potential to reduce the impact of war trauma on mental health both during and after military deployment. Additionally, BF has been shown to be an adaptive or learned process, which may be enhanced by personality factors such as ego resilience, social factors such as unit cohesion and leadership, and other factors such as meaningfulness of work. BF may also be useful for the identification of individuals who are vulnerable to mental health problems. In conclusion, benefit finding is a valuable coping mechanism that can augment a Soldier's psychological toolbox to better manage personal mental health.

References

Adler, A.B., Castro, C.A., McGurk, D., Bliese, P.D., Wright, K.M., and Hoge, C.W. (2006, November). *Post-deployment interventions to reduce the mental health impact of combat deployment to Iraq: Public health policies, psychological debriefing and battlemind training.* Paper presented at the International Society for Traumatic Stress Studies, Hollywood, CA.

Affleck, G., and Tennen, H. (1996). Construing benefits from adversity: Adaptational significance and dispositional underpinnings. *Journal of Personality* 64, 899–922.

Aldwin, C.M., Levenson, M.R., and Spiro III, A. (1994). Vulnerability and resilience to combat exposure: Can stress have lifelong effects? *Psychology and Aging* 9, 34–44.

Bartone, P.T., Adler, A.B., and Vaitkus, M.A. (1998). Dimensions of psychological stress in peacekeeping operations. *Military Medicine* 163, 587–593.

Bower, J.E., Kemeny, M.E., Taylor, S.E., and Fahey, J.L. (1998). Cognitive processing, discovery of meaning, CD4 decline, and AIDS-related mortality among bereaved HIV-seropositive men. *Journal of Consulting and Clinical Psychology* 66, 979–986.

Britt, T.W. (1998). Psychological ambiguities in peacekeeping. In H.J. Langholtz (ed.), *The psychology of peacekeeping* (111–128). Westport, CT: Praeger Publishers.

———. (2003). Can participation in peacekeeping operations be beneficial? The importance of meaning as a function of attitudes and identity. In T.W. Britt and A.B. Adler (eds.). *The psychology of the peacekeeper: Lessons from the field* (71–88). Westport, CT: Praeger Publishers.

Britt, T.W., Adler, A.B., and Bartone, P.T. (2001). Deriving benefits from stressful events: The role of engagement in meaningful work and hardiness. *Journal of Occupational Health Psychology* 6, 53–63.

Britt, T.W., Castro, C.A., McGurk, D., Herleman, H.A., and Odle-Dusseau, H.N. (2008, March 7). Benefit finding predicts dysfunctional and functional deployment outcomes. In T.W. Britt and J. Thomas (co-chairs), *Deriving benefits from stressful work: The case of combat veterans.* Paper presented at the American Psychological Association/U.S. National Institute for Occupational Safety and Health Work, Stress, and Health Conference, Washington, DC.

Britt, T.W., Dickinson, J.M., Moore, D.M., Castro, C.A., and Adler, A.B. (2007). Correlates and consequences of morale versus depression under stressful conditions. *Journal of Occupational Health Psychology* 12, 34–47.

Carver, C.S., and Antoni, M.H. (2004). Finding benefit in breast cancer during the year after diagnosis predicts better adjustment 5 to 8 years after diagnosis. *Health Psychology* 23, 595–598.

Dohrenwend, B.P., Neria, Y., Turner, J.B., Turse, N., Marshall, R., Lewis-Fernandez, R. et al. (2004). Positive tertiary appraisals and posttraumatic stress disorder in U.S. male veterans of the war in Vietnam: The roles of positive affirmation, positive reformulation, and defensive denial. *Journal of Consulting and Clinical Psychology* 72, 417–433.

Elder, G.H., and Clipp, E.C. (1989). Combat experience and emotional health: Impairment and resilience in later life. *Journal of Personality* 57, 311–341.

Fontana, A., and Rosenheck, R. (1998). Psychological benefits and liabilities of traumatic exposure in the war zone. *Journal of Traumatic Stress* 11, 485–503.

Helgeson, V.S., Reynolds, K.A., and Timich, P.L. (2006). A meta-analytic review of benefit finding and growth. *Journal of Consulting and Clinical Psychology* 74, 797–816.

Jennings, P.A., Aldwin, C.M., Levenson, M.R., Spiro III, A., and Mroczek, D.K. (2006). Combat exposure, perceived benefits of military service, and wisdom in later life: Findings from the normative aging study. *Research on Aging* 28, 115–134.

Lazarus, R.S., and Folkman, S. (1984). *Stress, appraisal, and coping.* New York: Springer.

Lewis, S.J. (2006). Combat stress control: Putting principle into practice. In Britt, T.W., Castro, C.A., and Adler, A.B. (eds.). *Military life: The psychology of serving in peace and combat* (121–140). Westport, CT: Praeger Security International.

McMillen, J.C., Smith, E.M., and Fisher, R.H. (1997). Perceived benefit and mental health after three types of disaster. *Journal of Consulting and Clinical Psychology* 65, 733–739.

Park, C.L., Cohen, L.H., and Murich, R.L. (1996). Assessment and prediction of stress-related growth. *Journal of Personality* 64, 71–105.

Park, C.L., and Helgeson, V.S. (2006). Introduction to the special section: Growth following highly stressful life events—current status and future directions. *Journal of Consulting and Clinical Psychology* 74, 791–796.

Ritzer, D.R., Campbell, S.J., and Valentine, J.N. (1999). Human dimensions research during operation "Joint Guard," Bosnia. *Army Medical Department Journal* 8, 5–16.

Schok, M.L., Kleber, R.J., Elands, M., and Weerts, Jos M.P. (2007). Meaning as a mission: A review of empirical studies on appraisals of war and peace keeping experiences. *Clinical Psychology Review* 28, 357–365.

Sledge, W.H., Boydstun, J.A., and Rabe, A.J. (1980). Self-concept changes related to war captivity. *Archives of General Psychiatry* 37, 430–443.

Southwick, S.M., Vythilingam, M., and Charney, D.S. (2004). The psychobiology of depression and resilience to stress: Implications for prevention and treatment. *Annual Review of Clinical Psychology* 1, 255–291.

Stouffer, S.A., Suchman, E.A., DeVinney, L.C., Star, S.A., and Williams, R.A. (1949). *The American soldier adjustment during army life* (vols. 1 and 2). Princeton: Princeton University Press.

Taylor, S.E. (1983). Adjustment to threatening events: A theory of cognitive adaptation. *American Psychologist* 38, 1161–1173.

Tedeschi, R.G., and Calhoun, L.G. (1996). The posttraumatic growth inventory: Measuring the positive legacy of trauma. *Journal of Traumatic Stress* 9, 455–471.

Thomas, J.L., Castro, C.A., McGurk, D.M., Cox, A., and Hoge, C.W. (2007, August). *The efficacy of Battlemind training at immediate post-deployment reintegration.* Paper presented at the American Psychological Association Conference, San Francisco, CA.

Tomich, P.L., and Helgeson, V.S. (2004). Is finding something good in the bad always good? Benefit finding among women with breast cancer. *Health Psychology* 23, 16–23.

Tugade, M.M., Fredrickson, B.L., and Barrett, L.F. (2004). Psychological resilience positive emotional granularity: Examining the benefits of positive emotions on coping and health. *Journal of Personality* 72, 1161–1190.

U.S. Army Surgeon General. (2008, February 14). Mental Health Advisory Team IV. Operation Iraqi Freedom 05–07. Retrieved June 26, 2008, from the U.S. Army Medical Command Web site, <www.armymedicine.army.mil/news/U.S. Army Surgeon General>.

Wood, M.D., Britt, T.W., Thomas, J.T., Klocko, R., Bliese, P.D., and Hoge, C.W. *Benefit finding in the war environment: A matter of time.* Unpublished manuscript.

Wood, M.D., Kim, P., Thomas, J.L., Bliese, P.D., and Hoge, C.W. (2008, March). *Active Duty Soldier's Postdeployment Benefit-Finding.* Paper presented at the Occupational Stress and Health Conference, Washington, DC.

Wood, M.D., Thomas, J.T., Klocko, R., Bliese, P.D., and Hoge, C.W. (2008, August). *Benefit finding in the war environment: A matter of time.* Poster presented at the Force Health Protection Conference, Albuquerque, NM.

Zapf, D., Dormann, C., and Frese, M. (1996). Longitudinal studies in organizational stress research: A review of the literature with reference to methodological issues. *Journal of Occupational Health Psychology* 2, 145–169.

Chapter 11

Enhancing Operational Readiness through Neuroimaging: Mapping the Pathophysiology of Mild Traumatic Brain Injury in Warfighters

Michael N. Dretsch

Estimates of mild traumatic brain injury (mTBI) in U.S. Soldiers as a result of blast injury during Operations *Iraqi Freedom* (OIF) and *Enduring Freedom* (OEF) are as high as 50 percent. Accordingly, there has been a surge of research to understand the pathophysiologic changes associated with mTBI, from inception to persistent and long-term sequelae. Technological advancement and innovative research in brain imaging techniques over the last two decades have greatly contributed to our current understanding of the pathophysiology of brain injury. Improved imaging techniques and sequences result in enhanced temporal and spatial resolution, validation, improved sensitivity and specificity, and utility of neuroimaging for assessing the degree of physical damage as well as predicting outcomes. Multimodal neuroscience approaches to neuroimaging will be the key to successful intervention of mTBI.

Background

Every Soldier understands the importance of placing the mission first. Yet when there is a rise in the number of Soldiers with impaired cognitive functioning and psychological well-being, whether transient or permanent and regardless of origin (for example, sleep deprivation, intoxication, combat stress conditions, or mTBI), then the risk of the mission failing greatly increases. In order to maintain operational readiness of our troops, leaders must establish guidelines and protocols to protect warfighters from exacerbation of their condition, while simultaneously achieving the mission objectives. Mild TBI cases present many difficulties for commanders because the injuries are predominantly "closed" head wounds, and diagnostic symptoms can overlap with psychological syndromes and malingering. Through advances in experimental design and integration of sophisticated techniques and disciplines, neuroscience research has been a pioneer in mapping the pathophysiology of mTBI. However, the application of neuroscience techniques for diagnoses and assessment of functioning has not yet been standardized. This chapter provides a general introduction to recent advances in neuroimaging techniques that have contributed to our knowledge of the pathophysiology of mTBI and explores the current and future trends in detection, diagnosis, and assessment of the injury fostered by the U.S. Army.

The brain, although remarkably resilient, is susceptible to transient and permanent dysfunction following an array of physical events. The activities in OIF and OEF have unfolded novel forms of physical insult to the central nervous system. Compared to previous wars, the current operations in southwest Asia have produced a greater number of brain-injured Soldiers, although this is accompanied by a greater survivability rate (Gawande, 2004; Ruff, 2005; Holcomb, Stansbury, Champion, Wade, and Bellamy, 2006). Trauma from the blast of improvised explosive devices is increasingly survivable due to improvements in armored vehicles, body armor, and medical care. However, medical providers are reporting staggering incidences of traumatic brain injury (TBI) among survivors. Reports from the Centers for Disease Control and Prevention (CDC, 2003) indicate that mTBI makes up 80 percent of all traumatic brain injuries in both military and civilian populations. Furthermore, some estimates indicate that 50 percent of Soldiers returning from OIF and OEF have sustained some level of mTBI (Jackson, Hamilton, and Tupler, 2008), with as many as 20 to 40 percent being redeployed with residual symptoms.

Similar to athletes who compete at a high level of intensity, Soldiers often have an ideology of warrior attributes such as resilience, invincibility, and risk-taking. What accompanies this ideology is an attitude to push through the pain of an injury and continue contributing to the team's goals. Such a virtue is laudable, especially when the injury is visible (for example, a wound to an extremity). However, mTBI is often invisible and consists of symptoms such as difficulty concentrating, headache, fatigue, irritability, sleep disturbances, and poor decision-making. Symptoms may persist or develop latently into transient, long-term, or permanent physical, psychological/behavioral, and cognitive symptoms. While headache is the most common physical symptom, a significant percentage of warriors develop sensory sensitivities and insomnia. Psychological/behavioral symptoms include depression, anxiety, fatigue, aggression, personality changes, emotional volatility, and reduced motivation (apathy). Cognitive changes include deficits in memory, attention, concentration, and judgment and decisionmaking (Ryan and Warden, 2003). The variation in symptom manifestation complicates diagnosis and leads to the need for advances in early detection and accurate diagnosis (for example, TBI versus acute stress disorder) and assessment of etiology (for example, asphyxia, focal lesions, contusion, intracranial hematoma/hemorrhage) and severity (mild, moderate, severe) of brain injury.

Classification

The terms *concussion* and *mTBI* are abundant within the literature and can be used interchangeably. *Post-concussion syndrome* (PCS) refers to the symptoms associated with the injury that persist for more than 3 months (Bigler, 2008). When residual symptoms are present past 6 months, the condition is referred to as *persistent post-concussion syndrome*. Currently, PCS is a proposed psychiatric disorder that is undergoing research for possible inclusion in future editions of the *Diagnostic and Statistical Manual of Mental Disorders* (American Psychiatric Association, 2000, Fourth Edition Text Revision). In an attempt to minimize confusion, the term *mTBI* is used throughout this chapter and refers to the injury and associated symptoms across time.

Currently, there is no solid evidence-based definition of mTBI. Criteria for the diagnosis vary by classification scheme. One of the most common screening tools for TBI is the Glasgow Coma Scale (GCS). It includes questions on motor responses (6 grades), verbal responses (5 grades), and eye-opening responses (4 grades). Lower scale scores indicate greater likelihood of more severe TBI (Teasdale and Jennett, 1974). Typically, a score of 13 to 15 suggests mild brain injury; 9 to 12 suggests moderate injury; and lower than 9, severe injury. There are various severity grading scales developed by different organizations. While each of these scales has shown a degree of success, experts have failed to reach a consensus. The Head Injury Interdisciplinary Special Interest Group of the American Congress of Rehabilitation Medicine has developed its own definition of mTBI (Esselman and Uomoto, 1995). For example, grade 1 is any alteration in mental state at the time of injury (such as feeling dazed, disoriented, or confused). Grade 2 is any loss of memory of the event immediately before or after the injury, with post-traumatic amnesia less than 24 hours. Finally, grade 3 is defined as any period of loss of consciousness (LOC) of less than 30 minutes followed by a GCS score of 13 to 15.7. The American Academy of Neurology guidelines (Kelly, 1999) require for a grade 1 that there be confusion, symptoms that last less than 15 minutes, and no loss of consciousness; grade 2, symptoms that last longer than 15 minutes and no loss of consciousness; and grade 3, loss of consciousness including coma, lasting either seconds (3a) or minutes (3b).

MTBI in military operational settings is defined by the Defense and Veterans Brain Injury Center's (DVBIC's) Clinical Practice Guideline and Recommendations (2006; http://dvbic.org/public_html/pdfs/clinical_practice_guideline_recommendations.pdf) as "an injury to the brain resulting from an external force and/or acceleration/deceleration mechanism from an event such as a blast, fall, direct impact, or motor vehicle accident which causes an alteration in mental status typically resulting in the temporally related onset of symptoms." DVBIC determines severity of TBI by set criteria based on duration of loss of consciousness, alternations of consciousness, and post-traumatic amnesia (see table 11–1).

Table 11–1. **Classification of Traumatic Brain Injury Severity**

Criteria	Mild	Moderate	Severe
Structural imaging	Normal	Normal or abnormal	Normal or abnormal
Loss of consciousness	0–30 minutes	> 30 minutes and < 24 hours	> 24 hours
Alteration of consciousness/ mental state*	A moment up to 24 hours	> 24 hours; severity based on other criteria	
Post-traumatic amnesia	0–1 day	> 1 and < 7 days	> 7 days
Glasgow Coma Scale (best available score in first 24 hours)	13–15	9–12	< 9

*Alteration of mental status must be immediately related to the trauma to the head. Typical symptoms would be looking and feeling dazed and uncertain of what is happening, confusion, difficulty thinking clearly or responding appropriately to mental status questions, and being unable to describe events immediately before or after the trauma event.

Source: VA/DOD Clinical Practice Guideline for Management of Concussion/Mild Traumatic Brain Injury, Version 1.0, March 2009, Table A–1, 8. Available at <www.dvbic.org/Providers/TBI-Screening.aspx>.

Experts have varying opinions of how accurate each TBI classification scheme is. However, all agree that clinical judgment is crucial in the majority of cases. This includes consideration of factors that frequently confound patient evaluation: preexisting conditions, intoxication, polypharmacy, and baseline functioning.

Impact on Physical, Metabolic, Vascular, and Endocrine Processes

Most cases of mTBI are acquired through mechanistic insult from explosion/blast, motor vehicle crash, blunt object, fragments, and gunshot wound. Brain injury may be from primary (exposure to overpressurization wave from blast), secondary (impact from blast, debris), tertiary (impact after displacement), quaternary (inhalation of toxins, hypoxia), or a combination of affects that results in stretching, lacerations, and contusions to grey and white matter, both directly (bony structures, foramen magnum, anterior and posterior fossa) and indirectly (intracranial hemorrhage or hematoma) and dysregulation of vascular, endocrine, and metabolic processes (Giza and Hovda, 2001; Bay and McLean, 2007; Korn, Golan, Melamed, Pascual-Marqui, and Friedman, 2005). To further complicate the process, the type of personal protective equipment a Soldier uses may accentuate the severity of the injury (Moore, Radovizky, Shupenko, Klinoff, Jaffee, and Rosen, 2008; Bhattacharjee, 2008). Although mild TBI can manifest as an open head wound, it more often is a closed wound, which has made this war injury easily misunderstood and often overlooked.

The brain's vulnerability to injury is a result of a number of structural and constitutional issues such as the surface location of arterial branches that nourish it and the soft arachnoid sinuses that filter cerebral sinus fluid and waste. These vessels can be impacted by coup-contra-coup impact with the skull (that is, the brain hitting the skull, then bouncing back and hitting the other side of the skull), resulting in degrees of contusion, intracranial hemorrhaging, and/or hematomas that can lead to increased intracranial pressure, and ultimately herniation of cortical mass. Herniation can stretch, tear, and compress cortical and subcortical nuclei, arteries, cranial nerves, and the brain stem, as well as produce secondary and tertiary hemorrhaging and swelling. Such structural alterations often result in an abnormal volume of cerebral spinal fluid and reduced circulation (Bay et al., 2007).

Individuals who experience repetitive concussions are at increased risk for developing long-term symptoms and even permanent brain damage (Mori, Katayama, and Kawamata, 2006). At a micro level, diffuse axonal injury and focal lesions impact biomechanical, hemodynamic, neurobiologic, and metabolic mechanisms leading to reduced or compromised axonal function, vascular blood flow, neurotransmitter and neuropeptide regulation, intra-axonal ion-gating, cerebral glucose metabolism, and cerebral immune response expression (Giza et al., 2001). Behan and colleagues (2008) identified pituitary hormone deficiencies in patients with previous brain injury in a study examining prevalence of neuroendocrine dysfunction in patients recovering from traumatic brain injury.

One of the medical community's biggest concerns is Soldiers prematurely returning to full duty following mTBI. Evidence suggests that premature return is associated with an increased susceptibility to a subsequent head injury with a prolonged neurological recovery and possible permanent neurobiologic and metabolic changes (Guskiewicz, McCrea, Marshall, Cantu, Randolph, Barr, Onate, and Kelly, 2003). Repetitive mTBI can produce increased vascular congestion and cerebral swelling, resulting in transtentorial herniation and subsequent death (Mendez, Hurley, Lassonde, Zhang, and Taber, 2005; Bailes and Hudson, 2001). It is of paramount importance to note that this vulnerability may exist even if the individual initially appears to be asymptomatic (CDC, 2002). Furthermore, regardless of severity, a repeat brain injury can be life threatening if experienced within hours or days of the first (Mori, Katayama, and Kawamata, 2006). The collateral loss from compromised cognitive functioning and the resulting human error includes decreased battlefield efficiency and increased risk of both fratricide and equipment loss.

Neuropsychological Functioning

The initial symptoms and persistent sequelae associated with mTBI can have deleterious effects on work performance. Individuals who report being asymptomatic after hours to days may still suffer from undiagnosed neurocognitive impairments, which can result in suboptimal long-term physical, mental, social, and occupational functioning (Kraus, Schaffer, Ayers, Stenehjem, Shen, and Afifi, 2005). In the context of operational combat settings, compromised functioning of the brain is particularly important when considering the job description. Initial mTBI-related symptoms reported often include fatigue, headache, nausea, amnesia, irritability, difficulties with memory and concentration, and affect/mood dysregulation (Creamer, O'Donnell, and Pattison, 2005; Stulemeijer, van der Werf, Bleijenberg, Biert, Brauer, and Vos, 2006; Bay et al., 2007). Furthermore, approximately 10 to 20 percent of individuals with mTBI suffer from long-term (greater than 6 months) neuropsychological sequelae that are often disabling (Nolin and Heroux, 2006; Kashluba, Paniak, Blake, Reynolds, Toller-Lobe, and Nagy, 2004).

The current literature provides good overviews of the various cognitive processes typically impacted by mTBI (Malojcic, Mubrin, Coric, Susnic, and Spilich, 2008; Kurca, Sivak, and Kucera, 2006; McAllistar, Sparling, Flashman, Guerin, Mamourain, and Saykin, 2001; McIntire, Langan, Halterman, Drew, Osternig, Chou, and van Donkelaar, 2006). Overall, the cognitive processes most often implicated include attention, memory, and speed of information processing.

Impairment in neuropsychological tasks has been observed in individuals with mTBI. For example, Nolin (2006) reported that patients with mTBI have executive dysfunction when engaged in memory tasks as assessed via the California Learning Test. The results revealed a specific deficit of impaired retrieval processes, but not of encoding or storage of information. Other studies concur with the notion that mTBI cognitive symptoms can be a result of a specific impairment such as inattention. However, disentangling the precise deficits associated with mTBI can sometimes be difficult.

Multiple studies have revealed attentional deficits in individuals with mTBI. Using single- and dual-task conditions, Azouvi and colleagues (2004) provided evidence that individuals with mTBI performed similar to controls, but reported significantly higher levels of effort. The results suggest that divided-attention deficits in individuals with mTBI are related to a reduction in available processing resources.

The concept that individuals with mTBI have reduced resources to apply to various tasks is supported by other studies. Tombaugh and colleagues (2007) reported a positive linear relationship between reaction time and complexity on computerized tests of information processing. A similar relationship between severity of injury and reaction time was also reported.

McIntire and colleagues (2006) provided evidence in accord with Azouvi et al. by using the rapid serial visual presentation task, which requires the participant to respond to a target letter and determine whether a probe letter was present or not in a stream of rapidly presented letters. Individuals with mTBI showed normal attentional blink as compared to matched controls. However, unlike health controls, the mTBI group showed impaired selective attention with significantly more omission errors. Based off results from a Gap Saccade task, Drew and colleagues (2007) reported that deficits in visuospatial attention associated with mTBI might be attributed to difficulties with the initial disengagement process. This was only present when the temporal gap was short, not when long. However, this deficit was resolved within 1 week. Other evidence suggests that mTBI results in deficits in cognitive-motor association (Sosnoff, Broglio, and Ferrara, 2008), which may be attributed to impairments in visuospatial attention.

Top-down processes often moderate general functions that we may take for granted. For example, reduced processing speed, attention, memory, and overall efficacy of cerebral functioning can dampen language ability. Aphasia batteries have been criticized for lack of sensitivity and specificity when used for assessment of subtle impairments associated with mTBI (Duff, Proctor, and Haley, 2002). In contrast, recent reports indicate that returning Soldiers being treated for mTBI show significant communication deficits that are not attributed to changes in hearing (personal communication, Dr. Anthony Salvatore, University of Texas–El Paso, and LTC Kathy Prue-Owens, USA, William Beaumont Army Medical Center, Fort Bliss, TX).

Most of the neuropsychology research has focused on "cool-headed" cognitive processes. Hence, there is a paucity of information concerning how mTBI impacts affective processes. This may be partially due to the complex affect-by-cognitive interplay that has been under investigation by psychological scientists from many fields of research (Dretsch and Tipples, 2008; Pecchinenda, Dretsch, and Chapman, 2006; Gray, Braver, and Raichle, 2002). Affective-based decisionmaking impairments have been associated with mTBI (Levine, Black, Cheung, Campbell, O'Toole, and Schwartz, 2005). Some evidence suggests reduced sensitivity of judgments and responding to stimulus-reinforcement contingencies following mTBI (Schlund and Pace, 2000).

An understanding of mTBI-related cognitive impairments, both transient and permanent, is necessary for determining treatment courses of action. Over the course of OIF and OEF, command leadership expressed a desire to know the cognitive status of the troops. However, understanding the impact of mTBI on neuropsychological functioning will require a multimodal research approach to account for limitations of behavioral measures.

Advances in Brain Imaging for Assessment of Neuropsychological Functioning

One of the greatest obstacles with mTBI is finding a nonintrusive way to see what is going on inside of the brain in order to evaluate the extent of damage and to provide appropriate treatment and rehabilitation interventions. While clinical neuropsychological assessment has been a chief contributor to capturing cerebral dysfunction following possible brain injury, the profession has acknowledged limitations. This being said, structural and functional brain biomarkers have advanced our understanding of the brain-mind interplay, which is one of the truly paramount enigmas of the 21st century.

Within the last decade, there has been a surge in the development of advanced neuroscience techniques for the assessment of mTBI. This is largely attributed to funding opportunities presented by the Department of Defense (DOD) for both independent and collaborative studies. Progressive advances in magnetic resonance (MR) techniques are currently being developed and applied, which has greatly improved our understanding of the impact of mTBI on multiple processes associated with physical, behavioral, psychological, and cognitive symptoms. The most common imaging findings include contusions, axonal shearing, hemorrhages, and herniations (Lizerbram and Moffit, 2001). The following sections will briefly outline some of the most recent advances in mTBI neuroimaging research. Emphasis will be on more recent imaging techniques that have only been available since the early 1990s (Belliveau, Kennedy, McKinstry, Buchbinder, Weisskoff, Cohen, Vevea, Brady, and Rosen, 1991; Basser and LeBihan, 1992). The reasons for primarily focusing on these more recent techniques are their noninvasive nature and potential for assessing mTBI symptoms.

Computed Tomography

Computed tomography (CT) is considered the gold standard for imaging in the acute management of TBI. However, less than 10 percent of patients with mTBI have positive CT findings (Saboori, Ahmadi, and Farajzadegan, 2007). Until recent advances, even the most sophisticated structural imaging analysis techniques lacked information about the functioning of the intact brain matter. This highlights the less than optimal predictive sensitivity and specificity of CT in comparison to other more advanced techniques (Lewine, Davis, Bigler, Thoma, Hill, Funke, Sloan, Hall, and Orrison, 2007; Niogi, Mukherjee, Ghajar, Johnson, Kolster, Sarkar, Lee, Meeker, Zimmerman, Manley, and McCandliss, 2008). This also emphasizes the importance of integrating functional imaging techniques, which may be more informative than structural imaging for milder cases of mTBI.

Magnetic Resonance Imaging

Unlike CT, magnetic resonance imaging (MRI) does not use ionizing radiation, but instead relies on a powerful magnetic field to align the nuclear magnetization of atoms in the water of the brain (Levine, Fujiwara, O'Connor, Richard, Kovacevic, Mandic, Restagno, Easdon, Robertson, Graham, Cheung, Gao, Schwartz, and Black, 2006). Although MRI has superior resolution to CT, small lesions associated with mTBI are often not visible to the naked eye and can result in underestimates of neuropathology. Approaches have been developed with combined field gradients and radiofrequency excitation to create higher resolution images. Structural imaging techniques such as diffusion tensor imaging (DTI), and functional techniques such as functional magnetic resonance imaging (fMRI), magnetic resonance spectroscopy (MRS), and positron emission tomography (PET), have contributed to the understanding of the impact of mTBI and show promise for diagnostic application to mTBI. Other imaging techniques such as fluid attenuated inversion recovery, arterial spin labeling, and magnetic resonance angiography have been shown to be superior to the conventional spin-echo images in detecting mTBI-related abnormalities. Yet studies have only shown weak correlations between abnormal findings on MRI and residual symptoms (Topal, Hakyemez, Erdogan, Bulut, Koksal, Akkose, Dogan, Parlak, Ozguc, and Korfali, 2008; Mendez et al., 2005). For this reason, we will focus on a few select MR techniques that appear to have the most potential at this time for use of diagnosis and clinical assessment of mTBI.

Diffusion tensor imaging. DTI is acquired with a standard MRI scanner, and has been shown to be much more sensitive to white matter injury than conventional T1- and T2-weighted anatomic MRI. DTI noninvasively represents myelin structural integrity and cellular membranes of white matter tracts. The technique is based on movement of water molecules or anisotropic diffusion (indexed as fractional anisotropy), which tend to move faster along, rather than perpendicular to, nerve fibers. Tractography is an advanced DTI application that has shown itself to be sensitive to detection of mTBI, especially in measuring the thickness of myelin sheaths and axons, distribution of directions, and density of white matter tracts (Kraus, Susmaras, Caughlin, Walker, Sweeney, and Little, 2007; Niogi, Mukherjee, Ghajar, Johnson, Kolster, Sarkar, Lee, Meeker, Zimmerman, Manley, and McCandliss, 2008).

DTI has shown promising results in its ability to detect subtle structural abnormalities associated with mTBI (Wilde, McCauley, Hunter, Bigler, Chu, Wang, Hanten, Troyanskaya, Yallamalli, Li, Chia, and Levin, 2008). Recently, Niogi and colleagues (2008) found that the number of damaged white matter structures, but not the number of traumatic microhemorrhages, was significantly correlated with mean reaction time on a simple cognitive task (Attention Network Task). These results are in accord with behavioral findings from other studies of deficits in speed of processing and attention following mTBI (Tombaugh et al., 2007; Drew et al., 2007).

The development of DTI is in the early stage, and therefore needs more validating. New methodological developments may lead to improved quantitative assessment of white-matter integrity and volume. This will greatly enhance clinicians' and researchers' ability to apply DTI to the diagnosis and assessment of mTBI.

Functional MRI. Functional MRI (fMRI) is a specialized form of MRI scan that measures the hemodynamic response related to neural activity in the brain or spinal cord. With fMRI, neuroimagers can map various types of physiological information, including baseline cerebral blood volume and changes in cerebral blood volume, baseline and cerebral perfusion, cerebral blood oxygenation, resting state cerebral oxygen extraction fraction, and cerebral metabolic rate for oxygen (Bandettini, 2006). Functional MRI studies typically incorporate blood oxygen level dependent (BOLD) changes that measure the hemodynamic response to neuronal activity. This response is a process in which blood releases oxygen to active neurons at a greater rate than to inactive neurons. The BOLD method measures the difference in magnetic susceptibility between oxyhemoglobin and deoxyhemoglobin.

Assessment of overall brain function and region(s) of interest using fMRI requires the use of experimental paradigms. For mTBI, fMRI imaging typically occurs during the performance of tasks that load frontal lobe functioning (for example, attention and working memory).

Tasks that rely on working memory neural circuitry have been helpful in many neuroimaging studies of mTBI. The impetus is that the psychological domains linked with frontal lobe functioning are inherently involved in many functional aspects of interest to humans, and they are the most anatomically vulnerable to cerebral insult as a result of coup-contra-coup contusions to the frontal and posterior (also temporal-temporal, which may include damage to lateral prefrontal cortex, somatosensory, insular cortices, and so forth) grey matter resulting from acceleration/deceleration impact against the skull.

Behavioral and imaging studies provide support for the use of working memory tasks as valid measures of frontal lobe functioning in mTBI patients (Lovell, Collins, Iverson, Field, Maroon, Cantu, Podell, Belza, and Fu, 2003; Chen, Johnston, Collie, McCrory, and Ptito, 2008; Tucker, 1999; Chen, Johnston, Petrides, and Ptito, 2008). Using various n-back conditions, McAllister and colleagues (2001) provided evidence that fMRI is sensitive to differential effects of memory load in individuals with mTBI. Using fMRI and a working memory task, Chen and colleagues (2008) replicated these findings and demonstrated atypical brain activation patterns in the dorsolateral prefrontal cortex (DLPFC) in all symptomatic concussed athletes compared to nonconcussed athletes. At a follow-up session, individuals that remained symptomatic continued to show atypical DLPFC activation. Jantzen, Anderson, Steinberg, and Kelso (2004) used a test battery that included mathematical, memory, and sensorimotor coordination tasks for comparing preseason baselines with postconcussion measures in athletes. Their results indicated that concussed players had marked within-subject increases in BOLD activation of parietal and lateral frontal and cerebellar regions during finger-sequencing. These studies plus findings from other studies (Laatsch, Thulborn, Krisky, Shobat, and Sweeney, 2004) reveal the potential value for the use of fMRI in developing return-to-duty standards for the military.

There is also evidence that fMRI may have predictive value for determining successful recovery from mTBI. Using athletes, Lovell and colleagues (2007) found that abnormal fMRI results during the first week of recovery predicted clinical recovery. The study suggested that some individuals with greater cerebral hyperactivation may recover more slowly than athletes with less hyperactivation.

Overall, compared to other functional imaging techniques such as PET and single photon emission computed tomography (SPECT), fMRI is preferred due to higher temporal resolution, better availability, lower costs, and the absence of ionizing radiation exposure. However, there are specific limitations of fMRI for assessing brain activation that can be attributed to the methods used to collect data and the relationship between neuronal activity and hemodynamic change. For example, compared to other underlying processes such as action potentials, the hemodynamic response is relatively slow with a latency range of 4 seconds, due to variation in the neuronal activity and oxygen and cerebral blood flow that accompanies brain activation.

Methodological considerations are also important for improving the diagnostic value of fMRI. Spatial resolution with BOLD fMRI is determined by spatial spread of hemodynamic and neurovascular coupling. Techniques allow for good resolution in the range of 1 to 3 cubic millimeters. In addition, BOLD can have a low signal sensitivity and a wide range of possible artifacts (for example, susceptibility artifact between air and tissue in ventral-frontal and temporal regions) that can pose difficulties. However, this issue has also been addressed by methodological strategies such as emphasizing task modulation to assess changes in latency and width of region activation (Bellgowan, Saad, and Bandettini, 2003). In addition, neurochemical and metabolic changes occur immediately following mTBI. These changes often disrupt various neuropsychological processes, but are typically ephemeral in that they resolve within days or weeks. For this reason, repetition of fMRI may be useful for patients who manifest symptoms of cognitive dysfunction after the acute phase has passed in order to tease apart transient physiological changes from residual cognitive deficits.

Brain Receptor Imaging

One aspect of cerebral trauma is an increased postinjury need for glucose (the brain's primary metabolic fuel), but due to the injury, there is hypoperfusion (Bonne, Gilboa, Louzon, Sherf, Katz, Fishman, Nahum et al., 2003). Thus, there is an imbalance between the brain's need for energy and the body's ability to meet that need on a metabolic basis. PET scans produce a 3–D image of functional processes in the brain by detecting pairs of gamma rays emitted indirectly by a radionuclide that is introduced into the body attached to an active biological molecule. Similarly, SPECT uses the detection of gamma rays to produce a valid 3–D image. However, the difference is that while the PET tracer emits positrons that collide with electrons, causing gamma photons to be emitted, the SPECT tracer (a radioactive tracer;

isotope) emits gamma radiation. Findings from PET and SPECT studies have shown frontal and temporal hypometabolism following mTBI at rest and during working memory tasks (Chen, Kareken, Fastenau, Trexler, and Hutchins, 2003; Umile, Sandel, Alavi, Terry, and Plotkin, 2002). PET scans can demonstrate precise regions of increased glucose needs, whereas SPECT can detect cerebral perfusion (Frankle, Slifstein, Talbot, and Laruelle, 2005).

Although SPECT scans are much less expensive than PET scans, both are considered to be invasive and risky because the procedures involve an intravenous injection of a radioactive tracer. Furthermore, due to the radioactive substance, imagers must limit the number of sessions/studies, therefore limiting serial testing, which has important clinical implications. Currently, there are an insufficient number of radiotracers available, making PET and SPECT less than optimal and not the most practical for serial assessment of mTBI. Overall, PET has better temporal resolution than SPECT (Frankle et al., 2005), which explains why it is often the preferred method.

Magnetic Resonance Spectroscopy

MRS is an advanced quantitative technique based on signals generated from the net spin of atomic nuclei from various molecules when disrupted by a radio frequency. Molecules utilized for MRS include hydrogen, phosphorus, lithium, fluorine, and sodium. These signals can be converted to a form of spectra to provide meaningful data on the chemical composition at the region of interest (Keshavan, Kapur, and Pettegrew, 1991).

This approach has shown promise for assessing brain metabolic abnormalities following cerebral insult. Using MRI with a group of mTBI patients, Cohen and colleagues (2007) were able to detect neuronal injury beyond the minimal focal lesions observed with MR using a whole-brain N-acetylaspartate analysis (NAA). Using proton magnetic resonance spectroscopy, Yeo and colleagues (2006) provided some evidence that neurometabolic ratios of NAA/creatine and choline/creatine predict neuropsychological performance in TBI patients. Changes in these ratios from 3 to 21 weeks post-injury were suggestive of neurometabolic recovery. In addition to Yeo et al., other studies have provided evidence, using athletes who experienced LOC, that brain metabolite ratios may be useful to support return-to-duty decisions (Vagnozzi,

Signoretti, Tavazzi, Floris, Ludovici, Marziali, Tarascio, Amorini, Di Pietro, Delfini, and Lazzarino, 2008).

MRS is limited by varying ranges of detectable chemicals, neurochemical noise, and varying metabolite to parenchyma (that is, brain matter) ratios for each receiver coil (Danielsen and Ross, 1999). Also, increased leakiness following cell membrane damage can result in a washout of metabolites, therefore making it more difficult to differentiate smaller areas that have been damaged from widespread cellular apoptosis and necrosis. Overall, MRS has good spatial resolution, but low sensitivity and specificity for optimal use in assessment of mTBI. MRS is best used in post-acute stages because cellular energy status is maintained until relatively late in the process.

Discussion

Roadblocks

With the exception of visible damage to brain matter, there is not one simple symptom or marker that can be reliably used to make a diagnosis of brain injury. Unless the TBI-related event is witnessed by an objective observer, or there is external damage to the head providing conclusive information detailing the mechanism of injury, orientation, and other relevant characteristics of the accident, health care providers must apply clinical judgment to bridge information gaps that diagnostic tools do not currently fill.

Neuroimaging techniques have greatly contributed to our understanding of the impact of mTBI. However, there are some pragmatic limitations of MR technology. With the use of ultra-high magnetic fields, there is better signal-to-noise ratio and improved temporal and spatial resolution. For example, scientists can map submillimeter columnar and laminar functional structures and detect neural activity in certain tasks to tens of milliseconds (Kim and Ogowa, 2002). However, strong magnetic fields such as those greater than 3 Tesla (3T) have greater technical demands, are more expensive, and have trade-off limitations such as increased noise within the scanner (Kim and Ogawa, 2002). The current standard is the use of the 3T in human research and clinical settings. Use of MRI is inadvisable for patients with shrapnel due to the inherent risks presented by the powerful magnet, which can be an obstacle when working with Soldiers.

Integration

Limitations in the spatial and temporal resolution of imaging techniques highlight the need to integrate neuroscience approaches (Lewine, Davis, Bigler, Thoma, Hill, Funke, Sloan, Hall, and Orrison, 2007; Niogi et al., 2008). There has been some work that combines structural and functional imaging to provide a more detailed analysis. For example, DTI and fMRI are complementary for use for assessing neuropsychological functioning. Nagy, Westerberg, and Klingerg (2004) provided evidence that DTI fractional anisotropy values in fronto-parietal white matter correlated with BOLD response in areas that could form a functional network underlying working memory function. Other evidence suggests that combined use of DTI and susceptibility weighted imaging (SWI) improve detection of mTBI (Kou, Benson, Gattu, and Haacke, 2008). These studies reported that SWI detected more hemorrhages than conventional MRI, and DTI detected more nonhemorrhagic lesions than observed with SWI. Concomitant use of various imaging approaches in assessment of cerebral functioning improves clinical utility for assessing mTBI (Schlosser, Nenadic, Wagner, Gullmar, Von Consbruch, Kohler, Schultz, Koch, Fitzek, Matthews, Reichenach, and Sauer, 2007).

Other physiological modalities have been simultaneously carried out during collection of fMRI data. Electroencephalography, eye tracking, optical imaging, skin conductance responses, magnetoencephalography, respiration, and heart rate have all been used concomitantly to enhance precision and alleviate uncertainty of imagers (Gosseries, Demertzi, Noirhomme, Tshibanda, Boly, de Beek, Hustinx et al., 2008). Improved methodological strategies, such as pre- and/or post-fMRI, have also minimized some uncertainty. Additionally, accurately reflecting physiological activity linked with mTBI requires continued sophistication of statistical analysis of single and combined techniques. Importantly, statistical models need to be developed to incorporate and integrate results from different laboratories across all the methods. Methodological and statistical analysis standardization would help homogenize efforts across the myriad of organizations and facilities involved with TBI research.

Imaging Biomarkers and Clinical Application

To date, CT and MRI are the only imaging techniques used for clinical diagnosis. As research progresses, there are prospects of the clinical application of functional imaging (fMRI, PET, and DTI) for mTBI diagnosis and evaluation. For the U.S. Armed Forces, this has beneficial implications when considering the need for return-to-duty guidelines. However, there is a need for validation of images to ensure face validity; standardization of imaging protocols, including upgraded scanners; development of sophisticated informatics to integrate with information collected from different imaging modalities, genetics, and protein expression; and the identification and development of new radiotracers and drugs capable of penetrating the blood-brain barrier.

In-theater Multimodal Technologic Approach

Improvements in helmet sensors for quantifying blast exposure and the addition of video instrumentation may provide better injury-event information for improving clinical diagnosis and assessment. In the future, if mTBI is suspected, new technologies will be used for multimodal testing, including neuropsychological testing (against pre-established baseline), structural and functional imaging, biomarker assaying, and advanced telemedicine capabilities, to get clinical interpretation of the results from experts. In addition, there is a need to develop algorithms based on models from neurocognitive, eye pupilometry, imaging, electrophysiology, genetic, and serum biomarkers measures for assessing the degree of neural damage sustained.

Conclusion

Recent brain imaging approaches and technologies have contributed to the medical and research communities' understanding of mTBI pathophysiological outcomes. Although there have been tremendous gains, limitations remain—produced not only by uncertainties that resulted in the injury, but also by the interplay between neurophysiology and imaging technology. To address the various gaps, enhance diagnostic accuracy, and improve imaging utility, DOD should emphasize basic projects that calibrate and validate the correlation of different imaging methods and physiological brain events.

References

Azouvi, P., Couillet, J., Leclercq, M., Martin, Y., Asloun, S., and Rousseaux, M. (2004). Divided attention and mental effort after severe traumatic brain injury. *Neuropsychologia* 42, 1260–1208.

Bailes, J.E., and Hudson, V. (2001). Classification of sport-related head trauma: a spectrum of mild to severe injury. *Journal of Athletic Training*, 36, 236–243.

Bandettini, P.A. (2006). Functional magnetic resonance imaging. In C. Senior, T. Russell, and M.S. Gazzaniga (eds.). *Methods in Mind* (193–235). Cambridge, MA: MIT Press.

Basser, P.J., and LeBihan, D. (1992). Fiber orientation mapping in an anisotropic medium with NMR diffusion spectroscopy [Abstract]. SMRM Proceedings 11:1221. Presented at Society for Magnetic Resonance in Medicine 11th Annual Meeting, Berlin, Germany.

Basser, P.J., Mattielo, J., and LeBihan, D. (1992) Diagonal and off-diagonal components of the self-diffusion tensor: Their relation to and estimation from the NMR spin-echo signal [Abstract]. SMRM Proceedings 11:1222. Presented at Society for Magnetic Resonance in Medicine 11th Annual Meeting, Berlin, Germany.

Bay, E., Hagerty, B.H., and Williams, R.A. (2007). Depressive symptomotology after mild-to-moderate traumatic brain injury: A comparison of three measures. *Archives of Psychiatric Nursing* 21, 2–11.

Bay, E., and McLean, S.A. (2007). Mild traumatic brain injury: An update for advanced practice nurses. *Journal of Neuroscience Nursing* 39, 43–51.

Behan, L.A., Phillips, J., Thompson, C.J., and Agha, A. (2008). Neuroendocrine disorders after traumatic brain injury. *Journal of Neurology, Neurosurgery, and Psychiatry* 79, 757.

Bellgowan, P.S.F., Saad, Z.S., and Bandettini, P.A. (2003). Understanding neural system dynamics through task modulation and management. *Proceedings of the National Academy of Sciences* 100, 1415–1419.

Belliveau, J.W., Kennedy, D.N., McKinstry, R.C., Buchbinder, B.R., Weisskoff, R.M., Cohen, M.S., Verea, J.M., Brady, T.J., and Rosen, B.R. (1991). Functional mapping of the human visual cortex by magnetic resonance imaging. *Science* 254, 716–719.

Bhattacharjee, Y. (2008). Shell shock revisited: Solving the puzzle of blast trauma. *Science* 319, 406–408.

Bigler, E.D. (2008). Neuropsychology and clinical neuroscience of persistent post-concussive syndrome. *Journal of the International Neuropsychological Society* 14, 1–22.

Bonne, O., Gilboa, A., Louzon, Y., Sherf, O.K., Katz, M., Fishman, Y., Nahum, Z.B., Krausz, Y., Bocher, M., Lester, H., Chisin, R., and Lerer, B. (2003). Cerebral blood flow in chronic symptomatic mild traumatic brain injury. *Psychiatry Research: Neuroimaging* 124, 141–152.

Centers for Disease Control and Prevention, National Center for Injury Prevention and Control. (2003). Report to Congress on mild traumatic brain injury in the United States: Steps to prevent a serious public health problem. Atlanta, GA: Centers for Disease Control and Prevention.

Centers for Disease Control and Prevention. (2002). Heads up: Facts for physicians about mild traumatic brain injury (MTBI). Retrieved March 28, 2008, from <www.cdc.gov/Migrated_Content/Brochures_and_Catalogs/tbi_mtbi_facts_for_physicians.pdf>.

Chen, J., Johnston, K.M., Petrides, M., and Ptito, A. (2008). Recovery from mild head injury in sports: Evidence from serial functional magnetic resonance imaging studies in male athletes. *Clinical Journal of Sports Medicine* 18, 241–175.

Chen, S.H.A., Kareken, D.A., Fastenau, P.S., Trexler, L.E., and Hutchins, G.D. (2003). A study of persistent post-concussion symptoms in mild head trauma using positron emission tomography. *Journal of Neurology, Neurosurgery, and Psychiatry* 74, 326–332.

Cohen, B.A., Inglese, M., Rusinek, H., Babb, J.S., Grossman, R.I., and Gonen, O. (2007). Proton MR spectroscopy and MRI-volumetry in mild traumatic brain injury. *American Journal of Neuroradiology* 28, 907–913.

Creamer, M., O'Donnell, M.L., and Pattison, P. (2005). Amnesia, traumatic brain injury, and posttraumatic stress disorder: A methodological inquiry. *Behaviour Research and Therapy* 43, 1383–1389.

Danielsen, E.R., and Ross, B. (1999). *Magnetic resonance spectroscopy diagnosis of neurological diseases*. New York: Marcel Dekker.

Dretsch, M., and Tipples, J. (2008). Working memory involved in predicting future outcomes based on past experiences. *Brain and Cognition* 66, 83–90.

Drew, A.S., Langan, J., Halterman, C., Osternig, L.R., Chou, L., and van Donkelaar, P. (2007). Attentional disengagement dysfunction following mTBI assessed with the gap saccade task. *Neuroscience Letters* 417, 61–65.

Duff, M.C., Proctor, A., and Haley, K. (2002). Mild traumatic brain injury (mTBI): Assessment and treatment procedures used by speech-language pathologists (SLPs). *Brain Injury* 16, 773–787.

Esselman, P.C., and Uomoto, J.M. (1995). Classification of the spectrum of mild traumatic head injury. *Brain Injury* 9, 417–424.

Frankle, W.G., Slifstein, M., Talbot, P.S., and Laruelle, M. (2005). Neuroreceptor imaging in psychiatry: Theory and applications. *International Review of Neurobiology* 67, 385–440.

Giza, C.C., and Hovda, D.A. (2001). The neurometabolic cascade of concussion: Sport-related concussion. *Journal of Athletic Training* 36, 228–235.

Gosseries, O., Demertzi, A., Noirhomme, O., Tshibanda, J., Boly, M., de Beeck, M.O., Hustinx, R., Maquet, P., Salmon, E., Moonen, G., Luxen, A., Laureys, S., and De Tiège, X. (2008). Functional neuroimaging (fMRI, PET and MEG): What do we measure. *Revue Medicale de Liege* 63, 231–237.

Gray, J.R., Braver, T.S., and Raichle, M.E. (2002). Integration of emotion and cognition in the lateral prefrontal cortex. *Proceedings of the National Academy of Sciences* 99, 4115–4120.

Guskiewicz, K.M., McCrea, M., Marshall, S.W., Cantu, R.C., Randolph, C., Barr, W., Onate, J.A., and Kelly, J.P. (2003). Cumulative effects associated with recurrent concussion in collegiate football players: The NCAA Concussion Study. *Journal of the American Medical Association* 19, 2549–2555.

Holcomb, J.B., Stansbury, L.G., Champion, H.R., Wade, C., and Bellamy, R.F. (2006). Understanding combat casualty care statistics. *Journal of Trauma* 60, 397–401.

Jackson, G.L., Hamilton, N.S., and Tupler, L.A. (2008). Detecting traumatic brain injury among veterans of Operations *Enduring* and *Iraqi Freedom*. *North Carolina Medical Journal* 69, 43–47.

Jantzen, K.J., Anderson, B., Steinberg, F.L., and Kelso, S. (2004). A prospective functional MR imaging study of mild traumatic brain injury in college football players. *American Society of Neuroradiology* 25, 738–745.

Johnston, K.M., Ptito, A., Chankowsky, J., and Chen, K. (2001). New frontiers in diagnostic imaging in concussive head injury. *Clinical Journal of Sport Medicine* 11, 166–175.

Kashluba, S. Paniak, C., Blake, T., Reynolds, S., Toller-Lobe, G., and Nagy, J. (2004). A longitudinal, controlled study of patient complaints following treated mild traumatic brain injury. *Archives of Clinical Neuropsychology* 19, 805–816.

Kelly, J.P. (1999). Traumatic brain injury and concussion in sports. *Journal of the American Medical Association* 282, 989–991.

Keshavan, M.S., Kapur, S., and Pettegrew, J.W. (1991). Magnetic resonance spectroscopy in psychiatry: Potential, pitfalls, and promise. *American Journal of Psychiatry* 148, 976–985.

Kim, S., and Ogawa, S. (2002). Insights into new techniques for high resolution functional MRI. *Current Opinion in Neurobiology* 12, 607–615.

Korn, A., Golan, H., Melamed, I., Pascual-Marqui, R., and Friedman, A. (2005). Focal cortical dysfunction and blood-brain barrier disruption in patients with postconcussion syndrome. *Journal of Clinical Neurophysiology* 22, 1–9.

Kou, Z., Benson, R., Gattu, R., and Haacke, M. (2008). Improving the detection of diffuse axonal injury by complementary use of advanced MRI: 0037. *Journal of Head Trauma Rehabilitation* 23, 351–352.

Kraus, J., Schaffer, K., Ayers, K., Stenehjem, J., Shen, H., and Afifi, A.A. (2005). Physical complaints, medical service use, and social and employment changes following mild traumatic brain injury: A 6-month longitudinal study. *Journal of Head Trauma Rehabilitation* 20, 239–256.

Kraus, M.F., Susmaras, T., Caughlin, B.P., Walker, C.J., Sweeney, J.A., and Little, D.M. (2007). White matter integrity and cognition in chronic traumatic brain injury: A diffusion tensor imaging study. *Brain* 130, 2508–2519.

Kurca, E., Sivak, S., and Kucera, P. (2006). Impaired cognitive functions in mild traumatic brain injury patients with normal and pathological magnetic resonance imaging. *Neuroradiology* 48, 661–669.

Laatsch, L.K., Thulborn, K.R., Krisky, C.M., Shobat, D.M., and Sweeney, J.A. (2004). Investigating the neurobiological basis of cognitive rehabilitation therapy with fMRI. *Brain Injury* 18, 957–974.

Levine, B., Black, S.E., Cheung, G., Campbell, A., O'Toole, C., and Schwartz, M.L. (2005). *Cognitive and Behavioral Neurology* 18, 45–54.

Levine, B., Fujiwara, E., O'Connor, C., Richard, N., Kovacevic, N., Mandic, M., Restagno, A., Easdon, R., Robertson, I.H., Graham, S.J., Cheung, G., Gao, F., Schwartz, M.L., and Black, S.E. (2006). In vivo characterization of traumatic brain injury neuropathology with structural and functional neuroimaging. *Journal of Neurotrauma* 23, 1396–1411.

Lewine, J.D., Davis, J.T., Bigler, E.D., Thoma, R., Hill, D., Funke, M., Sloan, J.H., Hall, S., and Orrison, W.W. (2007). Objective documentation of traumatic brain injury subsequent to mild head trauma: multimodal brain imaging with MEG, SPECT, and MRI. *Journal of Head Trauma Rehabilitation* 22, 141–155.

Lizerbram, E., and Moffit, B. (2001). Neuroimaging in acute brain injury. *Topics in Emergency Medicine* 23, 47–59.

Lovell, M.R., Collins, M.W., Iverson, G.L., Field, M., Maroon, J.C., Cantu, R., Podell, K., Powell, J.W., Belza, M., and Fu, F.H. (2003). Recovery from mild concussion in high school athletes. *Journal of Neurosurgery* 98, 295–301.

Lovell, M.R., Pardini, J.E., Welling, J., Collings, M.W., Bakal, J., Lazar, N., Roush, R., Eddy, W.F., and Becker, J.T. (2007). Functional brain abnormalities are related to clinical recovery and time to return-to-play in athletes. *Neurosurgery* 61, 352–360.

Macciocchi, S.N., Barth, J.T., Alves, W., Rimel, R.W., and Jane, J.A. (1996). Neuropsychological functioning and recovery after mild head injury in collegiate athletes. *Neuroscience Online* 39, 510–514. Retrieved from <http://ovidsp.tx.ovid.com/spb/ovidweb.cgi>.

Malojcic, B., Mubrin, Z., Coric, B., Susnic, M., and Spilich, G.J. (2008). Consequences of mild traumatic brain injury on information processing assessed with attention and short-term memory tasks. *Journal of Neurotrauma* 25, 30–37.

McAllistar, T.W., Sparling, M.B., Flashman, L.A., Guerin, S.J., Mamourain, A.C., and Saykin, A.J. (2001). Differential working memory load effects after mild traumatic brain injury. *Neuroimage* 14, 1004–1012.

McIntire, A., Langan, J., Halterman, C., Drew, A., Osternig, L., Chou, L., and van Donkelaar, P. (2006). The influence of mild traumatic brain injury on the temporal distribution of attention. *Experimental Brain Research* 174, 361–366.

Mendez, C.V., Hurley, R.A., Lassonde, M., and Taber, K.H. (2005). Mild traumatic brain injury: Neuroimaging of sports-related concussion. *Journal of Neuropsychiatry Clinical Neuroscience* 17, 297–303.

Moore, D.F., Radovizky, R.A., Shupenko, L., Klinoff, A., Jaffee, M.S., and Rosen, J.M. (2008). Blast physics and central nervous system injury. *Future Neurology* 3, 243–250.

Mori, T., Katayama, Y., and Kawamata, T. (2006). Acute hemispheric swelling associated with thin subdural hematomas: pathophysiology of repetitive head injury in sports. *Acta Neurochirurgica Supplement* 96, 40–43.

Nagy, Z., Westerberg, H., and Klingberg, T. (2004). Maturation of white matter is associated with the development of cognitive functions during childhood. *Journal of Cognitive Neuroscience* 16, 1227–1233.

Niogi, S.N., Mukherjee, P., Ghajar, J., Johnson, C., Kolster, R.A., Sarkar, R., Lee, H., Meeker, M., Zimmerman, R.D., Manley, G.T., and McCandliss, B.D. (2008). Extent of microstructural white matter injury in postconcussive syndrome correlates with impaired cognitive reaction time: A 3T diffusion tensor imaging study of mild traumatic brain injury. *American Journal of Neuroradiology* 29, 967–973.

Nolin, P. (2006). Executive memory dysfunctions following mild traumatic brain injury. *Journal of Head Trauma Rehabilitation* 21, 68–75.

Nolin, P., and Heroux, L. (2006). Relations among sociodemographic, neurologic, clinical, and neuropsychiatric variables, and vocational status following mild traumatic brain injury. *Journal of Head Trauma Rehabilitation* 21, 514–526.

Office of the Surgeon General and Office of Surgeon, Multi-National Force–Iraq. (2006). Mental health advisory team (MHAT) IV report, Operation *Iraqi Freedom* 05–07.

Ombaugh, T.N., Rees, L., Stormer, P., Harrison, A.G., and Smith, A. (2007). The effect of mild and severe traumatic brain injury on speed of information processing as measured by the computerized tests of information processing (CTIP). *Archives of Clinical Neuropsychology* 22, 24–36.

Pecchinenda, A., Dretsch, M., and Chapman, P. (2006). The effect of working memory load on emotion-based processes involved in choosing advantageously. *Experimental Psychology* 53, 191–197.

Ruff, R. (2005). Two decades of advances in understanding of mild traumatic brain injury. *Journal of Head Trauma Rehabilitation* 20, 5–18.

Ryan, L.M., and Warden, D.L. (2003). Post concussion syndrome. *International Review of Psychiatry* 15, 310–316.

Saboori, M., Ahmadi, J., and Farajzadegan, Z. (2007). Indication for brain CT scan in patients with minor head injury. *Clinical Neurology and Neurosurgery* 109, 399–405.

Schlosser, R.G.M., Nenadic, I., Wagner, G., Gullmar, D., Von Consbruch, K., Kohler, S., Christoph, S.C., Koch, K., Fitzek, C., Matthews, P.M., Reichenbach, J.R., and Sauer, H. (2007). White matter abnormalities and brain activation in schizophrenia: A combined DTI and fMRI study. *Schizophrenia Research* 89, 1–11.

Schlund, M.W., and Pace, G. (2000). The effects of traumatic brain injury on reporting and responding to causal relations: an investigation of sensitivity to reinforcement contingencies. *Brain Injury* 14, 573–583.

Schneiderman, A.I., Braver, E.R., and Kang, H.K. (2008). Understanding sequelae of injury mechanisms and mild traumatic brain injury incurred during the conflicts in Iraq and Afghanistan: Persistent postconcussive symptoms and posttraumatic stress disorder. *American Journal of Epidemiology* 15, 1446–1452.

Sivák, S., Kurca, E., Hladká, M., Zelenák, K., Turcanová-Koprusáková, K., and Michalik, J. (2008). Early and delayed auditory oddball ERPs and brain MRI in patients with mTBI. *Brain Injury* 22, 193–197.

Sosnoff, J.J., Broglio, S.P., and Ferrara, M.S. (2008). Cognitive and motor functions are associated following mild traumatic brain injury. *Experimental Brain Research* 187, 563–571.

Stulemeijer, M., van der Werf, S, Bleijenberg, G., Biert, J., Brauer, J., and Vos, P.E. (2006). Recovery from mild traumatic brain injury: A focus on fatigue. *Journal of Neurology* 253, 1041–1047.

Taber, K.H., Warden, D.L., and Hurley, R.A. (2006). Blast-related traumatic brain injury: What is known? *Journal of Neuropsychiatry and Clinical Neuroscience* 18, 141–145.

Teasdale, G. and Jennett, B. (1974). Assessment of coma and impaired consciousness: A practical scale. *Lancet* 2, 81–84.

Tombaugh, T.N., Rees, L., Stormer, P., Harrison, A.G., and Smith, A. (2007). The effects of mild and severe traumatic brain injury on speed of information processes as measured by the computerized tests of information processing (CTIP). *Archives of Clinical Neuropsychology* 22, 25–36.

Topal, N., Hakyemez, B., Erdogan, C., Bulut, M., Koksal, O., Akkose, S., Dogan, S., Parlak, M., Ozguc, H., and Korfali, E. (2008). MR imaging in the detection of diffuse axonal injury with mild traumatic brain injury. *Neurological Research* 30, 274–278.

Tucker, G. (1999). Functional MRI reveals dysfunction in the working brain. *Journal Watch Psychiatry*, 1201.

Umile, E., Sandel, E., Alavi, A., Terry, C., and Plotkin, R. (2002). Dynamic imaging in mild traumatic brain injury: Support for the theory of medial temporal vulnerabilities. *Archives of Physical Medicine and Rehabilitation* 83, 1506–1513.

Vagnozzi, R., Signoretti, S., Tavazzi, B., Floris, R., Ludovici, A., Marziali, S., Tarascio, G., Amorini, A.M., Di Pietro, V., Delfini, R., and Lazzarino, G. (2008). Temporal window of metabolic brain vulnerability to concussion: A pilot 1H-magnetic resonance spectroscopic study in concussed athletes—part III. *Neurosurgery* 62, 1286–1295.

Vanderploeg, R.D., Curtiss, G., and Belanger, H.G. (2005). Long-term neuropsychological outcomes following mild traumatic brain injury. *Journal of International Neuropsychological Society* 11, 228–236.

Warden, D. (2006). Military TBI during the Iraq and Afghanistan wars. *Journal of Head Trauma Rehabilitation* 21, 398–402.

Wilde, E.A., McCauley, S.R., Hunter, J.V., Bigler, E.D., Chu, Z., Wang, Z.J., Hanten, G.R., Troyanskaya, M., Yallampalli, R., Li, X., Chia, J., and Levin, H.S. (2008). Diffusion tensor imaging of acute mild traumatic brain injury in adolescents. *Neurology* 70, 948–955.

Yeo, R.A., Phillips, J.P., Jung, R.E., Brown, A.J., Campbell, R.C., and Brooks, W.M. (2006). Magnetic resonance spectroscopy detects brain injury and predicts cognitive functioning in children with brain injuries. *Journal of Neurotrauma* 23, 1427–1435.

Chapter 12

Why Do Soldiers Kill Themselves? Understanding Suicide in the Military

Sandra M. Escolas, Paul T. Bartone, Marek Rewers, Joseph M. Rothberg, and Joel Carter

Despite much research on suicide, we still do not have a clear understanding of what causes people to kill themselves. Recent upward trends in suicide in the U.S. military have raised concern among senior leaders and led to additional education and prevention efforts. Unfortunately, these programs to date have shown scant evidence of success (Kuehn, 2009). A better understanding of the causes of suicide is needed in order to take more effective preventive steps.

At least as far back as World War II, suicide rates have been lower in the U.S. military than in age-matched civilian groups (Cassimatis and Rothberg, 1997; Rothberg, Bartone, Holloway, and Marlowe, 1990). Historical trends also show that in general, suicide rates in the military tend to decline during wartime (for example, Oesterlen, 1865; Rothberg, Holloway, and Ursano, 1987). One possible reason is that wartime may impart a greater sense of meaning in soldiers' lives, thereby reducing any suicidal tendencies. Somewhat surprisingly, then, military suicide rates have increased in the years since 2003 when the United States began operations in Iraq. For example, in 2008, the U.S. Army suicide rate climbed to 20.2 (per 100,000), up from 12.7 in 2005 (see figure 12–1). For the first time, the Army suicide rate is higher than the comparable civilian rate, which was 19.5 for 2005 (most recent civilian data available; Centers for Disease Control [CDC], 2009). This alarming increase has led to redoubled efforts by the U.S. military to reduce or prevent suicides. Unfortunately, they continue to rise, as evidenced by continuing high figures for January and February of 2009 (Sheftick, 2009). Clearly, a better understanding of suicide in the military is needed in order to address it effectively.

This chapter will give a somewhat selective review of available studies and reports on suicide in the military, with the aim of clarifying the range of factors that can influence suicide in the military context. The research findings are diverse and at times inconsistent, but taken

Figure 12–1. **Suicide Rate among Active-duty U.S. Army**

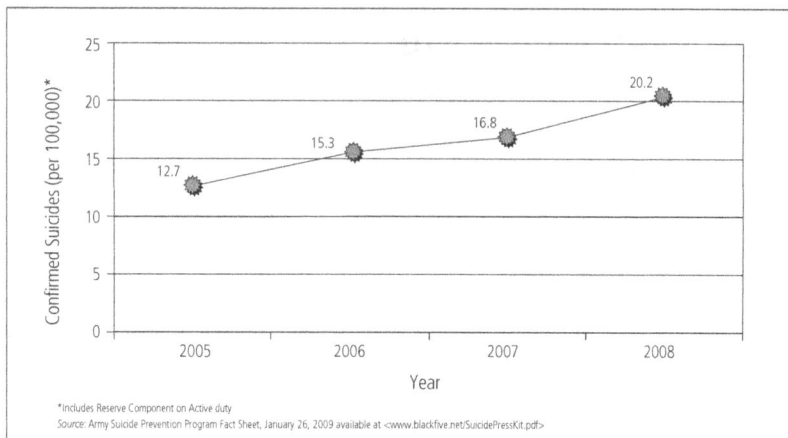

together, they can offer a better understanding of the factors that may lead to suicide. This understanding is a prerequisite to developing any truly effective prevention programs. The final section of this chapter will review some of the prevention programs that have been enacted by the military to date.

Definition of Suicide

Suicide is generally defined as the act of intentional, self-inflicted death. While this definition seems straightforward, in practice suicides are often difficult to determine with certainty. If the victim leaves no note or other clear evidence of intent, a suicidal death can easily be classified as an accident. For example, death by drug overdose, poisoning, motor vehicle accident, falling from a high place, or gunshot can be truly accidental or else suicide. In the U.S. military, all questionable deaths are carefully investigated to determine the cause. However, since many cases are ambiguous, investigations can last a year or more and do not always result in a clear determination. Most suicide rates are calculated using only clearly determined suicide cases, discarding ambiguous ones. Thus, the true prevalence of suicide in any group (including the military) is most likely underestimated.

This problem is underscored by a study of U.S. military casualties across all Services (Carr, Hoge, Gardner, and Potter, 2004). The authors examined deaths of Active-duty (including very recent retirees) Servicemembers in the Army, Navy, Marine Corps, and Air Force for 1998–

1999. All deaths (N = 124) that were classified in the Department of Defense Medical Mortality Registry as accidents (excluding motor vehicle deaths) or of undetermined cause were further examined as possible suicides. Suspicion of suicide was judged using Brent, Perper, and Allman's (1987) criteria, which include psychiatric history and evidence of intent to commit suicide communicated verbally or in writing. Based on these criteria, the authors found six "accidental" deaths that were likely suicides within their sample. The authors also identified four suspicious gun-shot-related deaths that were classified as accidents. Extrapolating from these numbers, correcting such misclassifications would increase the U.S. military suicide rate in 1998–1999 from 10.4 to 12.4 per 100,000. This means that suicides in the military could have been underreported by as much as 21 percent in that year. Similar studies in the civilian population have identified comparable rates of underreporting (Carr et al., 2004).

The problem of misclassification is compounded by the difficulties of calculating valid population-based suicide rates. Since suicide is a relatively low frequency event, even in large populations a small increase or decrease in incidence can have a dramatic effect on rates. To investigate this problem further, Eaton, Messer, Wilson, and Hoge (2006) reviewed suicides from 1990 to 2000 among all Active-duty Servicemembers aged 17 to 64 as compared to appropriate civilian population groups. Data were obtained from the Defense Medical Surveillance System, Centers for Disease Control (CDC), National Center for Health Statistics, and Vital Health Statistics System. After adjusting for demographic differences, the authors found that rates of suicide among the military were 25 to 33 percent lower than in the civilian population. However, with unaccounted variables and suicide rate fluctuations of 20 to 40 percent per year, it is extremely difficult to link suicide rate differences to membership in the military, or for that matter to any particular prevention program. One weakness of the study is that while it recognizes the problems inherent in comparing demographically disparate populations, the age range among the civilian population also includes 15- and 16-year-olds due to the Census Bureau's method of age stratification. Similar demographic challenges arise when adjusting for race and gender.

Theories of Suicide

Perhaps a less confounding introduction to the topic is a theoretical perspective on suicide. Durkheim (1951; originally published 1897) was one of the first to examine suicide from a population standpoint, noting a number of interesting trends and differences in suicide rates across various national and religious groups, as well as across seasons. In fact, Durkheim is credited with providing the first figures on suicide in the U.S. military (68 per 100,000 in the U.S. Army after the Civil War, from 1870 to 1884; Durkheim, 1951, as cited by Cassimatis and Rothberg, 1997). Durkheim describes four types of suicide: egotistical, altruistic, anomic, and fatalistic. Perhaps the most common is *egotistical* suicide, which occurs when people feel disconnected or separated from relevant social groups. This leads to a feeling of lonely desperation. Durkheim's conception of *altruistic* suicide is in some ways opposite to egotistical suicide. It is the result of overly deep integration of the individual with the group, to the point that the individual loses a sense of separate individuality and is willing to "die for the cause" of the group. Most suicide bombers would fall into this category. *Anomic* suicide occurs when the normal regulatory functions of society are loosened or absent, as, for example, in times of economic turbulence or any dramatic change. In theory, people feel lost or without guideposts under such conditions, leading to anomie and increased suicide. *Fatalistic* suicide, on the other hand, increases when individuals feel overregulated or controlled by society (for example, slaves or serfs). In this case, suicide may appear as a reasonable escape route.

Early studies by Menninger (1938) applied a Freudian psychoanalytic perspective, seeing suicide as the expression of basic self-destructive tendencies, or aggression turned inward. Another important theoretical perspective is that of Aaron Beck, for whom hopelessness is the common denominator for understanding suicide (Beck, Kovacs, and Weissman, 1979). In several studies, Beck and colleagues have shown that hopelessness is highly predictive of suicidal behavior (Beck, Brown, Berchick, Stewart, and Steer, 1990). Schneidman (1996) described suicide as mainly the result of frustrated or thwarted needs that result in intense psychic pain, or "psychache." This perspective is based largely on the personality theory of Henry Murray (1938), who emphasized the importance of satisfying a range of human needs, from primary biological ones to secondary psychological ones, including the need for affiliation, need for achievement and sense of competence, and the need for autonomy.

More recently, Joiner (2005) draws on Schneidman and Murray to provide his own theoretical explanation for suicide. Joiner posits three factors that must be present for suicide (or a serious attempt at suicide) to occur: the acquired capacity for lethal self-injury, which comes with practice and involves desensitization to pain and fear of death; thwarted effectiveness, which also increases the sense of being a burden on others; and thwarted belongingness (disconnection from social groups). Thwarted effectiveness refers to the growing sense that one is lacking competence, is ineffective at getting things done, and is a failure. Thwarted belongingness involves the perception of being isolated, not a part of the social world. Perhaps significantly for understanding military suicide, Joiner also suggests that the acquired capacity for lethal self-injury can come about indirectly, through repeated exposure to violence and death in the surrounding environment (Joiner, 2005). Joiner argues that these three factors align to form a "dangerous convergence zone," contributing to a greater risk for serious suicide behavior.

International Studies on Suicide

In this section, we review relevant studies on military suicide outside the United States. While not directly related to the topic of U.S. military suicide, these studies do examine the relationship between suicide and the trauma experienced by military populations exposed to wartime stressors. This relationship is clearly present among U.S. personnel deployed in Iraq and Afghanistan, and more broadly among all military personnel. After all, war and conflict represent the primary conditions for which soldiers continually train. Considered together, the works enumerated below may yield some insights into how people in general, and soldiers specifically, may slip into stress-related suicidal behaviors.

The available research has yielded inconsistent findings as to how suicide rates are affected by war. A study in Croatia before, during, and after the Croatian War of Independence found that the rate rose 20.9 percent during the wartime period itself over the prewar period (Bosnar, Stemberga, Coklo, Koncar, Definis-Gojanovic, Sendula-Jengic, and Katic, 2005). Conversely, the end of the war saw a drop of 26.2 percent. These findings found elevated wartime suicide rates as compared to peacetime rates. This trend was even more pronounced among the population under the age of 40, which saw a 45.3 percent rise in suicide rates from prewar to war, followed by a 56.6 percent drop in the 5 years after the war. However, because this study did not distinguish between military and civilian populations, the results are not readily generalizable to military forces. Also, the study does not explicitly use the standard method of reporting suicide rates as expressed per 100,000, making it difficult to interpret the results.

Despite these limitations, some valuable information can be taken from the study by Bosnar and colleagues (2005). They suggest that the availability of alcohol and firearms increases the risk of suicide during wartime, and that males are at greater risk than females. Results showed increases in completed suicides using firearms, and in the prevalence of intoxication among those committing suicide. Analysis also showed that the proportion of male to female suicides was significantly higher throughout all three time periods, which is consistent with other studies.

In a similar study of a wartime population, Henderson, Stark, Humphry, and Selvaraj (2006) reviewed suicide rates in Scotland before, during, and after World War II. Contrary to the Bosnar et al. (2005) study, these authors found that suicide rates for both females and males decreased during World War II. However, they observe that this overall trend conceals a sizable increase in suicides among certain subgroups. For example, the rate among young men (15–24 years old) rose to a peak of 14.8 per 100,000 during 1942, and declined to 3.9 by 1945 as the war wound down. Similarly, the rate among men 25–34 years old reached 19.9 per 100,000 in 1943 before falling to 6.6 by 1946. Throughout the same period there was little change in the age-specific suicide rates among females. This suggests that young men suffer an increased risk of suicide relative to women during times of war.

Henderson et al. (2006) also found an increase in the use of self-inflicted gunshot wounds. The authors attribute this in part to the increased availability of firearms, and the potential of becoming a combatant likely contributed to this rise. As in the Croatian study, occupation was not specifically considered, although the authors note that some of the observed suicides were likely members of the armed forces. Their conclusions are consistent with findings from the Bosnar et al. (2005) study in that males are more likely to commit suicide, and the ready availability of firearms is an added risk factor. This study also refines the understanding of changes in suicide rates during wartime by examining the rates among young males separately from the general population rates. This demographic group (which is roughly analogous to the population of the U.S. military) saw significant increases in suicide during wartime (World War II). While these studies of population suicide rates in Croatia and Scotland do not look specifically at military groups, they do suggest that the stress of war is associated with a rise in suicide rates generally, and especially among young males.

Perhaps a more relevant study from a military perspective is that of Mahon, Tobin, Cusack, Kelleher, and Malone (2005, 1688), who conducted a retrospective review of all deaths (N = 732) among regular-duty Irish Defence Forces between 1970 and 2002. The authors found an average suicide rate of 15.3 per 100,000, with most suicides carried out while soldiers were alone shortly after starting morning duty. Additionally, 53 percent were committed using a firearm. The authors conclude that the military occupation has an influence on incidence and manner of suicide, and that easy access to lethal weapons in the workplace is an inherent risk factor. In order to prevent suicides, the authors suggest that "moderating the opportunity for suicide is critically important," and that occupations that entail high exposure to lethal weapons should take greater measures for harm reduction centered on reducing the opportunity for self-injury (Mahon et al., 2005). This study once again confirms the risk associated with ready access to weapons. Recommendations for prevention are consonant with the "battle buddy" approach to suicide prevention that the U.S. military has recommended —that is, that small unit leaders and soldiers within the unit should be alert for any signs of suicide in their comrades, and move quickly to refer them for medical/professional support. While such steps are important and may indeed prevent some suicides, it must be recognized that they are in no way addressing the root causes of suicide in the military.

One study that sought to investigate the effectiveness of such preventive measures as well as determine the underlying causes of military suicide was conducted by the Russian Board of the Chief Military Prosecutor's Office. As reported by Ryvkin (2000), this study examined causes of death among Russian servicemen during peacetime and found that suicide rates had not diminished despite preventive measures. The authors reported that almost 75 percent of suicides were committed by enlisted soldiers or sergeants, 10 percent by warrant officers, and the remaining 15 percent by officers. While this suggests that enlisted soldiers are at greater risk, these figures may be skewed by the disproportionate distribution of various ranks—a fact that is not addressed in the study. Young males (age 20–24) made up the largest demographic group among those committing suicide, with 98.3 percent of suicides within this age group being enlisted men in the first year of service— although, again, proportionally adjusted rates were not provided. Similar to the studies already discussed, Ryvkin notes that enlisted servicemen generally commit suicide with firearms or by hanging, presumably because these are the most readily available methods. However, the authors also express concern about the growing use of explosives, which can have dangerous consequences for other military personnel or civilian bystanders. Like firearms, explosives (such as hand grenades) are readily available to military personnel and are highly lethal, which could make them an appealing means of suicide. High incidence of alcohol and narcotics use was also noted as contributing to suicide among Russian servicemen.

Ryvkin claims that conditions of Russian military life by themselves may increase the risk for suicide, citing "difficulties in adapting to army conditions, bullying and harassment, an unhealthy moral climate in military collectives, excessive workloads connected with under manning of military units and detachments, and failure to meet prescribed food-ration standards" (Ryvkin, 2000, 43). This conclusion differs from many other studies, which tend to focus on external factors such as relationship troubles or financial problems as primary causes of suicide in the military. Ryvkin also notes the important role played by psychological disorders that go undetected during the recruit screening process or that develop during military service. He states that the lack of a sense of meaning and the absence of effective coping mechanisms significantly increase a soldier's likelihood of committing suicide.

Bodner, Ben-Artzi, and Kaplan (2006) investigated suicide among 18- to 21-year-old Israeli soldiers in combat arms units, as compared to support units. The study examined 429 soldier suicides (388 males, 41 females) from 1979 to 1999, comparing them to a control group of 499 soldiers (450 males, 49 females) who did not commit suicide. Military service in the Israel Defense Force (IDF) is mandatory for all healthy men and women, and a large portion of the force is assigned to support units (noncombatants). Bodner et al. (2006) compared combat arms and noncombat arms units, using a distinction established in a previous study by Fishman, Morris-Dycian, and Kotler (1990), who said conditions in combat arms units are "physically tough and very demanding; their overall service conditions are harder and more often involve life-threatening situations than those of noncombatants" (226).

During the entry phase for Israeli military service, all recruits thought to have psychological or behavioral problems are referred for a mental health assessment interview (MHAI). In the nonsuicide control group, there were no significant differences between combat arms unit and noncombat arms unit soldiers sent for MHAIs. In contrast, among the suicide group, noncombat unit soldiers had been sent for mental health assessments significantly more often than combatant suicides. The authors suggest that noncombatant suicide may be related to preexisting dispositional factors. However, scores on the MHAI did not differentiate between soldiers who committed suicide and those who did not. Bodner and colleagues (2006) also found that combat unit soldiers who committed suicide showed a better behavioral adjustment to military service and expressed more motivation to serve than did noncombat unit soldiers committing suicide. High motivation, autonomy, and independence were seen as potential risk factors for combat unit member suicides because they may reflect perfectionist tendencies. The study also found that combat unit soldiers who committed suicide were referred for mental health services less frequently than noncombat unit soldier suicides. The combat unit suicides also had a stronger sense of duty, higher autonomy scores, and fewer unit changes. This suggests that combat unit suicides may be related to situational factors and firearms availability. The authors recommend further research into the personality and situational influences on soldier suicide, and more attention to identifying preexisting vulnerability characteristics in order to enable more effective prevention strategies.

Bodner et al. (2006) use combat-related service as a proxy for increased operational stress; however, they do not examine separately those troops who actually spent time in a war zone. Other studies have used peacekeeping operations as a proxy for increased operational stress, presumably because peacekeeping troops face many of the same difficult conditions and challenges as combat-deployed troops. A study conducted by Hansen-Schwartz, Jessen, Andersen, and Jørgensen (2002) looked at the frequency of suicide among Danish soldiers who participated in United Nations–mandated forces throughout the 1990s. Out of 3,859 soldiers deployed between 1995 and 1997, 4 male soldiers between the ages of 21 and 24 committed suicide. Of these four suicides, two were committed less than 1 month before deployment, and the other two soldiers committed suicide within 6 to 9 months of being repatriated. The study concluded that while suicide is not an overwhelming problem quantitatively, there are certain experiences "during a mission that may contribute to destabilizing the individual soldier psychologically and lead to suicidal ideation or even suicide" (Hansen-Schwartz et al., 2002, 57). The authors note that the soldiers studied had suffered severe artillery bombardment, and the resultant stress may have contributed to the suicides. The relative infrequency of suicide, however, calls into question whether deployment-related stress truly brings about higher incidence of suicide among soldiers than the general population.

This question is explored further in a study by Michel, Lundin, and Larsson (2007), which compared suicide rates among 39,768 former Swedish peacekeeping personnel with the general population in Sweden between 1960 and 1999. Data gathered from the Swedish National General Population Registry and the Cause-of-Death Registry demonstrated that rates were lower in peacekeeping personnel than among the general population. This is consistent with studies of U.S. military personnel, which (until very recently) exhibited lower rates of suicide than the general population. The authors suggest that the rigorous selection and screening process for peacekeepers may account for the lower suicide rate among former peacekeeping personnel. They also recommend that further studies be conducted to investigate the differences in suicide rates and selection criteria between countries.

Similar conclusions were reached in a Canadian study comparing 66 military suicides during the period 1990 to 1995 to 2,601 randomly selected military controls, and also to 66 matched controls containing complete personnel and medical data (Wong, Escobar, Lesage, Loyer, Vanier, and Sakinofsky, 2001). The study assessed whether military personnel who had served as peacekeepers demonstrated elevated rates of suicide. Of the 66 suicides examined, 95 percent were male, 91 percent were enlisted, and 46 percent were single persons. The mean age for those committing suicide was 30.4 years. While the authors did not find that peacekeeping increased overall suicide risk, they did conclude that "military lifestyles may strain interpersonal relationships, encourage alcohol abuse, and contribute to psychiatric illness and suicide in a minority of vulnerable individuals irrespective of peacekeeping assignment" (Wong et al., 2001, 103). They suggest that suicide risk may be reduced through a careful selection process and preparatory military training that encourages intra-group bonding and mutual support. In dealing with suicide prevention, the authors suggest there is an overarching problem of a "macho" soldier culture, which creates an environment that discourages soldiers from admitting to any emotional issues (Wong et al., 2001, 111).

In a large cohort study of suicide among 22,275 Norwegian former peacekeepers, Thoresen, Mehlum, and Moller (2003) found a moderate but statistically significant elevation in the standardized mortality ratio (SMR) for suicide among peacekeepers compared to the general population. However, when the figures were adjusted for marital status, the SMR became statistically nonsignificant. For this reason, the authors believed that lower rates of marriage among military peacekeepers may be an indication "that the personnel were characterized by certain vulnerability factors before entering peacekeeping service, resulting in a reduced ability to enter into and remain in stable love relations" and a greater susceptibility to suicide (Thoresen et al., 2003, 605). The authors also noted a significant increase in suicide by firearms and carbon monoxide poisoning among military peacekeepers, presumably because these means are readily available.

Thoresen conducted several other studies investigating suicide among military peacekeepers. Thoresen and Mehlum (2006) compared 43 suicide cases with a control group of 41 fatal accident cases among Norwegian peacekeepers from 1978 to 1995. This "psychological autopsy" study sought to discover risk factors for suicide among veterans of peacekeeping operations. Findings showed that some preexisting mental health problem was the major risk factor for suicide. Additionally, both living alone and the break-up of a love relationship contributed to suicide risk, even when controlling for known mental health problems. No specific peacekeeping-related factor was significantly correlated with suicide. A later study, also examining psychological autopsy results, demonstrated that involuntary repatriation (early return to home country) from peacekeeping service, negative life events prior to the mission, and marital status were all factors with unique impact on suicide risk for peacekeepers. Consistent with other studies, single soldiers were at greater risk (Thoresen, Mehlum, Røysamb, and Tønnessen, 2006).

Another study by Thoresen and Mehlum (2008) investigated the connection between war zone stress exposure and suicidal ideation in a 7-year followup of 1,172 Norwegian peacekeepers. The link between military stress exposure and suicidal ideation was stronger for those soldiers who also reported post-traumatic stress symptoms and general mental health problems. This suggests that post-traumatic stress disorder (PTSD) and other mental health problems may indicate greater risk for stress-related suicide. Thoresen and Mehlum (2004) also compared suicides with other types of death among peacekeepers. The authors observed that "alcohol-related fatal accidents were found to share many common features with the suicide group, such as depression, alcohol and substance abuse, and various social problems, and were also found to differ significantly from the other fatal accidents" (Thoresen and Mehlum, 2004, 988). These findings may indicate the misclassification of suicides as alcohol-related accidents, or at the least, a common disposition toward self-destruction between alcohol-related fatal accidents and suicides (Thoresen and Mehlum, 2004, 991).

The usefulness of these studies for understanding the impact of wartime stress exposure on outcomes such as suicide may be limited, however, because troops in peacekeeping operations are not generally exposed to the same acute stressors of combat. Several studies, such as Sareen, Cox, Afifi, Stein, Belik, Meadows, and Asumundson (2007), have attempted to understand the distinction between the two types of deployment as it pertains to troop mental health. Sareen et al. (2007) examined the perceived need for mental health care, usage of mental health services, and prevalence of suicidality among 8,441 active-duty Canadian military forces, comparing those deployed to combat zones with those on peacekeeping missions. They found that experiencing traumatic events (such as witnessing atrocities) during deployments on either combat or peacekeeping missions was related to increased mental health problems. However, in the absence of exposure to traumatic events, peacekeeping operations could be protective (related to decreased mental health problems). The study also demonstrated that perceived need for and use of mental health services and presence of common mental health diagnoses were not related to increased mental health problems. Few soldiers with a mental health diagnosis were receiving services even if they perceived a need for it. This could be a function of lack of confidence in the available services and the generalized stigma associated with admitting to any mental health problem.

A somewhat contradictory study by Belik, Stein, Gordon, and Sareen (2009) reported that "deployment-related traumatic experiences do not appear to confer additional likelihood of lifetime suicide attempt above and beyond the presence of development of a psychiatric disorder" (Belik et al., 2009, 93). Using data from the Canadian Community Health Survey, the authors derived a cross-sectional survey (N = 8,441) and concluded that "sexual and other interpersonal traumatic events are associated with suicide attempts in a representative sample of Canadian military men and women" (Belik et al., 2009, 93). This study thus points to the potential importance of previous history of exposure to traumatic events as a risk factor.

In 2004, the Centre for Military and Veterans' Health in Australia reviewed postdeployment health consequences, combining perspectives from American, British, Canadian, and Australian sources (Kitchener, 2005). The panel discussed the issues of poor understanding of health and well-being consequences, the positive and negative outcomes arising from deployment, and definitional distinctions between the many types of deployments associated with the topic of health and military deployments research. Their primary recommendation was to establish systems to better monitor Australian Defence Force personnel, starting with recruitment and continuing through service and following discharge. The panel also recommended that the health of servicemembers be closely monitored for each operational deployment, with systematic reviews of all personnel involved, and that systematic "hazard profiles" be created for each operational deployment to identify the primary occupational, environmental, operational, and psycho-social threats. While suicide is not addressed directly by this group, their recommendations for better surveillance would in principle lead to better documentation of factors associated with military suicide.

Specifically comparing mortality rates of military veterans to civilians, Crane, Barnard, Horsley, and Adena (1997) analyzed data from the Australian Department of Defence, Department of Veterans' Affairs, National Death Index, and Electoral Commission rolls. The authors compared Vietnam male veterans in Australia with the general male population. Even though the overall veteran death rate for all causes relative to the general population was lower (0.68), some specific categories of death were higher for veterans, including suicide (1.21). However, the authors indicate that this elevated risk for suicide in veterans was not statistically significant.

While the methods and findings of these international studies are quite diverse, they nevertheless provide useful insights. For example, a consistent finding across most studies is that young males make up the highest risk group for suicide during wartime. This parallels peacetime trends and is all the more relevant for the military since young males comprise its largest demographic group. However, the data are unclear as to whether this increased relative risk for young males in the military is a function of differential exposure to war-induced stress or to other factors. There is general consensus that increased availability of firearms and alcohol in the military environment elevates suicide risk. Thus, the inherent nature of serving in the armed forces, which usually demands ready access to weapons, can increase the risk. On the other hand, military selection standards may exclude those applicants from the general population most likely to commit suicide. Also, the stricter limitations on conduct imposed on soldiers in relation to their civilian counterparts may reduce the likelihood of alcohol or drug abuse, thereby protecting them from suicide. Successful interpersonal and intimate relationships also seem to play a role in mitigating risk. Some of the international studies reviewed here also indicate that increased operational and war-related stressors do have adverse effects on the mental health of soldiers, most notably with regard to the development of PTSD. However, the relationship between these mental health factors and suicide remains unclear. A number of studies recommend preventive measures such as more stringent selection criteria, greater supervision of at-risk soldiers, and limited access to weapons.

Suicide among Vietnam War Troops and Veterans

More systematic studies of suicide in the U.S. military appeared after the Vietnam War. The apparently high incidence of mental illness among those who served in Vietnam led to more in-depth studies, and subsequently to greater attention to suicide as a cause of death. For example, using data from the Southeast Asia Combat Area Casualties Database, Adams, Barton, Mitchell, Moore, and Einagel (1998) examined mortality in male U.S. ground troops in Vietnam from 1957 to 1973. Suicide deaths were compared to all other deaths in regard to military branch, age, race, marital status, religious preference, Service component, and year of death (Adams et al., 1998). No suicides were reported in the Navy, Coast Guard, or National Guard, so these branches were excluded. In contrast to other studies showing that young men were at the greatest risk, this study found that individuals who committed suicide in Vietnam were significantly older than those dying of other causes.

According to Adams and colleagues (1998), those who committed suicide had been in Vietnam approximately 2 months longer than those who died from other causes, suggesting a relationship between length of deployment and incidence of suicide. They also found that suicides peaked in early 1968 (coinciding with the Tet Offensive), while deaths from all other causes peaked in late 1967. The authors suggest that these two observations may indicate that combat-induced stress and depression were contributing factors to increased risk. Single men died more often from all causes than married men, which supports the conclusion of other studies that marriage may be a protective factor. Adams and colleagues (1998) discussed the difficulties of working with inconsistent mortality data and recommended further research to clarify risk factors such as history of suicidal behavior or depression and genetic factors. Interviews with surviving comrades, former girlfriends, and family members, and examination of documentary materials (letters, diaries, etc.) were also recommended.

As indicated previously, war-related mental health issues are a serious problem not only for Active-duty troops, but also among veterans. In 1987, the CDC published a study examining mortality figures for Vietnam veterans after their discharge from service (Boyle, Decoufle, Delaney, DeStefano, Flock, Hunter, Joesoef et al., 1987). Findings showed a 17 percent higher mortality rate for Vietnam veterans (N = 9,324) compared to veterans who served in Korea, Germany, or the United States (N = 8,989). The most dramatic difference between the two groups was during the first 5 years after discharge, when the mortality rate among Vietnam veterans was 1.45 times that of the non-Vietnam veterans. Death by motor vehicle accidents was a primary factor in the increased death rates. The study notes a 72 percent rise in the suicide rate over the first 5 years after deployment, followed by a drop in the rate. After the first 5 years, suicide rate ratios were not significantly different from 1.0, indicating that Vietnam veterans were no more likely than nonveterans to commit suicide.

Suicide in Soldiers Deployed during Operation
Desert Shield / Desert Storm

A number of studies examined soldier mental health and mortality during and after the first Gulf War. Writer, DeFraites, and Brundage (1996) sought to determine cause-specific mortality rates among U.S. troops stationed in the Persian Gulf during Operations *Desert Shield* and *Desert Storm* compared with those of U.S. troops serving elsewhere. The study looked at all Active-duty deaths between August 1990 and July 1991. Age-adjusted mortality rates among deployed soldiers were compared with rates projected from mortality rates among troops on Active duty elsewhere. Of the 1,769 Active-duty deaths, 372 occurred in the Persian Gulf region, while 1,397 happened elsewhere. Of the 372 deaths in the Persian Gulf region, 147 (39.5 percent) occurred as a direct result of combat during the war, 194 (52.2 percent) resulted from injuries not incurred in battle, and 30 (8 percent) resulted from illness. The

Table 12–1. **Comparison of Self-inflicted Deaths among Deployed and Nondeployed Soldiers**

Cause of Death	Deployed Number	Deployed Rate	Nondeployed Number	Nondeployed Rate	Total Number	Total Rate
Self-inflicted	10	3.78	206	10.82	216	9.96
Gunshot	10	3.78	120	6.30	130	5.99
Hanging/ asphyxia- tion	0	0.00	54	2.84	54	2.49
Other	0	0.00	32	1.68	32	1.48

Number and rate (expressed as per 100,000 person-years) of self-inflicted deaths among U.S. military personnel deployed to the Persian Gulf region and nondeployed forces, August 1, 1990, through July 31, 1991.

Source: Writer et al. (1996). Comparative Mortality among U.S. Military Personnel in the Persian Gulf Region and Worldwide during Operations Desert Shield and Desert Storm. *Journal of the American Medical Association* 275 (2), 118–121.

study found that there were 10 suicides among personnel stationed in the Persian Gulf region and 206 suicides outside of theater during this same 1-year period (see table 12–1).

Of the 10 suicides among those deployed in the Gulf region, all were by self-inflicted gunshot wounds. Meanwhile, of the 206 nondeployed suicide deaths, 120 were caused by gunshot, 54 by hanging or asphyxiation, and 32 by other means. The study found that deployed soldier suicide was lower than expected based on suicides elsewhere. Except for deaths from unintentional injury, deployed soldiers did not experience significantly higher mortality rates than soldiers serving elsewhere, nor were there any clusters of unexplained deaths (see table 12–2). The number and circumstances of nonbattle deaths among deployed soldiers were typical for the U.S. military population. This seems to indicate that deployment to a war zone does not necessarily increase the risk for suicide, although there may have been factors unique to the first Gulf War—including, for example, that casualties were relatively low, and the war was over in a matter of weeks.

A retrospective study by Macfarlane, Thomas, and Cherry (2000) compared all United Kingdom personnel who served in the Gulf at some time between September 1990 and June 1991 (N = 53, 416) to those serving elsewhere during the same period (N = 53, 450). They found that "the higher mortality rate from 'external' causes in the Gulf cohort was principally due to higher mortality rates from accidents"

Table 12–2. Observed and Expected Non-battle Deaths among Soldiers Deployed to Persian Gulf

Cause of Death	Observed Deaths	Expected Deaths	Difference	Standardized Mortality Ratio (95 percent confidence interval)
All non–battle-related deaths	225	202.7	22.3	111.5 (96.5–125.5)
All injury deaths	194	164.7	29.3	117.8 (101.2–134.3)
Unintentional Injury	183	118.6	64.4	154.3 (132.0–176.7)
Homicide	1	16.7	-15.7	6.0 (0.0–33.5)
Self-inflicted	10	29.4	-19.4	34.0 (16.3–62.6)
Deaths due to illness	30	31.9	-1.9	93.9 (63.3–134.2)
Cardiovascular, unexpected, or undefined	23	20.0	3.0	115.0 (72.9–172.6)

Number of observed and expected non-battle deaths (adjusted for age) in U.S. military personnel deployed to the Persian Gulf region, August 1, 1990, through July 31, 1991.

Source: Writer et al. (1996). Comparative Mortality among U.S. Military Personnel in the Persian Gulf Region and Worldwide during Operations Desert Shield and Desert Storm. *Journal of the American Medical Association* 275 (2), 118–121.

(MacFarlane et al., 2000, 17). Also, there were not more suicides in the Gulf cohort compared to the non-Gulf cohort. The authors concluded that Gulf War veterans had only a small increase in mortality compared to the non–Gulf War veterans. The groups differed somewhat in the causes of death, as Gulf War veterans had a slightly lower mortality from disease-related causes but higher mortality from external causes (due to an increased rate of accidental deaths). Suicide deaths in both groups were approximately equal. However, the study did note that suicide rates may have been higher than reported, since it is often difficult to differentiate between accidental deaths and suicides.

Bell, Amoroso, Wegman, and Senier (2001) reviewed elevated rates of injury among Persian Gulf War veterans in order to highlight the need for better data on suicide. They compared death rates of U.S. veterans of the Persian Gulf War with nondeployed veterans and the U.S. population at large. Findings showed that while death rates among veterans of the Gulf War were lower than for their counterparts, Gulf

War veterans were at significantly greater risk of mortality from injuries. The authors proposed five hypotheses as to why injury mortality was higher among the Gulf War veterans. First, the increased risk might be a consequence of depression, post-traumatic stress disorder, and symptoms of other psychiatric conditions developed after the war. Second, the development of unhealthy coping behaviors incident to physical and psychological trauma experienced during the war (for example, drinking) may be a risk factor. Third, increased risk may be an indirect consequence of the increased experience of ill-defined diseases and symptoms. Fourth, veterans may experience poorer survivability for a certain injury, resulting in a greater mortality but not morbidity. Finally, there may be a bias for selecting individuals for deployment who are inherently at greater injury risk. Bell et al. (2001) suggest that more research is needed so policymakers can better understand the link between deployment and increased risk of postwar injury.

Kang and Bullman (1996) concluded there was only a small increase in deaths from accidents among Gulf War veterans compared to non–Gulf War veterans, and no apparent increase in suicide. Their study examined mortality among 695,516 Gulf War veterans compared to 746,291 non–Gulf War veterans. Records indicated that out of 1,765 Gulf War veteran deaths, 261 were suicides, as compared to 277 suicides among 1,729 deaths of other veterans. This represents an almost identical rate of 15.3 per 100,000 for Gulf War veterans, and 15.4 per 100,000 for other veterans. A later report by Haley (1998) challenged these results, claiming the Kang and Bullman study was biased due to "errors in the calculation of confidence intervals for tests of statistical significance, a failure to appreciate a more pertinent application of the 'healthy-soldier effect,' and the unequal effects of excluding hospitalizations in nonmilitary hospitals" (Haley, 1998, 316). Kang and Bullman (2001) published a subsequent report that, though not a direct response to Haley, concluded that on further investigation there was still no significant difference between suicide rates of Persian Gulf deployed veterans and non–Persian Gulf veterans. Overall, then, most studies examining suicide among soldiers deployed in the first Gulf War have found no significant increase in risk for suicide.

More Recent U.S. Experience

In recent years, further studies have sought to identify factors that are inherent to military life and may therefore be helpful in explaining why suicide is on the rise. One of the foundations for this line of research, a study by Rothberg (1991), investigated the relationship of U.S. Army suicides and relocation from 1979 to 1983. Findings showed no correlation between suicide and geographic moves among total Active-duty Army personnel. However, in the 17 to 21 age group specifically, relocation was positively correlated with suicide (accounting for 10 percent of variance) from 1980 to 1983 (Rothberg, 1991). This implies that younger men may be more affected by relocation and thereby be more vulnerable to suicide.

In a more recent literature review by Allen, Cross, and Swanner (2005), the authors analyzed distal and proximal risk factors and problems with research, and provided some recommendations for preventing suicide in the military. Distal risk factors were found to include previous attempts (30 to 40 percent of suicides have made previous attempts), alcoholism, social isolation, and concurrent depression or other psychiatric condition (approximately 90 percent of suicides have had at least one diagnosable psychiatric condition). Proximal risk factors include having a specific plan to commit suicide, openly talking about suicide, and giving away one's possessions. Other events that seem to exacerbate risk include relationship problems, job loss, legal problems (including incarceration), and economic problems.

The authors provide several recommendations for preventing suicide in the military. These include educating medical and behavioral health specialists as to the distal, proximal, and exacerbating risk factors for suicide, improving coordination between the health care provider and the individuals' command structure, and establishing an environment in which individuals believe they are able to seek help without damaging their careers.

Regan, Outlaw, Hamer, and Wright (2005) examined trends among several different U.S. military groups. For example, they report that the suicide rate of Soldiers deployed to Kuwait and Iraq in 2003 was 17.3 per 100,000, while the overall rate in the Army was just 12.8 per 100,000. The Army suicide rate for the entire year of 2003 was even lower at 12.2 per 100,000. This suggests that deployment itself is a risk factor for suicide. The authors contend that suicide prevention is particularly important for U.S. forces in Iraq, since these forces have a higher percentage of males than the rest of the Army, and males historically exhibit higher rates of suicide.

The 2008 Mental Health Advisory Team V (MHAT–V), commissioned by the Office of the U.S. Army Surgeon General, reviewed

Table 12–3. **Demographics for Confirmed U.S. Army Suicides during Operation *Iraqi Freedom***

	2003 Iraq Suicides	2004 Iraq Suicides	2005 Iraq Suicides	2006 Iraq Suicides	2007 Iraq Suicides*	2007 Army Suicides
Suicide Rate	18.8	10.5	19.9	19.4	24.0	24.0**
Suicide by firearm/gun-shot (percent)	95	100	95	100	91	65
Male (percent)	91	100	95	82	100	94
Age 30 or younger (percent)	82	91	80	86	83	69
E–4 or below (percent)	68	82	65	77	61	46
Married (percent)	41	0	35	18	32	49
Minority/non-white (percent)	43	20	10	14	18	20

*Through October 15, 2007
** Estimated November 14, 2007
Source: Mental Health Advisory Team V (2008).

mental health effects associated with participation in Operation *Iraqi Freedom* (OIF) and Operation *Enduring Freedom* (OEF). While not primarily concerned with suicide, the issue was nevertheless included as one of a range of mental health problems meriting attention. The study found 113 confirmed U.S. Army suicides in Iraq since OIF began in 2003. The annual rates per 100,000 were 18.8 in 2003, 10.5 in 2004, 19.9 in 2005, and 19.4 in 2006 (MHAT–V, 2008). The estimated rate for 2007 was 24.0 per 100,000. The incidence of reported suicidal ideation was also elevated among troops deployed to Iraq. Soldiers committing suicide while deployed in OIF were mostly young, white, lower enlisted, single males who used a firearm. These demographic factors are summarized in table 12–3.

The MHAT–V team reported a curvilinear relationship between months deployed and suicide for the 2007 year. In the analysis, 15-month tours were divided into three 5-month deployment phases.

Findings showed significantly more suicides during the middle third of the deployment than in either the early or late phases. It is possible that once the Soldiers make it past the tenth month, they may see the end of their deployment is near and thus can deal with stresses that previously seemed overwhelming or that the Soldiers at risk had been identified and had received treatment that prevented a suicide. A similar curvilinear relationship was also noted for emergency psychiatric referrals.

Based on forensic investigations conducted by the Army's Criminal Investigation Division (CID), the MHAT–V identified failed intimate relationships as the single greatest precipitating factor (present in 68 percent of cases) for suicide while deployed in OIF. Other factors included pending Uniform Code of Military Justice (legal) actions (35 percent), other failed relationships (21 percent), job problems (17 percent), and financial debt (10 percent). The CID investigation also found that 60 percent of the cases demonstrated signs of depression or behavioral changes prior to suicide. In addition, a sizable percentage of victims had sought assistance for mental health problems within the 30 days prior to committing suicide. Of the soldier suicides in Iraq, 27 percent were seen at a medical treatment facility, 36 percent by a chaplain, 26 percent by mental health personnel, and 21 percent by both a chaplain and mental health personnel; and 13 percent were taking psychotropic medications. This finding suggests that the problem of stigma associated with seeking mental health care in the military may not be a major obstacle at least as regards suicidal persons. The reported data show that for U.S. soldiers who committed suicide in Iraq, up to one-third sought medical or mental health support immediately preceding the suicide. But the fact that so many soldiers committed suicide even though they were known to the military medical care system raises the disturbing possibility that current care and treatment approaches are ineffective. At any rate, these data suggest that simply making sure that suicidal soldiers are brought to the attention of the medical/mental health care system will not be sufficient to prevent a large proportion of military suicides. The rate of patients who sought assistance and did not later commit suicide was not reported.

The MHAT–V authors reported the most relevant risk factors to be a sense of loss (whether relational, social, work, or financial), isolation (defined as a feeling of being cut off from other people), barriers to accessing mental health treatment, and an unwillingness to seek help because of the stigma attached to mental health. Easy access to lethal weapons is also a risk factor. None of these factors, however, can account for the upward trends in suicide since they have all been present throughout the period of study, from 2003 through 2007.

The MHAT–V study group did suggest some possible protective factors for soldiers operating in the Iraqi theater. Although not yet studied in depth, these factors include the lack of intoxicants available in the theater; effective clinical care for mental, physical, and substance abuse disorders; easy access to a variety of clinical interventions and support for help seeking; family and community support; and teaching skills in problem solving and conflict resolution.

In another relevant study, Hoge, Auchterlonie, and Milliken (2006) investigated the mental health status of U.S. troops returning from combat deployments in Iraq and Afghanistan. Between May 2003 and April 2004, the authors conducted a population-based descriptive study of all U.S. Army Soldiers and Marines participating in routine mental health examinations as they came back from Iraq (N = 222,620), Afghanistan (N = 16,318), and other overseas deployments (N = 64,967). Findings showed that 19.1 percent of Servicemen returning from Iraq reported mental health problems—a higher proportion than those serving in Afghanistan (11.3 percent) or other locations (8.5 percent). The study also investigated the correlation between the postdeployment mental health screening results, the use of mental health services, and the rates of attrition from military service. The authors conclude that compared to deployments elsewhere, combat duty in Iraq was associated with higher utilization of mental health services (35 percent) and higher attrition from the military (17.3 percent) in the year following the deployment. Also, the high rate of utilization of mental health services after return from deployments (especially to Iraq) underscores the need to have adequate resources for returning veterans. While not specific to the problem of suicide, this study indicated that U.S. Service personnel seem to suffer disproportionate mental health problems as a result of deployment to a war zone.

Other studies have addressed the problem of suicide among veterans. Using data from the National Health Interview survey and the National Death Index, Kaplan, Huguet, McFarland, and Newsom (2007) compared veteran and nonveteran suicide rates between 1986 and 1994. The sample included 320,890 males, 104,026 of whom were veterans. Findings revealed that while causes of natural death were similar between both veterans and nonveterans, veterans were twice as likely to commit suicide. Veterans were also 58 percent more likely to use a firearm. This study was particularly interesting because it also included veterans who were not enrolled with the Department of Veterans Affairs (VA), who are often neglected when data are collected. As highlighted by the authors, relying only on data from veterans currently enrolled in the VA system is a critical research limitation, as this group comprises only about 25 percent of the total veteran population.

The Congressional Research Service also conducted a valuable study of suicide among veterans (Sundararaman, Panangala, and Lister, 2008). Their report states that "veterans have a number of risk factors that increase their chance of attempting suicide. These risk factors include combat exposure, PTSD and other mental health problems, traumatic brain injury (TBI), poor social support structures, and access to lethal means" (Sundararaman et al., 2008, 2). The authors stress the need to develop a more robust and reliable data system to track suicide issues among veterans. An improved data system would produce more accurate numbers for veteran suicides and also provide a means to analyze risk and protective factors. The system would also be useful in evaluating the effects of suicide intervention programs and to specifically identify which programs work and which ones do not.

These studies have pointed to a variety of risk factors for suicide, such as marital strife or general instability in a soldier's life. A few demonstrated some correlation between war deployment and increased likelihood of suicide. This does not necessarily mean that the current wars in Iraq and Afghanistan are to blame for rising suicide rates. There are many factors involved in suicide, and wartime deployment may have no real effect, or could be an aggravating factor, or even a protective factor as some older studies suggest (Oesterlen, 1865; Rothberg et al., 1987). Nevertheless, wartime deployment cannot be discounted as a potential cause of suicide. Many of the military's fledgling efforts to reduce suicide rates lay aside these questions and focus instead on more traditional explanations. For this reason, the resulting preventive approaches have tended to fall into a single mold: boosting suicide education.

Recent Prevention Efforts and Programs

The U.S. military is well aware of the increasing rates of suicide among its members and has initiated several measures aimed at curbing the rise. These efforts have focused on two aspects of the problem: gathering accurate data, and increasing awareness and education. By gathering more complete data on the suicides that occur, the military hopes to discover the root causes behind the rising rates. Meanwhile, increased awareness and education are intended to reduce the stigma associated with suicide and encourage preventive treatment of at-risk soldiers. While all branches have enacted similar measures, the Army has suffered the highest rates of suicide and been the most active in combating them. As such, this analysis will focus primarily on prevention programs enacted by the U.S. Army.

The Army's primary means of improving suicide data collection is the mandated use of the Army Suicide Event Report (ASER). These reports, which gather additional information regarding demographics, medical and psychological history, and event-specific information, are now required for all suicide-related behaviors that result in death, hospitalization, or evacuation. Such reporting should give researchers a better understanding of the factors involved in military suicide and help to reduce underreporting. In 2006, the Suicide Risk Management and Surveillance Office (SRMSO) received a total of 1,032 ASERs. Of these cases, 948 did not result in death, and 84 resulted in death (Active force only, excluding National Guard and Reserves; the total number of confirmed suicides in 2006 is 97, based on Armed Forces Medical Examiner's reports; SRMSO, 2007). The SRMSO found that suicide was more common among males (88 percent of suicides), Caucasians (64 percent), junior enlisted (63 percent), those under the age of 25 (49 percent), and unmarried (49 percent) (SRMSO, 2007). Other factors were failed marital relationships (present among 55 percent of completed suicides) and a history of at least one OIF/OEF deployment (62 percent) (SRMSO, 2007). The SRMSO report also concluded that within the U.S. Army population, marriage may be a factor related to prevention of suicide completion but unrelated to attempts.

The SRMSO released another ASER for 2007 with results quite similar to those of 2006. Of 108 confirmed suicides, the most common categories were again males (95 percent), Caucasians (67 percent), junior enlisted (55 percent), and Soldiers under the age of 25 years (45 percent) (SRMSO, 2008). Unlike the 2006 report, however, the 2007 statistics indicated that married Soldiers were disproportionately more likely to commit suicide (49 percent), raising some question as to the previous year's assessment of marriage as a preventive factor. Other factors that may have contributed were failed marital relationships (50 percent), a history of at least one mental disorder (44 percent), and a history of at least one OIF/OEF deployment (61 percent) (SRMSO, 2008).

While this demographic information is useful, it does little more than confirm the elevated risk for suicide among known at-risk groups. More relevant are the efforts to identify the causes of suicidal events. While in nearly half of committed suicides the primary motivation could not be identified, among those that could, emotion relief (stop bad feelings, self-hatred, or anxiety) made up the highest proportion (14 percent) (SRMSO, 2008). This was also reflected among attempted suicides, where 35 percent of cases were motivated primarily by emotion relief. Among unsuccessful suicide attempts, hopelessness, depression, and avoidance/escape also featured heavily (11 percent, 11 percent, and 10 percent respectively) (SRMSO, 2008). This trend shows that Soldiers attempting and committing suicide often sought to "stop bad feelings" or suffered from self-hatred and anxiety (SRMSO, 2008, 13). It is also interesting to note that suicidal activity was often not premeditated, and in most cases intent to commit suicide was not communicated to anyone.

The Army has also taken a broader look into fluctuations in the overall suicide rate. In January 2009, the Department of the Army released the "Army Suicide Prevention Program Fact Sheet." This report demonstrated the rising rates of suicide within the Army population. During the calendar year 2005, the Army suicide rate was 12.7 per 100,000; this rose to 15.3 in 2006, 16.8 in 2007, and 20.2 (estimated) in 2008 (Army Suicide Prevention Program, 2009). Interestingly, the study also listed the deployment status of those who committed suicide. The study found that 29 percent of suicides were committed while deployed, 39 percent were committed after deployment, and 30 percent had no deployment history.

The Army's suicide awareness and prevention efforts predated the recent uptick in military suicides. In 1997, the Army introduced the "Building Strong and Ready Families" program that encouraged relationship-building by Soldiers. The program held that strong intimate relationships were a key part of a Soldier's mental well-being and thereby a protective factor against suicide. In 2005, the program was renamed "Strong Bonds" and was expanded to address the full spectrum of those affected by military life, including single Soldiers, couples, and families with children. "Strong Bonds" seeks to establish common bonds and nurture friendships among Soldiers and their families. A particular effort has been made to connect those who are on the same deployment cycles. The program is led by chaplains who receive support from commanding officers and is fully funded through a grant managed by the

Army Chief of Chaplains (Strong Bonds, 2009). The American Forces Press Service reported on March 24, 2009, that more than 60,000 Soldiers from the Active and Reserve Components have gone through the "Strong Bonds" program (Quigley, 2009). While Soldiers and their leaders generally speak highly of such programs, it is unclear whether they are having any impact on the military's suicide rate.

There have also been several efforts aimed at preventing suicide among veterans. The Department of Veterans Affairs has recognized the risks for veterans and has responded with numerous initiatives including the establishment of a suicide prevention coordinator for each VA Medical Center to ensure veterans receive needed counseling and services; increases in the total number of VA mental health employees to 17,000 (400 of whom are specifically devoted to preventing suicide among veterans); the addition of 61 new veterans' centers (bringing the total to 268); continued education of veterans and family members on suicide risk factors and warning signs; screening of all new veterans from OEF/OIF to determine suicide risk; the development and testing of clinical and public health intervention standards for suicide prevention; and increased research into the clinical and neurobiological conditions that contribute to increased risk (U.S. Department of Veterans Affairs, 2009).

Congress has also taken note of the suicide risk faced by veterans. As a result, the 1st session of the 110th Congress passed H.R. 327—commonly referred to as the Joshua Omvig Veterans Suicide Prevention Act. This bill states that "suicide among veterans suffering from posttraumatic stress disorder is a serious problem" that must be rectified (U.S. Congress, 2007, 2). This comprehensive program mandates staff education, health assessments of veterans, designation of suicide prevention counselors, best practices research, sexual trauma research, 24-hour mental health care, a suicide hotline, outreach and education for veterans and families, a peer support counseling program, and "other components" that the Secretary of Veterans Affairs considers appropriate to reduce suicide among veterans. The act also required the Secretary of Veterans Affairs to submit a report to Congress addressing the status of program implementation, the time line and costs for complete implementation within 2 years, a plan for additional programs to reduce suicide among veterans, and recommendations for further legislation or administrative action deemed appropriate to improve suicide prevention (U.S. Congress, 2007).

All of these preventive efforts have helped to raise awareness regarding military suicide, and there is a growing body of literature on the subject. Unfortunately, there has not yet been a concerted effort to analyze the effectiveness of the programs described above and it is unclear whether they are having a positive effect. It is likely that a real reduction in the military suicide rate will require policy changes beyond the sphere of educational programs, and much more concerted study must be devoted to developing effective prevention measures.

Conclusion

The various studies reviewed in this chapter suggest that a variety of factors contribute to suicide among soldiers and recently discharged veterans. It is clear from the studies that certain demographic groups are in greater danger. Young males are clearly at the greatest risk. Caucasian ethnicity and lower enlisted ranks are also associated with higher suicide risk. There is near universal agreement that access to firearms and alcohol can be a lethal combination precipitating the danger. However, according to several studies, access to weapons alone does not seem to significantly raise the risk of suicide. Several studies suggest that recent exposure to war or deployment to conflict zones has an escalating effect on the rate of suicide among both civilian and military populations, most notably for young men. However, these conclusions are countered somewhat by studies showing lower than expected rates for soldiers in the first Gulf War. A better understanding of suicide during and following deployments is necessary in order to successfully address this pressing problem.

Notably, risk for suicide seems to be especially elevated after discharge from the military, rather than during service. However, difficulties in tracking and reporting suicide in the military thus far have greatly complicated efforts to identify the underlying causes. A history of mental illness as well as major life stressors (such as marital strife) increase the risk of suicide, but to what extent they are causal versus correlated but noncausal is unclear. Further research is needed to clarify risk factors such as history of suicidal behavior, depression, and genetic factors (Adams et al., 1998). Additional steps such as postmortem interviews with comrades, friends, and family members as well as examination of documentary materials (letters, diaries, etc.) can also help in identifying the causes of suicide, thus providing the knowledge to develop more effective prevention techniques.

A potentially important consideration for understanding suicide relates to soldier perceptions of how society regards their activities. If an already stressed deployed soldier also feels a lack of societal support for the war or mission, this could increase the sense of alienation and meaninglessness. Also, the usual beneficial effects of cohesion within the military (which may contribute to the lower peacetime military/civilian suicide ratio) could be diminished in that the soldier may feel he is part of a socially undesirable group. The potential impact of societal attitudes toward the military and toward particular operations merits further attention.

Under the current investigation and assessment system, it is still unclear who has real responsibility for determining if a death is a suicide. Investigations can take up to 2 or 3 years, and in some cases may never occur. This further complicates efforts to understand the causes and cures. Educational programs aimed at preventing suicide are being expanded, but this is being done without a real understanding of the factors at work. Not surprisingly, then, these prevention efforts have not worked. Despite a variety of awareness enhancement, stigma reduction, and relationship-building programs, the Army's suicide rate continues to rise. Suicide in the military can be prevented, or at least greatly reduced. But this will require selection, training, and treatment programs that are built upon a solid understanding of the range of factors, at multiple levels, that raise or lower the risk for such unfortunate outcomes.

References

Adams, D.P., Barton, C., Mitchell, G. L., Moore, A.L., and Einagel, V. (1998). Hearts and minds: Suicide among United States combat troops in Vietnam, 1957–1973. *Social Science and Medicine* 47 (11), 1687–1694.

Allen, J., Cross, G., and Swanner J. (2005). Suicide in the Army: A review of current information. *Military Medicine* 170 (7), 580–584.

Army Suicide Prevention Program. (2009). *Fact sheet.* Washington, DC: Army Public Affairs.

Beck, A., Brown, G., Berchick, R., Stewart, B., and Steer, R. (1990). Relationship between hopelessness and ultimate suicide: A replication with psychiatric outpatients. *American Journal of Psychiatry* 147 (2), 190–195.

Beck, A.T., Kovacs, M., and Weissman, A. (1979). Assessment of suicidal intention: The scale for suicide ideation. *Journal of Consulting and Clinical Psychology* 47, 343–352.

Belik, S., Stein, M., Gordon, J.G.A., and Sareen, J. (2009). Relation between traumatic events and suicide attempts in Canadian military personnel. *Canadian Journal of Psychiatry* 54 (2), 93–103.

Bell, N., Amoroso, P., Wegman, D., and Senier, L. (2001). Proposed explanations for excess injury among veterans of the Persian Gulf War and a call for greater attention from policymakers and researchers. *Injury Prevention* 7, 4–9.

Bodner, E., Ben-Artzi, E., and Kaplan, Z. (2006). Soldiers who kill themselves: The contribution of dispositional and situational factors. *Archives of Suicide Research* 10 (1), 29–43.

Bosnar, A., Stemberga, V., Coklo, M., Koncar, G., Definis-Gojanovic, M., Sendula-Jengic, V. and Katic, P. (2005). Suicide and the war in Croatia. *Forensic Science International* 147, S13–S16.

Boyle, C., Decoufle, P., Delaney, R., DeStefano, F., Flock, M., Hunter, M., Joesoef, M., Karon, J., Kirk, M., Layde, P., McGee, D., Moyer, L., Pollock, D., Rhodes, P., Scally, M., and Worth, R. (1987). *Postservice mortality among Vietnam veterans.* U.S. Department of Health and Human Services. Public Health Service Centers for Disease Control.

Brent, D.A., Perper, J.A., and Allman, C.J. (1987). Alcohol, firearms, and suicide among youth. *Journal of the American Medical Association* 257, 3369–3372.

Carr, J., Hoge, C., Gardner, J., and Potter, R. (2004). Suicide surveillance in the U.S. military: reporting and classification biases in rate calculations. *Suicide and Life-Threatening Behavior* 34, 233–242.

Cassimatis, E.G., and Rothberg, J.M. (1997) Suicide in the United States military. In A.J. Botsis, C.R. Soldatos, and C.N. Stefanis (eds.). *Suicide: Biopsychosocial approaches.* Amsterdam: Elsevier Science B.V. 23–32.

Centers for Disease Control. (2009). *Web-based Injury Statistics Query and Reporting System: Fatal Injury Reports.* Retrieved from <www.cdc.gov/injury/wisqars/index.html>.

Crane, P.J., Barnard, D.L., Horsley, K.D., and Adena, M.A. (1997) *Mortality of Vietnam veterans: The veteran cohort study. A report of the 1996 retrospective cohort study of Australian Vietnam veterans.* Canberra: Department of Veterans' Affairs.

Durkheim, E. (1951, originally published 1897). *Suicide: A Study in Sociology.* Trans. J.A. Spaulding and G. Simpson. New York: The Free Press.

Eaton, K., Messer, S., Wilson, A., and Hoge, C. (2006). Strengthening the validity of population-based suicide rate comparisons: An illustration using U.S. military and civilian data. *Suicide and Life-Threatening Behavior* 36 (2), 182–191.

Fishman, G., Morris-Dycian, A., and Kotler, M. (1990). Suicide in the Israeli army. *Suicide and Life-Threatening Behavior* 20 (3), 225–239.

Haley, R. (1998). Point: Bias from the "healthy-warrior effect" and unequal follow-up in three government studies of health effects of the Gulf War. *American Journal of Epidemiology* 148 (4), 315–323.

Hansen-Schwartz, J., Jessen, G., Andersen, K., and Jørgensen, H.O. (2002). Suicide after deployment in UN peacekeeping missions—A Danish pilot study. *Crisis* 23 (2), 55–58.

Henderson, R., Stark, C., Humphry, R.W., and Selvaraj, S. (2006). Changes in Scottish suicide rates during the Second World War. *BMC Public Health* 6, 167.

Hoge, C., Auchterlonie, J., and Milliken, C. (2006). Mental health problems, use of mental health services, and attrition from military service after returning from deployment to Iraq or Afghanistan. *The Journal of the American Medical Association* 295 (9), 1023–1032.

Joiner, T. (2005). *Why People Die by Suicide.* Cambridge: Harvard University Press.

Kang, H., and Bullman, T. (1996). Mortality among U.S. veterans of the Persian Gulf War. *The New England Journal of Medicine* 335 (20), 1498–1504.

———. (2001). Mortality among U.S. veterans of the Persian Gulf War: 7-year follow-up. *American Journal of Epidemiology* 154 (5), 399–405.

Kaplan, M., Huguet, N., McFarland, B., and Newsom, J. (2007). Suicide among male veterans: A prospective population-based study. *The Journal of Epidemiology and Community Health* 61, 619–624.

Kitchener, S. (2005). The health and wellbeing consequences of military deployment. *The Australian Defence Force Journal* 168, 45–53.

Kuehn, B.M. (2009). Soldier suicide rates continue to rise. *The Journal of the American Medical Association* 11, 1111–1113.

Macfarlane, G., Thomas, E., and Cherry, N. (2000). Mortality among UK Gulf War veterans. *The Lancet* 356, 17–21.

Mahon, M.J., Tobin, J.P., Cusack, D.A., Kelleher, C., and Malone, K.M. (2005). Suicide among regular-duty military personnel: A retrospective case-control study of occupation-specific risk factors for workplace suicide. *American Journal of Psychiatry* 162 (9), 1688–1696.

Menninger, K. (1938). *Man against himself.* New York: Harcourt, Brace and World.

Mental Health Advisory Team V. (2008). *Operation Iraqi Freedom 06-08: Iraq and Operation Enduring Freedom 08: Afghanistan.* Washington, DC: U.S. Army Medical Command, Office of the Surgeon General.

Michel, P.O., Lunden, T., and Larsson, G. (2007). Suicide rate among former Swedish peacekeeping personnel. *Military Medicine* 172 (3), 278–282.

Murray, H.A. (1938). *Explorations in personality*. New York: Oxford University Press.

Oesterlen, F. (1865). *Handbuch der medicinischen Statistik*. Tübingen.

Quigley, S. (2009). *Army program builds 'strong bonds' among couples, families*. Retrieved July 23, 2009, from American Forces Press Service Web site: <www.defenselink.mil/news/newsarticle.aspx?id=53621>.

Regan, J., Outlaw, F., Hamer, G., and Wright, A. (2005). Mental health series: Suicide in the military. *Tennessee Medicine* (July 2005), 400–401.

Rothberg, J., Bartone, P., Holloway, H., and Marlowe, D. (1990). Life and death in the U.S. Army: In corpore sano. *The Journal of the American Medical Association* 264 (17), 2241–2244.

Rothberg, J., Holloway, H., and Ursano, R. (1987). Suicide in the United States military. *Psychiatric Annals* 17, 545–548.

Rothberg, J. (1991). Stress and suicide in the U.S. Army: Effects of relocation on service members' mental health. *Armed Forces and Society* 3, 449–458.

Ryvkin, S. (2000). Suicidal phenomena among servicemen: Problem and ways of solution. *Military Thought* 6, 43–48.

Sareen, J., Cox, B.J., Afifi, T.O., Stein, M.B., Belik, S., Meadows, G., and Asumundson, G.J. (2007). Combat and peacekeeping operations in relation to prevalence of mental disorders and perceived need for mental health care. *Archive of General Psychiatry* 64 (7), 843–852.

Schneidman, E. (1996). *The suicidal mind*. New York: Oxford University Press.

Sheftick, G. (2009) *Army creates suicide prevention task force during 'stand down.'* American Forces Press Service. Retrieved from <www.defenselink.mil/News/newsarticle.aspx?id=53378>.

Strong Bonds. (2009). *The history of strong bonds*. Retrieved May 19, 2009, from <www.strongbonds.org/skins/strongbonds/display.aspx>.

Suicide Risk Management and Surveillance Office. (2007). *Army suicide event report (ASER) calendar year 2006*. Tacoma, WA: Suicide Risk Management and Surveillance Office.

———. (2008). *Army suicide event report (ASER) calendar year 2007*. Tacoma, WA: Suicide Risk Management and Surveillance Office.

Sundararaman, R., Panangala, S.V., and Lister, S.A. (2008). *Suicide prevention among veterans*. Washington, DC: Congressional Research Service.

Thoresen, S., and Mehlum, L. (2004). Risk factors for fatal accidents and suicides in peacekeepers: Is there an overlap? *Military Medicine* 169 (12), 988–993.

———. (2006). Suicide in peacekeepers: Risk factors for suicide versus accidental death. *Suicide and Life-Threatening Behavior* 36 (4), 432–442.

———. (2008). Traumatic stress and suicidal ideation in Norwegian male peacekeepers. *The Journal of Nervous and Mental Disease* 196 (11), 814–821.

Thoresen, S., Mehlum, L., and Moller, B. (2003). Suicide in peacekeepers: A cohort study of mortality from suicide in 22,275 Norwegian veterans from international peacekeeping operations. *Social Psychiatry and Psychiatric Epidemiology* 38 (11), 605–610.

Thoresen, S., Mehlum, L., Røysamb, E., and Tønnessen, A. (2006). Risk factors for completed suicide in veterans of peacekeeping: Repatriation, negative life events, and marital status. *Archives of Suicide Research* 10 (4), 353–363.

U.S. Congress. (2007). *Joshua Omvig veterans suicide prevention act*. (H.R. 327—1st Session, 110th Congress). Washington, DC: U.S. Government Printing Office. Retrieved May 20, 2009, from <http://thomas.loc.gov/cgi-in/query/D?c110:6:./temp/~c110hpFVh8:>.

U.S. Department of Veterans Affairs. (2009) *Suicide prevention*. Retrieved July 23, 2009, from <www.mentalhealth.va.gov/suicide_prevention/index.asp>.

Wong, A., Escobar, M., Lesage, A., Loyer, M., Vanier, C., and Sakinofsky, I. (2001). Are UN peacekeepers at risk for suicide? *Suicide and Life-Threatening Behavior* 31 (1), 103–112.

Writer, J.V., DeFraites, R.F., and Brundage, J.F. (1996). Comparative mortality among U.S. military personnel in the Persian Gulf region and worldwide during Operations Desert Shield and Desert Storm. *The Journal of the American Medical Association* 275 (2), 118–121.

Part V

Applying Psychology to
Chemical and Biological
Defense

Chapter 13

Psychological Effects of Chemical, Biological, Radiological, and Nuclear Weapons

Ross H. Pastel

Chemical, biological, radiological, and nuclear (CBRN) weapons have gained increased international attention in the last 20 years. Although the Soviet Union had long been suspected of developing biological warfare (BW) agents, it was not until 1992 that Russian President Boris Yeltsin admitted that the Soviet Union continued its offensive BW development program following its ratification of the Biological and Toxin Weapons Convention in 1972 (General Accounting Office, 2000). In 1994 and again in 1995, Japan suffered sarin attacks. It was not until after the second attack that the sarin was traced to the Aum Shinrikyo cult. Further investigation revealed that the cult had also attempted several attacks with BW agents—botulism and anthrax. The United States was rocked by the bombing of the Murrah Federal Building in Oklahoma City in April 1995. The new millennium was greeted with the September 11, 2001, al Qaeda attacks on the Pentagon and the twin towers of the World Trade Center in New York City. Although anthrax attacks in the mail followed within the week, the first realization of them was not until October when the first victim fell ill. In 2002–2003, Mother Nature released a new contagious disease: severe acute respiratory syndrome, better known as SARS. Although not spread by terrorists, the disease caused widespread death and illness in multiple countries including China, Taiwan, Hong Kong, Singapore, Vietnam, and Canada.

CBRN weapons, which have been a high priority for al Qaeda, are no longer weapons possessed only by states; they have become available to terrorists as well. Many experts believe that a large-scale terrorist attack with CBRN weapons is not a matter of *if*, but of *when*. Therefore, it is critical that mental health practitioners become aware of the possible psychological consequences following a CBRN attack.

The psychological effects of a CBRN attack differ from other medical effects in that personnel do not need to be physically exposed to these agents in order to exhibit symptoms. Psychological effects can cause symptoms that may mimic the prodromal or early symptoms of CBRN weapons. Either psychological stress or CBRN weapons can cause such ubiquitous symptoms as fatigue, headache, nausea, joint and muscle ache, dizziness, and shortness of breath. Many of these symptoms can also be induced by respiratory alkalosis caused by hyperventilation. Fortunately, the acute and long-term psychological effects after CBRN attacks demonstrate no apparent unique psychological disorders, but rather seem to exist on a continuum with effects seen after exposure to natural disasters or high explosives (Scharf, Vaught, Kidd, Steiner, Kowalski, Wiehagen et al., 2001). Therefore, currently available treatments for mental disorders will help victims of CBRN attacks (Ritchie, Watson, and Friedman, 2006). Psychophysiologic effects, typically syndromes of medically unexplained physical symptoms (MUPS), will likely dominate the long-term picture (Hyams, Murphy, and Wessely, 2002). Historically, this has been seen in war syndromes as well as civilian disasters (Hyams et al., 2002; Vasterman, Yzermans, and Dirkzwager, 2005). However, treatment of MUPS may be difficult, due to patient resistance and difficulties with doctor-patient relationships (Engel, Adkins, and Cowan, 2002).

This review will begin with a discussion of risk and risk perception, which have important impacts on the psychological effects of CBRN. Both risk communication and the mass media influence risk perception. Potential acute and long-term effects of CBRN agents will be discussed briefly, followed by a study of historical examples of the psychological effects of chemical, biological, radiological, and nuclear weapons. In addition to discussion of the anthrax attacks in the United States, this study will also discuss the pneumonic plague outbreak in India in 1994 and SARS in 2003, because the psychological effects accompanying these two natural outbreaks of contagious disease will likely resemble the psychological effects seen following a biological weapons attack with a contagious agent such as smallpox or plague.

Risk, Risk Communication, Risk Perception, and the Mass Media

In a CBRN event, it is likely that the extent of the danger will not be known immediately. Public health authorities and public officials will need time to survey the damage and determine the location of the victims before they can calculate the extent of the threat and inform the media and the public.

Risk Communication

Risk communication has been defined as the "interactive process of exchange of information and opinions among individuals, groups, and institutions concerning a risk or potential risk to human health or the environment" (National Research Council, 1989). Although risk communication can be used in a variety of contexts, this review is most concerned with crisis communication in response to situations of extreme, sudden danger as seen in a CBRN incident. Risk communication plays an important role in how people react to the threat. New York Mayor Rudolph Giuliani was extremely effective following the events of September 11. He demonstrated the value of daily or twice-daily scheduled information briefings with the media and the public. In times of crisis, people have an overwhelming need to hear what is happening from a credible, trustworthy person, well placed in the administration, who can give up-to-date information. Risk communication following a CBRN attack should be based on existing principles of health communication. Some of the more important principles include having a consistent message delivered by a knowledgeable and credible official, listening and responding to the concerns of the public, and avoiding the appearance of defensiveness or concealment.

After any toxic industrial accident or terrorist CBRN attack, many people will feel anxious about the health effects of any release. Such anxieties may be multiplied by the news media, which will be full of devastating descriptions of the event and the potential consequences. Public officials will need to provide accurate hazard communication and information on workable measures that individuals can take to protect themselves and their families.

According to one risk communication approach, risk equals hazard plus outrage (Sandman, 2003). Hazard is the scientifically based risk assessment, but outrage is made up of nonquantifiable factors related to the public's concern about and perception of the event. Outrage following a CBRN attack may significantly influence both acute and long-term psychological effects. If people focus on the idea that the event should not have happened, they are more likely to respond with feelings of hopelessness and helplessness. On the other hand, effective leaders may be able to use risk communication to promote a more helpful resilience in the community. For example, research on hardiness has suggested that hardy individuals face crises as challenges to overcome, feel a greater sense of control, and respond with high commitment (Bartone, 2006). Studies in the military have demonstrated that leaders can influence hardiness in their personnel (Bartone, 2006).

Risk Perception

Risk perception is an important driver of the outrage component in risk communication. There are a number of factors involving CBRN weapons that increase the perception of risk. Many of them are invisible and odorless (radiation, biological agents, and some chemical agents), which leads to uncertainty about whether exposure occurred and if so, the amount. In many cases, exposure is not known until the patients become symptomatic. However, these agents may initially induce nonspecific symptoms—for example, fatigue, headache, nausea, difficulty breathing, dizziness, and muscle/joint ache. Regardless of illness induced, chemical agent and radiation exposures will also increase the fear of long-term effects of the exposure.

In the risk literature, a number of factors have been shown to increase the perception of risk, including potentially fatal illness, involuntary exposure (lack of control), catastrophic event, presence of an unknown perpetrator, delayed detection and reaction by authorities, and potential effect on future generations (Brecher and Flynn, 2002; Covello, Peters, Wojtecki, and Hyde, 2001). Fear of radiation in particular is prevalent, largely due to ignorance and misinformation. The thoughts and images typically associated with radiation are death, cancer, sterility, and fear for future generations. A number of factors—scapegoating, distrust of governmental and industrial experts, and news media hype and misinformation—may further amplify risk perception.

Mass Media

The acute and long-term consequences of terrorism and CBRN are shaped by risk perception, which itself is shaped, at least in part, by the mass media. The media have played an important role in various outbreaks of multiple unexplained symptoms (Vasterman, Yzermans, and Dirkzwager, 2005). The media are an important risk perception amplifier because they select and frame risk messages to inform the public, and ensuing media hype can create waves of news due to intensive reporting (Covello et al., 2001).

The power of the media can be seen in studies that followed both the Oklahoma City bombing and the events of September 11, 2001. In a study done 7 weeks after the Oklahoma City bombing, post-traumatic stress symptomatology was higher in nonexposed children (those with no direct or emotional exposure to the bombing) who had high television exposure than in those who did not (Pfefferbaum, Nixon, Tivis, Doughty, Pynoos, Gurwitch et al., 2001). A telephone survey study done 3 to 5 days following September 11 found that 44 percent of the people surveyed had one or more substantial stress symptoms, including sleep difficulties, irritability and anger, difficulty concentrating, and disturbing thoughts, memories, and dreams (Schuster, Stein, Jaycox, Collins, Marshall, Elliott et al., 2001). It is important to note that people responding to the survey were not present at the event. Therefore, much of what they knew was presumably based on media reporting.

Similarly, in a contagious disease outbreak, information becomes extremely important. The public is eager for information and needs to know what precautionary measures should be taken. In Hong Kong, 90 percent of respondents to a telephone survey reported actively seeking SARS information on a daily basis, and relied more on mass media (television, newspapers, and radio), than on medical professionals, friends, or the Internet (Lau, Yang, Tsui, and Kim, 2003). Substantial misinformation and false beliefs persisted even at an advanced stage of the SARS epidemic, despite constant media and public service announcements (Leung, Lam, Ho, Ho, Chan, Wong et al., 2003). For example, recommended measures were not practiced uniformly. Many people did not understand transmission routes—only one-third of respondents avoided direct contact by touch with contaminated objects (fomites), and less than half practiced at least five of the seven recommended precautions.

Acute and Long-term Psychological Effects

Many symptoms commonly seen following a CBRN incident (fatigue, nausea, vomiting, headaches, and anorexia) are common in combat (Stouffer, Lumsdaine, Lumsdaine, Williams, Smith, Janis et al., 1949) and can be induced by acute radiation sickness (ARS), chemical agent exposure, and during the prodromal syndrome of various biological agents and toxins (Anno, Baum, Withers, and Young, 1989; U.S. Army Medical Research Institute of Chemical Defense [USAMRICD], 2000; U.S. Army Medical Research Institute of Infectious Diseases [USAMRIID], 2005). Because many CBRN agents are invisible, Soldiers may experience symptoms that they may blame on CBRN exposure, regardless of actual exposure or dose of exposure. These patients are not "worried well." They are worried—possibly with good reason—but they are not well if they are in distress and pain. The term *worried well* has a pejorative connotation. The doctor-patient relationship may become strained if patients feel that their symptoms are being dismissed by health care providers. Thus, the term should be avoided.

Some CBRN agents may directly induce psychological effects, in addition to medical effects—for example, nerve agents can induce anxiety (DiGiovanni, 1999). In other cases, symptoms may precede signs— pulmonary agents may initially be present with respiratory distress without measurable physical signs (USAMRICD, 2000). Symptomatic ambulatory cases with mild or perceived exposures will present difficulties for CBRN event triage.

Based on historic experience in World War II, military medical planners can get a rough estimate of battle fatigue or combat stress casualties (CSC), based on the number of the wounded in action (WIA) expected from different types of battles (Levin, 1991; Vineberg, 1965). In World War II, the ratio of CSC to WIA was in the range of 1:10 to 1:2 (Levin, 1991; Vineberg, 1965). An analogous measure among a civilian population would be the ratio of psychological casualties (PC) to wounded. During the first Israeli Scud missile attack, this ratio was 16:1 (172 PC plus 171 unjustified atropine injections to 22 wounded) or 8:1 (if atropine injections are excluded). In the 1985 radiological contamination accident in Goiania, Brazil, where no explosion occurred, the ratio of PC to wounded was 500:1 (if wounded is defined as the 249 people contaminated either externally or internally) or 2,500:1 (if wounded is defined as the 50 people requiring close medical surveillance). This implies that CBRN events are significantly more psychologically traumatic than normal combat, especially on civilian populations.

While neither of these cases occurred with American military populations, the available data do suggest that it is unlikely that the PC to wounded ratio following a CBRN attack or incident will resemble the 1:10 to 1:2 range seen in World War II battles. The low end of the range may resemble World War II statistics, but the high end could go much higher, depending on the characteristics of the CBRN attack. Unfortunately, most CBRN exercises account for relatively few psychological casualties. These exercises should include a minimum of 25 percent psychological casualties and would be more realistic with ratios of 3:1 or greater PC to wounded ratios.

Mass Panic

The popular expectation for behavior during or after a disaster is mass panic, described as "highly disorganized flight by hysterical individuals who have stampeded at the sight of actual or potential danger" (Janis, 1951). During the Cold War, many civil defense planners feared that a mass panic would follow a nuclear attack. However, studies of disasters and wars over the last 50 years show that disorganized flight (mass panic) is, in fact, very rare (Caldwell, Ranson, and Sacks, 1951; Glass and Schoch-Spana, 2002; Quarantelli, 1960; Smelser, 1962). The few occasions when it did occur were very circumscribed and were characterized by limited escape routes with the possibility of entrapment, a perception of collective powerlessness, and a feeling of individual isolation (Caldwell et al., 1951; Quarantelli, 1960; Smelser, 1962). The most frequent historical examples of mass panic are in cases of fires, mine collapses, and sinking ships. Mass anxiety is not mass panic. Mass anxiety is an expression of fear. If public officials use effective risk communication, they can help people deal with their fears and guide them to more helpful responses.

Distress and Outbreaks of Multiple Unexplained Symptoms (OMUS)

Based on published literature concerning effects of terrorist bombing attacks, psychological distress should be common following a CBRN attack. Acute psychological effects were reported in 50 percent of bomb-injured patients in one study (Hadden, Rutherford, and Merrett, 1978). Another study reported that 12 percent of the casualties presented with emotional distress, with another 6 percent presenting with medical problems (such as angina, diabetes, headache, or asthma) (Carley and Mackway-Jones, 1997). Thus, it seems likely that acute psychological effects will also be seen with CBRN attacks, although the scale of response probably would be greater due to fear and the uncertainty of both exposure and its effects.

Perceived exposure to a CBRN agent can also result in the appearance of symptoms that may be hard to differentiate from mild symptoms expected from actual exposure to a CBRN agent. Thus, OMUS can occur independently or concurrently with a CBRN event. Consequently, in a CBRN event, not all symptomatic casualties have been exposed to a toxic agent (Singer, 1982). Symptoms of psychological origin can also occur in casualties actually exposed to a CBRN agent and may make treatment more difficult. Regardless of the degree of exposure, it is important to address the patient's symptoms of pain and distress while attempting to discern actual exposure.

An example of perceived exposure leading to OMUS was seen in 1988, when 1,800 male military recruits were evacuated from military barracks when an epidemic of coughing, difficulties with breathing, and chest pain broke out at a training center (Struewing and Gray, 1990). These symptoms were consistent with exposure to a pulmonary chemical agent. Although recruits and medical personnel suspected an airborne toxin, none was detected. Various OMUS have demonstrated that perceived exposure can induce symptoms resembling an actual toxic exposure.

Although acute OMUS has been widely studied, the possibility of long-duration and large-scale OMUS syndromes has only recently been suggested (Showalter, 1997). In recent years, a number of different chronic syndromes (chronic fatigue syndrome, environmental somatization syndrome, multiple chemical sensitivity syndrome, and sick building syndrome) have appeared, all characterized by multiple nonspecific symptoms (fatigue, headaches, sleep disturbances, nausea, dizziness, muscle and joint pains, and difficulties with memory and concentration) that are not connected with specific infectious or toxic agents (Gothe, Molin, and Nilsson, 1995; Rothman and Weintraub, 1995; Weiss, 1998). Each of these syndromes is characterized by an invisible contaminant or infectious agent that is blamed for the symptoms.

The military has seen several chronic OMUS syndromes, including Atomic Veterans Syndrome, Agent Orange Syndrome, and Gulf War Illness (Hyams, Wignall, and Roswell, 1996). Agent Orange Syndrome began when the media publicized an association between exposure of Vietnam veterans to Agent Orange and a reported epidemic of cancer and children born with birth defects. Epidemiological studies done by the Centers for Disease Control (CDC) found no evidence for an increased incidence of cancer or birth defects in this population (CDC, 1998b; CDC, 1998c). However, an increased prevalence of depression, anxiety, alcohol abuse or dependence, and post-traumatic stress disorder (PTSD) was demonstrated in Vietnam veterans compared with subjects who had not fought in Vietnam (CDC, 1998a). Another study found that symptoms of psychological distress were strongly associated with self-reported herbicide exposure (Decoufle, Holmgreen, Boyle, and Stroup, 1992). This group presented with more symptoms than were found in Air Force personnel actively involved in aerial spraying of herbicides. This suggests that Agent Orange Syndrome might be more related to a perception of exposure than to actual exposure.

Mental Disorders

Long-term psychological consequences after terrorist bomb attacks have also been reported. After the Oklahoma City bombing, 45 percent of the survivors suffered a post-disaster psychiatric disorder, including 34 percent with PTSD (North, Nixon, Shariat, Mallonee, McMillen, Spitznagel et al., 1999). Another study reported PTSD in 50 percent of the patients 6 months after a bombing (Curran, Bell, Murray, Loughrey, Roddy, and Rocke, 1990). Nearly one in five (18 percent) civilian survivors of terrorist attacks in another study suffered from PTSD, while another 13 percent suffered from major depression (Abenhaim, Dab, and Salmi, 1992). When broken down by severity of injury, PTSD was present in 31 percent of the severely injured, but in only 11 percent of the uninjured and 8 percent of the moderately injured. The adjusted prevalence ratio for PTSD (severely injured/others) was 4.2. Similarly, major depression occurred in 22 percent of the severely injured but in only 9 percent of the moderately injured or uninjured.

Chemical Warfare Agents

World War I Chemical Warfare

Although World War I chemical warfare agents (CWAs) (predominantly chlorine, phosgene, and mustard) caused 31 percent of battle injuries, they were responsible for only 2 percent of deaths in U.S. forces (Gilchrist, 1928). Nevertheless, CWAs had an enormous psychological effect. One contemporary report described "the most horrible and ghastly sight of the war. . . to see a hundred or more men, hale and hearty a few hours before, slowly strangling to death from pulmonary edema, with gradually increasing dyspnea, cyanosis and pallor, making futile efforts to expectorate" (Norris, 1919).

In the initial use of chlorine gas on the Western Front by the Germans in 1915, "a full-blown, blind, contagious panic swept portions of the line" (Hammerman, 1987). However, there was no panic farther out on the line where there was little or no gas. In the next six gas attacks over the next 2 months, there were no mass panics, although protective equipment was rudimentary and not widely available. In total, only four other gas panics were documented during World War I.

The American military experienced several OMUS with relation to chemical weapons. In World War I, outbreaks of gas neurosis (gas hysteria) occurred, in which some soldiers experienced symptoms of gas poisoning (dyspnea, coughing, and burning of skin) without clinical exposure to gas (Hulbert, 1920). In one incident, 500 battle-tested troops drifted into medical aid stations over a 1-week period following desultory gas shelling. They suffered from chest pain, fatigue, dyspnea, coughing, husky voice, and indefinite eye symptoms—all consistent with chemical exposure (Salmon and Fenton, 1929). However, the divisional gas officer found no evidence of gas inhalation or burning.

Three years after World War I, approximately one-half of gassed veterans claimed subjective complaints in medical examinations (Wachtel, 1941). When there were no objective findings, no compensation or pensions were paid, nor were these included in statistics of permanent disabilities. There were reports of large numbers of men who had recovered from acute gas poisoning and had good physical examinations, but suffered from serious sequelae, most particularly of easy fatigability and difficulty breathing on exertion (Haldane, 1919). This condition was variously known as effort syndrome, disordered action of the heart, and neurocirculatory asthenia. In chronic gas cases, there were often acute attacks of breathlessness at night accompanied by nightmares, and patients usually reported insomnia and unrefreshing sleep (Haldane, 1919).

Chemical Warfare Attacks on Sardasht, Iran

During the 1980–1988 Iran-Iraq War, Iraq attacked both military and civilian targets with a variety of chemical warfare agents. In June 1987, Iraq attacked Sardasht, Iran, with four 250-kilogram warheads filled with sulfur mustard (SM), a vesicant chemical warfare agent first used in World War I (Hashemian, Khoshnood, Desai, Falahati, Kasl, and Southwick, 2006). Vesicants produce delayed effects—blisters (vesicles)—with a latent period of hours following exposure. Depending on exposed areas, the most common effects are on the skin, eyes, and upper respiratory system (Chemical Casualty Care Division, 2000).

Approximately 4,500 civilians, over one-third of the Sardasht population of 12,000 residents, were exposed to SM (Ghanei, Aslani, Khateri, and Hamadanizadeh, 2003). Most sustained mild exposure and were treated on an outpatient basis, but 1,500 developed moderate to severe medical complications that required hospitalization (Khateri,

Ghanei, Keshavarz, Soroush, and Haines, 2003). Many suffered injuries from the explosions in addition to their exposure to SM.

One brief report looked at the long-term psychosocial impact on Sardasht civilians 16 years after the attack (Khateri, Ghanei, and Palmer, 2004). They administered a questionnaire to 400 individuals (69 percent male, 31 percent female) with confirmed exposure to SM in the June 1987 attack. Nearly all (94 percent) remembered the deaths from the attack as more horrifying than death by conventional weapons. Many (54 percent) reported long-term feelings of depression. Although 76 percent reported improved family cohesion and stronger friendships, 83 percent reported that relationships with some family and friends deteriorated significantly. Women were reported as being more vulnerable, and those who were younger than 19 years old at the time of the attack suffered more severe psychological effects. The authors also reported that 69 percent of males and 89 percent of females suffered from PTSD, based on response to a standard PTSD assessment instrument. The results are interesting, but it is hard to evaluate them due to the brevity of the report and the lack of detail on the methodology. Although all subjects had confirmed SM exposure, no mention is made of any long-term health effects. In a review article, mention is made of a neuropsychiatric evaluation of 1,428 Iranian veterans 3 to 9 years following SM exposure (Balali-Mood and Hefazi, 2005). The study found high rates of depression (46 percent), personality disorders (31 percent), and anxiety (15 percent). Unfortunately, no diagnostic detail is given in the review.

Recently, a cross-sectional randomized survey study was published that looked at anxiety, depression, and PTSD in three towns exposed to warfare in northwestern Iran (Hashemian et al., 2006). One town was exposed to low-intensity conventional warfare (LIW), one to high-intensity warfare (HIW), and one (Sardasht) to high-intensity warfare in addition to exposure to SM (HIW/SM). Those from Sardasht had a verified SM exposure–related disability, with a median disability score of 30 (range 5–70). Seventeen years after exposure, 56 percent of the Sardasht sample experienced mild respiratory, ocular, and cutaneous symptoms, 22 percent had at least 1 moderate symptom, and 11 percent had at least 1 severe symptom. The combination of HIW/SM was associated with higher rates of Clinician-Administered PTSD Scale–diagnosed PTSD (lifetime and current; partial and full) as well as anxiety and depressive symptoms. There was no current full PTSD for those exposed to LIW or HIW compared to 12 percent for HIW/SM.

For lifetime full PTSD, the corresponding rates were 0 percent (LIW), 6 percent (HIW), and 24 percent (HIW/SM). With regard to anxiety symptoms, 47 percent of the LIW group reported none compared to 29 percent (HIW) and 6 percent (HIW/SM). Reporting of no depressive symptoms showed a similar progression: LIW, 41 percent; HIW, 31 percent; and HIW/SM, 14 percent. Multivariate analysis showed that compared to the HIW group, the HIW/SM group were at elevated risk for lifetime PTSD (OR, 3.4; 95 percent CI, 1.5–7.4), current PTSD (OR, 6.2; 95 percent CI, 2.0–20.1), increased anxiety symptoms (OR, 5.6; 95 percent CI, 2.5–12.6), and increased depressive symptoms (OR, 3.7; 95 percent CI, 1.8–7.2).

Scud Missile Attacks on Israel, 1991

The Israeli experience of suffering 18 Scud missile attacks during the 1991 Gulf War involved both the effects of missile explosions and, at least initially, the perception of a possible nerve agent attack. One study of patients arriving in the emergency departments of 11 local hospitals in Israel (Bleich, Dycian, Koslowsky, Solomon, and Wiener, 1992) found that 332 (43 percent) of the 773 casualties were psychological casualties and an additional 209 (27 percent) had injected themselves with atropine because they thought the missiles contained nerve agent. After the first Scud attack, there were 365 casualties: 172 (47 percent) psychological casualties, 171 (47 percent) unjustified atropine injections, and only 22 (6 percent) physical injuries. Another study looked at patients reporting to the emergency department of a Tel Aviv hospital within 8 hours of a Scud attack (Rotenberg, Noy, and Gabbay, 1994). Of the 103 patients admitted, 70 (68 percent) were for psychological distress and 19 (18 percent) were for unjustified atropine injections; only 9 (9 percent) had direct injuries. The reported findings for the Israeli Scud missile experience are for civilians, not for soldiers. Both of these studies clearly demonstrate that psychological casualties should be expected to be at least as frequent as physical injuries, if not more frequent.

Tokyo Sarin Attack, 1995

In March 1995, terrorists released sarin, a nerve agent, in the Tokyo subway system during the morning rush hour (Asukai and Maekawa, 2002; Ohbu, Yamashina, Takasu, Yamaguchi, Murai, Nakano et al., 1997). Most of the victims were office workers commuting to central Tokyo. Despite the crowded conditions of the morning rush hour and the limited escape routes, there were no reports of mass panic. One fireman reported a "perplexing silence" at the accident scene—no talking, just the coughing of the victims as they awaited medical assistance (Asukai and Maekawa, 2002).

Over 5,500 people visited 280 medical facilities the following week and 1,046 were admitted as patients. St. Luke's International Hospital saw the most patients: 641 on the first day and 349 in the following week (Ohbu et al., 1997). Of the 641 patients admitted to the emergency department on the first day, 111 were admitted to the hospital for further treatment (4 severe cases, 107 moderate cases), and 530 mild cases were observed for 6 hours and then released from the emergency department. The mild cases suffered mainly from eye problems. It is difficult to determine from the literature how many of the mild cases may have actually been psychological casualties.

Most admitted patients from the Tokyo sarin attack were hospitalized for a few days. Some reported sleep disturbances, nightmares, and anxiety. Whether these were due to acute stress disorder or to exposure to nerve agent is not known. In studies conducted 1 month after the event, nearly 60 percent of casualties reported suffering from post-incident symptoms including fear of using the subway, sleep disturbances, flashbacks, depression, nightmares, irritability, headaches, malaise, physical tension, and emotional lability (Asukai and Maekawa, 2002; Ohbu et al., 1997). Follow-up questionnaires at 3- and 6-month intervals showed little decrease in the percentage reporting symptoms. Unfortunately, it is difficult to determine the extent to which these symptoms were either psychological effects or sequelae to the cholinergic effects of sarin exposure. Given that most of the casualties from St. Luke's were mild cases (suffered mainly eye symptoms), it is possible that many of the post-incident symptoms were psychological effects.

A long-term study of sarin patients who had been hospitalized at St. Luke's found that somatic and psychological symptoms continued for 5 years after the incident (Kawana, Ishimatsu and Kanda, 2001). There was a high rate of reporting of medically unexplained physical symptoms. Eye symptoms, fatigue, muscle stiffness, and headache were all reported by more than 10 percent of the study population.

According to follow-up studies done 3 and 5 years after the incident by the National Police Agency and the National Research Institute for Police Science, reporting of somatic complaints—eye strain, weakened eyesight, and easy fatigability—remained relatively stable from the acute stage through both follow-up periods (Asukai and Maekawa, 2002). Post-traumatic stress symptoms still reported by 14 to 18 percent of studied survivors included flashbacks, fear of the subway, intense distress at exposure to reminders of the attack, and avoidance of thinking about the attack (Asukai and Maekawa, 2002).

The casualties seen at St. Luke's Hospital were surveyed at 2, 3, and 5 years, using a questionnaire asking about 14 physical symptoms, 8 eye symptoms, and 11 psychological symptoms (avoidance, hyperarousal, and re-experiencing) (Kawana et al., 2001). The most common symptoms across all time periods were eye symptoms, broken down into eye strain (33–39 percent), dim vision (23–26 percent), and difficulty focusing (17–21 percent). Physical symptoms (tiredness, fatigue, muscle ache, headache) were also common. Most of the psychological symptoms remained stable over the three time periods, with rates of 10 to 16 percent still being reported at the 5-year point for memory difficulties, depressed mood, avoidance of accident reminders, flashbacks, and fear in the subway or at the attack site. PTSD by DSM–IV criterion remained stable with 2 to 3 percent meeting criteria during the three time points. The incidence of partial PTSD (one symptom from each category) ranged from 7 to 9 percent. Due to the persistence of physical symptoms, a modified set of PTSD criteria (adding at least one medically unexplained physical symptom to the diagnosis) were developed. Using these criteria, 10 to 14 percent met criteria for modified PTSD. The physical symptoms were reported to deteriorate following flashbacks and to improve during psychiatric therapy. The victims continued to be stressed by the lack of support from government systems, limits on resources available for medical follow-ups, and a feeling of stigmatization.

Radiological Agents

Three Mile Island

The Three Mile Island (TMI) nuclear power plant accident in 1979 demonstrated the importance of psychological effects in a CBRN-type event. According to the President's Commission that studied the accident, the only medical effect documented was mental distress (Kemeny, Babbitt, Haggerty, Lewis, Marks, Marrett et al., 1979). There were no cases of acute radiation sickness (ARS), and the estimated dose for people living within 10 miles of TMI was approximately that of an average chest x-ray and much lower than the annual background radiation dose. Populations exhibiting the most distress were TMI workers, families with preschool-age children, and those living within 5 miles of TMI (Fabrikant, 1983). Studies of TMI workers reported no long-term effects, only short-term acute effects. TMI personnel reported nausea, stomach troubles, headaches, diarrhea, sleep disturbances, and loss of appetite in greater frequency than did control group personnel. These symptoms are also common to the acute radiation sickness prodrome, but TMI personnel were not exposed to such doses.

No long-term psychological stress in TMI workers has been reported (Parkinson and Bromet, 1983). However, TMI residents, compared to controls, displayed a significant amount of stress on several measures (performance, self-report measures of anxiety, depression and somatic complaints, physiological measures of urinary norepinephrine, epinephrine, and cortisol, disturbed sleep, and changes in immune system parameters) for up to 6 years after the accident (Baum, Gatchel, and Schaeffer, 1983; Davidson, Fleming, and Baum, 1987; McKinnon, Weisse, Reynolds, Bowles, and Baum, 1989). It would appear that the TMI symptoms were not the result of exposure to radiation but to perceived radiation threat. Therefore, TMI demonstrates that fear of exposure to radiation can cause significant distress and stress symptoms that can mimic some of the symptoms of actual radiation exposure.

Chernobyl

Unlike the TMI accident, the Chernobyl accident in 1986 did release significant amounts of radiation. Approximately 135,000 people were evacuated from a 30-kilometer zone in the first 2 weeks after the accident. Most of these people had to be permanently relocated. In addition, an estimated 600,000 to 800,000 liquidators (workers who were brought in to handle the emergency situation and for the subsequent cleanup operations) were also exposed to significant levels of radiation (Nuclear Energy Agency [NEA], 2002). Although there were over 200 cases of ARS, the primary health effect was widespread psychological distress (NEA, 2002).

A variety of psychoneurological syndromes have been reported as sequelae of Chernobyl in the Russian literature (Novikov, Tsygan, Borisova, and Rybina, 1997; Pastel, 2002; Torubarov, 1991). These syndromes are characterized by medically unexplained physical symptoms (MUPS) including fatigue, sleep and mood disturbances, headaches, impaired memory and concentration, and muscle and/or joint pain.

These syndromes were reported in both liquidators who had suffered ARS and those who had not (Novikov et al., 1997; Torubarov, 1991). No significant correlations were found among physical symptoms, radiation dose, and physical examination data (Torubarov, 1991).

A study of over 1,400 Latvian liquidators found that 44 percent had an International Classification of Disease, 9th edition (ICD–9)–coded mental-psychosomatic disorder (depression, physiologic malfunction arising from mental factors, or unspecified disorders of the autonomic nervous system) (Viel, Curbakova, Dzerve, Eglite, Zvagule, and Vincent, 1997). Due to lack of ICD–9 codes, anxiety, PTSD, and sleep disturbances were not diagnosed. PTSD and PTSD symptoms have been found in two studies of people exposed to Chernobyl (Cwikel, Abdelgani, Goldsmith, Quastel, and Yevelson, 1997; Tarabrina, Lazebnaya, Zelenova, and Lasko, 1996).

An epidemiologic study of over 4,700 Estonian liquidators found an increase in suicide, but no increases in cancer, leukemia, or overall mortality (Rahu, Tekkel, Veidebaum, Pukkala, Hakulinen, Auvinen et al., 1997). Suicide accounted for almost 20 percent of mortality in the liquidator cohort. Reasons for the increased suicide rate are not currently known. However, data from Vietnam veterans with PTSD have demonstrated an increased risk for traumatic deaths, including suicide (Bullman and Kang, 1997). Given other studies demonstrating a variety of mental health disorders in Chernobyl liquidators, one could speculate that fear of radiation in these liquidators might cause depression, PTSD, and other disorders that are associated with increased rates of suicide. As such, the primary toxic agent appears to be fear, not radiation.

Goiania

In September 1987, two scavengers removed a cesium-137 (Cs-137) teletherapy unit from an abandoned radiotherapy institute in Goiania, the capital of Goias State in Brazil (International Atomic Energy Agency [IAEA], 1988). While dismantling the unit, they accidentally ruptured the source capsule, which contained radioactive Cs-137 powder. When the radiation accident became public, the perceived threat of radiation exposure caused over 120,000 people (approximately 10 percent of the city of 1.2 million people) to be screened over a 6-month span for possible contamination (Lipsztein, Cunha, and Oliveira, 1991). Residents and others who happened to be in the city at that time felt sufficiently at risk that they took time off from work or came on weekends to wait in line to be scanned (Petterson, 1988; Rosenthal, de Almeida, and Mendonca, 1991). Approximately 5,000 (8 percent) of the first 60,000 people screened presented with symptoms that mimicked acute radiation sickness (for example, rash around neck and upper body, vomiting, diarrhea), but none of these individuals were contaminated (Petterson, 1988). Only 249 people had measurable radiological contamination.

The 11 most seriously affected victims at Goiania had previously been hospitalized, but were all moved to one hospital. Due to immunosuppression, patients were kept confined and isolated and medical personnel wore protective masks. Both measures increased stress in patients. Most of the 20 hospitalized patients suffered from depression and anxiety (Brandao-Mello, Oliveira, and de Carvalho, 1991). Uncertainty about future health effects also increased stress, as did the lack of information concerning the duration of their treatment and the long-term prognosis.

The large number of people (approximately 10,000) who lived or worked within 300 meters of the contaminated area in Goiania exhibited fear, psychosomatic reactions, apprehension about the future, insecurity, and disbelief regarding the effectiveness of remedial measures the government took (Curado, Costa Neto, and Helou, 1991). A public opinion poll found that two-thirds of both affected Goiania residents and a control group living away from the contamination believed that Goiania was still contaminated. Research conducted 3 years later showed that stress parameters were still increased and performance was decreased both in nonirradiated individuals with perceived exposure (those living within 1 kilometer of Abadia, the area where contaminated waste from the incident had been stored) and in irradiated individuals from Goiania (Collins and de Carvalho, 1993).

Nuclear Weapons

Of all the CBRN agents, nuclear weapons have the greatest destructive impact; they are the quintessential weapons of mass destruction. The atomic weapons dropped on Hiroshima and Nagasaki caused incredible devastation, outbreaks of fires, and large numbers of dead, dying, and injured people (Janis, 1951). In interviews done after the war, approximately two-thirds described psychological disturbances of intense fear, emotional upset, or depression. Nevertheless, there was only a single incident of an apparent mass panic reported at Hiroshima:

a large group of frightened people in a park pressed some victims into a river, resulting in several deaths (Hersey, 1946).

During the following weeks, survivors continued to witness severely injured people suffering from burns and blast injuries. In addition, there were outbreaks of acute radiation sickness. The continued exposure to the devastation and human suffering served as a constant reminder to survivors and reinforced the psychological impact of the original event (Janis, 1951).

An estimated 200,000 Department of Defense personnel, both military and civilian, observed the early U.S. above-ground nuclear tests (Johnson, Goetz, and McRaney, 1986). The external doses received by these "atomic veterans" averaged about 0.5 rem, with many receiving no dose and only 1 percent receiving a dose greater than 5 rem (5 rem = maximum annual occupational dose). Several case studies of atomic veterans reported long-term psychological distress (Garcia, 1994; Vyner, 1983). Initially, troops at Desert Rock V seemed to go through the test-shot experience with equanimity (White, 1953), but many years later, an anecdotal study found that veterans reported vivid recollections of an atmosphere of tension and fear at the test sites and thought they had been ill prepared (Garcia, 1994).

Survivors of Hiroshima and Nagasaki were stigmatized, especially those with severe burns that resulted in scarring and keloids. Lifton (1967) described a "Neurasthenic Survivor Syndrome" characterized by "persistence of symptoms of withdrawal from social life, insomnia, nightmares, chronic depressive and anxiety reactions and far-reaching somatization.... in addition, fatigue, emotional lability, loss of initiative, and generalized personal, sexual and social maladaptation." A study of over 7,000 Nagasaki atomic bomb patients conducted 15 years later showed long-term psychological effects in approximately 7 percent, with the majority complaining of fatigue, lack of spirit, poor memory, and introversion (Nishikawa and Tsuiki, 1961). These symptoms were twice as common in survivors who had shown ARS symptoms and were related to severity of ARS symptoms.

A cluster of functional somatic symptoms reported in atomic veterans was dubbed the Radiation Response Syndrome (RRS) (Vyner, 1983). RRS has two components: a core belief that radiation had caused physical harm, and functional somatic symptoms that appeared to be an expression of this belief. The RRS belief system included the views "that men were dying, that doctors are of little help, that one doctor may exist who could help, that the government is to blame for their illness, and that people think they are crazy for blaming exposure to ionizing radiation for their illnesses" (Vyner, 1983). RRS resembles a delayed-onset PTSD, but rather than a re-experiencing of the trauma, the veteran is preoccupied with radiation and its impact on his life.

Biological Agents

U.S. Anthrax Attacks

In the fall of 2001, the United States was shocked by the mailing of anthrax-laced letters leading to 23 cases of anthrax (CDC, 2001). Following September 11, 2001, and prior to the first case of anthrax, there had already been media reports of increased purchases of gas masks and ciprofloxacin (commonly called cipro), which is an antibiotic used to treat anthrax. After the anthrax mail attack was discovered, there were increased patient requests for cipro as well as anecdotal reports of increased prescriptions for cipro (Steinhauer, 2001). Hospitals reported their already busy emergency rooms were filled with people anxious about anthrax, many demanding treatment.

Puzzling long-term effects were seen in the survivors of anthrax. Newspapers reported that survivors suffered symptoms of fatigue, shortness of breath, chest pains, memory problems, nightmares, and rage 6 to 12 months after their illnesses (Stolberg, 2002). Only one of the inhalational anthrax survivors was well enough to return to work. In the one published study of anthrax survivors one year after the attack, many reported reduced health-related quality of life (HR QOL) and psychological distress (Reissman, Whitney, Taylor, Hayslett, Dull, Arias et al., 2004).

1994 Plague Outbreak in India

In 1994, there were two outbreaks of plague in India: a bubonic plague outbreak in Maharashtra State followed a month later by a pneumonic plague outbreak 500 kilometers away in Surat (John, 1994). Of the 5,000 suspected cases of plague, there were 167 confirmed cases and 55 deaths (Mavalankar, 1994). Unfortunately, no data is readily available on psychological reactions or rates of such reactions. However, there were observable effects on behavior. The local media helped fuel the anxiety with exaggerated reports (Mavalankar, 1994; Madan, 1995). An estimated 400,000 to 600,000 people fled Surat, including hospital staff, private medical practitioners, and municipal workers (Kumar, 1994; Ramalingaswami, 1995). In Delhi, 1,200 kilometers from Surat, people fashioned masks from available materials and many bought and hoarded tetracycline, an antibiotic used to treat plague (Ramalingaswami, 2001).

Severe Acute Respiratory Syndrome

The outbreak of SARS, a new and emerging infection, likewise created much fear and anxiety. In Beijing, schools and universities were shuttered, hundreds of companies closed their doors, and some surrounding villages shut themselves off (Pomfret, 2003). Rumors of neighborhoods being quarantined led to people stockpiling food (Eckholm, 2003). Although officials asked people to avoid travel, thousands of businessmen, migrant workers, and college students left Beijing. In Taiwan, 160 doctors and nurses quit work at various hospitals out of fear that infection control measures were inadequate (McNeil, 2003).

SARS patients often spent hours in isolation between contacts with staff and were deprived of family visits, leading to complaints of sadness, anxiety, boredom, loneliness, and nonspecific anger and frustration (Avendano, Derkach, Swan, 2003; Maunder, Hunter, Vincent, Bennett, Peladeau, Leszcz et al., 2003). Fear and anxiety often waxed and waned with fever (Maunder et al., 2003).

One study used questionnaires to measure the psychosocial effects of SARS on staff in a Toronto hospital (Nickell, Crighton, Tracy, Al Enazy, Bolaji, Hanjrah et al., 2004). Almost two-thirds of the respondents reported concerns for their own or their family's health. Factors associated with increased concerns were perception of a greater risk of death from SARS, living with children, personal or family lifestyle affected by SARS outbreak, and being treated differently because of employment in a hospital. Emotional distress was found in almost 30 percent of all responders but was present among 45 percent of nurses, who were most at risk for infection. Factors identified for significant association with emotional distress were being a nurse, part-time employment status, lifestyle affected by SARS outbreak, and the ability to do one's job affected by precautionary measures.

Several infectious disease outbreaks have been reported to cause both PTSD and a decreased HR QOL. For example, the majority of survivors of an outbreak of Legionnaire's disease reported fatigue, neurologic symptoms, and neuromuscular symptoms 17 months after diagnosis (Lettinga, Verbon, Nieuwkerk, Jonkers, Gersons, Prins et al., 2002).

HR QOL was impaired in seven out of eight dimensions, and 15 percent of patients experienced PTSD. Similarly, survivors of acute respiratory distress syndrome have also reported PTSD and decreased HR QOL (Schelling, Stoll, Haller, Briegel, Manert, Hummel et al., 1998; Weinert, Gross, Kangas, Bury, and Marinelli, 1997). Because most of the most severe (Centers for Disease Control Category A) BW agents cause acute respiratory distress syndrome, similar long-term effects should be expected.

Summary

Risk perception will be a critical variable that mediates the psychological effects following a CBRN attack. Effective risk communication by authorities may decrease psychological effects while the mass media is likely to increase risk perception and possible psychological effects. Although mass panic can occur in situations involving limited escape routes, it will still likely be a rare event. Psychological effects are likely to cause large numbers of casualties following attacks using CBRN weapons. Initial presentation may resemble combat stress casualties or may present with a variety of nonspecific symptoms including difficulty breathing, dizziness, fatigue, headache, and sleep disturbances. Triage and differential diagnosis may present problems in the initial stages. The number of psychological casualties could increase based on possible amplification of risk perception by mass media reporting. Unlike most physical injuries or illnesses caused by CBRN agents, psychological effects can be contagious. People anxious about a CBRN attack may suffer from psychologically induced nonspecific symptoms such as fatigue, headache, muscle ache, and dizziness.

Given the history of postcombat war syndromes (Hyams et al., 1996), long-term effects are also very likely and will be difficult to diagnose and treat. These chronic OMUS syndromes will be greatly influenced by risk perception and mass media reporting.

References

Abenhaim, L., Dab, W., and Salmi, L.R. (1992). Study of civilian victims of terrorist attacks (France 1982–1987). *Journal of Clinical Epidemiology* 45, 103–109.

Anno, G.H., Baum, S.J., Withers, H.R., and Young, R.W. (1989). Symptomatology of acute radiation effects in humans after exposure to doses of 0.5–30 Gy. *Health Physics* 56, 821–838.

Asukai, N. and Maekawa, K. (2002). Psychological and physical health effects of the 1995 sarin attack in the Tokyo subway system. In J.M. Havenaar, J.G. Cwikel, and E.J. Bromet (eds.). *Toxic turmoil: Psychological and societal consequences of ecological disasters* (149–162). New York: Kluwer Academic/Plenum Publishers.

Avendano, M., Derkach, P., and Swan, S. (2003). Clinical course and management of SARS in health care workers in Toronto: A case series. *Canadian Medical Association Journal* 168, 1649–1660.

Balali-Mood, M. and Hefazi, M. (2005). The pharmacology, toxicology, and medical treatment of sulphur mustard poisoning. *Fundamental and Clinical Pharmacology* 19, 297–315.

Bartone, P. (2006). Resilience under military operational stress: Can leaders influence hardiness? *Military Psychology* 18 (Suppl.), S131–S148.

Baum, A., Gatchel, R.J., and Schaeffer, M.A. (1983). Emotional, behavioral, and physiological effects of chronic stress at Three Mile Island. *Journal Consulting and Clinical Psychology* 51, 565–572.

Bleich, A., Dycian, A., Koslowsky, M., Solomon, Z., and Wiener, M. (1992). Psychiatric implications of missile attacks on a civilian population: Israeli lessons from the Persian Gulf War. *Journal of the American Medical Association* 268, 613–615.

Brandao-Mello, C.E., Oliveira, A.R., and de Carvalho, A.B. (1991). Psychological effects of the Goiania radiation accident on the hospitalized victims. In R.C. Ricks, M.E. Berger, and F.M. O'Hara (eds.). *Medical basis for radiation-accident preparedness III: The psychological perspective* (121–129). New York: Elsevier.

Brecher, R.W., and Flynn, T. (2002). Principles of risk communication: building trust and credibility with the public. In W.M. Haschek, C.G. Rousseaux, and M.A. Wallig (eds.), *Handbook of toxicologic pathology.* New York: Academic Press.

Bullman, T.A., and Kang, H.K. (1997). Posttraumatic stress disorder and the risk of traumatic deaths among Vietnam veterans. In C.S.Fullerton and R.J. Ursano (eds.), *Posttraumatic stress disorder. Acute and long-term responses to trauma and disaster* (175–189). Washington, DC: American Psychiatric Press, Inc.

Caldwell, J.M., Ranson, S.W., and Sacks, J.G. (1951). Group panic and other mass disruptive reactions. *United States Armed Forces Medical Journal* 2, 541–567.

Carley, S.D., and Mackway-Jones, K. (1997). The casualty profile from the Manchester bombing 1996: A proposal for the construction and dissemination of casualty profiles from major incidents. *Journal of Accident and Emergency Medicine* 14, 76–80.

Centers for Disease Control (1988a). Health status of Vietnam veterans I. Psychosocial characteristics. The Centers for Disease Control Vietnam experience study. *Journal of the American Medical Association* 259, 2701–2707.

———. (1988b). Health status of Vietnam veterans II. Physical health. The Centers for Disease Control Vietnam experience study. *Journal of the American Medical Association* 259, 2708–2714.

———. (1988c). Health status of Vietnam veterans III. Reproductive outcomes and child health. The Centers for Disease Control Vietnam experience study. *Journal of the American Medical Association* 259, 2715–2719.

———. (2001). Update: Investigation of bioterrorism-related inhalational anthrax—Connecticut, 2001. *Morbidity and Mortality Weekly Report* 50, 1049–1051.

Collins, D.L., and de Carvalho, A.B. (1993). Chronic stress from the Goiania 137Cs radiation accident. *Behavioral Medicine* 18, 149–157.

Covello, V.T., Peters, R.G., Wojtecki, J.G., and Hyde, R.C. (2001). Risk communication, the West Nile virus epidemic, and bioterrorism: Responding to the communication challenges posed by the intentional or unintentional release of a pathogen in an urban setting. *Journal of Urban Health* 78, 382–391.

Curado, M.P., Costa Neto, S.B., and Helou, S. (1991). Psychological aspects of the radiation accident in Goiania: a general overview on victims and population. In R.C. Ricks, M.E. Berger, and F.M. O'Hara (eds.), *Medical basis for radiation-accident preparedness III. The psychological perspective* (143–154). New York: Elsevier.

Curran, P.S., Bell, P., Murray, A., Loughrey, G., Roddy, R., and Rocke, L.G. (1990). Psychological consequences of the Enniskillen bombing. *British Journal of Psychiatry* 156, 479–482.

Cwikel, J., Abdelgani, A., Goldsmith, J.R., Quastel, M., and Yevelson, I.I. (1997). Two-year follow up study of stress-related disorders among immigrants to Israel from the Chernobyl area. *Environmental Health Perspectives* 105, 1545-1550.

Davidson, L.M., Fleming, R., and Baum, A. (1987). Chronic stress, catecholamines, and sleep disturbance at Three Mile Island. *Journal of Human Stress* 13, 75-83.

Decoufle, P., Holmgreen, P., Boyle, C.A., and Stroup, N.E. (1992). Self-reported health status of Vietnam veterans in relation to perceived exposure to herbicides and combat. *American Journal of Epidemiology* 135, 312-323.

DiGiovanni, C. (1999). Domestic terrorism with chemical or biological agents: Psychiatric aspects. *American Journal of Psychiatry* 156, 1500-1505.

Eckholm, E. (2003, April 24). Illness psychological impact in China exceeds its actual numbers. *The New York Times*, A13.

Engel, C.C., Jr., Adkins, J.A., and Cowan, D.N. (2002). Caring for medically unexplained physical symptoms after toxic environmental exposures: Effects of contested causation. *Environmental Health Perspectives* 110, 641-647.

Fabrikant, J.I. (1983). The effects of the accident at Three Mile Island on the mental health and behavioral responses of the general population and nuclear workers. *Health Physics* 45, 579-586.

Garcia, B. (1994). Social-psychological dilemmas and coping of atomic veterans. *American Journal of Orthopsychiatry* 64, 651-655.

Ghanei, M., Aslani, J., Khateri, S., and Hamadanizadeh, K. (2003). Public health status of the civil population of Sardasht 15 years following large-scale wartime exposure to sulfur mustard. *Journal of Burns and Surgical Wound Care* 2, 7-18.

Gilchrist, H.L. (1928). *Comparative study of world war casualties from gas and other weapons.* Washington, DC: U.S. Government Printing Office.

Glass, T.A., and Schoch-Spana, M. (2002). Bioterrorism and the people: How to vaccinate a city against panic. *Clinical Infectious Disease* 34, 217-223.

Gothe, C.J., Molin, C., and Nilsson, C.G. (1995). The environmental somatization syndrome. *Psychosomatics* 36, 1-11.

Hadden, W.A., Rutherford, W.H., and Merrett, J.D. (1978). The injuries of terrorist bombing: A study of 1532 consecutive patients. *British Journal of Surgery* 65, 525-531.

Haldane, J.S. (1919). Lung-irritant gas poisoning and its sequelae. *Journal of the Royal Army Medical Corps* 33, 494-507.

Hammerman, G. (1987). The psychological impact of chemical weapons on combat troops in World War I. In R.W. Young and B.H. Drum (eds.), *Proceedings of the defense nuclear agency symposium/workshop on the psychological effects of tactical nuclear warfare* (87-209). Washington, DC: Defense Nuclear Agency.

Hashemian, F., Khoshnood, K., Desai, M.M., Falahati, F., Kasl, S., and Southwick, S. (2006). Anxiety, depression, and posttraumatic stress in Iranian survivors of chemical warfare. *Journal of the American Medical Association* 296, 560-566.

Hersey, J. (1946). *Hiroshima.* New York: Knopf.

Hulbert, H.S. (1920). Gas neurosis syndrome. *American Journal of Insanity* 77, 213-216.

Hyams, K.C., Murphy, F.M., and Wessely, S. (2002). Responding to chemical, biological, or nuclear terrorism: The indirect and long-term health effects may present the greatest challenge. *Journal of Health Politics, Policy and Law* 27, 273-291.

Hyams, K.C., Wignall, F.S., and Roswell, R. (1996). War syndromes and their evaluation: From the U.S. Civil War to the Persian Gulf War. *Annals of Internal Medicine* 125, 398-405.

International Atomic Energy Agency (1988). *The radiological accident in Goiania.* Vienna: International Atomic Energy Agency.

Janis, I.L. (1951). *Air war and emotional stress. Psychological studies of bombing and civilian defense.* New York: RAND Corporation.

John, T.J. (1994). Learning from plague in India. *Lancet* 344, 972.

Johnson, A.A., Goetz, J.L., and McRaney, W.K. (1986). *For the record—A history of the nuclear test personnel review program, 1978-1986* (Rep. No. DTIC AD No. ADA190410). Washington, DC: U.S. Government Printing Office.

Kawana, N., Ishimatsu, S., and Kanda, K. (2001). Psycho-physiological effects of the terrorist sarin attack on the Tokyo subway system. *Military Medicine* 166, 23-26.

Kemeny, J.G., Babbitt, B., Haggerty, P.E., Lewis, C., Marks, P., Marrett, C.B. et al. (1979). *Report of the president's commission on the accident at Three Mile Island*. New York: Pergamon Press.

Khateri, S., Ghanei, M., Keshavarz, S., Soroush, M., and Haines, D. (2003). Incidence of lung, eye, and skin lesions as late complications in 34,000 Iranians with wartime exposure to mustard agent. *Journal of Occupational and Environmental Medicine* 45, 1136-1143.

Khateri, S., Ghanei, M., and Palmer, C. (2004). Long-term psycho-social impact of a chemical weapons attack on the population of Sardasht, Iran. *Applied Science and Analysis Newsletter*.

Kumar, S. (1994). Plague in India. *Lancet* 344, 941-942.

Lau, J.T., Yang, X., Tsui, H., and Kim, J.H. (2003). Monitoring community responses to the SARS epidemic in Hong Kong: From day 10 to day 62. *Journal of Epidemiology and Community Health* 57, 864-870.

Lettinga, K.D., Verbon, A., Nieuwkerk, P.T., Jonkers, R.E., Gersons, B.P., Prins, J.M. et al. (2002). Health-related quality of life and posttraumatic stress disorder among survivors of an outbreak of legionnaires disease. *Clinical Infectious Disease* 35, 11-17.

Leung, G.M., Lam, T.H., Ho, L.M., Ho, S.Y., Chan, B.H., Wong, I.O. et al. (2003). The impact of community psychological responses on outbreak control for severe acute respiratory syndrome in Hong Kong. *Journal of Epidemiology and Community Health* 57, 857-863.

Levin, S.G. (1991). *Estimating battle fatigue casualties in tactical nuclear combat* (Rep. No. DNA–TR–90–114). Washington, DC: Defense Nuclear Agency.

Lifton, R. (1967). *Death in life: Survivors of Hiroshima*. New York: Random House.

Lipsztein, J.L., Cunha, P.G., and Oliveira, C.A. (1991). The Goiania accident: Behind the scenes. *Health Physics* 60, 5-6.

Madan, T.N. (1995). The plague in India, 1994. *Social Science and Medicine* 40, 1167-1168.

Maunder, R., Hunter, J., Vincent, L., Bennett, J., Peladeau, N., Leszcz, M. et al. (2003). The immediate psychological and occupational impact of the 2003 SARS outbreak in a teaching hospital. *Canadian Medical Association Journal* 168, 1245-1251.

Mavalankar, D.V. (1994). Plague in India. *Lancet* 344, 1298.

McKinnon, W., Weisse, C.S., Reynolds, C.P., Bowles, C.A., and Baum, A. (1989). Chronic stress, leukocyte subpopulations, and humoral response to latent viruses. *Health Psychology* 8, 389-402.

McNeil, D.G. (2003, May 21). SARS fears shake Taiwan medical staffs. *The New York Times*, A14.

Nickell, L.A., Crighton, E.J., Tracy, C.S., Al Enazy, H., Bolaji, Y., Hanjrah, S. et al. (2004). Psychosocial effects of SARS on hospital staff: Survey of a large tertiary care institution. *Canadian Medical Association Journal* 170, 793-798.

Nishikawa, T., and Tsuiki, S. (1961). Psychiatric investigations of atomic bomb survivors. *Nagasaki Medical Journal* 36, 717-722.

Norris, G.W. (1919). Some medical impressions of the war. *American Journal of the Medical Sciences* 157, 628-634.

North, C.S., Nixon, S.J., Shariat, S., Mallonee, S., McMillen, J.C., Spitznagel, E.L. et al. (1999). Psychiatric disorders among survivors of the Oklahoma City bombing. *Journal of the American Medical Association* 282, 755-762.

Novikov, V.S., Tsygan, V.N., Borisova, E.D., and Rybina, L.A. (1997). Changes in cerebral bioelectric activity in the Chernobyl NPP accident liquidators. *Human Physiology* 23, 542-546.

Nuclear Energy Agency (2002). *Chernobyl. Assessment of radiological and health impacts. 2002 update of Chernobyl: Ten years on*. Paris: Organization for Economic Co-operation and Development.

Ohbu, S., Yamashina, A., Takasu, N., Yamaguchi, T., Murai, T., Nakano, K. et al. (1997). Sarin poisoning on Tokyo subway. *Southern Medical Journal* 90, 587–593.

Parkinson, D.K., and Bromet, E.J. (1983). Correlates of mental health in nuclear and coal-fired power plant workers. *Scandinavian Journal of Work, Environment and Health* 9, 341–345.

Pastel, R.H. (2002). Radiophobia: Long-term psychological consequences of Chernobyl. *Military Medicine* 166, 134–136.

Petterson, J.S. (1988). Perception vs. reality of radiological impact: The Goiania model. *Nuclear News* 31, 84–90.

Pfefferbaum, B., Nixon, S.J., Tivis, R.D., Doughty, D.E., Pynoos, R.S., Gurwitch, R.H. et al. (2001). Television exposure in children after a terrorist incident. *Psychiatry* 64, 202–211.

Pomfret, J. (2003, May 1). SARS inciting "mass panic" in Beijing. *The Washington Post.*

Quarantelli, E.L. (1960). Images of withdrawal behavior in disasters: Some basic misconceptions. *Social Problems* 8, 68–79.

Rahu, M., Tekkel, M., Veidebaum, T., Pukkala, E., Hakulinen, T., Auvinen, A. et al. (1997). The Estonian study of Chernobyl cleanup workers: II. Incidence of cancer and mortality. *Radiation Research* 147, 653–657.

Ramalingaswami, V. (1995). Plague in India. *Nature Medicine* 1, 1237–1239.

———. (2001). Psychosocial effects of the 1994 plague outbreak in Surat, India. *Military Medicine* 166, 29–30.

Reissman, D.B., Whitney, E.A., Taylor, T.H., Hayslett, J.A., Dull, P.M., Arias, I. et al. (2004). One-year health assessment of adult survivors of bacillus anthracis infection. *Journal of the American Medical Association* 291, 1994–1998.

Ritchie, E.C., Watson, P.J., and Friedman, M.J. (2006). *Interventions following mass violence and disasters.* New York: Guilford Press.

Rosenthal, J.J., de Almeida, C.E., and Mendonca, A.H. (1991). The radiological accident in Goiania: The initial remedial actions. *Health Physics* 60, 7–15.

Rotenberg, Z., Noy, S., and Gabbay, U. (1994). Israeli ED experience during the Gulf War. *American Journal of Emergency Medicine* 12, 118–119.

Rothman, A.L., and Weintraub, M.I. (1995). The sick building syndrome and mass hysteria. *Neurology Clinics* 13, 405–412.

Salmon, T.W., and Fenton, N. (1929). Neuropsychiatry in the American expeditionary force. In T.W. Salmon and N. Fenton (eds.), *Volume 10. Neuropsychiatry* (217–474). Washington, DC: U.S. Government Printing Office.

Sandman, P.M. (2003). Bioterrorism risk communication policy. *Journal of Health Communication* 8 (Suppl. 1), 146–147.

Scharf, T., Vaught, C., Kidd, P., Steiner, L., Kowalski, K., Wiehagen, B. et al. (2001). Toward a typology of dynamic and hazardous work environments. *Human Ecology and Risk Assessment* 7, 1827–1841.

Schelling, G., Stoll, C., Haller, M., Briegel, J., Manert, W., Hummel, T. et al. (1998). Health-related quality of life and posttraumatic stress disorder in survivors of the acute respiratory distress syndrome. *Critical Care Medicine* 26, 651–659.

Schuster, M.A., Stein, B.D., Jaycox, L., Collins, R.L., Marshall, G.N., Elliott, M.N. et al. (2001). A national survey of stress reactions after the September 11, 2001, terrorist attacks. *New England Journal of Medicine* 345, 1507–1512.

Showalter, E. (1997). *Hystories: Hysterical epidemics and modern culture.* New York: Columbia University Press.

Singer, J.E. (1982). Yes Virginia, there really is a mass psychogenic illness. In M.J. Colligan, J.W. Pennebaker, and L.R. Murphy (eds.), *Mass psychogenic illness* (127–135). Hillsdale, NJ: Lawrence Erlbaum Associates, Inc.

Smelser, N.J. (1962). *Theory of collective behavior.* New York: The Free Press.

Steinhauer, J. (2001, October 21). Hysteria can be hazardous. *The New York Times*, C9.

Stolberg, S.G. (2002, May 7). For anthrax survivors, a halting, painful recovery. *The New York Times*, F1.

Stouffer, S.A., Lumsdaine, A.A., Lumsdaine, M.H., Williams, R.M., Jr., Smith, M.B., Janis, I.L. et al. (1949). *The American soldier: combat and its aftermath.* Princeton: Princeton University Press.

Struewing, J.P., and Gray, G.C. (1990). An epidemic of respiratory complaints exacerbated by mass psychogenic illness in a military recruit population. *American Journal of Epidemiology* 132, 1120–1129.

Tarabrina, N., Lazebnaya, E., Zelenova, M., and Lasko, N. (1996). Chernobyl clean-up workers' perception of radiation threat. *Radiation Protection and Dosimetry* 68, 251–255.

Torubarov, F.S. (1991). Psychological consequences of the Chernobyl accident from the radiation neurology point of view. In R.C. Ricks, M.E. Berger, and F.M. O'Hara (eds.), *The medical basis for radiation-accident preparedness III: The psychological perspective* (81–91). New York: Elsevier Science Publishing.

U.S. Army Medical Research Institute of Chemical Defense (2000). *Medical Management of Chemical Casualties Handbook* (3ᵈ ed.). Aberdeen Proving Grounds, MD: USAMRICD.

Vasterman, P., Yzermans, C.J., and Dirkzwager, A.J. (2005). The role of the media and media hypes in the aftermath of disasters. *Epidemiolic Reviews* 27, 107–114.

Viel, J.F., Curbakova, E., Dzerve, B., Eglite, M., Zvagule, T., and Vincent, C. (1997). Risk factors for long-term mental and psychosomatic distress in Latvian Chernobyl liquidators. *Environmental Health Perspectives* 105, 1539–1544.

Vineberg, R. (1965). *Human factors in tactical nuclear combat* (Rep. No. HumRRO TR 65-2). Alexandria, VA: George Washington University, Human Resources Research Office.

Vyner, H.M. (1983). The psychological effects of ionizing radiation. *Culture, Medicine and Psychiatry* 7, 241–261.

Wachtel, C. (1941). *Chemical warfare.* Brooklyn, NY: Chemical Publishing Co., Inc.

Weinert, C.R., Gross, C.R., Kangas, J.R., Bury, C.L., and Marinelli, W.A. (1997). Health-related quality of life after acute lung injury. *American Journal of Respiratory and Critical Care Medicine* 156, 1120–1128.

Weiss, B. (1998). Neurobehavioral properties of chemical sensitivity syndromes. *Neurotoxicology* 19, 259–268.

Chapter 14

Mitigating the Medical Consequences of Exposure to Chemical Weapons: Contributions of U.S. Army Research Psychologists to Chemical Defense

Maurice L. Sipos and Matthew G. Clark

U.S. Army research psychologists play a key role in evaluating the behavioral and functional impact of all sorts of chemical and pharmaceutical toxicants, including chemical warfare agents and potential therapeutics. Additionally, they are essential to the mission of developing safe and effective therapies for warfighters, particularly medical countermeasures against chemical warfare agents. This chapter provides a brief survey of the contributions of research psychologists in support of the Medical Chemical Defense Program with emphasis on the effects of chemical warfare nerve agents (CWNAs) like sarin, soman, tabun, and VX and the development of medical countermeasures.

Introduction

Chemical warfare agents (CWAs) are extremely toxic chemicals that are intended to kill, seriously injure, or incapacitate military and civilian personnel. The threat of CWAs on the battlefield dates back to ancient times. The American perspective on the history of chemical warfare is summarized in the *Textbook of Military Medicine (Part 1): Medical Aspects of Chemical and Biological Warfare* (Smart, 1997).

In brief, World War I marked the beginning of the modern era of chemical warfare. Its horror was captured by Lieutenant Wilfred Owen of the Royal Army in the poem entitled "Dulce et Decorum Est" (Owen, 1963). The image of a soldier "guttering, choking, drowning" permeated the narrator's dreams, forcing him to relive the horror of hearing "the blood come gargling from the froth-corrupted lungs." Chlorine, phosgene, and sulfur mustard were common CWAs during World War I, and they had a devastating psychological impact on Allied forces that the Germans were not prepared to exploit.

Following World War I, there was a rapid escalation in the science surrounding all aspects of chemical warfare, to include research and development of decontamination methods, chemical detection and alarms, treatment strategies, protective devices, and mechanisms to deliver CWAs to the enemy. The discovery by German chemist Dr. Gerhard Schrader in the 1930s of tabun, an extremely toxic organophosphorus insecticide that affected the nervous system, led to the development of a new class of CWA—nerve agents including sarin, soman, and VX (Smart, 1997).

Although CWAs have not been used extensively since World War I, several incidents involving them in the last half-century underscored the need to continue toxicological research on the threats and develop countermeasures for defending military and civilian populations from their indiscriminate effects (Salem, Ternay, and Smart, 2008). For example, Iraq used sulfur mustard and nerve agents against Iran during the Iran-Iraq War in the 1980s. In the mid-1990s, Aum Shinrikyo, a Japanese religious sect, used the nerve agents sarin and VX in a series of terrorist incidents that specifically targeted civilians.

The fact that CWAs are relatively easy and inexpensive to synthesize on a large scale, coupled with the fact that rogue nations and militant groups rely on asymmetric warfare, has led pundits to state that it is not a matter of *if*, but *when*, CWAs will be used on American soil or against U.S. military and civilians. The Defense Threat Reduction Agency and the National Institutes of Health continue to invest in research directed toward mitigating and eliminating the medical consequences of exposure to CWAs.

Outside of counterterrorist activities, the U.S. Army is aggressively expanding preparations to conduct full-spectrum operations worldwide as evidenced by new doctrine on counterinsurgency and stability operations (U.S. Army Field Manual [FM] 3–07, *Stability Operations* [Army, 2008b]; FM 3–24, *Counterinsurgency* [Army, 2006]; FM 7–0, *Training for Full Spectrum Operations* [Army, 2008c]; FM 3–0, *Operations* [Army, 2008a]). These military operations emphasize the increased potential for unconventional attacks on conventional U.S. forces involving asymmetric threats. In addition to the use of improvised explosive devices, these asymmetric threats will likely include CWAs ranging from toxic industrial compounds (such as chlorine and bromine) to deadly nerve agents (for example, sarin and soman), as has been exhibited to some degree in current operations. Thus, understanding the functional psychological and toxicological impact of CWAs and potential therapies is critical for both U.S. citizens and military personnel.

U.S. Army research psychologists are essential for supporting the mission of evaluating the behavioral and functional impact of CWAs and potential therapeutics. More specifically, as uniformed military scientists, they are trained to employ and understand holistic functional assessments in the areas of psychobiology, toxicology, and neuroscience, while also receiving extensive training and field experience in military operational medicine. As presented in this chapter, Army research psychologists conducted much of the early research on the performance impacts of nerve agents and their pharmacological countermeasures, and are ideally suited to investigate nerve agent–induced seizure activity and the resulting brain injury.

While this chapter has a limited focus, it must be noted that the breadth and volume of work done by Army neuroscientists, psychobiologists, and neurotoxicologists could fill a separate volume. This chapter will survey the contributions of U.S. Army research psychologists in support of the Medical Chemical Defense Program, with emphasis on the effects of CWNAs such as sarin, soman, tabun, and VX.

Defining the Physiological Toxicity of Chemical Warfare Nerve Agents

Research on the neurobehavioral and physiological impact of CWNAs has been limited due to the highly toxic nature of these compounds. U.S. Army and allied research psychologists provided much of the seminal data on these threats. They conducted research on the physiological and behavioral toxicity of cholinesterase-inhibiting nerve agents.

Classical CWNAs are extremely toxic organophosphorus cholinesterase inhibitors that cause a rapid progression of clinical signs that include hypersecretions, fasciculations, generalized convulsions, coma, and eventually death due primarily to respiratory failure (Taylor, 1990). The physiological consequences of exposure to CWNAs have been defined in part by research psychologists over the past three decades.

In the 1980s, research psychologists helped characterize the effects of CWNAs on general health, lethality, and brain function (McDonough, Hackley, Cross, Samson, and Nelson, 1983; Churchill et al., 1984; Rickett, Glenn, and Beers, 1986; McDonough, McLeod, and Nipwoda, 1987). For example, McDonough et al. (1983) suggested that the brain damage seen in rats following soman exposure was likely due to the excessive neuronal stimulation that occurs during prolonged soman-induced seizures. Other research psychologists determined that the predominant cause of CWNA-induced respiratory failure was related to the effects of CWNAs on the central respiratory drive rather than to their peripheral effects in the diaphragm (Rickett et al., 1986).

To further elucidate the central effects of CWNAs, McDonough et al. (1987) directly infused soman or VX into discrete brain regions and found that neuropathology was only observed in animals that developed repetitive convulsions. Their results indicated that CWNAs were not directly neurotoxic, that peripherally induced hypoxia or anoxia was unlikely to be the mechanism of the neuropathology, and that the brain damage produced by these compounds was primarily seizure-mediated (McDonough et al., 1987). These studies identified the need to include an anticonvulsant in the standard treatment regimen, as described below.

Age-related differences in soman toxicity, in signs of soman intoxication, and in blood and brain regional cholinesterase activity have also been examined by military research psychologists and their civilian colleagues (Shih, Penetar, McDonough, Romano, and King, 1990). Shih et al. (1990) found that older rats were more susceptible than younger rats to the lethal and nonlethal effects of soman. These findings had important implications for subsequent neurobehavioral, toxicological, and pharmacological studies of CWNAs across the lifespan.

More recently, U.S. Army research psychologists and their colleagues have assessed the consequences of repeated sublethal exposure to CWNAs (Moffett, Price, Anderson, Sipos, Moran, Tortella, and Dave, 2003; Dave, Connors, Genovese, Whipple, Chen, DeFord, Moran, and Tortella, 2007; Johnson, Daugherty, Gallagher, Moran, and DeFord, 2008). Repeated exposures to sublethal doses of soman (Moffett et al., 2003) and sarin (Dave et al., 2007) significantly increased DNA fragmentation in blood leukocytes, suggesting that assays measuring leukocyte DNA fragmentation may provide a sensitive biomarker for low-dose CWNA exposure. Repeated sublethal exposure to soman has also been found to affect glutamate receptor subunit expression (Johnson et al., 2008) that may contribute to other cognitive deficits observed following sublethal nerve agent exposure.

Defining the Behavioral Toxicity of Chemical Warfare Nerve Agents

Military research psychologists have conducted a significant number of studies to understand the functional manifestations of neurotoxicants including CWNAs. By understanding the functional impact of exposure to CWNAs, research psychologists gained insight into their mechanism of action and were thus able to evaluate whether therapeutic strategies were effective countermeasures devoid of untoward behavioral side effects if administered in the absence of exposure.

Functional assessments of toxicity in rodents have involved a variety of techniques ranging from very simple to complicated measures of behavior. For example, Romano and Landauer (1986) studied the effects of tabun on motor performance and balance using a rotarod and were able to determine the dose range that disrupted performance on this task. Similarly, Koplovitz, Romano, and Stewart (1989) used a very simple inverted screen test to assess motor function in mice following soman poisoning to rapidly screen the effectiveness of pretreatment and treatment compounds against CWNA toxicity. In their study, Koplovitz and colleagues (1989) found that pretreatment with pyridostigmine bromide combined with therapy of atropine and pralidoxime chloride (2-PAM) after soman exposure significantly improved performance on the inverted screen test. All three of these compounds are currently fielded as components of the nerve agent antidote kit and are discussed more below.

A variety of other assessment techniques have also been used to define the functional impact of CWNAs, including measuring acute signs of toxicity and weight loss (Shih et al., 1990), conditioned taste aversions (Romano and Landauer, 1986), nociception using tail flick and hot plate tests (Romano, King, and Penetar, 1985), conditioned avoidance (Romano, Terry, Murrow, and Mays, 1991), fixed ratio operant responding and extinction (Harris, McDonough, Stitcher, and Lennox, 1984; Modrow and Jaax, 1989), acoustic startle response (Sipos, Moran, Kahler, Lukefahr, and DeFord, 2005), active avoidance learning (Sipos et al., 2005), and spatial memory via the Morris water maze (Johnson et al., 2008).

Research psychologists have used more complex behavioral assessments to determine the correlation between brain damage and performance in some of these tasks. For example, McDonough, Smith, and Smith (1986) demonstrated that exposure to high doses of soman can lead to neural damage as well as significant, long-term decrements in certain learned behaviors. Raffaele, Hughey, Wenk, Olton, Modrow, and McDonough (1987) demonstrated that rats that performed abnormally in behavioral tests following soman exposure were more likely to have abnormal brain pathology than rats that performed normally in the same behavioral tests.

Behavioral assessments of toxicity in nonhuman primates have included assessments of cognition and short-term memory using serial probe recognition (Castro, Gresham, Finger, Maxwell, Solana, Lenz, and Broomfield, 1994), balance and equilibrium (Doctor, Blick, Caranto, Castro, Gentry, Larrison et al., 1993), and the rate and incidence of convulsions (Castro, Larsen, Finger, Solana, and McMaster, 1992), to name only a few. All of these studies were useful in establishing the neurotoxic effects of CWNAs, and they served as a battery of tests that were used subsequently to evaluate the effectiveness of therapeutics in preventing or ameliorating the consequences of exposure.

Therapeutic Development

In general, research on therapeutics against CWNAs has focused on four types of compounds: anticholinergics (for example, atropine) that reduce the impact of cholinergic hyperactivity; cholinesterase reactivators (such as 2-PAM); short-acting cholinesterase inhibitors (for example, pyridostigmine bromide) that serve as a prophylactic to physically block CWNAs; and anticonvulsants (such as diazepam and its likely replacement midazolam). The current standard therapy for nerve agent exposure in the military now includes one of each type of these compounds. Research psychologists helped characterize the impact of these compounds and contributed to the development of the new and novel therapeutics detailed below.

Regarding assessments of general health and toxicity, research psychologists demonstrated that the therapeutic regimen of atropine and 2-PAM provides protection against soman-induced lethality and behavioral incapacitation of a conditioned avoidance response in guinea pigs (Romano et al., 1991). More specifically, McDonough, Jaax,

Crowley, Mays, and Modrow (1989) demonstrated that atropine and diazepam provide dose-dependent protection against soman-induced epileptiform seizure activity and cardiac and neural pathology. However, while this regimen reduced mortality, it did not eliminate most of the acute signs of CWNA toxicity.

In pharmaceutical development, scientists must understand the mechanism of action, the functional behavioral impact, and the potential side effects of a given compound. Research psychologists helped demonstrate that while atropine increases the core body temperature, repeated exposure to the prophylactic pyridostigmine offsets this increase and improves heat tolerance and thermoregulation (Matthew, Glenn, Bowers, and Navara, 1994). Given the environmental range within which warfighters operate (from sweltering deserts to frigid tundra), any effect on thermoregulation must be understood for the effective planning of military operations.

Romano and Shih (1983, 1984) examined anticholinesterase physostigmine as a potential therapeutic and demonstrated that the cholinergic system is likely involved in centrally mediated analgesia. They also identified key targets in the brain for the analgesic/antinociceptive effects of physostigmine (Romano and Shih, 1984). Similarly, Romano et al. (1985) identified the limits of nociception and conditioned taste aversion produced by physostigmine. This mechanistic understanding provided important information about potential effects and unintended side effects of CWNAs and therapeutics. Furthermore, this understanding helps military scientists identify targets for future drug development.

In addition to more basic biological assessments, research psychologists bring a specific expertise in toxicological assessments of whole animals (including humans), particularly in the area of behavior and performance. Research psychologists used behavioral screening of large numbers of therapeutic compounds including various anticholinergics in search of a potential replacement for atropine. Sipos, Burchnell, and Galbicka (1999, 2001) screened eight different anticholinergic compounds using an assessment of locomotor activity and the acoustic startle reflex. These behavioral assessments provided an excellent way to quickly evaluate overt peripheral and central nervous system function. While still in development, this research provided data that proved useful in down-selecting potential drugs before testing with more complex behavioral assessments. For example, these data collectively demonstrated that biperiden, benactyzine, and procyclidine might serve as candidate drugs for further development.

Beyond basic physiological and simple behavioral assessments, research psychologists also employ and evaluate higher order cognitive assessments including coordinated fine motor performance, memory, cognition, and neurobehavioral performance. These assessments serve as very sensitive neurobehavioral and psychological measures that can detect changes in the central and peripheral nervous systems. In addition to enhanced sensitivity, results from these assessments correlate well across species, thus allowing screening of candidate drugs using less expensive species. Lastly, these assessments examine the impact of CWNAs on the entire organism and go well beyond more common biological assessments that tend to focus on specific systems or portions of biological systems.

Various anticholinergics including atropine, benactyzine, scopolamine, and biperiden and the anticholinesterase physostigmine have been evaluated using complex behavioral assessments that included evaluations of delayed reinforcement of low rate schedules, repeated acquisition, delayed match to sample, and serial probe recognition. These sensitive assessments have examined fine and gross motor learning and performance, and memory tasks including short-term memory, primacy and recency, and serial list recall (Castro, 1997; McDonough, 1982; Myers, Galbicka, Sipos, Varadi, Oubre, and Clark, 2002; Penetar, 1985; Penetar and McDonough, 1983).

Research on anticholinergics using complex behavioral assessments has characterized the dose-dependent impact of these drugs on sensitive and highly militarily relevant behaviors and cognition. This characterization helps combat and medical leadership and advisors understand the relative benefit and potential risks of current and future therapeutics. Clearly, the standard therapy against CWNAs significantly reduces mortality. However, the research indicates that when therapeutic anticholinergics are given alone, they too may have a deleterious effect on motor performance, cognition, and short-term memory. The research also indicates that the cholinergically mediated impact on memory and centrally mediated performance deficits might be moderated by the development and use of a centrally active anticholinesterase prophylactic as a possible future replacement for pyridostigmine bromide.

As previously mentioned, the standard therapy against CWNAs now includes the anticonvulsant diazepam (brand name Valium).

Research psychologists helped determine that the cessation of convulsions and epileptiform seizure activity is critical for enhanced survivability, reduced cardiac and brain pathology, and long-term outcome (McDonough et al., 1989).

Behaviorally, it was also demonstrated that when given alone, diazepam impacts primacy and recency as measured in a serial probe recognition assessment (Castro et al., 1992). Moreover, Castro and colleagues (1992) demonstrated that when combined with the standard therapy, diazepam improved the outcome following exposure to CWNAs. The addition of diazepam reduced the incidence of convulsions and attenuated CWNA-induced behavioral impairments in a serial probe recognition task. Functionally, diazepam-treated nonhuman primates exposed to soman recovered in only 6 days compared to non-diazepam-treated control subjects that required 15 days for recovery. Additionally, only one of five diazepam-treated subjects exhibited the excitotoxic epileptiform convulsions whereas all five control subjects that received only the standard therapy demonstrated this deleterious activity. Thus, research psychologists played a key role in demonstrating that diazepam is a valuable addition to the standard therapy against CWNA exposure.

Beyond the standard therapy of pyridostigmine, atropine, 2-PAM, and diazepam, research psychologists contributed to the development of novel bioscavengers of CWNAs (Broomfield, Maxwell, Solana, Castro, Finger, and Lenz, 1991; Castro et al., 1994; Clark, Sun, Myers, Bansal, Doctor, and Saxena, 2005; Doctor et al., 1993; Saxena, Luo, Bansal, Sun, Clark, Ashani, and Doctor, 2004). CWNA bioscavengers are primarily enzymes (for example, acetylcholinesterase or butyrylcholinesterase) that permanently bind CWNAs before they can cause physiological or behavioral damage.

Broomfield and colleagues (1991) demonstrated that nonhuman primates pretreated with butyrylcholinesterase showed no signs of CWNA poisoning after exposure to 2 LD_{50} of soman, a dose that is typically fatal to those exposed without prophylaxis or postexposure therapy. Serial probe recognition, a complex assessment of memory and motor behavior, was briefly depressed following CWNA exposure but returned to its baseline within only 8 hours. Similar results were obtained with even larger doses of CWNA.

Doctor and colleagues (1993) demonstrated similar results with the bioscavenger acetylcholinesterase by showing protection against lethality of up to 5 LD$_{50}$ of soman. As before, complex motor and short-term memory performance, as measured by serial probe recognition, were preserved in nonhuman primates treated with bioscavenger and then exposed to soman. Complex motor, vestibular equilibrium, and balance performance were preserved as assessed using a primate equilibrium platform, a highly challenging neurobehavioral assessment. The Broomfield and colleagues (1991) and Doctor and colleagues (1993) studies demonstrated protection against significant levels of CWNAs. The protection against the highly toxic CWNA soman by these novel bioscavengers is notable because it preserved both life and militarily relevant neurobehavioral performance with limited side effects.

Collectively, the research by uniformed and civilian psychologists on therapeutics against CWNAs demonstrated that novel therapies and prophylactics can be developed against the most toxic CWNAs known to man. It also demonstrated the importance of understanding and protecting immediate and long-term behavior and cognition as medical science develops to the point where protection against lethality is surpassed as the main objective. Uniformed research psychologists provide the capability to examine not only the threat, but also the neurobehavioral response of organisms, and potential therapies as a whole.

Safety Assessments

After animal testing has identified candidate drugs, two questions must be answered before the Food and Drug Administration (FDA) approves a drug for use in humans. First, is the drug safe (are there any secondary or tertiary effects and what is the margin of safety)? Second, is the drug effective at treating the identified disease, symptom, or condition? Research psychologists helped characterize the safety of both the standard and developing therapies against CWNAs (see reviews by Headley, 1982, and Penetar, 1990).

Penetar and Beatrice (1986) and Penetar, Haegerstrom-Portnoy, and Jones (1988) evaluated the effects and duration of atropine and 2-PAM on pursuit tracking, visual function, general health measures (such as heart rate and blood pressure), and cognitive function in humans. In brief, they found that the standard therapy had the greatest impact on the visual system, while psychological and cognitive functions were relatively unaltered over time. Thus, this research suggested that the standard therapy is relatively innocuous when given in the absence of CWNAs.

Research psychologists further examined the psychoactivity and abuse potential of atropine in humans (Penetar and Henningfield, 1986). Atropine was administered intramuscularly to normal volunteers using the same route of administration Soldiers would use in the field. This research established that while atropine alone had a noticeable effect immediately following administration in humans, the drug was not likely to be abused as indicated by a drug addiction assessment. This determination is important because it indicates that Soldiers in the field can safely carry several autoinjectors of this compound so it is available for immediate use and maximal effectiveness. Understanding the potential risks of a given capability helps military leaders develop a strategy to mitigate potential operational risks during combat.

The research with atropine represents a type of "postmarketing" safety assessment (that is, after the FDA has approved a therapy for use). However, safety assessments are also important as a pharmaceutical is under development or being considered for an alternate indication than originally approved because they help to identify potential risks and possible mitigation strategies early. For example, Myers and Clark (2006) recently examined the functional impact of the next generation anticonvulsant midazolam, which will likely replace diazepam. They examined memory and fine motor performance in a serial probe recognition task following administration of the likely dose of midazolam that will be included in future autoinjectors. The data indicated that at that dose, midazolam has minimal impact on memory and fine motor performance. This is notable, because a common surgical practice involves the use of a much larger dose of midazolam to reduce the likelihood of surgery-related memories.

Similar research with potential bioscavengers has also demonstrated limited adverse impact. Saxena and colleagues (2004) and Clark and colleagues (2005) demonstrated that a dose 30 times greater than that required for protection against 2 LD_{50} of soman had no negative physiological impact and minimal to no alteration in avoidance behavior in rodents. Together with other bioscavenger data involving nonhuman primates (Broomfield et al., 1991; Doctor et al., 1993), the data suggest that cholinesterase bioscavengers will be safe for both prophylactic and therapeutic use and have limited physiological and neurobehavioral side effects. The work also demonstrates research psychologists' efforts to characterize the safety of both the standard and developmental therapies against CWNAs.

Administrative and Operational Support

Beyond the laboratory, uniformed research psychologists provide essential information on the functional employment of therapies and their potential impact on operational military units in the field. They have also managed large-scale chemical defense research and acquisition programs, commanded military medical research units focused on chemical defense, supported military leadership in chemical defense and pharmaceutical development, and even supported recent combat operations in the area of chemical defense. More specifically, in addition to commanding the primary Army medical chemical defense research laboratory, research psychologists have commanded around the Army in deployed field medical units; helped to develop employment doctrine for chemical and biological protective shelters in Korea; served in a valuable advisory role to combatant commanders and key medical and chemical officer staffs in the Multi-National Corps and Multi-National Force–Iraq during recent combat operations; and served as Army Headquarters Staff in support of strategic level initiatives involving chemical and various asymmetric warfare threats.

Discussion

Research psychologists have played a key role in measuring the short- and long-term impact of toxicants and demonstrating the efficacy and safety of therapies against CWNAs. More specifically, because they employ behavioral and holistic assessments at the systems and organism levels, they provide a valuable evaluation and screening of candidate therapeutics. Research psychologists have provided data essential for pharmaceutical development and approval by the FDA. However, valuable research on nerve agent toxicity and the development of medical chemical defense countermeasures are not the only contributions of uniformed research psychologists in the Army.

Through their contributions both in the laboratory and in the field, uniformed research psychologists help to mitigate the horrific image so graphically illustrated in Lieutenant Wilfred Owen's poem. We must continue to recruit and retain the expertise of neuroscientists, psychobiologists, and neurotoxicologists in uniform as we face the challenges of unconventional warfare in the future.

References

Broomfield, C., Maxwell, D., Solana, R., Castro, C., Finger, A., and Lenz, D. (1991). Protection by butyrylcholinesterase against organophosphorus poisoning in nonhuman primates. *Journal of Pharmacology and Experimental Therapeutics* 259 (2), 633–638.

Castro, C. (1997). Primacy and recency effects in rhesus monkeys (Macaca mulatta) using a serial probe recognition task: II. Effects of atropine sulfate. *Behavioral Neuroscience* 111 (4), 676–682.

Castro, C., Gresham, V., Finger, A., Maxwell, D., Solana, R., Lenz, D., et al. (1994). Behavioral decrements persist in rhesus monkeys trained on a serial probe recognition task despite protection against soman lethality by butyrylcholinesterase. *Neurotoxicology and Teratology* 16 (2), 145–148.

Castro, C., Larsen, T., Finger, A., Solana, R., and McMaster, S. (1992). Behavioral efficacy of diazepam against nerve agent exposure in rhesus monkeys. *Pharmacology Biochemistry and Behavior* 41 (1), 159–164.

Churchill, L., Pazdernik, T., Jackson, J., Nelson, S., Samson, F., and McDonough, J.H. (1984). Topographical distribution of decrements and recovery in muscarinic receptors from rat brains repeatedly exposed to sublethal doses of soman. *Journal of Neuroscience* 4 (8), 2069–2079.

Clark, M., Sun, W., Myers, T., Bansal, R., Doctor, B., and Saxena, A. (2005). Effects of physostigmine and human butyrylcholinesterase on acoustic startle reflex and prepulse inhibition in C57BL/6J mice. *Pharmacology Biochemistry and Behavior* 81 (3), 497–505.

Dave, J., Connors, R., Genovese, R., Whipple, R., Chen, R., DeFord, S., et al. (2007). DNA fragmentation in leukocytes following repeated low dose sarin exposure in guinea pigs. *Cellular and Molecular Life Sciences* 64 (21), 2823–2828.

Doctor, B., Blick, D., Caranto, G., Castro, C., Gentry, M., Larrison, R., et al. (1993). Cholinesterases as scavengers for organophosphorus compounds: protection of primate performance against soman toxicity. *Chemico-Biological Interactions* 87 (1–3), 285–293.

Harris, L., McDonough, J.H., Stitcher, D., and Lennox, W. (1984). Protection against both lethal and behavioral effects of soman. *Drug and Chemical Toxicology* 7 (6), 605–624.

Headley, D. (1982). Effects of atropine sulfate and pralidoxime chloride on visual, physiological, performance, subjective, and cognitive variables in man: a review. *Military Medicine* 147 (2), 122–132.

Johnson, E., Daugherty, K., Gallagher, S., Moran, A., and DeFord, S. (2008). Glutamate receptor pathology is present in the hippocampus following repeated sub-lethal soman exposure in the absence of spatial memory deficits. *Neurotoxicology* 29 (1), 73–80.

Koplovitz, I., Romano, J., and Stewart, J. (1989). Assessment of motor performance decrement following soman poisoning in mice. *Drug Chemical Toxicology* 12 (3–4), 221–235.

Matthew, C., Glenn, J., Bowers, W.J., and Navara, D. (1994). Cholinergic drug interactions and heat tolerance. *Life Sciences* 54 (17), 1237–1245.

McDonough, J.H. (1982). Effects of anticholinergic drugs on DRL performance of rhesus monkeys. *Pharmacology Biochemistry and Behavior* 17 (1), 85–90.

McDonough, J.H., Hackley, B.J., Cross, R., Samson, F., and Nelson, S. (1983). Brain regional glucose use during soman-induced seizures. *Neurotoxicology* 4 (2), 203–210.

McDonough, J.H., Jaax, N., Crowley, R., Mays, M., and Modrow, H. (1989). Atropine and/or diazepam therapy protects against soman-induced neural and cardiac pathology. *Fundamentals of Applied Toxicology* 13 (2), 256–276.

McDonough, J.H., McLeod, C.J., and Nipwoda, M. (1987). Direct microinjection of soman or VX into the amygdala produces repetitive limbic convulsions and neuropathology. *Brain Research* 435 (1–2), 123–137.

McDonough, J.H., Smith, R., and Smith, C. (1986). Behavioral correlates of soman-induced neuropathology: deficits in DRL acquisition. *Neurobehavioral Toxicology and Teratology* 8 (2), 179–187.

Modrow, H., and Jaax, N. (1989). Effect of soman exposure on the acquisition of an operant alternation task. *Pharmacology Biochemistry and Behavior* 32 (1), 49–53.

Moffett, J., Price, R., Anderson, S., Sipos, M., Moran, A., Tortella, F., and Dave, J.R. (2003). DNA fragmentation in leukocytes following subacute low-dose nerve agent exposure. *Cellular and Molecular Life Sciences* 60 (10), 2266–2271.

Myers, T., and Clark, M. (2006). Serial-probe recognition in rhesus macaques: effects of midazolam. *Pharmacology Biochemistry and Behavior* 85 (3), 555–561.

Myers, T., Galbicka, G., Sipos, M., Varadi, S., Oubre, J., and Clark, M. (2002). Effects of anticholinergics on serial-probe recognition accuracy of rhesus macaques (Macaca mulatta). *Pharmacology Biochemistry and Behavior* 73 (4), 829–834.

Owen, W. (1963). "Dulce et Decorum Est." In C.D. Lewis (ed.), *The Collected Poems of Wilfred Owen*. New York, NY: New Directions Publishing.

Penetar, D. (1985). The effects of atropine, benactyzine, and physostigmine on a repeated acquisition baseline in monkeys. *Psychopharmacology (Berl)* 87 (1), 69–76.

———. (1990). Brief review of atropine effects on physiology and performance. *Drug Development Research* 20, 117–121.

Penetar, D., and Beatrice, E. (1986). Effects of atropine on human pursuit tracking performance. *Aviation, Space, and Environmental Medicine* 57 (7), 654–658.

Penetar, D., Haegerstrom-Portnoy, G., and Jones, R. (1988). Combined atropine and 2-PAM Cl effects on tracking performance and visual, physiological, and psychological functions. *Aviation, Space, and Environmental Medicine* 59 (12), 1125–1132.

Penetar, D., and Henningfield, J. (1986). Psychoactivity of atropine in normal volunteers. *Pharmacology Biochemistry and Behavior* 24 (4), 1111–1113.

Penetar, D., and McDonough, J.H. (1983). Effects of cholinergic drugs on delayed match-to-sample performance of rhesus monkeys. *Pharmacology Biochemistry and Behavior* 19 (6), 963–967.

Raffaele, K., Hughey, D., Wenk, G., Olton, D., Modrow, H., and McDonough, J.H. (1987). Long-term behavioral changes in rats following organophosphonate exposure. *Pharmacology Biochemistry and Behavior* 27 (3), 407–412.

Rickett, D., Glenn, J., and Beers, E. (1986). Central respiratory effects versus neuromuscular actions of nerve agents. *Neurotoxicology* 7 (1), 225–236.

Romano, J., King, J., and Penetar, D. (1985). A comparison of physostigmine and soman using taste aversion and nociception. *Neurobehavioral Toxicology and Teratology* 7 (3), 243–249.

Romano, J., and Shih, T.M. (1983). Cholinergic mechanisms of analgesia produced by physostigmine, morphine and cold water swimming. *Neuropharmacology* 22 (7), 827–833.

———. (1984). The effect of adrenalectomy and dexamethasone on the antinociceptive effects of physostigmine. *Psychopharmacology (Berlin)* 84 (3), 426–430.

Romano, J. and Landauer, M. (1986). Effects of the organophosphorus compound, O-ethyl-N-dimethyl-phosphoramidocyanidate (tabun), on flavor aversions, locomotor activity, and rotarod performance in rats. *Fundamentals of Applied Toxicology* 6 (1), 62–68.

Romano, J., Terry, M., Murrow, M., and Mays, M. (1991). Protection from lethality and behavioral incapacitation resulting from intoxication by soman (pinacolyl methylphosphonofluoridate) and treatment with atropine sulfate and 2-PAM chloride in the guinea pig, cavia porcellus. *Drug and Chemical Toxicology* 14 (1–2), 21–44.

Salem, H., Ternay, A.L.J., and Smart, J.K. (2008). Brief history and use of chemical warfare agents in warfare and terrorism. In J.A. Romano, Jr., B.J. Lukey, and H. Salem (eds.), *Chemical*

Warfare Agents: Chemistry, Pharmacology, Toxicology, and Therapeutics (2ᵈ ed.), 1–20. Boca Raton: CRC Press.

Saxena, A., Luo, C., Bansal, R., Sun, W., Clark, M.G., Ashani, Y., et al. (2004). Human butyrylcholinesterase: A future generation antidote for protection against organophosphate agents. In S.J.S. Flora, J.A. Romano, S.I. Baskin, and K. Sekhar (eds.), *Pharmacological Perspectives of Toxic Chemicals and Their Antidotes*. New York: Springer.

Shih, T.M., Penetar, D., McDonough, J.H., Romano, J., and King, J. (1990). Age-related differences in soman toxicity and in blood and brain regional cholinesterase activity. *Brain Research Bulletin* 24 (3), 429–436.

Sipos, M., Burchnell, V., and Galbicka, G. (1999). Dose-response curves and time-course effects of selected anticholinergics on locomotor activity in rats. *Psychopharmacology (Berlin)* 147 (3), 250–256.

Sipos, M., Burchnell, V., and Galbicka, G. (2001). Effects of selected anticholinergics on acoustic startle response in rats. *Journal of Applied Toxicology* 21 (Suppl. 1), S95–101.

Sipos, M., Moran, A., Kahler, D., Lukefahr, H., and DeFord, S. (2005). Development of a behavioral assessment model in guinea pigs to study the effects of low dose chronic exposure to chemical warfare nerve agents. In *Low Level Chemical Warfare Agent Toxicology Research Program FY04 Report and Analysis*. AFRL-HE-WP-TR-2005-0054. Air Force Research Laboratory, Aberdeen Proving Ground, MD.

Smart, J.K. (1997). History of chemical and biological warfare: An American perspective. In F.R. Sidell, E.T. Takafuji, and D.R. Franz (eds.), *Medical Aspects of Chemical and Biological Warfare*, vol. 1, 9–86. Washington, DC: Department of the Army, Office of the Surgeon General.

Taylor, P. (1990). Anticholinesterase agents. In A. Gilman, T. Rall, A. Nies, and P. Taylor (eds.), *The Pharmacological Basis of Therapeutics*, 131–149. New York: Pergamon Press.

U.S. Army (2006). Field Manual 3-24, *Counterinsurgency*. Washington, DC: Headquarters Department of the Army.

———. (2008a). Field Manual 3-0, *Operations*. Washington, DC: Headquarters Department of the Army.

———. (2008b). Field Manual 3-07, *Stability Operations*. Washington, DC: Headquarters Department of the Army.

———. (2008c). Field Manual 7-0, *Training for Full Spectrum Operations*. Washington, DC: Headquarters Department of the Army.

Chapter 15

New Biodosimetry Tools for Mental Health Support in Nuclear/Radiological Accidents or Terrorism

Charles A. Salter, David J. Sandgren, Ira H. Levine, and William F. Blakely

Terrorist attacks or disasters involving radiological materials can cause severe stress. After an accidental release of radioactive material in Goiania, Brazil, the ratio of noncontaminated people to those actually contaminated reporting for voluntary radiological screening was about 500 to 1. This ratio was derived from a total of approximately 125,000 people seeking screening. The public health community would be overwhelmed in such incidents. Rapid biodosimetry tools may help distinguish those needing medical treatment for radiation injuries from those needing psychological support alone. The Armed Forces Radiobiology Research Institute (AFRRI) has prepared a rapid biodosimetry tool in a CD-ROM package entitled "Radiation Training and Assessment Tools" (4th edition, December 2007) that includes:

- Casualty Management Guidance, including the *Terrorism with Ionizing Radiation Guidance Pocket Guide* and *Medical Management of Radiological Casualties Handbook*
- Medical Data Forms, including the AFRRI Adult/Pediatric Medical Field Record and the AFRRI Biodosimetry Worksheet
- Exposure Assessment Software, including the Biodosimetry Assessment Tool (BAT), which was developed to record diagnostic information and estimated dose in suspected radiological exposures based on symptoms reported.

The First-responders Radiological Assessment Triage (FRAT), which will be available for a future edition of the AFRRI CD-ROM, is similar to the BAT but is intended for use on hand-held personal digital assistant (PDA) devices. With the beta version released August 2008, it will provide "triage" dose assessments and estimates of hospitalization and mortality outcomes. The contents of the AFRRI CD-ROM are also available to download from AFRRI's Web site (www.afrri.usuhs.mil).

Introduction

Although the world has not experienced nuclear warfare since 1945, there have been a large number of nuclear/radiological accidents over the years and a recent increase in the perceived threat of nuclear terrorism. Most people fear a nuclear/radiological threat even more than a conventional explosion due to their inability to perceive the presence of radiation with the ordinary human senses and their concerns about long-lasting radiation effects.

Public fears about radiation are often excessive, unrealistic, and persistent. Attitudes have been shaped over time by viewpoints expressed in the media and in the culture at large. While there was considerable American pride in nuclear weaponry after World War II, decades of exaggeration and misconceptions in books, movies, television, and other media led primarily to fear regarding anything related to radiation. Following Chernobyl, for instance, some people who actually had little exposure risk became obsessed with avoiding contamination. Some became so depressed due to their fears that they gave up hope entirely and committed suicide (Drottz-Sjoberg and Persson, 1993). Similarly, the 1979 Three Mile Island nuclear power plant accident in Pennsylvania—generally considered America's worst nuclear disaster to date—caused no actual radiation injuries or deaths and only a minor release of radioactive gas into the environment (Collins, 1991). Few people received a detectable dose of radiation above the general background level, and even of those, the largest possible exposures were roughly equivalent to the dose of an ordinary chest X-ray. Yet the concern and anxiety were very real. Psychophysiologic reactions were common among the citizens of the area, and many fled the region in the mistaken belief that a major nuclear detonation might occur at any moment.

More typical stress reactions may take many forms, including psychophysiologic reactions, adjustment disorders, acute stress disorder, and post-traumatic stress disorder. For example, data from the Hiroshima and Nagasaki attacks revealed widespread acute reactions such as psychic numbing, severe anxiety, and disorganized behavior, and there were later chronic effects including survivor guilt and psychosomatic reactions (Mickley, 1989).

Studies of severe radiological accidents have revealed that the number of people frightened and/or demonstrating psychophysiologic symptoms may far outweigh the number of victims actually contaminated. The ratio of people with medically unexplained physical symptoms to radiation injured may be as high as 500 to 1, as occurred in the radiation accident in Goiania, Brazil, the worst radiological incident in the Western Hemisphere to date (Petterson, 1988). A deliberate terrorist attack with a radiological dispersal device, such as a so-called dirty bomb, would use conventional explosives to spread radiological material. In such an event, the psychological impact may prove far worse than the medical effects. The potential for additional attacks might also contribute to the fear.

Providing early response diagnostic assessment information about radiation dose and injury received by each victim may help prevent or reduce anxiety and other adverse psychological reactions following a radiological incident. In addition, facilitating accurate exposure assessment would support medical triage and treatment of those actually exposed to significant levels of radiation and reassure minimally and non-exposed persons that they have not received medically significant doses.

The Concept of Biodosimetry

Workers at nuclear facilities who may be exposed to radiological materials must wear physical dosimetry devices by regulation. These devices directly measure radiation received in the event of an accident. The public at large, however, does not typically have access to such devices. Thus, in the event of a surprise attack or accident, the vast majority of those affected will have no idea about their potential radiation exposure—whether it was minimal, enough to sicken but not kill them, or lethal. This uncertainty regarding exposure and its consequences adds to the victim's psychological distress.

Biodosimetry can provide a quantitative assessment of potential exposure. Years of research on the victims of past exposures have revealed a number of biological indicators of the dose received. Simply put, the greater the radiation dose, the greater the biomedical effect on the body. Therefore, the assessment of physiological effects in those present at an incident can be compared to historic data and the approximate radiation dose estimated. These biological indicators include a decline in lymphocytes with increasing dose, an increase in dicentric chromosomes, and an increasing severity of various prodromal signs and symptoms such as nausea, vomiting, and erythema (Blakely, Brooks, Lofts, van der Schans, and Voisin, 2002; Waselenko, MacVittie, Blakely, Pesik, Wiley, Dickerson et al., 2004; Blakely, Salter, and Prasanna, 2005). Some of these dose indicators, such as the dicentric assay, require sophisticated equipment and too much time to be practical in a mass casualty situation where immediate triage is necessary. We have developed tools that can provide a reasonably accurate estimation of dose within a few minutes per victim, using only laptop computers or even hand-held computer devices.

A Dynamic Package of Biodosimetry Tools for First Responders

The fourth edition of the Radiation Training and Assessment Tools includes guidance for managing casualties, forms for recording medical information, and software tools for assessing exposure doses. This CD can be obtained by taking the Medical Effects of Ionizing Radiation course held routinely by the AFRRI's Military Medical Operations department. More information is available at the AFRRI Web site.

Casualty Management Guidance

The CD-ROM emphasizes what treatment providers should administer in the medical management of casualties, providing a complete handbook, an abbreviated summary for quick reference, and a brief overview. The *Medical Management of Radiological Casualties Handbook* (2ᵈ edition, 2003) helps prepare medical care providers to treat injuries complicated by ionizing radiation exposure and radioactive contamination. The handbook provides concise reading material for health care professionals in the management of uncontrolled ionizing radiation exposure. The AFRRI *Emergency Radiation Medicine Response Pocket Guide* can be quickly consulted in a radiation emergency for assistance in assessing and controlling the situation. It includes such topics as diagnosis, treatment considerations, decontamination procedures, and public health reporting procedures. This document was originally developed by the U.S. Department of Veterans Affairs as a two-page pocket guide. Both the handbook and the pocket guide are available in print form or downloadable at the AFRRI Web site.

Medical Data Forms

The second subset of biodosimetry tools on the CD-ROM deals with the actual recording of medical data relevant to radioactive contamination, exposure, and the resulting patient symptoms. Included are tools such as a brief field record and a more comprehensive biodosimetry worksheet. The AFRRI Adult/Pediatric Field Medical Record is a convenient one-page form for recording emergency medical information in the field. It was adapted from the U.S. Army's Field Medical Card into an electronic form applicable to both military personnel and civilians, for adult and pediatric cases alike. For instance, it has both an adult and a pediatric body map for labeling areas of contamination or injury. It also includes sections on personal data, types of injury, other symptoms, treatments administered, and disposition of the case. The AFRRI Biodosimetry Worksheet provides a six-page data entry worksheet for gathering facts about a case of radiation exposure, including the source and type of radiation, the extent of exposure, and the nature of the resulting injuries. It also is applicable to both adult and pediatric cases. This worksheet was adapted by AFRRI from the North Atlantic Treaty Organization Standardization Agreement 2474 Appendix 1, "Medical Record of Ionising Radiation Dose and Contamination," and incorporates a version of the Medical Treatment Protocols for Radiation Accident (METREPOL) grading scale (Fliedner, Friesecke, and Beyrer, 2001). It includes the basic items from the one-page Field Medical Record, and quite a bit more as well. Additional features include a detailed section on types of radiation sources present, radioactive exposure, external and internal contamination, and consequent symptoms. There are two sets of adult and pediatric body maps, one for contamination distribution and the other for estimated dose distribution. There is also a section for tracking changes in counts of various blood cell lines over time. Using a modified METREPOL grading system, the Acute Radiation Syndrome Responses Assessment based on the neurovascular, cutaneous, gastrointestinal, and hematopoietic systems can be determined. Finally, there is an entire page for additional notes and comments. The Biodosimetry Worksheet is in PDF format and can be filled out on the computer, or printed to be filled out by hand. Information on this Biodosimetry Worksheet can be used to complete many sections of the BAT software program.

Exposure Assessment Software

The current edition of the CD-ROM includes one complete software program and will soon include a hand-held application that is currently in beta testing phase.

Biodosimetry assessment tool. The BAT software program was developed by AFRRI in collaboration with the Radiation Emergency Assistance Center/Training Site for recording diagnostic information in suspected radiological exposures. Version 1.03 is included on the CD-ROM and information on downloading BAT is also available on the AFRRI Web site. The minimum system requirements for BAT are Microsoft Windows XP SP2 operating system and a Pentium 200 with 256 MB of RAM and 10 MB of free disk space. A typical patient's complete record will require about 1 MB of disk space.

The BAT software equips health care providers with diagnostic information (for example, physical dosimetry and clinical signs and symptoms) germane to the management of human radiation casualties. Designed primarily for prompt use after a radiation incident, the user-friendly program facilitates collection, integration, and archiving of data obtained from exposed persons. Data collected in templates are compared with established radiation dose responses obtained from the literature to provide multiparameter dose assessments. The program archives clinical information (such as the extent of contamination, wounds, and infection) that is useful for casualty management, displays relevant diagnostic information in a concise format, and can be used to manage both military and civilian radiation accidents (see figure 15–1). In addition, the use of this program to monitor individual patient diagnostic information could minimize the severity of psychological casualties to

Figure 15–1. **Biodosimetry Assessment Tool Flowchart**

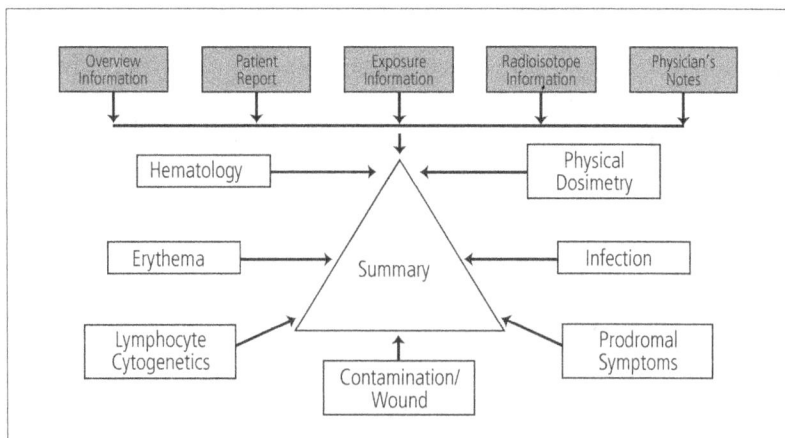

radiation victims and the concerned public by influencing how they view their exposure, dose, and potential risk for the development of disease.

Background information about the patient and the circumstances of the suspected radiological incident (date of exposure, time of exposure, location) are entered first. Next, medical data on various symptom categories are entered under the appropriate tabs (physical dosimetry, contamination/wound, prodromal symptoms, hematology, lymphocyte cytogenetics, erythema, infection) on the master screen. Only the categories for which data are available would be used for a given patient. In some cases, that may mean only two or three sections would be used, and the other tabs could be ignored.

Clinical signs and symptoms associated with radiation exposure are varied and numerous (see table 15–1) (Anno et al., 1989). The BAT's prodromal symptom screen allows data entry of this diagnostic information. While all the symptom categories for which data are available would be useful in treating a patient, three are especially relevant to estimating dose received. These are latency to vomiting in the prodromal symptoms category and the absolute number of lymphocytes and rate of lymphocyte depletion in the hematology category. These data are used by the BAT program to estimate exposure dose based on human accident databases and algorithms (see table 15–2). In the BAT prodromal symptom screen there are cells for entry of vomiting data, including time of onset and duration, while an "Obtain Dose Assessment" button can then be clicked to display an estimated dose from onset of emesis.

The hematology data entry screen has room for up to 13 serial measures of lymphocytes over time (see figure 15–2). Note that in this screen, dose can be estimated from a single lymphocyte count simply by clicking on the "SHOW DOSE" button after entering the number. A better estimate, however, can be derived from lymphocyte kinetics showing changes over time as revealed by multiple measurements. A click on "Get Multi-Sample Dose Estimate" will reveal a feedback screen estimating dose and 95 percent confidence level. The dose estimate based on lymphocyte kinetics is determined by one of two methods based primarily on data from Goans, Holloway, Berger, and Ricks (1997, 2001) and Goans (2001) or Guskova, Barabanova, Baranov, Gruszdev, Pyatkin, Nadezhina et al. (1988) (see table 15–2). While other patient data like cytogenetics can be used to calculate dose later, the advantage of the onset of vomiting and lymphocyte counts or depletion kinetics is that the physician or other user of the BAT program can get immediate

Table 15–1. **Early Clinical Responses Following Radiation Exposure**

Nausea	Fatigue
Vomiting	Weakness
Headache	Abdominal pain
Fever	Parotid pain
Tachycardia	Erythema

Table 15–2. **Human Databases Used in BAT Dose Assessment Algorithms**

Onset of Vomiting	Hematology	Radiation Quality	References
X	X	Photon	Anno et al., 1989 Dr. George Anno (personal communication)
X	X	Photon, Mixed Photon, and Neutron	Goans et al., 1997; 2001; Goans, 2001 Dr. R.E. Goans (personal communication)
X	X	Photon, Mixed Photon, and Neutron	REAC/TS Registry
	X	Photon, Mixed Photon, and Neutron	Guskova et al., 1988 Dr. A.K. Guskova (personal communication)
	X	N/A (control)	AFRRI studies (BD–02)

feedback on estimated dose, bound by error bars or upper and lower confidence limits.

The BAT summary screen captures all the most relevant information entered or computed earlier for the given patient (see figure 15–3). In this screen the various symptom categories present are displayed, along with information on dose estimates. Note in this example

Figure 15–2. **Hematology Data Entry Screen**

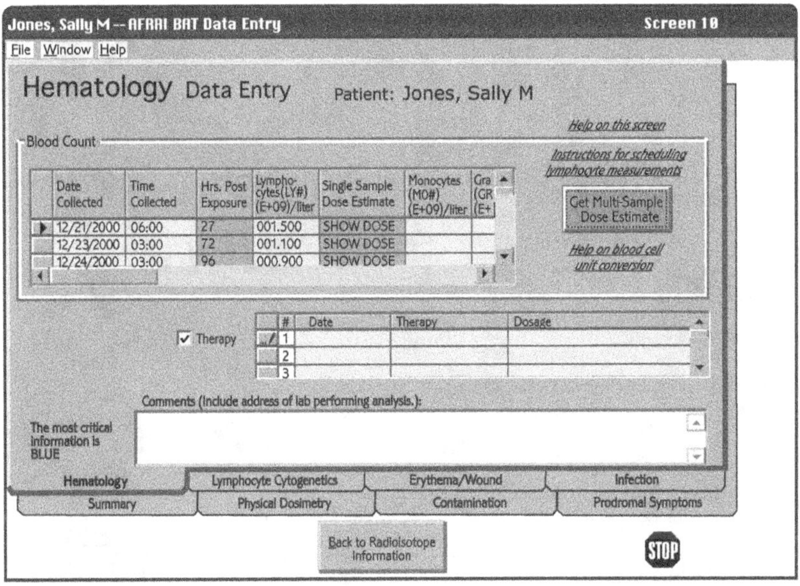

that the dose estimated by onset of vomiting is slightly different from that estimated using hematology. An expert can view the information and assign a dose in the specified box. If more information or additional diagnostic indices became available later, of course, the operator could add that at any time.

First-responders radiological assessment triage. A beta version FRAT was released in August 2008, and the final version will be available in a later edition of the CD-ROM. While the two programs are somewhat related, BAT requires a desktop or laptop computer, while FRAT works on a hand-held PDA device using the Palm operating system. BAT computers could be used in a far-forward command post, but the size and portability of FRAT make it ideal for first responders or others who will enter a radiological incident zone and need to make patient data entry or dose estimates while operations are under way.

The FRAT was developed by surveying experts in radiobiology and medical effects of ionizing radiation. A group of experts responded to a questionnaire asking them to provide their judgments about the possible contributions of several diagnostic factors to predict radiation dose received. First, they were asked to estimate how useful each factor

Figure 15–3. **Summary Screen**

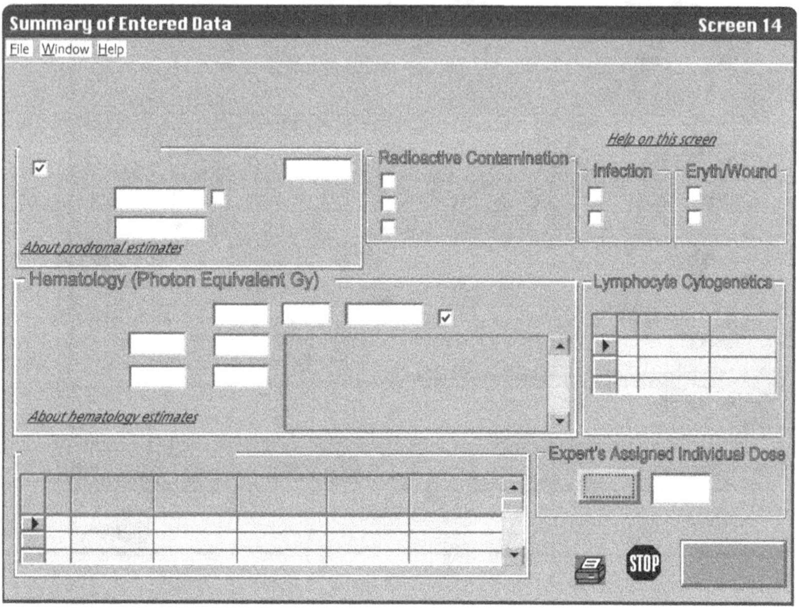

would be in predicting whether radiation exposure had occurred. Then they were asked to rate how reliable that factor might be in estimating the dose if exposure had indeed occurred. Answers were to be provided on a 100-point scale with 0 indicating that the item had no value in such judgments. Ratings on each item were ranked and then the median score was used to weigh each sign and symptom factor in the multiparameter-based determination of a triage dose.

The FRAT software allows first responders to triage suspected radiation casualties based on the initial or prodromal features as listed in the AFRRI *Emergency Radiation Medicine Response Pocket Guide.* An additional feature of FRAT is the inclusion of a digital version of the AFRRI Pocket Guide that can be accessed quickly from the help menu. FRAT was developed initially to use the Palm operating system due to the portability of PDA devices. The program was written using NS Basic Palm and is compatible with Palm OS3 or higher.

The FRAT program permits convenient entry of signs and symptoms, blood lymphocyte counts, and dosimetry data, all with minimum text entry requirements (see figure 15–4 for a schematic of possible inputs to FRAT). As with BAT, only the program components for which data are available are used. Otherwise, they can be ignored until such time as pertinent data are gathered.

Figure 15–4. **Schematic Outline for First-responder Radiological Assessment Triage**

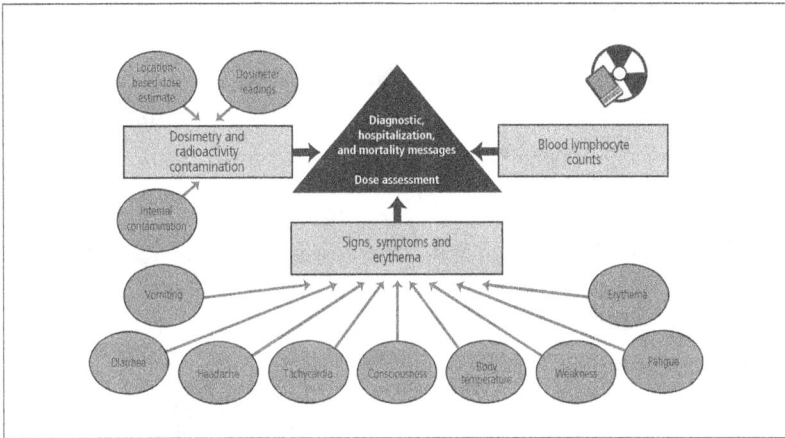

Figure 15–5. **Sample First-responder Radiological Assessment Triage Entry Screens**

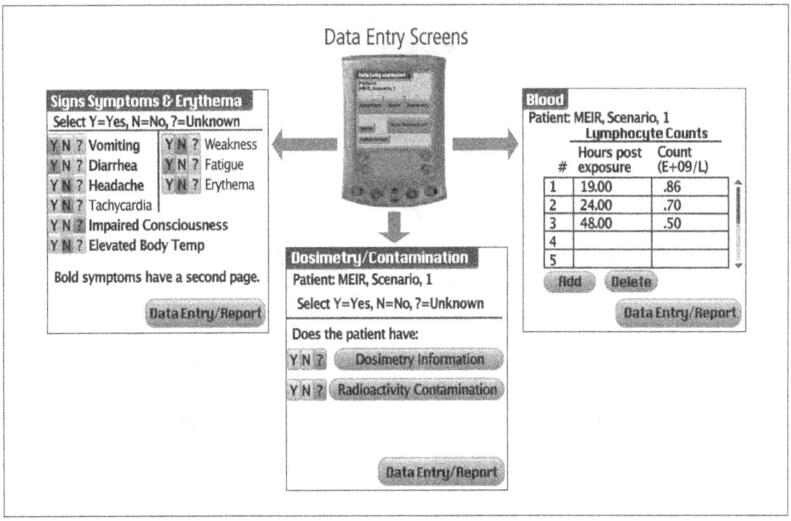

The program comes equipped with a variety of input screens (see figure 15–5). The main input screen allows the user to link to the three subsections mentioned above. The "Signs Symptoms and Erythma" screen allows the responder to select from the nine symptomatic conditions indicated in the input schematic. The symptoms highlighted in bold, such as vomiting, have a sub-screen where additional information on time of onset and duration can be logged. The "Blood" screen is capable of saving up to 10 lymphocyte counts. The "Dosimetry/Contamination" screen allows the responder to store exposure information. Note that no writing is required for most data entry since in most cases the options are built in and one must simply touch the on-screen menus with a stylus to proceed. Once familiarity with the program has been achieved, it is possible to enter all the known facts about a given case quite rapidly.

At the conclusion of data entry, the user taps the "Dose Assessment" button on the main screen to get immediate feedback in the form of five triage dose assessment screens to advise the first responder about triage and prognosis on the case (see figure 15–6). "Triage Dose Assessment 1" indicates whether the patient has received no dose, a dose of inestimable magnitude, or a measurable degree of overexposure. The second dose assessment screen supplies messages related to hematology, such as warnings related to dose estimate or to draw serial blood samples.

Figure 15–6. **First-responder Radiological Assessment Triage Dose Assessment Screens**

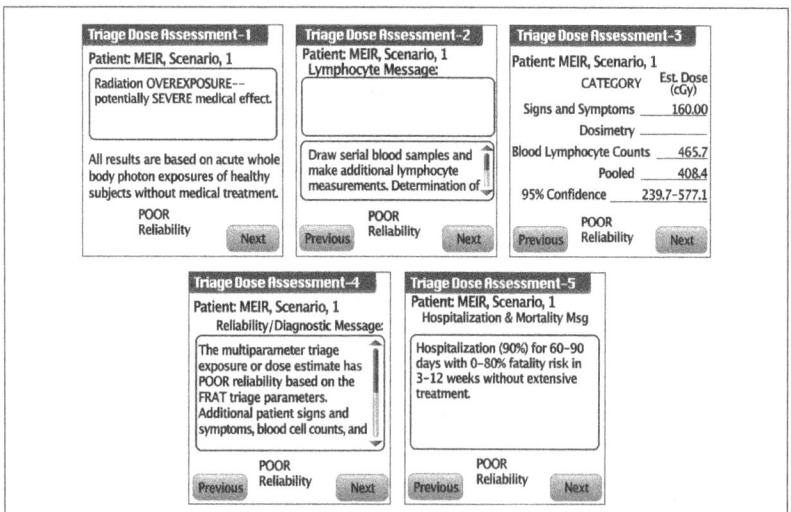

The third assessment screen provides a dose estimate for each of the three categories (if available) as well as a weighted multiparameter triage dose in centi-Gray (cGy), along with confidence limits. The fourth assessment screen provides reliability and diagnostic information. Finally, the fifth screen reports hospitalization estimations and mortality projections. All this information has great utility for triage in a mass casualty situation as well as for treatment.

Conclusion

The tools included on AFRRI's CD-ROM can greatly assist first responders at the scene of a nuclear/radiological mass casualty accident or attack and medical treatment providers at all subsequent echelons back to tertiary care. The doctrinal guidelines and articles provided offer an enormous amount of useful information pertaining to both radiation and casualty care management. The medical data forms offer a means to collect the necessary information that physicians can use to help triage a patient and decide on the best course of treatment. The BAT and FRAT software can provide almost immediate radiation received dose estimates for each individual with a tremendous potential to triage large groups of victims rapidly. Such expedited triage can differentiate patients who actually have been exposed from those showing psychosomatic symptoms, thereby allowing medical treatments to be rapidly focused where they are needed most. With the truly exposed being taken care of, the likely hundreds of non-seriously exposed persons can be reassured that they are in no medical danger from radiation. The psychodynamics causing persons to be less likely to trust such "government/official pronouncements" are beyond the scope of this chapter, as is the interaction between such distrust and continuing psychological distress. Not everyone will be reassured, but many might experience a significant relief of stress and impetus toward sustained mental health through the use of biodosimetry tools to assess radiation exposure.

References

Anno, G.H., Baum, S.J., Withers, H.R., and Young, R.W. (1989). Symptomatology of acute radiation effects in humans after exposure to doses of 0.5–30 Gy. *Health Physics* 56 (6), 821–838.

Blakely, W.F., Brooks, A.L., Lofts, R.S., van der Schans, G.P., and Voisin, P. (2002). Overview of low-level radiation exposure assessment-biodosimetry. *Military Medicine* 167 (2), 20–24.

Blakely, W.F., Salter, C.A., and Prasanna, P.G. (2005). Early-response biological dosimetry: Recommended countermeasure enhancements for mass-casualty radiological incidents and terrorism. *Health Physics* 89 (5), 494–504.

Collins, D.L. (1991). Stress at Three Mile Island: Altered perceptions, behaviors, and neuroendocrine measures. In R. Ricks, M. Berger, and F. O'Hara (eds.). *The medical basis for radiation-accident preparedness III: The psychological perspective* (71–79). New York: Elsevier.

Drottz-Sjoberg, B.M., and Persson, L. (1993). Public reaction to radiation: Fear, anxiety, or phobia? *Health Physics* 64, 223–231.

Fliedner, T.M., Friesecke, I., and Beyrer, K., eds. (2001). *Medical management of radiation accidents—manual on the acute radiation syndrome.* London: The British Institute of Radiology.

Goans, R.E., Holloway, E.C., Berger, M.E., and Ricks, R.C. (1997). Early dose assessment following severe radiation accidents. *Health Physics* 72 (4), 513–518.

———. (2001). Early dose assessment in criticality accidents. *Health Physics* 81 (4), 446–449.

Goans, R.E. (2001). Clinical care of the radiation-accident patient: patient presentation, assessment, and initial diagnosis. In R. Ricks, M. Berger, and F. O'Hara (eds.). *The medical basis for radiation-accident preparedness: The clinical care of victims, proceedings from the fourth international conference REAC/TS conference on the medical basis of radiation-accident preparedness, March 2001, Orlando, Florida* (11–22). Boca Raton, FL: The Parthenon Publishing.

Guskova, A.K., Barabanova, A.V., Baranov, A.Y., Gruszdev, G.P., Pyatkin, Y.K., Nadezhina, N.M. et al. (1988). Acute radiation effects in victims of the Chernobyl nuclear power plant accident. In United Nations Scientific Committee on the Effects of Atomic Radiation (UNSCEAR). *Sources, effects and risks of ionizing radiation* (613–647). New York: United Nations.

Mickley, G.A. (1989). Psychological factors in nuclear warfare. In R. Zajtchuk (ed.). *Medical consequences of nuclear warfare* (153–169). Falls Church, VA: U.S. Army Office of the Surgeon General.

Petterson, J.S. (1988). Perception vs. reality of radiological impact: The Goiania model. *Nuclear News* 31, 84–90.

Waselenko, J.K., MacVittie, T.J., Blakely, W.F., Pesik, N., Wiley, A.L., Dickerson, W.E. et al. (2004). Strategic national stockpile radiation working group medical management of the acute radiation syndrome: Recommendations of the strategic national stockpile radiation working group. *Annual of Internal Medicine* 140 (12), 1037–1051.

Part VI

Psychology and
Military Families

Chapter 16

Families Facing the Demands of Military Life: New Research Directions

Lolita Burrell, Gary A. Adams, Doris Durand, and Carl Andrew Castro

Military families, whether Active-duty, Reserve, or National Guard, face a multitude of demands in times of both peace and war, and these demands will shift throughout a Servicemember's career. Our research at the Walter Reed Army Institute of Research (WRAIR), as well as research done at other institutions, has shown that the demands experienced by military families have both negative and positive effects in terms of health, marital satisfaction, and satisfaction with the Army. Appraisal of these demands and the ability to obtain the necessary resources to deal with them are important determinants of a variety of well-being-, family-, and Army-related outcomes. This chapter will focus on the findings of family studies conducted by researchers at WRAIR and examine the road ahead with studying military families based on the outcomes discussed.

Introduction

As of September 2006, a total of 275,016 Active-duty (54 percent of the Active force), 189,975 Army Reserve (47 percent of the Reserve force), and 160,921 Army National Guard (46 percent of the National Guard force) Servicemembers were married (Maxfield, 2006). As we can see from the demographics, many in the Army have families that will experience a multitude of demands while part of the military, some of which have always existed. However, the war on terror has brought unexpected additional challenges. The long-term impact of these recurring demands within a military and society that continue to grow and change remains to be seen. Segal discusses a number of military lifestyle demands that may negatively impact family members. These include geographic mobility, residence in foreign countries, periodic separations from family, and risk of Servicemember injury/death. These demands may be experienced individually or collectively. They are also connected to another demand, which is the ability to integrate with the military community. Of course, these demands are not all-inclusive, but they may often be closely tied to one another whether through creating conditions that lead to other demands and/or through their effects on various health-, family-, and Army-related outcomes.

Geographic Mobility and Residence in Foreign Countries

The relocations that are part of military life can disrupt family life and existing supportive relationships. Although moving can be a positive experience for some spouses (Warner, 1983), many report experiencing difficulty adjusting to their most recent move (Croan, Levine, and Blankinship, 1992; Orthner, 2002). In the Survey of Army Families IV, recently moved spouses reported less positive adjustments on 13 out of 18 family, personal, and Army adjustment indicators (Orthner, 2002). Research in this area suggests that geographic mobility is related to lower psychological well-being (Croan et al., 1992; Gaylord and Symons, 1986; Norell and Copeland, 2002) and physical health (Jensen, Lewis, and Xenakis, 1986; Andersen and Arnetz, 1999), as well as with lower marital happiness (Schneider and Gilley, 1984; Ozkaptan, Sanders, and Holz, 1986; Makowsky, Cook, Berger, and Powell, 1988) and lower retention (Lewis, 1985; Mohr, Holzbach, and Morrison, 1981).

Moving across town may be difficult enough for some, so moving to an entirely different country, complete with its own customs and language, is likely to place a strain on even those families looking forward to the new assignment. As of 2006, approximately 186,393 of the 1,381,401 Active-duty Servicemembers were stationed overseas (Bruner, 2007), not including those who were deployed. Like geographic mobility, residence in a foreign country can be a positive experience (Ozkaptan, Sanders, and Holz, 1983); however, research also suggests that it can lead to lower well-being (Burnam, Meredith, Sherbourne, Valdez, and Vernez, 1992; Puskar, 1990).

Separation from Family

Moving and living overseas may be tied to separation from family when the Soldier is assigned to a unit that is likely to deploy and/or frequently travel for field training exercises, school, peacekeeping, or combat missions. These separations can place additional demands on family members in terms of managing the household and taking on the role of single parent. They can also lead to feelings of isolation and loneliness.

Some of the research on periodic separations has examined its impact by focusing on the frequency of separations (Rohall, Segal, and Segal, 1999), while other research has focused on the hardships imposed on the family by separations (Rosen, Teitelbaum, and Westhius, 1993; Marshall-Mies, 2001). Generally, the research has shown a negative relationship between separations and well-being (Adler, Bartone, and Vaitkus, 1995; Bell and Schumm, 1999, Frankel, Snowden, and Nelson, 1992) and separations and marital satisfaction (Schumm, Bell, and Gade, 2000; Roschman, Patterson, and Schofield, 1989). While the effects of deployment on marital and familial relationships reported are generally negative, positive effects may also occur. Positive outcomes may include the opportunity for the spouse to develop independence and self-sufficiency (Coolbaugh and Rosenthal, 1992; Hunter and Hickman, 1981; Schwartz, Braddy, Griffith, and Wood, 1988). Generally speaking, research has indicated that there are spillover effects between job and family that can affect marital quality and job satisfaction (Gal and Syna, 1998; Rogers and May, 2003).

Risk of Soldier Injury or Death

While the risk of Servicemember injury or death is most obvious during wartime, it is also present during humanitarian and peacekeeping missions, as well as during field training exercises. Approximately 500 civilian wives of enlisted Soldiers were surveyed regarding their spouses' deployment to Somalia in 1993. Fears regarding Soldier safety were some of the most cited problems regarding the deployment along with loneliness and lack of knowledge of the Somalia situation (Schumm, Bell, and Knott, 2001). Research conducted during the Gulf War (Operation *Desert Storm*) also indicated that spouses feared for the safety and welfare of their Soldiers (Rosen and Durand, 1995). This and other studies of wartime fears have shown that spouse reports of such fears were related to their reports of health symptoms (Scurfield and Tice, 1992; Cohen and Dotan, 1976). Other studies have found that negative health symptoms such as anxiety and depression can also occur during peacekeeping missions (Adler, Bartone, and Vaitkus, 1995; Van der Kloet and Moelker, 2002).

Integration with the Military Community

Such fears may be exacerbated if spouses feel isolated from the military community. Reserve and National Guard Soldiers have taken on duties and responsibilities closely reflecting those of the Active Component, to include deployments. Yet it is unclear how well the spouses of these Soldiers have become integrated into the military way of life and how that may compare to the integration of Active-duty spouses, especially considering that they are more likely to be geographically isolated from military communities. In McClure and Broughton's (1988) study of military communities, they note that the term *community* refers to constructs, such as satisfaction with and attachment to the community, social networks, and a psychological sense of community, which tap into the construct cohesion. Regardless of component, if spouses feel they are not integrated into the military community, findings indicate that they may experience less marital satisfaction and have a negative attitude toward their Soldier continuing in the organization (Bourg and Segal, 1998; McClure and Broughton, 1988).

The focus of this chapter will be on health-, family-, and Army-related findings, and those demands spouses experienced that either positively or negatively contributed to those outcomes. The findings presented below reflect earlier studies conducted by researchers at Walter Reed that examine the impact of both deployed (peacekeeping) and nondeployed environments on Active-duty, Reserve, and National Guard spouses who were primarily civilian.

Research Hypotheses

The findings presented are from two survey studies we conducted in the late 1990s to early 2000s timeframe. Each spouse study was based on a study that was conducted separately with their Soldiers. Before discussing the findings, a brief overview of the study questions and methodology will be presented.

Study 1

1. Active-duty spouses will report greater levels of integration within the military community than will Reserve or National Guard spouses.

2. Spouses who are less integrated into the military way of life will a) report poorer physical health; b) report higher depression levels; c) report increased drinking and smoking; and d) want their Soldier to leave the military.

Study 2

1. Hypothesis 1a-d: Fear for the Soldier's safety will be negatively related to (a) psychological well-being, (b) physical well-being, (c) Army life satisfaction, and (d) marital satisfaction.
2. Hypothesis 2a-d: Impact of moving will be negatively related to a-d above.
3. Hypothesis 3a-d: Impact of separations will be negatively related to a-d above.
4. Hypothesis 4a-d: Impact of foreign residence will be negatively related to a-d above.

Study 1 Methods

Participants and Procedures

During 2000, a convenience sample of Soldiers within the III Corps area completed a questionnaire that was given in person. A subsample of all participants who indicated they were married was identified (N = 1,384). This subsample included 444 Active-duty, 417 Guard, and 523 Reserve Soldiers ranging in age from 18 to 63 years (M = 33.44, SD = 9.97). Of these, 92 percent were men, 74 percent had children living in the home, and 70 percent had working spouses.

Questionnaires were then mailed or delivered, via the units, to the spouses of the Soldiers who previously participated. Of the 709 spouses who participated, 90 percent were female and 8 percent were male. The response rate for Active-duty spouses was 31 percent, while the rates for National Guard and Reserve spouses were 76 percent and 42 percent, respectively. The mean age of the participants was 35. With regard to education, only 1 percent had less than a high school education. Twenty-two percent had a high school diploma or its equivalent; 66 percent had some college or a college degree; and 10 percent had some graduate training or a graduate degree. The majority of participants (89 percent) were Caucasian; 4 percent were Hispanic, 2 percent were African-American, 2 percent were either Asian, Native American, or other, and 2 percent were multiracial.

Measures

Integration. Two programs that the Army offers to assist spouses in adjusting to the military way of life include Family Support Groups (FSGs), which are now known as Family Readiness Groups (FRGs), and

Army Family Team Building (AFTB). These programs provide assistance regardless of deployment status. Spouses were asked if their units had FSGs, if they attended meetings, and if they had received AFTB training. Spouses also answered a two-part question regarding where they would go if they needed support: "Many spouses need support at some time—whether it is someone to talk to or someone to help with a specific problem. Below are listed persons or agencies where one might go for support." The first part of the question was: "Please indicate how likely you are to seek support from each." The resource choices were: 1) Your spouse, 2) Your children, 3) Your extended family (parents, siblings, etc.), 4) A clergyman from the civilian community, 5) Members of your church/mosque, synagogue, 6) Your boss, 7) Co-workers, 8) Friends/neighbors not associated with your Soldier's unit, 9) A Soldier from your Soldier's unit, 10) A spouse from your Soldier's unit, 11) Army Community Service (ACS), 12) Army Emergency Relief (AER), 13) An Army Chaplain, and 14) Your Family Support Group. The second part of the question was: "Please indicate how often you have gone to that person or agency for support during the past year." The four response categories were: 1) not at all, 2) a few times, 3) moderately, and 4) very often. We had two other questions that looked at military associations: "Are you friends with the members of your Soldier's unit?" and "Are you friends with the spouses of unit members?"

An "integration score" was developed based upon spouses answering "yes" to attending FSG meetings, taking AFTB training, having friends in the unit, or being friends with spouses of unit members. Additionally, spouses who answered that they "moderately/very often" sought support from Army resources (AER, ACS, Chaplain) were included in the score. A score of 1 was given for each "yes" or "moderately/very" answer. Therefore, scores could range from 0 to 7, with 7 showing the highest degree of integration with the military.

Physical and psychological health. Physical health was measured through a self-assessment of one question: "How would you rate your current state of physical health?" Responses ranged from "excellent" to "very poor." Questions also addressed the individual's smoking and drinking habits. We used the four-item CAGE scale (Mayfield, McLeod, and Hall 1974) to determine if there was an alcohol abuse problem (key words, as italicized below, form the acronym CAGE). 1) Have you ever felt you should *cut* down on your drinking? 2) Have people *annoyed* you by criticizing your drinking? 3) Have you ever felt *guilty* about your drinking? 4) Have you ever had a drink first thing in the morning (an *eye-opener*) to steady your nerves or get rid of a hangover? A score of > 2 on the CAGE indicates a problem with alcohol. Participants were also asked how much alcohol they consumed in a typical week and scores were categorized from 1 (none) to 5 (10 or more). Smoking behavior was determined by the question: "Which statement best describes your smoking habits in the last year?" The statements were: 1) I have never been a smoker, 2) I smoked but quit, 3) I smoke 10 or fewer cigarettes a day, 4) I smoke 11–19 cigarettes a day, and 5) I smoke more than 20 cigarettes a day.

The individual's psychological health was evaluated by the Center for Epidemiological Studies for Depression Scale, which consists of 20 questions chosen to reflect various aspects of depression including depressed mood; feelings of guilt and worthlessness, helplessness, and hopelessness; psychomotor retardation; loss of appetite; and sleep disturbance (Radloff, 1977). Respondents reported the frequency of occurrence for each item during the previous week on the following 4-point scale: 0 (rarely, less than 1 day), 1 (some of the time, 1 to 2 days), 2 (a moderate amount of the time, 3 to 4 days), or 3 (most or all of the time, 5 to 7 days). Summary scores can range from 0 to 60 with a score of 16 or above indicating depression.

Retention intentions. We asked the spouses what their Soldiers' retention intentions were. Options included: 1) My Soldier wants to make the Army a career, 2) My Soldier wants to stay in the Army after his/her current enlistment/tour, 3) My Soldier wants to get out at the end of his/her current enlistment/tour, 4) My Soldier wants to get out of the Army before the end of his/her current enlistment/tour. Spouses were asked if they agreed with their Soldiers' retention intentions and, if not, what they would prefer their Soldiers to do.

Study 2 Methods

Participants and Procedures

During 2002, approximately 3,900 questionnaires were either mailed directly to spouses or delivered via units that were located within Germany and Italy. A total of 506 questionnaires were returned for a response rate of 13 percent. From this sample, a subsample of spouses who reported at least one deployment was drawn. Deployment was defined as the spouse being away from home for 30 days or more on a combat, peacekeeping, or humanitarian mission. This subsample consisted of 346 (7 men and 339 women) persons ranging in age from 18 to 51 years (M = 30.41, SD = 6.36), the majority of whom were Caucasian (77 percent). African-Americans accounted for 9 percent of the sample, Hispanics 6 percent, Asians 3 percent, and others 5 percent. Most were well educated, with 25 percent indicating that they graduated from college and another 15 percent reporting either some graduate training or a graduate degree.

Measures

Military lifestyle demands. Fear for Soldiers' safety was measured with four items: "I worry about my spouse being injured while on deployment," "I worry about my spouse being killed on deployment," "There is a strong possibility that my spouse will be involved in combat during deployment," and "Given my spouse's job in the military, there is a higher risk for injury or death during deployment." Impact of moving was measured with seven items: "Moving has had a positive impact on my family," "We move more frequently than I would like," "Moving has provided me with many positive opportunities," "Moving has allowed me to make new friends," "We have moved to exciting places," "Moving is difficult on our children," and "One of the benefits of being a military spouse is getting to move." Impact of separations was measured with four items: "The separations from my spouse are stressful," "The number of deployments has put a strain on our family," "The number of deployments has hurt the stability of our marriage," and "I worry about the effects of my spouse's deploying on our children." Impact of foreign residence was measured by 11 items: "I like living in Europe," "I find the people in this country are very friendly to our Soldiers and families," "Since I have been living here, I feel isolated," "I am comfortable using the local language," "It has been difficult for me to make friends with others in the military community," "I like learning about this country," "I like to shop on the economy," "It has been difficult to make friends with the Germans/Italians," "I try to participate in this country's activities," "I can hardly wait to get back to the States," and "Being away from friends and relatives back home is very hard for me." All items for these measures were rated on a five-point, Likert-type scale ranging from 1 (strongly agree) to 5 (strongly disagree), and then scored/reverse-scored such that a higher score indicates a higher/more negative standing on the variable of interest. That is, higher scores indicated greater fear and a greater negative impact of moving, separations, and foreign residence.

Physical and psychological health. Well-being was measured in two ways. The Brief Symptom Inventory was used to assess psychological well-being. Participants were asked to indicate how much they had been bothered or distressed by each of 18 items over the past 7 days (Derogatis, 2000). Sample items included, "Feeling no interest in things" and "Nervousness or shakiness inside." The response scale ranged from 1 (none) to 5 (extreme). Walter Reed Army Institute of Research's physical health symptom checklist was used to assess physical well-being. Participants were asked to indicate how often they experienced each of 22 health symptoms during the past month. Sample items included "headaches," "eye/ear/nose problems," "stomach/intestinal problems," and "shortness of breath." The response scale ranged from 1 (not at all) to 5 (very often). For both measures, items were scored/reverse-scored such that a higher value represented higher levels of well-being (fewer symptoms). These scores were then averaged to create a composite score.

Army-related attitudes. Participants were asked to indicate how satisfied they were with each of five items related to Army life attitudes. Items included, "The Army as a way of life," "The concern your spouse's unit has for families," "The respect the Army shows spouses," "How you would feel if your spouse were to make the Army a career," and "The kind of family life you can have in the Army." The response scale ranged from 1 (very satisfied) to 5 (very dissatisfied). Items were scored/reverse-scored such that a higher value represented a more favorable value on the variable and then averaged to create a composite score.

Marital satisfaction. Participants were asked to indicate how much they agreed with each of six items from Norton's Quality of Marriage Index (Norton, 1983). Sample items included, "I have a good marriage," "I am happy in my marriage," and "I feel like I am part of a team with my partner."

Results of Studies 1 and 2

Various analyses, including correlations, analyses of variance, chisquares, and regressions, were conducted to assess the impact of separations, moving, residence in a foreign country, fear for a Soldier's safety, and community integration on health, Army, and family-related outcomes.

Predictors

Using our indicators of integration in study 1, the data suggested that the United States Army Reserve (USAR) and Army National Guard

(ARNG) spouses are not integrated into the military way of life to any great degree. Table 16–1 shows that only about half of the Reserve Components had FSGs and, where they were available, only about one-fifth of spouses attended meetings. Similarly, only about 20 percent of them had taken AFTB training. Approximately half of ARNG spouses claimed no friends in the unit, as did 70 percent of USAR spouses.

There were significant differences between the Reserve and Active Component spouses on all these measures. Significant results emerged on whether or not they attended FSG meetings ($X^2 = (2, N = 494) = 67.62$, $p = .000$) and if they attended AFTB training ($X^2 = (2, N = 663) = 77.87$, $p = .000$). Likewise, there were significant differences on where they had friends in the unit: with regard to having Soldiers as friends ($X^2 = (2, N = 697) = 34.70$, $p = .000$) and with regard to having unit spouses as friends ($X^2 = (2, N = 693) = 34.95$, $p = .000$). Thus, hypothesis 1 was supported. However, while there were significant differences between Active-duty spouses and the Reserve Component spouses, even Active-duty spouses were not strongly integrated into the military way of life;

Table 16–1. **Indicators of Army Spouse Integration into Military Way of Life (in percent)**

Integration Indicator	Active-duty spouses	Reserve spouses	National Guard spouses
Does your Soldier's unit have a Family Support Group (FSG)? **"Yes" (N = 691)**	88	52	54
If yes, do you attend FSG meetings? **"I do not attend" (N = 494)**	43	86	79
Have you taken Army Family Team Building (AFTB) training? **"I have not taken AFTB training" (N = 663)**	57	77	83
Are you friends with members of your Soldier's unit? **"No, I have no friends in the unit" (N = 697)**	34	68	49
Are you friends with the spouses of unit members? **"No, I have no friends who are unit spouses" (N = 693)**	44	76	53

a large percentage of them did not attend FSGs, take AFTB training, or have friends in the unit, as indicated in table 16–1.

Another indicator of integration was the use of unit or formal Army support programs for problems. We asked spouses how often they have sought support from the unit, Army formal support programs, or others for their problems. As can be seen in table 16–2, these resources were used by very few of the spouses. Integration scores ranged from 0 to 6, with 74 percent of responses in the 0 to 2 range, suggesting little overall integration into the military. A score of 3 was indicated by 16 percent, approximately 9 percent scored 4, and only 1 percent scored in the 5–6 range. Active-duty spouses were significantly more integrated than either Reserve or National Guard spouses ($p < .05$). The mean score for Reserve spouses was .84, while for National Guard spouses the mean was 1.42; for Active-duty spouses, the mean integration score was 2.38.

Military Lifestyle Demands

In the second study, different factors were assessed with regard to the outcomes described above. The impact of four different demands was analyzed, and descriptive findings are presented in table 16–3. Mean

Table 16–2. **Moderate or Frequent Army Spouse Use of Military Resources for Support (in percent)**

Resource	Active-duty	Reserve	National Guard
Unit Resources:			
A Soldier from your Soldier's unit (N = 663)	2	1	1
Another spouse from your Soldier's unit (N = 663)	11	2	2
Family Support Group (N = 667)	4	1	< 1
Army Resources:			
Army Community Service (N = 666)	< 1	2	1
Army Emergency Relief (N = 665)	< 1	2	1
Army Chaplain (N = 667)	3	2	1

scores show that fear for Soldier safety was the greatest concern, followed by the impact of separations, of moving, and of foreign residence.

Outcomes

Physical and Psychological Health

Although the first study indicated significant differences in terms of integration across the three components, level of integration within the military community was not significantly correlated with, or predictive of, any of the physical or psychological health measures studied; this

Table 16–3. **Descriptive Statistics (N = 346)**

Variable	Mean	Standard Deviation	Possible Range	Observed Range	Cronbach's Coefficient Alpha
Age	30.41	6.36	18-	18–51	N/A
Soldier rank	8.91	6.78	(2–29) E1–O10	(2–28) E2–O8	N/A
Number of moves	3.43	3.10	0-	0–33	N/A
Number of separations	3.53	5.10	1-	1–40	N/A
Fear for Soldier safety	3.76	.81	1–5	1.25–5	.81
Impact of moving	2.89	.69	1–5	1.38–5	.82
Impact of separations	3.28	.75	1–5	1–5	.70
Impact of foreign residence	2.63	.68	1–5	1–5	.84
Psychological well-being	4.59	.44	1–5	2–5	.88
Physical well-being	4.25	.52	1–5	1.5–5	.86
Army life satisfaction	3.01	.81	1–5	1–5	.82
Marital satisfaction	4.49	.83	1–5	1–5	.97

indicates a lack of support for hypotheses 2a–c in study 1. However, some of the demands in the second study did yield significant results. After controlling for age, rank, number of moves, and number of separations the family had experienced, a series of two-step hierarchical regressions were calculated. Control variables were entered on the first step and the four lifestyle demand variables were entered on the second step.

As can be seen in table 16–4, on the first step, the control variables accounted for 7 percent of the variance in psychological health ($R^2 = .07$, F [4, 341] = 6.04, p <. 01). The addition, the lifestyle demand variables accounted for an additional 12 percent of the variance ($\Delta R^2 = .12$, F [4, 337] = 11.79, p < .01). An examination of the betas for the individual variables at this second step indicated that impact of separation and impact of foreign residence each had significant negative relationships with psychological health. In terms of physical health (see table 16–5), the control variables accounted for 1 percent of the variance in physical well-being ($R^2 = .01$, F [4, 341] = .37, ns), while the addition of the lifestyle demand variables accounted for an additional 6 percent of the variance ($\Delta R^2 = .06$, F [4, 337] = 5.23, p <. 01). An examination

Table 16–4. **Regression of Psychological Health on Control and Lifestyle Demand Variables (N = 346)**

	Step 1 ($R^2 = .07$*) Beta	Step 2 ($R^2 = .18$*, $\Delta R^2 = .11$*) Beta
Control Variables		
Age	.19*	.11
Rank	.11	.05
Number of moves	.03	.05
Number of separations	.01	.04
Lifestyle Demand Variables		
Fear for Soldier safety	—	-.07
Impact of moving	—	.09
Impact of separations	—	-.22*
Impact of foreign residence	—	-.24*

*p < .01

Table 16–5. **Regression of Physical Health on Control and Lifestyle Demand Variables (N = 346)**

	Step 1 (R² = .01) Beta	Step 2 (R² = .06**, Δ R² = .05**) Beta
Control Variables		
Age	-.02	-.06
Rank	.04	.04
Number of moves	-.04	-.02
Number of separations	.03	.00
Lifestyle Demand Variables		
Fear for Soldier safety	—	-.12*
Impact of moving	—	.14*
Impact of separations	—	-.12*
Impact of foreign residence	—	-.16**

*p < .05; **p < .01

that fear for Soldiers' safety, impact of moving, impact of separation, and impact of foreign residence each had significant relationships with physical well-being. Contrary to expectations, the relationship between impact of moving and physical well-being was actually positive in sign and significant (p < .05).

Army-related Outcomes

Although integration into the military community was not correlated with health, it was significantly correlated with the spouse's desire for their Soldier to remain in the military (r = .12, p < .05). Spouses who felt more integrated indicated a desire for their Soldiers to remain in the military even when Soldier age and component were partialed from integration scores (see table 16–6). When assessing satisfaction with Army life, mixed results occurred in the second study (see table 16–7). The control variables in the regression analysis accounted for 13 percent of the variance in Army life satisfaction (R² = .13, F [4, 341] = 12.87, p < .01). The addition of the four lifestyle demand variables accounted for an additional 14 percent of the variance (ΔR² = .14, F [4, 337] = 16.47,

Table 16–6. **Logistic Regression of Integration into the Military Community and Retention Preference**

Variable	b	SE	Wald	p	Estimated Odds Ratio	Confidence Interval (Odds)
Age	.105	.018	33.889	.000	1.111	1.072–1.151
Active-duty spouse	-.058	.419	.019	.891	.944	.415–2.146
National Guard spouse	.411	.344	1.426	.232	1.508	.768–2.962
Integration into military	.302	.116	6.748	.009	1.353	1.077–1.699
Constant	-2.691					

Table 16–7. **Regression of Army Life Satisfaction on Control and Lifestyle Demand Variables (N = 346)**

	Step 1 $(R^2=.13^*)$ Beta	Step 2 $(R^2=.27^*, \Delta R^2=.14^*)$ Beta
Control Variables		
Age	.26*	.18*
Rank	-.02	.02
Number of moves	.01	.04
Number of separations	.20*	.13*
Lifestyle Demand Variables		
Fear for Soldier safety	–	.02
Impact of moving	–	-.20*
Impact of separations	–	-.25*
Impact of foreign residence	–	-.10

*$p < .01$

$p < .01$). An examination of the betas for the individual variables at this second step indicated that the impact of moving and impact of separation had significant negative relationships with Army life satisfaction.

Residence in a foreign country and fear for their Soldiers' safety were not significantly related to satisfaction with the Army.

Family-related Outcomes

Again, mixed findings occurred for the second study (see table 16–8). On the first step, the control variables accounted for 3 percent of the variance in marital satisfaction ($R^2 = .03$, F [4, 341] = 2.37, ns). The addition of the lifestyle demand variables accounted for an additional 3 percent of the variance ($\Delta R^2 = .03$, F [4, 337] = 3.04, p < .05). An examination of the betas for the individual variables at this second step indicated that only one of the four lifestyle demands, impact of separation (beta = -.17), had a significant relationship with marital satisfaction.

Conclusion

The results of the analyses from the two studies yielded mixed support for the hypotheses. While we did find that Active-duty spouses indicated a greater degree of integration into the military community than did Reserve or National Guard spouses, a high degree of integration was not experienced by any of the three components. Additionally,

Table 16–8. **Regression of Marital Satisfaction on Control and Lifestyle Demand Variables (N = 346)**

	Step 1 ($R^2 = .03$) Beta	Step 2 ($R^2 = .06$**, $\Delta R^2 = .03$*) Beta
Control Variables		
Age	-.14*	-.19**
Rank	-.01	.02
Number of moves	.03	.04
Number of separations	.13*	.12*
Lifestyle Demand Variables		
Fear for Soldier safety	–	.04
Impact of moving	–	.02
Impact of separations	–	-.17**
Impact of foreign residence	–	-.07

*p < .05; **p < .01

integration was not significantly related to any of the health measures; however, a greater degree of integration was associated with wanting the Soldier to remain in the Army. One possible explanation for the low levels of integration is that the Soldiers of the spouses surveyed were not deployed and deployment would likely result in a greater need to use such resources to obtain information about their Soldiers and their safety. A second possible explanation may be due to children and employment, which may not leave time for friends or involvement within the military community. In our study, 65 percent of the spouses were working either part- or full-time and 80 percent had at least 1 child living at home. A third possibility is that the spouses may indeed feel integrated but their behavior suggests otherwise.

The impact of separations was the one demand variable that was predictive of all of the outcomes reported in study 2, which suggests that it may play a more critical role in determining how spouses are affected. Moving was associated with physical health and Army life satisfaction. Living in a foreign residence was associated with both physical and psychological health, and fear concerning Soldier safety was associated with physical health. Based on these findings, physical health was the outcome most susceptible to the influence of the military life demands as it was the only variable predicted by all four demands. However, in study 1, integration was not predictive of physical health outcomes.

One possible explanation for the difference in findings between study 1 and study 2, beside the different predictors measured, is that study 1 included nondeployed Soldiers while study 2 included deployed Soldiers. Deployment status has been linked to health. Adler, Bartone, and Vaitkus (1995) assessed the effects of deployment on spouses whose Soldiers were on a peacekeeping mission in Croatia and found that over half of the spouses reported sleep problems and more than a third reported a loss of appetite. In another study of health outcomes that included Navy wives, spouses whose husbands were deployed were not significantly different in terms of physical or psychological symptoms when compared to wives whose husbands were not deployed; the only exception was depression (Nice and Beck, 1980). Those wives whose husbands were deployed had significantly higher depression scores during the pre- and mid-deployment phases than did the nonseparated wives. During the deployment, separated wives also went to the doctor significantly more than the nonseparated wives. As mentioned previously, integration was not predictive of psychological health. However, in study 2, impact of separations, which is consistent with deployment, was predictive of psychological health.

Based on the findings from the two studies we can see that, regardless of deployment status, Army-related outcomes such as retention preferences and Army life satisfaction may be impacted by integration and a variety of demands. However, deployment status may play a role in health outcomes. In the first study, level of integration in a nondeployed environment was not related to health outcomes; however, this might have been different if studied in a deployed situation. It is possible that integration may serve as a buffering source of social support rather than a main source. The buffering effect suggests that support is beneficial under stressful conditions such as deployment, but is neither helpful nor harmful in the absence of stress (Griffith, 1985).

Several topics within military family research deserve further study. While the effects of deployment on families is certainly at the forefront of our minds and is another demand that extends beyond its ties to separation and integration, it is also important to continue to study the effects on those families whose spouses have not deployed. It should be noted if there are differences in health-, family-, and Army-related outcomes between the two groups as well as the potential unique challenges these families may face. For example, Soldiers who do not deploy may not have enough manpower back in garrison to efficiently accomplish the mission, or perhaps these families feel somewhat ostracized/isolated and find it difficult to relate to other military families who have been through a deployment experience. Although we do not know just how different or similar the experiences of deployed and nondeployed spouses are, the findings presented hopefully shed some light on that issue.

Additionally, understanding the ties that spouses in the various components have to the military community, and the impact these ties have on a broad range of outcomes, should continue to be investigated. This is especially important as spouses are an integral part of the Army as an organization. Typically, it is the spouse that is called upon to hold fundraisers or participate in community service programs that are used to ensure a better quality of life for the military family. In many cases, it is the spouse who notifies other spouses about what is happening with a Soldier's unit through means such as FRGs and, if this support system is not in place, this may place an even greater burden on the organization to inform spouses of what the units are currently doing. Even if the spouse does not like the deployments, feeling supported may at least buffer the stressful effects of the separation; this would give them a more positive view of the organization for its sense of community and, perhaps, a subsequent desire for the Soldier to stay in the Service.

Along those same lines, a better understanding of coping strategies families use, and the appraisal of the demands the military places on its families, is critical to reducing the negative effects that may occur in response to these demands. Fortunately, programs such as FRGs, AFTB, Morale Welfare and Recreation, and others exist within the Army to help families have a better quality of life. However, not everyone knows, uses, or has easy access to these programs. This suggests a need to explore other coping strategies as well.

The variables described above, which include deployment status, military community integration, and coping, are just a few that will provide better insight into how families are affected by a wide range of demands. Thus, new opportunities for resolving the stress associated with these demands may be discovered in the process. Opportunities may be reflected in programs and/or policies, and may exist within the individual, the unit, and the Army as an organization where, ideally, all three will collaborate to develop family resilience.

Portions of this chapter were previously published in Burrell, L., Durand, D.B., and Fortado, J. (2003). Military community integration and its effect on well-being and retention. *Armed Forces and Society*, 30, 7–24; and Burrell, L., Adams, G.A., Durand, D.B., and Castro, C.A. (2006). The impact of military lifestyle demands on well-being, Army, and family outcomes. *Armed Forces and Society*, 33, 43–58.

References

Adler, A.B., Bartone, P.T., and Vaitkus, M.A. (1995). *Family stress and adaptation during a U.S. Army Europe peacekeeping deployment* (USAMRU–E Technical Report 95–1). Fort Detrick, MD: U.S. Army Medical Research Unit–Europe.

Allen, T.D., Herst, D.E., Bruck, C.S., and Sutton, M. (2000). Consequences associated with work to family conflict: A review and agenda for future research. *Journal of Organizational Health Psychology* 5, 278–308.

Anderzen, I., and Arnetz, B.B. (1999) Psychophysiological reactions to international adjustment: Results from a controlled longitudinal study. *Psychotherapy and Psychosomatics* 68, 67–75.

Bedeian, A.G., Burke, B.G., and Moffett, R.G. (1988). Outcomes of work-family conflict among married male and female professionals. *Journal of Management* 14, 475–491.

Bell, D.B., and Schumm, W.R. (1999). Family adaptation to deployments. In Peggy McClure (ed.), *Pathways to the future: A review of military family research*. Scranton, PA: Military Family Institute, Marywood University.

Bourg, C., and Wechsler Segal, M. (1999). The impact of family supportive policies and practices on organizational commitment to the Army. *Armed Forces and Society* 25 (4), 633–652.

Bruner, E.F. (2007). *U.S. military dispositions: Fact sheet*. Retrieved July 27, 2008, from <http://digital.library.unt.edu/govdocs/crs/permalink/meta-crs-10172:1>.

Burnam, M.A., Meredith, L.S., Sherbourne, C.D., Valdez, R.B., and Vernez, G. (1992). *Army families and soldier readiness* (R–3384–A). Santa Monica, CA: RAND.

Cohen, A.A., and Dotan, J. (1976). Communication in the Family as a Function of Stress during War and Peace. *Journal of Marriage and the Family* 38, 141–148.

Coolbaugh, K.W., and Rosenthal, A. (1992). *Family separations in the Army* (ARI Technical Report 964). Alexandria, VA: U.S. Army Research Institute for the Behavioral and Social Sciences.

Copeland, A.P., and Norell, S.K. (2002). Spousal adjustment on international assignments: The role of social support. *International Journal of Intercultural Relations* 26 (3), 255–272.

Croan, G.M., Levine, C.T., and Blankinship, D.A. (1992). *Family adjustment to relocation* (ARI Report 968). Alexandria, VA: U.S. Army Research Institute for the Behavioral and Social Sciences.

Derogatis, L.R. (2000). *The brief symptom inventory 18 (BSI 18): Administration, scoring and procedures manual.* Minneapolis, MN: National Computer Systems.

Finkel, L.B., Kelley, M.L., and Ashby, J. (2003). Geographic mobility, family, and maternal variables as related to the psychosocial adjustment of military children. *Military Medicine* 168 (12), 1019–1024.

Frankel, H., Snowden, L.R., and Nelson, L.S. (1992). Wives' adjustment to military deployment: An empirical evaluation of a family stress model. *International Journal of Sociology of the Family* 22, 93–117.

Frone, M.R., Yardley, J., and Markel, K.S. (1997). Developing and testing an integrative model of the work-family interface. *Journal of Vocational Behavior* 50, 145–167.

Gal, R., and Syna, H. (1988). *Relationship between the soldier in the military unit and family* (No. ADA 198585). Israeli Institute for Military Studies Zikhron Ya'akov.

Gaylord, M., and Symons, E. (1986). Relocation stress: A definition and a need for services. *Employee Assistance Quarterly* 2 (1), 31–36.

Griffith, J. (1985). Social support providers: Who are they? Where are they met? And the relationship of network characteristics to psychological distress. *Basic and Applied Social Psychology* 6 (1), 41–60.

Hunter, E.J., and Hickman, R.A. (1981). *Family separation and reunion: A style of life in the military.* San Diego, CA: United States International University.

Jensen, P.S., Lewis, R.L., and Xenakis, S.N. (1986). The military family in review: Context, risk, and prevention. *Journal of the American Academy of Child Psychiatry* 25, 225–234.

Kopelman, R.E., Greenhaus, J.H., and Connolly, T.F. (1983). A model of work, family, and inter-role conflict: A construct validation study. *Organizational Behavior and Human Performance* 32, 198–215.

Lewis, P.M. (1985). *Family factors and the career intent of Air Force enlisted personnel* (LMDC Technical Report 85–9). Washington, DC: USAF Office of Scientific Research.

Makowsky, P.P., Cook, A.S., Berger, P.S., and Powell, J. (1988). Womens' perceived stress and well-being following voluntary and involuntary relocation. *Lifestyles: Family and Economic Issues* 9 (2), 111–122.

Marshall-Mies, J.C. (2001). *Civilian spouses of soldiers: Comparison of the results for male and female spouses* (ARI Technical Report 2001–04). Alexandria, VA: U.S. Army Research Institute for the Behavioral and Social Sciences.

Maxfield, B.D. (2006). *Army Demographics: FY06 Army Profile.* Retrieved July 26, 2008, from <www.2k.army.mil/downloads/FY06Tri-Fold.pdf>.

Mayfield, D., McLeod, G., and Hall, P. (1974). The CAGE questionnaire: Validation of a new alcoholism instrument. *American Journal of Psychiatry* 131, 1121–1123.

McClure, P., and Broughton, W. (1988). *Military community cohesion* (MFI Technical Report 98–4). Scranton, PA: Marywood University.

Mohr, D.A., Holzbach, R.L., and Morrison, R.F. (1981). *Surface warfare junior officer retention: Spouses' influence on career decisions* (NPRDC Technical Report 81–17). San Diego, CA: Navy Personnel Research and Development Center.

Netermeyer, R.G., Boles, J.S., and McMurrian, R. (1996). Development and validation of work-family conflict and family-work conflict scales. *Journal of Applied Psychology* 81, 400–410.

Nice, D.S., and Beck, A. (1980) *Feelings of depression in Navy wives prior to family separation.* San Diego, CA: Navy Personnel Research and Development Center.

Norell, S., and Copeland, A.P. (2002).*US–UK moves: Cross cultural adjustment in American and British expatriate accompanying spouses.* Retrieved November 16, 2003, from <www.interchangeinstitute.org/html/ research.htm>.

Norton, R. (1983). Measuring marital quality: A critical look at the dependent variable. *Journal of Marriage and the Family* 45 (1), 141–151.

Orthner, D.K. (2002). *Deployment and separation adjustment among Army civilian spouses.* SAF IV Survey Report retrieved September 13, 2005, from <www.armymwr.com/corporate/operations/planning/surveys.asp>.

Ozkaptan, H., Sanders, W., and Holz, R. (1986). *A profile of Army families in USAREUR: Results of the 1983 families in Europe survey* (ARI Technical Report 925). Alexandria, VA: U.S. Army Research Institute for the Behavioral and Social Sciences.

Puskar, K.R. (1990). International relocation: Womens' coping methods. *Health Care for Women International* 11, 263–276.

Radloff, L.S. (1977). The CES-D Scale: a self-report depression scale for research in the general population. *Applied Psychological Measurement* 1, 385–400.

Raschman, J.K., Patterson, J.C., and Schofield, G.L. (1989). *A retrospective study of marital discord in pilots* (Technical Report USAFSAM–JA–90–5). Brooks AFB, TX: The USAF School of Aerospace Medicine.

Rogers, S.J., and May, D.C. (2003). Spillover between marital quality and job satisfaction: Long term patterns and gender differences. *Journal of Marriage and Family* 65 (2), 482–495.

Rohall, D.E., Segal, M.W., and Segal, D.R. (1999). Examining the importance of organizational supports on family adjustment to Army life in a period of increasing separation. *Journal of Political and Military Sociology* 27, 49–65.

Rosen, L.N., and Durand, D.B. (1995). The family factor and retention among married soldiers deployed in Operation Desert Storm. *Military Psychology*, 7 (4), 221–234.

Rosen, L.M., Teitelbaum, J.M., and Westhuis, D. (1993). Stressors, stress mediators, and emotional well-being among spouses of soldiers deployed to the Persian Gulf during Operation Desert Shield/Storm. *Journal of Applied Social Psychology* 23, 1587–1593.

Schneider, R., and Gilley, M. (1984). *Family adjustment in USAREUR.* U.S. Army Medical Research Unit–Europe Final Report. Heidelberg, Germany.

Schumm, W.R., Bell, D.B., and Gade, P. (2000). Effects of a military overseas peacekeeping deployment on marital quality, satisfaction, and stability. *Psychological Reports* 87 (3), 815–821.

Schumm, W.R., Bell, D.B., and Knott, B. (2001). Predicting the extent and stressfulness of problem rumors at home among Army wives of soldiers deployed overseas on a humanitarian mission. *Psychological Reports* 89 (1), 123–134.

Schwartz, J.B., Braddy, B.A., Griffith, J.D., and Wood, L. (1988). *The employment status of Army spouses* (ARI Technical Report 14). Alexandria, VA: U.S. Army Research Institute for the Behavioral and Social Sciences.

Scurfield, R.M., and Tice, S.N. (1992). Interventions with medical and psychiatric evacuees and their families: From Vietnam through the Gulf War. *Military Medicine* 15 (2), 88–97.

Van der Kloet, I., and Moelker, R. (2002). *Overcoming stressful experiences: Military families in the frontline.* Paper presented at the 38th IAMPS. Amsterdam, Netherlands.

Warner, W.M. (1983). *The Air Force wife: Her perspective.* Unpublished manuscript, Maxwell Air Force Base, AL: Air Command and Staff College.

Chapter 17

To Stay or Not to Stay? Family-friendly Unit Climate and Career Intentions

Edward N. Edens, Lyndon A. Riviere, Charles W. Hoge, and Paul D. Bliese

For decades, the Army has invested substantial amounts of money into training Soldiers. Therefore, it is in our best interest to retain as many of them as possible. However, many Soldiers voluntarily separate from the Army before retirement. According to a Congressional Budget Office report, since 2000, over 15 percent of Active-duty Soldiers have left the Army annually.

In this study, the outcome of interest is career or turnover intentions. The relationship between turnover intentions and behavior has been established (Huffman, Adler, Dolan, and Castro, 2005). They found that among Soldiers who were undecided as to whether they planned to stay in the Army or separate early, more than 75 percent actually remained in the Army. Approximately 96 percent of those who indicated that they would stay actually did, while a plurality (59 percent) of those who planned to leave actually left.

When analyzing why Soldiers decide to leave the Army or remain in until retirement, many possible reasons emerge. These vary on an individual basis, but one theme that often appears is family ties. With family ties come family demands, which have been associated with Soldier attrition, particularly among those who are married with children (Bedeian, Burke, and Moffett, 1988; Carlson and Perrewe, 1999; Duxbury and Higgins, 1991; Frone, Yardley, and Markel, 1997; Major, Klein, and Ehrhart, 2002). As of 2004, over half of all Active-duty Soldiers were married; 58 percent of those had an average of two children, which adds up to roughly 1.17 million children belonging to Active-duty military families (Ender, 2006; National Council on Family Relations, 2004). At the same time, the military is seeing an increase in dual-career couples, many of whom have dependents (Griffith, Rakoff, and Helms, 1992; Huffman and Payne, 2006; Martin and McClure, 2000; Schumm, Bell, Rice, and Sanders, 1996). In 2004, there were over 47,000 dual-Service couples with children in the military (National Council on Family Relations, 2004).

The purpose of this study is to examine the relationship between a family-friendly unit climate and the decision of married Soldiers with children to remain or leave the Army beyond their present commitment. Examining the relationship between a family-friendly unit climate and married Soldiers' decisions to remain in the Army or leave early can ultimately benefit the military as a whole and help the Army extend Soldiers' enlistment beyond their present time commitment.

Work-Family Conflict

Like people employed in other careers, Soldiers sometimes experience the conflicting demands of work and family (Kopelman, Greenhaus, and Connolly, 1983). Work-family conflict (WFC) is likely experienced most acutely by those Soldiers who have dependent children (Bedeian et al., 1988; Bruck and Allen, 2003; Eagle, Icenogle, Maes, and Miles, 1998; Frone, 2003; Grywacz and Marks, 2000; Kinnunen and Mauno, 1998; Major et al., 2002). Further, military personnel have obstacles that are not common to their civilian counterparts, such as multiple deployments into life-threatening situations for long periods and increasing demands to begin training for the next deployment as soon as they complete their current deployment.

In recent years, there has been an increase in the number of deployments as the United States continues to send military personnel to Afghanistan as well as Iraq. During these deployments and the concomitant training for them, Soldiers are separated from their families. Several studies have been conducted on how separation between military personnel and their families increases the likelihood that they will not remain in the military beyond their current obligation (Schumm, Jurich, Stever, Sanders, Castelo, and Bollman, 1998; Smith, 1988).

Other demands more common to military personnel than the typical civilian workforce include residence in foreign countries, frequent moves, physical separation, long working hours, and demanding shift work. These demands increase WFC for all Soldiers, but especially for those with children at home. Implementing a family-friendly working environment may help alleviate the intensity of WFC (Weiss, 2003).

In the past, the military has been in the forefront of providing medical care for dependents. In recent years, the military has also followed the lead of many civilian organizations by incorporating family-friendly working environments. One example is instituting flexible work hours so Soldiers can be available to take care of their families in emergencies. Again like many civilian companies, the military has also embraced on-site Childhood Development Centers where parents can leave their children, with one or more located on many posts and bases throughout the United States and around the world. The Department of Defense also provides youth services to Soldiers' school-age children in order to focus on their particular needs (Office of Family Policy, 1999).

Recently, the Department of Defense has begun to use the Internet as a medium to help military family members cope with unique situations. A particular example is the creation of the "Military Teens on the Move" Web site (<www.defenselink.mil/mtom>), which has coined the catch phrase, "Moving can be cool . . . and sometimes challenging." It provides information to children of military personnel about posts they are moving to, as well as other tips on how to handle the move (Ender, 2006).

Other family-friendly policies and programs implemented in recent years include the "family/Soldier time" policy where Soldiers are allowed to leave work early to tend to personal business or spend time with family. On-site military family service centers have been built to offer support groups for family members and host Army Family Team Building, which trains family members on such topics as "family and military expectations." In addition, mentoring programs have been advanced where a married Soldier is teamed up with another married Soldier, or a married Soldier with children is teamed up with another Soldier who is married with children, to offer each other comfort, support, and insight into programs and policies they can use to assist themselves and their dependents (Huffman and Payne, 2006).

The significant increase in dual-military career couples over the past decade introduces a unique set of issues that the Army must work on to prevent losing qualified personnel. In this case, it can mean the loss of not just one but two Soldiers if both spouses leave the Army as soon as their enlistments are concluded.

One of the biggest issues the Army faces with this group of Soldiers is what will happen if two married Soldiers cannot be reassigned to the same new location due to their career fields or branch of service (Huffman and Payne, 2006). The Married Army Couples Program was created to assist dual-career Army couples in being relocated to the same place (U.S. Army Human Resource Command, 2008). Yet it is sometimes necessary that dual-career couples accept less than desired locations in order to be stationed together (Williams, 1978).

The old adage "A happy family makes for a happy soldier" appears to be true. It has been shown that military spouses have a significant impact on the career intentions of Soldiers (Rosen and Durand, 2000). One of the major causes for spouses giving their active support is that the military environment is considered conducive for raising a family (Ender, 2006).

While there has been a substantial amount of research focusing on family-friendly working environments and their impact on WFC, very little has been done examining the relationship between a family-friendly unit climate and the decision of married Soldiers with children to remain or leave the Army beyond their present commitment (Bourg and Segal, 1999; Huffman, Culbertson, and Castro, 2008).

Method

Sample

The sample for this study is comprised of 492 Active-duty Soldiers from three brigades, who were surveyed 3 months after returning from a year-long deployment to Iraq between May and June of 2004. The response rate was 66 percent. The sample only included junior enlisted or noncommissioned officer (NCO) married male Soldiers with children.

Thirty-eight percent of the respondents were between 30 and 39 years of age, 30 percent between 25 and 29 years of age, 26 percent between 20 and 24, 1 percent under 20, and 5 percent over 39. The average time of military service was more than 10 years. Almost two-thirds of the respondents were NCOs and, of those, 16 percent were senior NCOs (E8–E9). More than half of the respondents had some college education, while almost 40 percent had a high school diploma or less. Ethnically, the sample was segregated into two groups: 67 percent were Caucasian, and 33 percent were minorities.

Females were removed from this study because they represented a very small sample of the total population (<1 percent). Officers and warrant officers were also excluded because their perception of what constitutes a family-friendly unit climate is different enough from those of junior enlisted soldiers and NCOs to warrant a separate study in the future.

Measures

Career intentions of Soldiers. This was a single-item, six-point Likert scale: "Which best describes your current Active-duty Army career intentions?" The response options were (1) definitely stay in until retirement; (2) probably stay in until retirement; (3) definitely stay in beyond my present obligation but not until retirement; (4) undecided about whether to stay after completion of my current obligation; (5) probably leave upon completion of my current obligation; or (6) definitely leave upon completion of my current obligation. If participants answered (4), undecided, they were removed from the sample. The distribution of the sample is displayed in figure 17–1. This item has been used previously in military research to measure career intent (Tremble et al., 2003). The mean for this variable was 3.17. This result is displayed in table 17–1.

Demographic variables. These include age, education, race/ethnicity, rank/grade, and years in the military. The distributions of the demographic variables are displayed in table 17–2.

Family-friendly unit climate. This measure includes six items, each with a five-point Likert scale ranging from strongly disagree (1) to strongly agree (5). An example of an item for this measure is, "My unit makes allowances for Soldiers to drop off and pick up children." Two of the items

Figure 17–1. **Career Intentions of Soldiers (N = 492)**

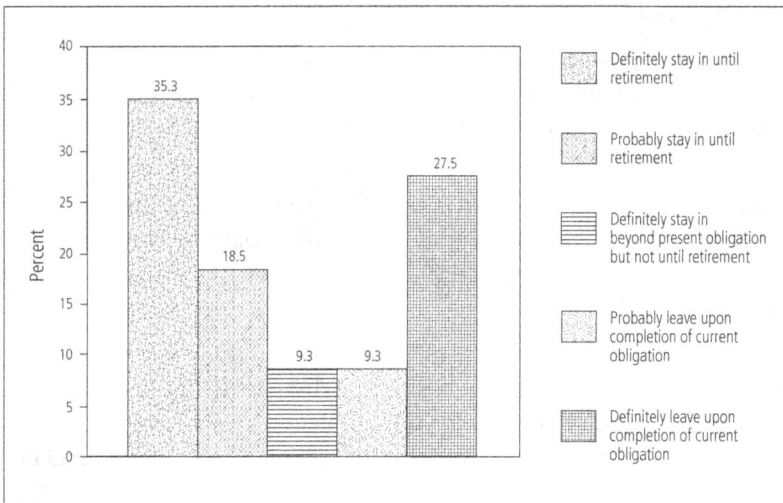

Table 17–1. **Measures Affecting Soldiers' Decision to Remain in Army**

Scales (possible range of response in brackets)	Mean	SD	α	n
Career intentions (1 item [1–6])	3.17	2.09		492
Family-friendly unit climate (6 items [6–30])	15.69	4.70	.80	492
Individual morale (1 item [1–5])	2.97	1.15		487
Unit morale (1 item [1–5])	2.63	0.93		485
Unit cohesion (3 items [3–15])	9.97	2.75	.88	492
Combat experiences (29 items [29–145])	67.15	18.89	.89	492

were negative ("Attending to family needs such as taking time off for sick children is frowned upon") and were recoded so all questions would have a positive scale. Scores were calculated by summing up the items with the higher numbers representing a greater belief in a positive family-friendly unit climate. The mean for this scale was 15.69, and the Cronbach's alpha coefficient was .80 (see table 17–1).

Individual morale. This was a single-item, five-point Likert scale that asked the respondent to rate his personal morale. Response choices ranged from very low (1) to very high (5). The mean for this variable was 2.97 (see table 17–1).

Unit morale. This was a single-item, five-point Likert scale that asked the respondent to rate the morale of his unit. Response choices ranged from very low (1) to very high (5). The mean for this variable was 2.63 (see table 17–1).

Unit cohesion. This contains three different items, each with a five-point Likert scale ranging from strongly disagree (1) to strongly agree (5). One example of an item for this measure is, "The members of my unit know that they can depend on each other." Scores were calculated by summing the results from each of the three items. All items with a higher number represent a greater belief in positive unit cohesion. The

Table 17-2. **Sample Demographic Variables**

Variable	Sample (N = 492) (in percent)
Age	
18–19	1
20–24	26
25–29	30
30–39	38
40 and over	5
Education	
GED	8
High school diploma	31
Some college	58
Bachelor's degree	2
Advanced	1
Race/Ethnicity	
Caucasian	67
Minority	33
Rank/Grade	
E1–E4	39
E5–E6	45
E7–E9	16
Years in Military	
1 or less	5
2	12
3–4	19
5–9	28
10+	36

mean for this scale was 9.97, and the Cronbach's alpha coefficient was .88 (see table 17–1).

Mental health status (screened positive for depression, anxiety, or post-traumatic stress disorder [PTSD]). This was measured using the Post-traumatic Stress Disorder Checklist, an extremely useful assessment tool for conducting mental health screenings on a large population where a structured interview is not feasible due to time and manpower constraints (Blanchard, Jones-Alexander, Buckley, and Forneris, 1996).

It was therefore well suited for the current sample of a large group of Soldiers returning from a military theater at the same time. The 17-item checklist yielded a total of 96 Soldiers (19.5 percent of the sample) who screened positive for a mental health problem.

Combat experience/exposure. This measure includes 29 items, each with a five-point Likert scale ranging from never (1) to 10 or more times (5). An example of an item for this measure is, "Being attacked or ambushed." Scores were calculated by adding together all of the items. Results with a higher number represented a higher exposure to combat. The mean for this scale was documented as 67.15, and the Cronbach's alpha coefficient was .89. This particular index has been used in previous military research, and its results can be seen in table 17–1 (Hoge et al., 2004).

Analysis

Logistic regression was used to determine the relationship between career intention (outcome) and family-friendly unit climate, individual and unit morale, unit cohesion, mental health status, combat exposure, age, rank, education, race, and years in service (predictors).

Results

Demographics. A cross-tabulation was conducted for age and career intentions for this group of Soldiers. As the age of the Soldier increased, he was more inclined to remain in the Army beyond his current enlistment. Only 49 percent of Soldiers under 30 years of age intended to remain in the Army, while almost 80 percent over 30 years of age reported that they were going to continue beyond their current enlistment time.

Similarly, when number of years in service was compared to career intention in a cross-tabulation, an increase in likelihood to remain in the Army can be seen as the number of years in the Army increases. While 51 percent of the Soldiers with less than 10 years of service intended to stay in the Army, 83 percent with 10 or more years of service planned to remain.

When cross-tabulation between race and career intentions is conducted based on ethnic differences with regard to intended career choice, there was only a slight difference, with 64 percent of Caucasians and 62 percent of minorities intending to remain in the Army.

Additionally, there was only a small increase in intention to make the Army a career when education was examined in a cross-tabulation with the career variable. Whether the Soldier had only obtained a general equivalency diploma (GED) or an advanced degree, the likelihood that he would make the Army a career was only minimally different (GED, 56 percent; high school diploma, 57 percent; some college, 68 percent; college graduate, 69 percent).

The results of the logistic regression analysis in table 17–3 revealed that, among the demographics, the variables of age (β = .458, $p < .05$) and rank (β = .806, $p < .05$) significantly predicted the career intentions for this group (0 = probably or definitely leave the Service after current obligation, 1 = definitely stay in the Service at least beyond current obligation). However, ethnicity, education, and years in service were not significant predictors of whether these Soldiers would remain or leave after their current commitment.

It is not surprising that rank was the strongest significant predictor of career intentions in this study. Sixty-one percent of the respondents were NCOs and, of those, 16 percent were senior NCOs.

Table 17–3. **Logistic Regression Analysis on Career Intention to Stay in Service (N = 485)**

Predictors	B	Odds-Ratio	CI
Family-friendly unit climate (low = 0, high = 1)	.085*	1.09	(1.03–1.15)
Unit cohesion (low = 0, high = 1)	.157	1.17	(.97–1.42)
Unit morale (low = 0, high = 1)	.219	0.80	(.36–1.79)
Individual morale (low = 0, high = 1)	.687*	1.99	(1.10–3.60)
Combat exposure	.035	1.04	(.99–1.08)
Mental health status (no problem = 0, problem = 1)	-.027*	0.97	(.96–.99)
Race/ethnicity (Caucasian = 0, minority = 1)	.115	1.12	(.69–1.83)
Age	.458*	1.58	(1.10–2.28)
Education	-.238	0.79	(.59–1.05)
Rank	.806*	2.24	(1.34–3.74)
Years in service	.005	1.01	(.94–1.07)
Constant	3.706*		

*$p < .05$

As Soldiers rose in rank, especially to the senior levels, their motivation to stay in the Army shifted strongly; more than likely, they would make a career out of serving.

Only 38 percent of junior enlisted Soldiers reported that they wanted to remain in the Army beyond their current enlistment. On the other hand, almost three-quarters of the lower level NCOs (E5–E6) stated that they wish to it make a career. This number increased to 88 percent for senior NCOs (E7–E9).

Mental health status and combat exposure. There was a significant relationship between career intentions and mental health status (β = -.027, p < .05). This indicated that married soldiers with children who screened positive for depression, anxiety, or PTSD were more likely to intend to leave the Army. This argument was strengthened by running a cross-tabulation on career intentions and mental health issues. Soldiers who screened positive for mental health issues were much more likely to get out of the Army when their enlistment was completed than those who had not screened positive. In fact, 58 percent of Soldiers with mental health issues intended to leave, as opposed to the 33 percent without the issues.

While mental health status was a significant predictor of the career intentions for this group of Soldiers, combat exposure was not. In an attempt to examine this further, combat exposure was divided into three categories: low, moderate, and high. These categories were determined by breaking the overall combat exposure index into quartiles, with the lowest quartile representing low combat exposure, the middle two representing moderate combat exposure, and the highest representing high combat exposure. When career intention was examined with these three combat exposure categories by way of cross-tabulation, the level of combat exposure did not appear to be a driver of career intention. Soldiers in this study with low combat exposure were just as likely to leave the Army upon completion of their contract as Soldiers who had experienced moderate or even high levels of combat exposure (low, 37 percent; moderate, 35 percent; high, 42 percent). While there was a distinct relationship between the amount of combat exposure a Soldier experienced and mental health problems, combat exposure was not a significant predictor of career intentions within this study group, whereas mental health was. This suggested that Soldiers with high levels of combat exposure were no more likely to plan to leave or stay in the Army than those with low levels.

Family-friendly unit climate. When cross-tabulation was conducted on career intention and family-friendly unit climate, it became clear that family-friendly unit climate was a strong driver of the career intentions of those in this study. Soldiers who believed that their unit climate was family-friendly were much more likely to be interested in making a career out of the Army. Eighty-five percent of them chose to remain, compared to the 54 percent who chose to leave because they did not find their unit to be family-friendly. To conduct the cross-tabulation, the family-friendly unit index was divided at the median into two categories: low and high.

The logistic regression analysis displayed in table 17–3 revealed that family-friendly unit climate was found to be a significant predictor of a Soldier's career intention (β = .085, p < .05). Thus, we can argue that married Soldiers with children who perceived their unit climate to be family-friendly were more likely to stay in the Army than those who believed their unit was not family-friendly. This suggests that unit leadership should provide a family-friendly unit climate for the Soldiers directly under them. This may be accomplished through giving appropriate amounts of time off to take care of dependents when needed, or a flexible work schedule for Soldiers who have children in daycare. Failure to address these needs may result in greater turnover of qualified Soldiers.

Morale and cohesion. In 1999, morale was identified by the Department of the Army as "the state of the spirits of an individual or group as shown by confidence, cheerfulness, discipline, and willingness to perform tasks" (Department of the Army, 1999). When a dichotomous cross-tabulation was conducted on these variables and career intention, the findings indicated that those with high individual morale were more likely to report they would remain in the Army (83 percent) than those reporting high unit morale (79 percent) or high unit cohesion (71 percent). To examine unit cohesion in the cross-tabulation, this index was divided at the median into two categories: low and high. Unit level morale and individual morale were dichotomized based on their own lower ("very low," "low," and "medium") and higher ("high" and "very high") categories.

The logistic regression analysis displayed in table 17–3 shows that at the unit level, both morale and cohesion were found to be nonsignificant predictors of career intentions; however, high individual morale was found to significantly predict the career intentions among these Soldiers (β = .687, p < .05).

While unit morale and unit cohesion were not significant indicators of career intention for this group, married Soldiers with children that have high individual morale were significantly more likely to remain in the Army. After studying these results, the question becomes why unit cohesion and unit morale were not significant predictors in the model, while individual morale was a significant indicator of career intentions. Our research suggests that, while individual morale has a more direct influence on the personal lives of Soldiers, unit cohesion and unit morale does not have the same direct influence. While a family-friendly unit climate is not directed at Soldiers on an individual level per se, its influences are very personal in the lives of Soldiers and their families; therefore, it is clear that the dynamics measured by unit cohesion and morale differ from those of family-friendly unit climate.

Discussion

Given the relationship between turnover intent and behavior, and despite the fact that the study used a peacekeeping sample, the percentages give us reason to expect that a large proportion of Soldiers who indicate that they will leave the Army actually will.

From this study, we know that career intentions of married Soldiers with children are influenced by factors including family-friendly unit climate, individual morale, mental health issues, age, and rank. While the effect is modest, Soldiers who see their units as family-friendly are more likely to remain in the Army.

Kelley, Hock, Bonney, Smith, Jarvis, and Gaffney (2001) found support among Navy personnel with dependents that their commitment to the military was the main reason they were making a career out of the Navy. Additionally, many respondents reported that military benefits were a huge incentive in continuing their careers. Among dual-member Army families, items that contributed to Soldiers wanting to remain involved their spouse's military career intention, family income, family size, and job satisfaction (Lakhani and Gade, 1992). In 1998, however, Pierce conducted a study on Active-duty Air Force women with young children. These women believed that civilian work conditions were better than those in the military, and they were much more likely to leave. Furthermore, it was found that women who left the military after giving birth perceived their role as mothers to be conflicting with their job behavior. Kelley and colleagues (2001) also reported that a significant percentage of Navy mothers identified the conflict of balancing their naval career and family responsibilities as the reason for planning on not renewing their contract after their current obligation.

In this study, it is not being suggested that Soldiers with dependents be given preferential treatment or a reduction in their workload to accommodate their family needs; however, given the amount spent on Soldiers' training, the Army has a vested interest in ensuring that attrition is minimized. This study points to the importance of unit leadership, both at the command level, as well as the immediate supervisor, to ensure that their Soldiers are given at least some flexibility to take care of emergency family needs. Additionally, work predictability is important to Soldiers, especially those with dependents. It is in the unit leadership's best interest to work with their Soldiers who have preschool-age children in childcare facilities to try to match the daycare schedule.

Available programs and policies should also be a reflection of the population of the Army; therefore, as the number of single parents increases, so should the development of leadership programs focusing on that population's specific needs. Currently, many Army programs are geared toward the traditional family (Kelley, 2006). Failure to adapt to the changing needs of Army members will result in reduced productivity and the possible loss of excellent workers who are successfully contributing to the mission. Neither the Army nor unit leadership can be expected to not deploy Soldiers with dependents; however, they can put supportive practices and policies in place that assist these Soldiers at all times, whether prior to deployment or during it. Failure may result in a continued loss of qualified Soldiers who would otherwise remain till retirement (Bourg and Segal, 1999).

Leaders should strive to create a culture and climate that are conducive to the needs of Soldiers who have outside commitments due to their dependents, instead of one where these individuals are viewed as less effective in the workplace. As the Army population has changed, many family-friendly programs and policies have been implemented within the Service to adapt. However, not all of these policies and programs have been embraced by military leaders or those that they were created to assist (Huffman and Payne, 2006).

Our findings show that, while there is a dose-response relationship between combat exposure and mental health problems, mental health status is a significant direct predictor of career intentions among

Soldiers who are married and are parents, and combat exposure is not. This indicates that Soldiers with high levels of combat exposure are no more likely to plan to leave or stay in the Army than those with low levels, unless mental health problems have developed as a result.

Interestingly enough, while unit cohesion and unit morale are not significant predictors in the model, individual morale is. This indicates that even though unit cohesion and morale are often reflective of indicators of unit dynamics like the family-friendly unit climate measure, they measure distinct aspects of those dynamics that are not additive when it comes to individual career intentions.

Overall, it appears as though certain factors, like unit family-friendliness and individual morale, may have a more direct influence on the personal lives of Soldiers and their career intentions, while factors that operate at a more distal level do not.

References

Bedeian, A.G., Burke, B.G., and Moffett, R.G. (1988). Outcomes of work-family conflict among married male and female professionals. *Journal of Management* 14, 475–491.

Blanchard, E.B., Jones-Alexander, J., Buckley, T.C., and Forneris, C.A. (1996). Psychometric properties of the PTSD checklist (PCL). *Behaviour Research and Therapy* 34, 669–673.

Bourg, C., and Segal, M.W. (1999). The impact of family supportive policies and practices on organizational commitment to the Army. *Armed Forces and Society* 25, 633–652.

Bruck, C.S., and Allen, T.D. (2003). The relationship between big five personality traits, negative affectivity, type A behavior, and work-family conflict. *Journal of Vocational Behavior* 63, 457–472.

Carlson, D.S., and Perrewe, P.L. (1999). The role of social support in the stressor-strain relationship: An examination of work-family conflict. *Journal of Management* 25, 513–540.

Congressional Budget Office (2006). *Recruiting, retention, and future levels of military personnel* (Publication No. 2777). Washington, DC: Government Printing Office.

Department of the Army (1999). *Army leadership: Be, know, do* (Field Manual 22–100). Washington, DC: Headquarters Department of the Army.

Duxbury, L.E., and Higgins, C.A. (1991). Gender differences in work-family conflict. *Journal of Applied Psychology* 76, 60–73.

Eagle, B.W., Icenogle, M.L., Maes, J.D., and Miles, E.W. (1998). The importance of employee demographic profiles for understanding experiences of work-family inter-role conflict. *The Journal of Social Psychology* 138, 690–709.

Ender, M.G. (2006). Voices from the backseat: Demands of growing up in military families. In T.W. Britt, A.B. Adler, and C.A. Castro (eds.). *Military life: The psychology of serving in peace and combat. Vol. 3: The military family* (138–166). Westport, CT: Praeger Security International.

Frone, M.R. (2003). Work family balance. In J.C. Quick and L.E. Tetrick, *Handbook of occupational health psychology* (143–162). Washington, DC: American Psychological Association.

Frone, M.R., Yardley, J., and Markel, K.S. (1997). Developing and testing an integrative model of the work-family interface. *Journal of Vocational Behavior* 50, 145–167.

Grzywacz, J.G., and Marks, N.F. (2000). Reconceptualizing the work-family interface: An ecological perspective on the correlates of positive and negative spillover between work and family. *Journal of Occupational Health Psychology* 5, 111–126.

Griffith, J., Rakoff, S.H., and Helms, R.F. (1992). *Family and other impacts of retention* (Technical Report 951). Alexandria, VA: U.S. Army Research Institute for the Behavioral and Social Sciences.

Hoge, C.W., Castro, C.A., Messer, S.C., McGurk, D., Cotting, D.I., and Koffman, R.L. (2004). Combat duty in Iraq and Afghanistan, mental health problems, and barriers to care. *The New England Journal of Medicine* 351, 13–22.

Huffman, A.H., Adler, A.B., Dolan, C.A., and Castro, C.A. (2005). The impact of operations tempo on turnover intentions of Army personnel. *Military Psychology* 17 (3), 175–202.

Huffman, A.H., and Payne, S.C. (2006). The challenges and benefits of dual-military marriages. In T.W. Britt, A.B. Adler, and C.A. Castro (eds.). *Military life: The psychology of serving in peace and combat. Vol. 3: The military family* (115–137). Westport, CT: Praeger Security International.

Huffman, A.H., Culbertson, S.S., and Castro, C.A. (2008). Family-friendly environments and U.S. Army soldier performance and work outcomes. *Military Psychology* 20 (4), 251–270.

Kelley, M.L. (2006). Single military parents in the new millennium. In T.W. Britt, A.B. Adler, and C.A. Castro (eds.). *Military life: The psychology of serving in peace and combat. Vol. 3: The military family* (93–114). Westport, CT: Praeger Security International.

Kelley, M.L., Hock, E., Bonney, J.F., Smith, K.M., Jarvis, M.S., and Gaffney, M.A. (2001). Enlisted Navy mothers experiencing and not experiencing deployment: reasons for staying or leaving the military. *Military Psychology* 13, 17–20.

Kopelman, R.E., Greenhaus, J.H., and Connolly, T.F. (1983). A model of work, family, and inter-role conflict: A construct validation study. *Organizational Behavior and Human Performance* 32, 198–215.

Kinnunen, U., and Mauno, S. (1998). Antecedents and outcomes of work-family conflict among employed women and men in Finland. *Human Relations* 51, 157–177.

Lakhani, H., and Gade, P.A. (1992). Career decisions of dual military career couples: A multidisciplinary analysis of the U.S. Army. *Journal of Economic Psychology* 13, 153–166.

Major, V.S., Klein, K.J., and Ehrhart, M.G. (2002). Work time, work interface with family, and psychological distress. *Journal of Applied Psychology* 87, 427–436.

Marshall-Mies, J.C., Seligson, T.B., and Martin, J.A. (2000). *Differences in Job Satisfaction of Soldiers in Dual Military and Traditional Marriages* (Technical Report 2000–04). Alexandria, VA: United States Army Research Institute for the Behavioral and Social Sciences.

Martin, J.A., and McClure, P. (2000). Today's active-duty military family: The evolving challenges of military family life. *The military family. A practice guide for human service providers,* 3–24.

National Council on Family Relations (2004). *Building strong communities for military families.* Minneapolis, MN. Retrieved from <www.parenting.umn.edu/militaryKids/buildingStrongCommunities.pdf>.

Office of Family Policy (1999). *Strategic youth action plan.* Arlington, VA: Office of Family Policy, Department of Defense Office of Personnel Support, Families and Education.

Pierce, P.F. (1998). Retention of Air Force women serving during Desert Shield and Desert Storm. *Military Medicine* 10, 195–213.

Rosen, L.N., and Durand, D.B. (2000). Coping with the unique demands of military family life. *The military family: A practice guide for human service providers,* 55–72.

Schumm, W.R., Bell, D.B., Rice, R.E., and Sanders, D. (1996). Trends in dual military couples in the U.S. Army. *Psychological Reports* 78, 1287–1298.

Schumm, W.R., Jurich, A.P., Stever, J.A., Sanders, D., Castelo, C., and Bollman, S.R. (1998). Attitudes of Reserve component service members regarding the consequences of frequent overseas deployments. *Psychological Reports* 83, 983–989.

Smith, A.L. (1988). *A multivariate analysis of determinants of re-enlistment: A decision-making model for enlisted personnel* (ARI Technical Report No. 797). Alexandria, VA: U.S. Army Research Institute.

Tremble, T.R., Jr., Payne, S.C., Finch, J.F., and Bullis, R.C. (2003). Opening organizational archives to research: Analog measures of organizational commitment. *Military Psychology* 15, 167–190.

U.S. Army Human Resource Command. (2008). U.S. Army Human Resource Command. Fort Hood, TX. Retrieved from <https://www.hrc.army.mil/site/index-flash.aspx>.

U.S. General Accounting Office. (1982). *Army needs better data to develop policies for sole and in-service parents* (GAO/FPCK–92–50). Washington, DC: Government Printing Office.

Weiss, H., MacDermid, S., Strauss, R., Kurek, K., Le, B., and Robbins, D. (2003). *Retention in the Armed Forces: Past approaches and new research directions.* Purdue University, Military Family Research Institute.

Williams, J.W., Jr. (1978). Dual-career military families. In E.J. Hunter and D.S. Nice, *Military families: Adaptation to change* (103–110). New York: Praeger Publishers.

Part VII

Perspectives on Army
Research Psychology

Chapter 18

Science in Action: Meeting Urgent Field Requirements through Science and Technology on the Ground in Iraq

Matthew G. Clark

The uniformed Services currently identify capability require-ments using a deliberate process that is ineffective in emergency situa-tions. Similarly, operational needs statements take time to validate and even more time to fund and address. The Army Materiel Command/ Research Development and Engineering Command Field Assistance in Science in Technology (AMC/RDECOM FAST) Team and the Rapid Equipping Force lead efforts that capture and address current and ur-gent requirements on the fluid battlefield. Since early 2005, these pro-grams have been augmented by personnel from the U.S. Army Medical Command (USAMEDCOM), who work with operational units to iden-tify capability gaps for both current and future operations and provide commanders with immediate access to the development centers with-in the RDECOM and USAMEDCOM. As a result, potential technol-ogy solutions can be tested by the warfighter, and performance in the field can be evaluated for immediate and long-term effectiveness. Uni-formed Army research psychologists have served on these AMC/RDE-COM FAST teams and have enabled significant equipment programs such as the armored ambulance initiative.

Uniformed Army research psychologists serve in roles that span the spectrum of military operations, from the more traditional fields of psychological and biological laboratory and field research, to operation-al assessments and systems acquisition. In addition to their field of ex-pertise, many Army medical and scientific research officers can also be acquisition professionals. One nonstandard acquisition assignment op-portunity for research psychologists and other Army Medical Depart-ment (AMEDD) officers involves identifying urgent warfighter require-ments for full-spectrum operations as part of the four-man AMC/RDE-COM FAST team in Iraq.

Background: Requirements Process Limitations during War

Generally, addressing warfighter capability requirements in the Department of Defense (DOD) involves three key components: the requirements process, the acquisition process, and the budgeting and funding process. Previously referred to as "the requirements generation and identification process," the requirements process now focuses on the broader capability needs of all Services and is called the Joint Capabilities Integration and Development System (JCIDS). Per the Chairman of the Joint Chiefs of Staff Instruction 3170.01F (2007): "The JCIDS process was created to support the statutory requirements of the JROC [Joint Requirements Oversight Council] to validate and prioritize joint warfighting requirements. JCIDS is also a key supporting process for DOD acquisition and PPBE [Planning, Programming, Budgeting, and Execution] processes. The primary objective of the JCIDS process is to ensure that the joint warfighter receives the capabilities required to successfully execute the missions assigned to them."

The current JCIDS process was initiated in 2003 and updated in December 2008 to overcome shortfalls in joint Service capability integration, as required by the JROC. An outgrowth of the Goldwater-Nichols Department of Defense Reorganization Act of 1986, the JCIDS is a top-down process in which joint capability requirements (that is, military needs) evolve through a lengthy and deliberate developmental process. Through JCIDS, equipment and capability requirements are validated. With a validated capability requirement, Services can acquire systems to address the specific need. Accordingly, the chosen system is funded and budgeted until the item is removed from the military inventory. The JCIDS is contrasted against the previous requirements generation system, in which Service-unique requirements were captured and developed in Service-specific "silos." In the old system, joint interoperability and shared utility were an afterthought at best. In total, JCIDS, the acquisition system, and the PPBE processes are intended to effectively control resources to maximize capability development across and between the Services in DOD.

However, as highlighted in a December 2006 article in *Defense AT&L*, "Recent reporting has also brought to light the United States' deficiencies in getting innovative solutions to our warfighters rapidly enough to adjust to the changing tactics and techniques of our enemies" (Buhrkuhl, 2006). Simply put, the DOD acquisition and budget systems and JCIDS are not agile enough to respond to the demands of war when rapid adaptation is required on the battlefield. A change in emphasis from maneuver warfare to full-spectrum operations (as defined by Army Field Manual 3–0, *Operations*), and specifically the security, stability, transition, and reconstruction missions in Iraq and Afghanistan, revealed significant shortfalls in doctrine, training, and equipment across the Army. Therefore, the Army needed a means to identify and address current shortfalls and unforeseen capability gaps on a fluid battlefield because the capabilities identification and acquisition systems were incapable of meeting the need.

Out of necessity, the U.S. Army and DOD developed and continue to employ various approaches to identify and address critical shortfalls for current military operations. Some of these approaches included the Rapid Equipping Force, the Joint Improvised Explosive Device Defeat Organization, the Asymmetric Warfare Group, the Army Asymmetric Warfare Office, and the AMC/RDECOM FAST team, to name just a few.

FAST Teams in the Iraqi Theater and Need for AMEDD Input

The broader FAST Activity started in 1985 to "bring Army Materiel Command Laboratories and Research, Development and Engineering Centers into closer contact with their 'customers'—the major Army commands throughout the world" (<www.rdecom-fast.army.mil/about_fast.htm>). Regarding Operation *Iraqi Freedom* wartime efforts, the first three-member AMC/RDECOM FAST teams, also called Science and Technology Assistance Teams, deployed in 2003. These teams comprised an acquisition corps major or lieutenant colonel (officer in charge), a government civilian scientist or engineer, and a senior noncommissioned officer (staff sergeant to sergeant major). The team was an extension of the RDECOM in AMC. Initially, during the 4-month rotations (extended to 6 months in 2007) they focused largely on issues related to weapons, vehicles, and various other unit, individual, and force protection systems.

In 2005, the AMC FAST Headquarters recognized their inability to address requests for new medical capabilities related to the reality on the ground. To meet the growing need, they invited the USAMEDCOM and the U.S. Army Medical Research and Materiel Development Command (USAMRMC) to participate in their battlefield rotations.

Accordingly, the science and technology assistance teams were expanded to include an AMEDD officer (captain, major, or lieutenant colonel).

As is often the case with new initiatives, the initial rotations that included a medical professional produced mixed results. Sometimes they were well received as a value-added additional staff member across various levels of command, from combat battalions through the Multi-National Corps–Iraq (MNCI) and Multi-National Force–Iraq (MNFI). Occasionally, however, the officer was viewed with skepticism or resistance in the theater as medical and nonmedical leaders learned and tested the capabilities and limitations of the program.

Early FAST successes involving the AMEDD were primarily medical-specific shortfalls such as the need for hypothermia management in combat casualties. This was addressed particularly for those evacuated with truncal or abdominal insult or recently controlled hemorrhage via air medical evacuation (MEDEVAC). This large-scale effort led to broad employment of the hypothermia management kit (North American Rescue Products, Inc., Greenville, SC) in both medical and nonmedical vehicles and settings. These are now used in Humvees, Mine Resistant Ambush Protected (MRAP) vehicles, and with combat, engineer, and explosive ordnance disposal teams.

The early FAST teams also assisted by addressing numerous requests for information and leading fielding and feedback efforts on the Improved First Aid Kit. Notably, these teams assisted with initial efforts in the evaluation, deployment, and even developmental feedback of body cooling and ventilation systems, "golden hour" blood shipment containers, and even new and novel tactical communication headsets with active noise reduction for hearing protection. These efforts were urgently required because of the unique nature of irregular combat, and the need for improved Soldier effectiveness and health in desert and urban environments. More relevant to the traditional role of medical researchers, these efforts also contributed to the long-term health and immediate safety of American Soldiers fighting in an extreme and atypical environment.

In early 2006, the medical FAST effort started to build momentum by tackling military medical and materiel issues that required a coordinated AMEDD and AMC effort. One of the major efforts was the analysis of personal protective equipment (PPE). This effort required the shared resources and skills of operationally minded medical, research, acquisition, and materiel development professionals.

These professionals included the AMEDD and AMC FAST team on the ground in Iraq and various individuals from the USAMRMC, RDECOM, and Program Executive Offices (for example, PEO Soldier) under the Assistant Secretary of the Army for Acquisition, Logistics, and Technology (ASA[ALT]).

The PPE project began as a proof of concept conducted by the AMC/RDECOM FAST team where PPE was collected from Soldiers who were wounded or killed in action. Following a combat incident, the FAST team collected Interceptor Body Armor and Advanced Combat Helmets. The equipment was then shipped to PEO Soldier in the United States for analysis. The PPE study included an analysis of operational data from each combat incident co-analyzed with casualty medical data. Effectively, this was the first effort to combine operational information with medical and materiel analyses to help develop the next generation of protective equipment and improve tactics, techniques, and procedures (TTPs) as quickly as possible. This "fusion analysis" approach, which included the examination of operational, intelligence, medical, and materiel data, eventually developed into the Joint Trauma Analysis for the Prevention of Injury in Combat program. This program is now progressing toward becoming an enduring capability for support to combat operations, materiel developers, and the military intelligence community.

After the first three rotations, the USAMRMC prepared a selection process to ensure that the AMEDD provided the most qualified officers for the team. Besides the appropriate rank, command support, and letters of recommendation, the team's only eligibility requirement was completion of the Captain's Career Course. Other career experiences that improved selection for this opportunity included field or deployment experience, certification as a Defense Acquisition Workforce Professional, completion of the Combat Casualty Care Course, and possession of the Expert Field Medical Badge.

My FAST Team Experience and the Armored MEDEVAC Challenge

In 2006, the first officer selected and deployed through the USAMRMC process was a research psychologist. The FAST mission in Iraq at that time was four-fold:

- to provide expert technical advice and support to commanders and staff in the theater of combat operations

- to query Operation *Iraqi Freedom* units on science and technology gaps at the user level for both combat and stability
  and support operations
- to provide operational commanders immediate access to the
  development centers within the RDECOM and USAMED-
  COM
- to expedite technology solutions to the warfighter and act as
  an RDECOM/AMEDD forward element to evaluate and integrate solutions.

As the research psychologist who was selected for this team, I used a
fusion analysis approach similar to that used in the PPE analysis to address various issues including a countersniper effort, burn mitigation,
and the need for armored ground MEDEVAC on the rapidly adaptive
battlefield. While changes in materiel systems did not occur immediately, TTPs were altered and rapid battlefield feedback and education to
warfighters were completed without delay.

During this rotation, the FAST team tackled the need for improved ground medical evacuation, arguably the most challenging issue to date because it spanned various sectors of the Army and had to
counter widely held and incorrect beliefs and assumptions. Like the
PPE analysis, the issues required the shared resources of the AMEDD
and AMC FAST team on the ground in Iraq, with the support of stateside activities. Various individuals from the USAMRMC, the U.S. Army
Medical Materiel Development Activity (USAMMDA), the AMEDD
Center and Schools Directorate of Combat and Doctrine Development
(DCDD), and Program Management shops (for example, PM–Light
Tactical Vehicle) under ASA(ALT) were involved in providing this im-
mediate warfighter need.

In the summer of 2006, the heightened threat and record use of
improvised explosive devices (IEDs) created several significant challenges for medical care in the austere environment of urban Iraq and
Afghanistan. Most notably, the prevalence of IEDs, rocket-propelled
grenades, and certain small arms resulted in restrictions that kept all
unarmored vehicles to forward operating bases, particularly in Iraq.
As reported in January 2007, this restriction applied to all unarmored
Humvees, including M996 and M997 ambulances (McDonnell, 2008).
Medical support units were severely impacted by the restriction because
a comprehensive patient evacuation and medical treatment plan must

include effective ground and air MEDEVAC (FM 8–10–6, *Medical Evacuation in a Theater of Operations*, 2000).

Addressing this particular issue required the presence of the FAST team, which could evaluate and address shortfalls across all levels of command, from the company through MNCI and MNFI levels. This issue also required rapid collection, organization, and analysis of data from a highly adaptive and seemingly disjointed battlefield. The data included information about combat incidents, the type, number, and method of threats, the TTPs and equipment used, the method and time of evacuation needed, and the outcomes for Soldiers involved. In total, the data set that had to be synthesized was not simply maintained in one location in a usable or readily accessible fashion. Instead, the data was diffuse and variously collected across several levels of command and staff. One of the key contributions of having a research psychologist on the team was the capability to address the lack of a defined data source. With my research background, I brought methodological integrity, knowledge, and a broader understanding of the analysis of a poorly defined but clearly evident problem. Additionally, the collective efforts of the FAST team identified related needs and provided potential solutions.

Addressing the ground MEDEVAC issue required clear and open communication with development centers in the United States. Data was collected from various command, control, communications, computers, intelligence, surveillance, and reconnaissance systems. It was also drawn from Soldiers and Marines on the ground and from generic patient information collected by various staff surgeons for units including MNCI and the Joint Theater Trauma Registry. The analysis required a clear understanding of doctrinal and current medical and combat operations and required support from new, nondoctrinal units like the Combined Joint Task Force, which addressed asymmetric threat and IED defeat issues.

The first challenge was that no data had been collected, collated, or analyzed regarding the requirement and magnitude of the ground MEDEVAC problem. Once collected, an initial assessment and discussions with the commander of the 30th Medical Brigade revealed that the first approach had to be the development of nonstandard vehicles for casualty evacuation (CASEVAC) use. Usually conducted by nonmedical personnel, CASEVAC is the movement of casualties in a combat zone without en route medical care to an initial treatment location or facility (FM 8–10–6, *Medical Evacuation in a Theater of Operations*, 2000). This was the most expeditious way to start to address the problem because CASEVAC provided at least some form of armored ground evacuation in an environment with zero ground MEDEVAC capacity due to the threat and a potentially limited air MEDEVAC capability.

Stateside, the USAMMDA and DCDD began working on CASEVAC conversion kits for all up-armored Humvees, an issue that eventually became a reality in late 2007 to early 2008. In theater, we designed two CASEVAC configurations for the Rhino Runner armored buses and prepared rapid conversion kits with the 134th Medical Company (Ground Ambulance) that could be used in a mass casualty situation. This CASEVAC alternative allowed these buses to be used to evacuate from 5 to 12 litter patients or from 1 to 23 ambulatory patients. However, there were a very limited number of these vehicles, they did not travel on combat missions, and they could only be used in a catastrophic mass casualty situation. These vehicles were designed and employed to keep Servicemembers, government workers, and contractors somewhat safe as they traversed the battlefield between bases much like a very small public transit system. Due to the availability, developing a CASEVAC configuration for the Rhino Runner would only serve as a potential Band-Aid for the greater need that might only be available for the worst cases.

As the use and effectiveness of IEDs expanded to record levels, combatant commanders and staff throughout the Iraqi theater of operations began to acknowledge the magnitude of the ground MEDEVAC shortfall. After initially working with a brigade combat team and brigade support battalion commander from the 10th Mountain Division in the Baghdad area, various division and corps leaders and medical advisors started to realize and accept that air MEDEVAC could not address all urgent evacuation situations. This was particularly true for some urban areas. The need was clearly evident and growing when a resourceful combat medic cut a hole in the back blast wall of his up-armored Humvee to accommodate a casualty as a CASEVAC vehicle (Powell, 2006). In support of the definition of MEDEVAC from *Medical Evacuation in a Theater of Operations* (FM 8–10–6, 2000), the simple reality was that a timely means of moving patients with en route care by medics on medically equipped vehicles was needed both to enhance casualties' potential for recovery and to reduce the potential for long-term disability.

An armored wheeled ground MEDEVAC vehicle (that is, an armored ambulance) was badly needed.

Unfortunately, the misunderstanding of the MEDEVAC problem was perpetuated by poor communication across the combat theater. Incorrect information existed at all levels of command and even within the stateside military medical community. They believed that the problem was already being addressed and that a solution was on its way. To gather the facts on the issue, the FAST team constantly communicated with PEO and PM shops in the United States. The team learned that the reason that M996 and M997 Humvee ambulances were not available was because it was not technologically possible to up-armor these specific vehicles. That is, the solution many believed was already in production could not be implemented because these vehicles could not support the weight of armor in an ambulance configuration along with the weight of all personnel and related equipment. Despite repeated efforts to communicate this reality, a widespread misunderstanding of the situation made it very difficult to address the issue within the Army.

Adding to the state of affairs, the MEDEVAC problem was viewed as unique to Army units. This incorrect and widely held belief limited the possibility of any joint Service request because all joint requests had to address an "inherently joint" problem that involved other Services (for example, the Marines). Despite the FAST team's work with Soldiers and Marines at the MNCI and MNFI levels, U.S. Central Command (USCENTCOM) did not understand the situation until the end of 2006. Eventually, with the support of the MNFI Surgeon, the late Colonel Brian Allgood, I presented the issue at the USCENTCOM Surgeon's conference in November 2006, which educated and energized the appropriate levels of leadership across the military Services. The result was that the Navy surgeon representing the Marine Corps Central Command and surgeons and staff covering the USCENTCOM area of responsibility finally understood the full nature of the situation. This realization, along with the help of Frank Van Syckle, a key science and technology advisor in USCENTCOM headquarters, led to the identification of and required support for a joint immediate warfighter need request for armored ambulances.

As the AMEDD FAST representative on the ground, I collected the data, analyzed the issue, and then wrote the urgent requirement request that was initiated with an MNCI endorsement and forwarded to the Joint Staff through USCENTCOM. When we returned stateside at the end of 2006, I worked through USCENTCOM, the Joint Staff, and up to the Joint Rapid Acquisition Cell and Office of the Deputy Secretary of Defense to procure funding approval for the first 16 heavy armored ground ambulances (HAGA) for Soldiers and Marines in Iraq and Afghanistan. This modest request was approved by the Deputy Secretary of Defense in May 2007. However, delivery of the urgent request for 16 vehicles was only the beginning of addressing a much larger requirement. Accordingly, in a follow-on appointment I served as one of the primary subject matter experts in the design of the HAGA and MRAP ambulance along with representatives from the AMEDD DCDD and USAMMDA. The development and production of the vehicles was achieved through work with the Marine Corps Systems Command and the MRAP Joint Program Office. The vehicle that met the identified need was produced by BAE Systems. By the end of 2008, nearly 700 armored ambulances were fielded for use in current operations.

Progress on a materiel solution to the armored ambulance requirement had been stalled prior to fall 2006. Therefore, the recognition and fielding of this urgent warfighter requirement was a major success for the AMEDD. Even with increased interest in armored vehicles and the larger MRAP effort that developed at the beginning of 2007, an armored ambulance was not included as one of the first vehicles until the FAST team entered the process. This success required the skills of personnel who could rapidly collect, analyze, and merge data from diverse levels of command, locations, and sources. It required personnel who could think strategically, yet understand the people and processes so that they could coordinate action and move between tactical and operational levels through a highly convoluted, constantly changing bureaucratic process. Lastly, the number of ambulances and the seemingly low density of the requirement meant that its success was dependent on personnel who could speak across levels of combat leadership, champion the effort through to the highest levels of the Department of Defense, conduct and translate the analysis of data employed, and communicate the strategic and psychological impact of medical support to American forces. Collectively, these qualities were key in accelerating the timeline to make armored ambulances a reality. These are also qualities required for success as an Army research psychologist.

Summary: Uniformed Army Research Psychologists in Combat Operations

It is important to understand the value that this assignment provided to both research psychologists and the Army. First, the FAST position provided scientific and acquisition officers the opportunity to work in an area that is immediately relevant to current operations, instead of the more traditional roles of researching post-traumatic stress disorder, battle stress, and mild traumatic brain injury. It also allowed research psychologists to work on medical integration across functional areas, branches, and Services, thereby expanding medical lessons learned and improving the understanding of medical research shortfalls. This understanding will enhance future medical research and development.

Similarly, several research psychologists have completed research on medical, chemical, and biological defense issues as is presented in other chapters in this publication. They have knowledge and expertise that can serve as a force multiplier on the current battlefield for current operations. While serving on the FAST team, I served on the MNCI Corps Chemical Officer's chemical, biological, radiological, nuclear, and explosive working group. Thus, a medical chemical defense researcher was able to impact and assist current chemical defense issues in real time. The combatant command benefited from relevant and timely research expertise on urgent and unexpected events. In short, research psychologists were immediately relevant to the current fight rather than just future or investment operations through research and development.

Additionally, the Army and the research psychology specialty shared the advantage of applying the core acquisition abilities of a highly skilled and motivated workforce to some of the more challenging analytical and operational challenges facing combat and military medical operations. This is particularly apparent because the FAST assignment was accomplished one-on-one with Soldiers and Marines from the level of combat and the battalion aid station, through combat support hospitals, or even through the combatant command, Department of the Army, and joint levels of command. Thus, the core research and acquisition capabilities were applied and extended to the benefit of both the Army and the individual.

Groups like the FAST team provide the Army immediate relevance and utilization of research and analytical subject matter experts for current operations. This flexibility of assignments raises questions about the future role and placement of research psychologists across the

Army. A full mission analysis of the requirements for this specialty in full-spectrum operations will likely lead to the conclusion that research psychologists can and should play a greater role in defense acquisition, both medical and nonmedical research and development, and military intelligence analysis and fusion. This conclusion may be particularly compelling when considering irregular and asymmetric warfare, and it is amplified by the recent placement of a research psychologist in the Army Asymmetric Warfare Office of the Army Operations Center (under the Deputy Chief of Staff for Operations, Headquarters Department of the Army G-3/5/7). While a full consideration of the future assignments for this specialty is beyond the scope of this article, the success of research psychologists in the field illustrates the need for a comprehensive evaluation of their role, particularly for positions outside of the USAMEDCOM.

In conclusion, research psychologists serve in various assignments that include roles as acquisition and analytical professionals. They must continue to be prepared to meet general medical, Army, or even joint requirements as a commissioned officer of the United States. Uniformed research psychologists of all backgrounds must be prepared to modify, develop, create, and execute efforts outside more traditional assignments in laboratories. The success of research psychologists in various relevant, nontraditional assignments like the AMC/RDECOM FAST team in Iraq underscores the potential future role this specialty can play across the U.S. Army. These accomplishments clearly show that the analytical and leadership skills of research psychologists are useful well beyond the medical research realm.

References

Buhrkuhl, R.L. (2006, November/December). When the warfighter needs it now. *Defense AT&L*, 28–31.

Joint Chiefs of Staff. (2007). *Chairman of the Joint Chiefs of Staff instruction 3170.01F*. Washington, DC: U.S. Government Printing Office.

McDonnell, J.J. (2007). Instilling innovation in Iraq. *Army Logistician*, 39.

Powell, A. (2006, July 27). Medics humvee design not much of a stretch. *Stars and Stripes, Mideast Edition*.

RDECOM–FAST. (n.d.) *About FAST*. Retrieved December 8, 2008, from <www.rdecom-fast.army.mil/about_fast.htm>.

U.S. Army. (2008). Field Manual 3–0, *Operations*. Washington, DC: Headquarters Department of the Army.

U.S. Army. (2000). Field Manual 8–10–6, *Medical operations in a theater of operations*. Washington, DC: Headquarters Department of the Army.

Chapter 19

Looking Back: From the Lab to Iraq... and to the Clinic

Melba C. Stetz

In this chapter, I share some of my experiences as a uniformed Army research psychologist (71F). My primary goal is to interest and invite more researchers to join our group. I was raised in a bicultural and medical household (with psychiatrist and dietitian parents), which probably helped prepare me for the eclectic Army Medical Department (AMEDD) life.

In 1986, I became the first military Servicemember in my family. Because I already had completed 2 years toward a Bachelor's degree, I was able to start in the Army as a private first class. After training at Fort Sam Houston, in San Antonio, Texas, and at the Walter Reed Army Medical Center, in Washington, DC, I served as an orthopedic specialist in the 8th Evacuation Hospital at Fort Ord, California. After this tour, I went back to Puerto Rico to finish college. While working full-time in my parents' medical center and part-time in the Puerto Rico Army National Guard, I finished a bachelor's degree with a double minor in Clinical Psychology and Industrial/Organizational (I/O) Psychology.

Once promoted to a second lieutenant in the National Guard, I went back to Fort Sam Houston to the Officer Basic Course (OBC). In the Guard I had held jobs ranging from an ambulance platoon leader in the medical company to a Total Quality Management (TQM) Advisor for the 92[d] Infantry Brigade (Separate). After finishing a Masters of Science degree in I/O Psychology, I received the TQM assignment, following a meeting with my Brigade General, Felix Ocasio, to whom I offered a few win-win solutions based on insights from my I/O Psychology background. At the time, I worked as a research assistant during the day at Carlos Albizu University in San Juan, Puerto Rico, and studied for my master's and doctorate degrees at night and during the weekends. After a decade in "the National Guard system" (Reserve promotions are slow!), I became a first lieutenant and was sent back to Fort Sam Houston to attend the Officer Advanced Course (OAC). A few months after graduation, I volunteered for a short tour in Panama as a medical plans officer (70H). I spent 9 months in the Civil Military Operations section of Fort Clayton, where I coordinated medical readiness training exercises (MEDRETEs) for the U.S. Army South General Office. Specifically, my mission was to bring Reserve Component (Guard or Reserve) medical units to remote areas in the U.S. Southern Command (USSOUTHCOM) to provide free medical services to the local populace. Units would typically participate during their annual 2-week training. It was a tough environment for a junior female officer in a male-dominated environment. One highlight for me was meeting the president of Panama, Ernesto Pérez Balladares, who invited me to one of his special inaugurations after being impressed by our MEDRETEs.

After Panama and a logistics course at Fort Sam Houston, I transferred to the Active Guard Reserve as a Medical Logistician (70K). In my new role, I led the 424th Medical Logistics Battalion (Rear), a Reserve unit in Pedricktown, New Jersey. We supported five states during their continental U.S. (CONUS) and deployment (that is, Bosnia) missions. While on that job, I finished my dissertation. At that time, sexual harassment was becoming a more commonly recognized social problem, with cases against VIPs such as President Bill Clinton and Sergeant Major Gene McKinney, USA. Therefore, I decided to study sexual harassment, and administered the validated and standardized "Military Equal Opportunity Climate Survey" to Army, Marine, and Navy Reserve units in the area. I presented my findings during the Defense Equal Opportunity Management Institute conference at Patrick Air Force Base in Florida and published part of the data. Validation analysis suggested that a shorter, 17-item version of the original 50-item scale could be used when studying military equal opportunity climate (Estrada, Stetz, and Harbke, 2007). While defending my study in Puerto Rico, I also passed the general psychology licensing board exams.

In 1999, I transferred to the Individual Ready Reserve (IRR) to be part of an Army group that made it easier to accept a 71F slot when one became available. For a year, I worked as a Government civilian in the Office of Personnel Management in Washington, DC. As a personnel research psychologist, I worked on I/O-related issues (such as organizational diagnoses and training). I met my husband, a fellow I/O psychologist and researcher in the local area, while working there.

Walter Reed Army Institute of Research

Once commissioned as a 71F captain, I attended the longer active versions of both the OBC and the OAC, and reported to the Walter Reed Army Institute of Research (WRAIR) on graduation. At that lab, I helped with survey design and administration, unit coordination, participant recruitment, and data analysis, and gave presentations to commanders and at professional conferences. One of my first military research opportunities was analyzing data recently acquired by the WRAIR on Army Rangers. Our main results suggested that Rangers with work-relevant values may experience a stronger relationship between work stressors and outcomes such as job satisfaction and distress than Rangers with stronger affiliation values (Britt, Stetz, and Bliese, 2004).

While at WRAIR, we also studied Reservists activated after the terrorist attacks on September 11, 2001 (Stetz, Castro, and Bliese, 2007). The data suggested that with high deactivation uncertainty, workload, and organizational constraints, Reservists reported low psychological well-being and high turnover intentions. Some of these activated Reservists also shared their concern about the security of the jobs they were leaving behind. We also interviewed civilian supervisors and, interestingly, found that supervisors generally supported their employees' military activation (Allison-Aipa, De la Rosa, Stetz, and Castro, 2005).

In other research, we found moderating effects of social support on a few stressor-strain relationships (Stetz, Stetz, and Bliese, 2006). Specifically, we found that this moderating effect depended on one's self-efficacy or sense of preparedness. Following the importance of self-efficacy, we developed a theoretical model of Reservists activated after 9/11 on the impact of their perceptions of procedural justice during the early months of their mobilization on their job satisfaction and well-being (Bliese and Stetz, 2007). Procedural justice, self-efficacy, and intragroup conflict showed a 3-way interaction where self-efficacy served only as a buffer when accompanied by procedural justice.

A few of us (Killgore, Stetz, Castro, and Hoge, 2006) also analyzed the relationship between deployment experience and somatic symptoms in Soldiers. Consistent with the literature, previously deployed Soldiers reported less affective, but more somatic complaints than those without any deployment experience. Following these research efforts at WRAIR, I attended the Combined Arms and Services Staff School at Fort Leavenworth, Kansas, where I learned more about military history, cohesion, and leadership. More importantly, I was able to interact with non-AMEDD military officers.

U.S. Army Medical Research and Materiel Command

My next assignment was with the U.S. Army Medical Research and Materiel Command (USAMRMC) Headquarters (HQ) at Fort Detrick, Maryland. I worked as the Deputy Director for the Research Area Directorate 3—Medical Operational Medicine. I screened and supported research from throughout the USAMRMC laboratories and collaborating parties. I was able to help others while still continuing to analyze and publish my own data. For example, we analyzed 5,671 records of Servicemembers who had been medically evacuated (MedEvac) out of theater during 2003 (Stetz, McDonald, Lukey, and Gifford, 2005). Following the International Statistical Classification of Diseases and Related Health Problems, it seemed that 13 percent of this sample was classified as either under "top level mental disorders" or "persons without reported diagnosis encountered during examination and investigation of individuals and populations." Sadly, further analysis of that data (Stetz, Thomas, Russo, Stetz, Wildzunas, McDonald et al., 2007) suggested that most of these MedEvac cases had psychiatric problems prior to deploying to the battlefield (Operations *Iraqi Freedom* and *Enduring Freedom*).

An advantage of not being in a specific research laboratory was being able to work different facets of research psychology. Thus, I was also able to contribute to an article about the relationship between cognitive fitness and decisionmaking in the increasingly complex network-centric military world (Russo, Stetz, and Thomas, 2005). I believe all of my opportunities and early career achievements were validated when in 2004 I received the Arthur W. Melton Award, which the American Psychological Association, Division 19, gives in recognition of early career achievement in military psychology.

I also developed as an officer by working with Brigadier General Lester Martínez-López, whom I coincidentally had escorted during a ceremony at WRAIR. While at HQ, I coordinated the 2004 71F meeting. I also took courses at the Defense Acquisition University and became Level III–certified in Science and Technology Management and Level II–certified in Program Management.

During this tour, I attended the Aeromedical Psychology Course at Fort Rucker, Alabama (2004), where I learned about the cognitive demands that pilots and their crews face while flying (for example, eye-hand coordination demands while fatigued and under foggy vision). We had to take a battery of tests inside a hyperbaric chamber, swim with flying gear, tread water for an extended period, and get spun around in a simulated helicopter cabin under water while finding "the nearest exit." Another important experience was being able to work with clinical psychologists.

Colonel James McGhee, the commander of the U.S. Army Aeromedical Research Laboratory (USAARL), also at Fort Rucker, asked me to come to the laboratory to do research. When he discovered that I had always wanted to be a company commander, however, he offered me a detachment commander position to be followed by a research psychology slot.

Upon getting my orders to Fort Rucker, I decided to apply to the Airborne School. I thought I should earn another pair of wings (in addition to the flight psychology ones) before reporting to Fort Rucker, the well-known "U.S. Army Aviation Center of Excellence." Jump school was physically demanding. However, I enjoyed jumping out of "perfectly functioning" airplanes (such as C–130s and C–17s) with Soldiers who were almost young enough to be my own kids.

U.S. Army Aeromedical Research Laboratory

I was, and still am, proud to be the first laboratory officer, the first 71F, and the first woman ever to serve as the Detachment Commander and Adjutant of USAARL. I helped open a door for other scientists to follow in my footsteps, adding diversity to the typical group of very smart but not research-minded pilots. Keeping the holistic approach to health, I was able to maintain my physical training badge while training with my Soldiers. In fact, one of my contributions to the lab was bringing the German Armed Forces Proficiency Badge certification training there. Considering my new role in life as a commander, an adjutant, and a mom, earning the gold was probably as challenging as my first day learning SPSS syntax. I also took the opportunity to organize the 2006 biannual 71F meeting.

After my year in command, I reported to Fort Belvoir, Virginia, to attend the 4-month Command and General Staff College, Intermediate-level Education. Upon graduation, I returned to USAARL and became the Chief of Combat Stress Research. I ran several studies that were funded by USAMRMC's Telemedicine and Advanced

Technologies Research Center (TATRC). While working in that lab, I had access to other types of scientists (such as physiologists and engineers) and was able to discuss additional ways to measure stress that were not limited to the typical paper-and-pencil surveys.

In my first research project at USAARL, I brought in a 71F Individual Mobilization Augmentee, Lieutenant Christopher Long, who helped me with virtual reality (VR) technology during his 2-week annual training. Our research team ran a few pilot studies testing this technology (Stetz, Long, Schober, Cardillo, and Wildzunas, 2007; Stetz, Wildzunas, Wiederhold, and Hunt, 2006), including a study in which participants played a game where they had 3 minutes to save, triage/assess, and treat a casualty.

We used monitors and biofeedback equipment to measure participants' reactions while they played a triage and treat casualties video game. Our three main experimental groups comprised: one that did only VR (play a game), one that did VR and relaxation techniques, and one that did only relaxation techniques. Our main findings suggested that the group only exposed to VR showed higher levels of hostility than the other groups. The participants in the VR group that practiced relaxation techniques showed higher levels of sensation-seeking. Interestingly, further analyses showed higher levels of both anxiety and dysphoria in previously deployed participants in the VR and relaxation groups (Stetz, Long, Wiederhold, and Turner, 2008).

For the second TATRC-funded study, my team examined relaxation techniques using VR. Our participants were Forward Surgical Teams training at the Ryder Trauma Center in Miami. In this study, we tested the usefulness of VR for relaxation by translating an existing relaxation environment being used in a laboratory in Milan, Italy, and adding simultaneous progressive muscular relaxation and controlled breathing techniques. Participants took the Combat Stress Assessment test (a Stroop test adapted for military populations) after practicing relaxation techniques on a laptop. They practiced the relaxation techniques again at night via portable play stations and wore an actigraph (a wristwatch-sized instrument to measure movement and activity) to measure sleep (related to relaxation). Early results suggest that VR may be a promising deployable technology for relaxation, and not just for entertainment.

Field Assistance in Science and Technology Team

In 2008, I volunteered to serve in Operation *Iraqi Freedom* for 6 months as part of the Field Assistance in Science and Technology (FAST) Team, where I followed in the footsteps of Major Matt Clark, another 71F who was the first to serve on a FAST Team. The other members of our four-person team worked for the Army Materiel Command (AMC)/Research and Development Command (RDECOM), while I represented the U.S. Army Medical Command and the USAMRMC. It was difficult being away from my family, but we did have access to a phone, as well as a web camera to help us maintain communication.

My mission was to identify the needs of the various medical levels of care in the whole theater of operations. I quickly recognized some gaps and worked with CONUS USAMRMC personnel to identify what could be accomplished at the Medical Research and Materiel Command level versus what should be done at the unit level (that is, logistical or command issues). In one project, we provided the veterinary community with intravenous (IV) fluid warmers that were needed to prevent dogs from experiencing hypothermia during surgery. Another project developed a smaller safety harness to be used as a restraint system for injured children and smaller adults who would otherwise slide off the chopper seats when traveling to the nearest medical care facility. The U.S. Army's Rapid Equipping Force helped fund 20 civilian air-worthy seats to test in both Operation *Iraqi Freedom* and Operation *Enduring Freedom*. During my trips, we also provided feedback to developers about already-in-use items, such as the combat tourniquet and the Talon litters.

I also visited a few combat stress control clinics. Since the FAST team supporting Operation *Enduring Freedom* in Afghanistan had no medical representative, I also flew there to meet the corps surgeon. We discussed many of the ongoing projects in Operation *Iraqi Freedom* (such as up-armored ambulances), and I sent them a few of the IVs we had given to the veterinarians in Iraq.

As I was still on "flight status" at USAARL, I was able to engage with pilots and crewmembers—and collected more data as part of a study to identify the main stressors and coping methods used by aviators. Preliminary results found that 59 out of 283 deployed aviators were having problems sleeping. A further analysis of 22 (8 percent) of those with a self-reported head injury suggested that the group scored higher than eight points on the Epworth Sleepiness Scale (the threshold for excessive daytime sleepiness and referral to a sleep specialist (Stetz, Russo, and Stetz, forthcoming).

Tripler Army Medical Center

I am currently Director of Research for the Psychology Department at Tripler Army Medical Center (TAMC) in Hawaii. I recently finished collecting data for an anger study in which we compared technologies (VR head-mounted display [HMD] versus flat-screen computer monitors) and their effectiveness in prompt behavioral reactions (that is, anger) to video vignettes in participants. In the study, 30 participants viewed video anger vignettes through a VR HMD and 30 on a flat-screen computer monitor. Preliminary results (Folen, Miyahira, and Stetz, 2009) suggest that the VR HMD technology might positively affect presence and reactivity to anger stimuli (psychological, emotional, and physiological).

Presently, we are continuing to investigate VR applications to important warfighter and family matters including Post-traumatic Stress Disorder, pain tolerance (for example, during cystoscopy), biofeedback, and nicotine cravings. Furthermore, we are engaging in collaborative efforts with USAMRMC as the TATRC's Pacific Telehealth and Technology Hui and fellows in TAMC's departments of Neurosurgery and Urology.

In summary, the field of research psychology has been a very exciting military career path. In this profession, individuals can contribute to the fighting force by generating scientific findings in a laboratory, by commanding research units, and even by branching out and carving new paths for others to follow at home or abroad.

References

Allison-Aipa, T.S., De la Rosa, G.M., Stetz, M.C., and Castro, C.A. (2005). The impact of National Guard activation for homeland defense: Employer's perspective. *Military Medicine* 170 (10), 846–850.

Bliese, P.D., and Stetz, M.C. (2007). Modeling the effects of efficacy, justice, and conflict among reservists activated for homeland defense. *Military Psychology* 19 (1), 27–43.

Britt, T.A., Stetz, M.C., and Bliese, P. (2004). Work-relevant values strengthen the stressor–strain relation in elite army units. *Military Psychology* 16 (1), 1–17.

Estrada, A.X., Stetz, M.C., and Harbke, C.R. (2007). Further examination of the psychometric properties of the military equal opportunity climate survey in a sample of reserve component personnel. *International Journal of Intercultural Relations* 31, 137–161.

Folen, R.A., Miyahira, S.D., and Stetz, M.C. (2009, June). *Immersive panoramic video display compared to flat screen display: Psychological and physiological reactions to anger stimuli.* Presentation at the 14th Annual International CyberTherapy and CyberPsychology Conference, Verbania-Intra, Italy.

Killgore, W., Stetz, M.C., Castro, C.A., and Hoge, C.W. (2006). Somatic and emotional stress symptom expression prior to deployment by soldiers with and without combat experience. *Journal of Psychosomatic Research* 60, 379–385.

Russo, M.B., Stetz, M.C., and Thomas, M.L. (2005). Monitoring and predicting cognitive state and performance via physiological correlates of neuronal signals. *Aviation, Space, and Environmental Medicine*, C59–63.

Stetz, M.C., Castro, C.A., and Bliese, P. (2007). The impact of deactivation uncertainty, workload, and organizational constraints on reservists' psychological well-being and turnover intentions. *Military Medicine* 172 (6), 576–580.

Stetz, M.C., Long, C.P., Schober, W.V., Cardillo, C.G., and Wildzunas, R.M. (2007). Stress assessment and management while medics take care of the VR wounded. *Annual Review of Cybertherapy and Medicine*, 191–204.

Stetz, M.C., Long, C.P., Wiederhold, B.K., and Turner, D.D. (2008). Combat scenarios and relaxation training to harden medics against stress. *Journal of CyberTherapy & Rehabilitation*, 239–246.

Stetz, M.C., McDonald, J.J., Lukey, B.J., and Gifford, R.K. (2005). Psychiatric diagnoses as a cause of medical evacuation. *Aviation, Space, and Environmental Medicine* 76 (7), C15–20.

Stetz, M.C., Russo, M.B., and Stetz, T.A. (2009, June). *An accessible sleep tool for military personnel: The post traumatic stress disorder checklist.* Presentation at the 23d SLEEP annual meeting, Seattle, WA.

Stetz, M.C., Thomas, M.L., Russo, M.B., Stetz, T.A., Wildzunas, R.M., McDonald, J.J., Wiederhold, B.K., and Romano, J.A. (2007). Stress, mental health, and cognition: A brief review of relationships and countermeasures. *Aviation, Space, and Environmental Medicine* 78 (5), B252–260.

Stetz, M.C., Wildzunas, R.M., Wiederhold, B.K, and Hunt, M.P. (2006). The usefulness of virtual reality stress inoculation training for military medical females: A pilot study. *Annual Review of Cybertherapy and Medicine*, 51–58.

Stetz, T.A., Stetz, M.C., and Bliese, P.D. (2006). The importance of self-efficacy in the moderating effects of social support on stressor-strain relationships. *Work & Stress* 20 (1), 49–59.

Chapter 20

Notes from a Research Psychologist at U.S. Special Operations Command

Craig A. Myatt

This chapter describes my role as the first military research psychologist assigned to the headquarters of U.S. Special Operations Command (USSOCOM). This descriptive account of the research psychologist duties in USSOCOM headquarters provides new perspectives on the application of military psychology. The research psychology position at USSOCOM is new and was created to meet command leadership concerns articulated by the commander. As the first incumbent in this position, my contributions rely on research skills and expertise in biopsychosocial (BPS) applications, as well as leadership perspectives gained from previous assignments as a platoon leader, company commander, and battalion commander.

The Special Operations Forces (SOF) environment, further described below, is a somewhat unique one for the command research psychologist to work within. He or she can play a critical role in SOF command consultation by bringing specialized knowledge on the psychology of leadership and on enhancing program development. A good understanding of the principles of operational psychology is also important for effective consultation and program development in this environment (Staal and Stephenson, 2006; Williams, Picano, Roland, and Banks, 2006).

The U.S. military Services as a whole currently number approximately 1.8 million military and civilian personnel (U.S. Department of Defense, 2009); USSOCOM has only 56,000. As a relatively small and specialized force, USSOCOM aims to develop a well-structured, properly resourced, culturally attuned, and adaptive SOF. The assignment of a research psychologist to the command is a seminal step toward reaching this goal. This chapter will also highlight the SOF leadership perspectives that a research psychologist must appreciate in order to support his work at the headquarters of USSOCOM.

USSOCOM and Research Psychology

SOF Characteristics

The USSOCOM is a unified combatant command that maintains authorized oversight for special operations of the U.S. Armed Forces. There are two parts to its mission. First, the mission established by U.S. Code, Title 10 (Section 167), is to organize, train, and equip SOF and to command and control continental United States (CONUS)-based SOF. The second part of the mission, directed by the Unified Command Plan, is to synchronize planning of global operations against terrorist networks.

USSOCOM is unique in that it functions as a combatant command with Service-like functions and distinct acquisition authorities. The commander's funding, however, differs from that for the other combatant commanders because Congress granted USSOCOM its own budget authorities. The Major Force Program 11 (MFP–11) budgetary control established for USSOCOM enables it to modernize SOF and provide SOF-peculiar program support to maintain SOF characteristics (see table 20–1).

Table 20–1. **Special Operations Forces Characteristics**

Leadership	Culture and language familiarity
Warrior ethos	Innovation and creativity
Integrity	Intellect and problemsolving skills
Teamwork	Adaptability to changing environments
Physical and moral courage	Savvy and diplomatic awareness
Maturity and reliability	Physical readiness

Consultation and Program Development

The duties of a research psychologist assigned to USSOCOM involve consultation and program development tailored to the unique characteristics of SOF. For example, if a military Servicemember at USSOCOM exhibits disruptive behavior in the workplace that leads to behavioral interventions, the research psychologist is available to consult supervisors on the use of counseling, social work services, and other available programs. In such instances, the military psychologist serves as an additional source of research-based information. For example, the research psychologist assigned to USSOCOM helps to formulate metrics that reveal shortfalls in program support for both military and civilian personnel to complete the assessment and analysis required for the Planning, Programming, and Budgeting System (PPBS) process. The relevance of outlining leadership perspectives in SOF psychology at USSOCOM arises from the commander's unique institutional authority under Title 10, the budget process with MFP–11, and a command-focused interest in developing programs that effectively mitigate potentially adverse behavioral effects stemming from an increased operations tempo (OPTEMPO).

My own perspectives on leadership were developed through previous training and experiences as a platoon leader, as well as a commander at the company and battalion levels (Department of the Army, 2003). These experiences have proven highly relevant to the decision-making processes exercised by the leadership at USSOCOM, in part because they enhance multidisciplinary approaches in supporting SOF and their families. Though SOF is a relatively small force, it is faced with many of the same OPTEMPO challenges encountered by conventional forces due to accelerated rotational deployment cycles and training exercises in support of overseas contingency operations. SOF's smaller force structure may, however, enhance its ability to mitigate potentially adverse behavioral effects resulting from a high OPTEMPO.

As USSOCOM's current research psychologist, I place leadership at the forefront of my approach to SOF psychology. While I also focus on developing programs that support the well-being of Servicemembers and their families, from a leadership standpoint it is also important to recognize how MFP–11 funding leverages opportunities to develop SOF-specific programs.

Strategic Perspective

The strategic perspective of a research psychologist in a combatant command with MFP–11 funding is different from that of the operational psychologist functioning at the component or unit level. In the USSOCOM setting, the research psychologist must pay attention to the strategic dynamics of leadership and the delivery of leadership perspectives unique to SOF (see table 20–1). Program development, even if streamlined by supplemental funding for rapid execution, is strategic in nature. Therefore, as the research psychologist attends to strategic approaches for program planning, his support for rapid implementation of USSOCOM command-direct initiatives reinforces the unique leadership ethos for warriors trained in the application of irregular warfare (U.S. Department of Defense, 2008).

Stress on the Forces

Concern about stress on the forces is due, in part, to the contemporary global operating environment. As a result, an increasing number of units now have stronger institutional support to address combat and operational stress (Mangelsdorff, 2006). While not used by SOF, Combat Stress Control teams are an example of this type of support. Operational psychologists supporting SOF in the Army, Air Force, and Navy are typically assigned at the SOF group level and provide direct support to subordinate units based on mission requirements.

OPTEMPO Behavioral Effects

The research psychologist assigned to USSOCOM receives direct guidance from the USSOCOM commander on the development of a centralized SOF-peculiar program with decentralized execution. A research psychologist was chosen to develop the program in order to allow component level operational psychologists to mitigate the potential adverse effects stemming from the operational demands on Servicemembers and their families generated by a high OPTEMPO. This is feasible since potential strain in SOF and family member resilience, based on operations tempo behavioral effects (OTBE), can be assessed at every level throughout an organization. Thus, the USSOCOM research psychologist functions in a consultative capacity at the headquarters level, providing leaders at every level with additional tools to manage these demands.

Multidisciplinary Approach

The USSOCOM research psychologist offers a central role in addressing OTBE and works closely with clinical psychologists throughout the component commands. The use of formal working groups supports this collaboration. Although the research psychologist is not an operational psychologist, both research and clinical psychologists counsel unit commanders on issues related to combat and operational stress control (COSC). As consulting psychologists, they offer commanders an additional mechanism to reinforce the favorable interpersonal aspects of leadership, such as sustaining the dynamics of caring, support, and growth.

Consulting psychologists function as staff officers trained to provide a multidisciplinary approach to COSC (Kennedy and Zillmer, 2006). The USSOCOM research psychologist offers organizational consultation and supports a multidisciplinary approach to COSC that involves other USSOCOM personnel—teams of psychologists, mental health specialists, psychiatrists, psychiatric nurses, chaplains, and social workers—also serving the needs of SOF warriors and their families. From a consultation standpoint, the teams work outside of a clinic setting to support command programs. The USSOCOM research psychologist works with the multidisciplinary teams to address what the U.S. Army Training and Doctrine Command (2008) describes as combat and operational stress responses (COSR) and combat and operational stress injury (COSI).

Leadership Focus

Previous command experience helps the research psychologist maintain timely awareness of command-focused issues for program development that is responsive to the USSOCOM commander's focus. Mission-directed leadership focuses on achieving the mission's objective. Concerns for personnel exist, but not at the cost of the mission. Interpersonal leadership also places a high value on the mission's objective, but often prioritizes issues of personnel. The balance between mission-directed and interpersonal leadership may vary, depending on national strategic, operational, and tactical goals and the effects of the different types of goals on warriors, their friends, and families. Ensuring clarity across the entire spectrum from strategic to tactical is a challenge for effective leaders at all levels (Wass de Czege, 2009).

With experience in the applied knowledge of the BPS perspective (Engel, 1980), the research psychologist at USSOCOM is familiar with research perspectives that assess behavior using a systems approach, along a continuum from individual physiologic subsystems to social macrosystems. Previous command experience in deployable units and a working knowledge of the BPS perspective provide the USSOCOM research psychologist with innovative insights on how to integrate a BPS systems approach into program development. That insight is tailored to maximize what effective leaders do to improve their organizations across all applicable systems, both individual and social.

Effective leaders use their staff and subordinate leaders to establish balance throughout the spectrum of strategic, operational, and tactical interests. In elite and high-OPTEMPO organizations and units, the use of multidisciplinary staff components (such as consulting psychologists) aids in leadership focus.

When leadership is focused, organizations, units, and teams perform to standard or better, regardless of the complexity or ambiguity of a given situation. Central to that focus is an element of caring that defines a leader not only as a person in a position of authority and power, but also as one who, through experience and support, channels all required energies and resources toward the achievement of specified goals. Caring is an aspect of leadership focus that is observed not only in the outcomes of mission-directed effort and leader interpersonal skills, but also in the types of programs supported and endorsed by the leadership.

Screening and Selection

The availability of military psychology consultants allows command leaders to obtain practical and expert input on the potential fragility of forces based on assessments of OTBE, COSR, COSI, and decisionmaking. While the military utilizes extensive screening and selection criteria for personnel decisions, personality factors in potential selectees are only one of several psychological assessments that consulting psychologists can provide to institutional leaders and subordinate commanders. Although personality affects leadership focus, dynamics, and perspectives, the degree to which personality influences each is variable and insufficient to transcend the overall dynamics of leadership. Historically, personality factors are measured along several domains.

The Domains of Personality

Personality taxonomy contains three to five well-defined domains. The three basic domains, as defined by Eysenck and Eysenck (1985), are neuroticism, extraversion, and psychoticism. Neuroticism expresses negative affect, sadness, anger, guilt, and shame. Extraversion is characterized by sociability, ambition, dominance, positive emotionality, and excitement-seeking. Psychoticism involves aggressive, impersonal, and impulsive behavior with underlying thought patterns that can be antisocial, unempathetic, creative, and tough-minded (Hall, Lindsey, and Campbell, 1997).

Derivations of those three basic domains offer a different taxonomy with five domains: neuroticism, extraversion, openness, agreeableness, and conscientiousness (McCrae and Costa, 1996). Costa and McCrae (1997) correlate openness (to experience) with creativity, unconventionality, and broad-mindedness. Agreeableness reflects an interpersonal stance of cooperation, trustfulness, and compliance. The five-factor model removes psychoticism from the taxonomy, assigning many of its factors to one of three different domains: openness, agreeableness, and conscientiousness.

When one analyzes each of these domains as multiple or single factors for measure, analyses can support choices about which factors are inherently relevant to leadership and unit performance. Personality patterns in distinct performing groups, such as elite warriors, deviate from the general population. For example, in military pilots who completed the five-factor Neuroticism Extroversion Openness Personality Inventory–Revised (Bernard and Walsh, 2004), those in initial military flight training showed higher levels of extraversion and openness with lower levels of agreeableness compared to the normal population (Callister, King, Retzlaff, and Marsh, 1999). Those with substantial military flight experience showed lower levels of neuroticism and some extraversion—but not significantly high compared to the normal population—and lower levels of openness (Grice and Katz, 2006).

Hardiness and Resilience

Like the more mature and experienced pilot, the SOF warrior displays the hardiness and mental toughness common among groups that undergo rigorous physical and cognitive screening. Hardiness as a personality construct provides the SOF warrior with initial advantages that promote selection, but leadership focus helps to direct performance enhancement observed through the biobehavioral construct resilience. The focus of leadership applied through its dynamics is what governs the delivery of leadership perspectives in elite warrior units. Understanding the role that personality plays in the SOF screening and selection process and how hardiness as a personality construct is a trait characteristic (Bartone, 1999; Maddi, 2002) helps the USSOCOM research psychologist appreciate how important leadership focus is in the development of SOF resilience (Lukey and Tepe, 2008).

Dynamics of Leadership

If one accepts that a leadership focus is important in the development of SOF resilience, he may then acknowledge that the dynamics of leadership also develop resilience. The research psychologist offers an essential role at USSOCOM; this is seen in his interactions with leadership in the capacity of a consultant and program manager to reinforce mastery over the dynamics of leadership for stronger resilience development in SOF and their families.

No leader is ever fully isolated from OTBE, though the inherent demands of leadership can at times remove a leader from direct involvement with execution. Awareness of OTBE through a reporting process may be less than ideal, but an effective leader will employ mechanisms within an organization to strengthen awareness and mitigate the adverse effects of OTBE.

The role of leadership can also separate a key decisionmaker from full participation in COSC, but well-developed multidisciplinary teams can circumvent that alienation. Leaders of elite warriors, regardless of how much they become isolated from OTBE, can effectively participate in COSC with the aid of well-trained consulting psychologists who can assist leadership in making appropriate strategic, operational, or tactical mission-directed decisions to reinforce mastery over the dynamics of leadership. Regardless of the leadership traits and characteristics that one exhibits as a leader, three dynamics are universal: caring, support, and growth.

Caring

There must be an observable methodology for acknowledging care to Servicemembers, including well-defined programs and clearly understood leadership behaviors. For example, when there are mishaps such as loss of life during training exercises, effective leaders should demonstrate care for their personnel by adhering to the basic aspects of grieving, such as maintaining a continued presence among Servicemembers, attending to the immediate concerns of family members, and showing active participation at memorial and funeral services.

In the small USSOCOM community, failure to show presence and care in the midst of often hectic and time-consuming incident investigations would qualify as a failure of leadership contrary to the SOF culture. The USSOCOM research psychologist must appreciate longstanding military traditions that reinforce the dynamic of caring and the value added in providing ancillary participation as a staff consultant. Leaders who have access to consultative services build an atmosphere of caring that strengthens command effectiveness. Multidisciplinary consultative services offer responsive delivery for both leadership and other Servicemembers.

Although uncommon, consultative mental health services, such as those provided by a clinical psychologist or psychiatrist working in a medical treatment facility, can be therapeutic in nature. More commonly, nontherapeutic consultative mental health services for the leadership are available within the chain of command and can be offered by psychologists, mental health specialists, psychiatrists, psychiatric nurses, chaplains, social workers, and authorized support elements outside of the chain of command. Used effectively, each addresses the dynamic of caring. A command-directed program with a formal working group authorized to review best business practices within each discipline can address the dynamic of caring on behalf of a commander and then make recommendations for improvement.

Nontherapeutic consultative mental health services provided to the leadership can occur in the scope of operational psychology (Williams, Picano, Roland, and Banks, 2006). In the USSOCOM setting, the research psychologist may provide nontherapeutic consultative mental health services depending on the situation and preliminary assessments. For example, the research psychologist may be summoned to the command suite to provide insight to senior leaders who are attempting to address a civilian employee's family-related problem.

Called into a meeting to review the situation on behalf of the commander and to provide appropriately constructive input, the research psychologist might ask if the employee had been referred to the Employee Assistance Program (EAP) on the supporting installation. If the military psychologist discovered that there was no EAP at the supporting installation, he or she would provide short-term recommendations to address the immediate family-related problem. In a consultative role, the research psychologist could also provide long-term recommendations to assess the overall feasibility of developing an EAP to offer counselors to civilian employees and their family members

Support

The dynamic of support, like caring, requires mechanisms of human intervention. Unlike caring, support as a mechanism of human intervention can be augmented by technological systems, such as online access to information and referral services. The application of adequate telecommunications techniques, graphic user interfacing, and workstation ergonomics cannot be ignored in the process of reaching out to prospective warriors and families in need of command-sponsored support. Technology-based programs are better received if the applications meet identified end-user technology requirements. Familiarity with acquisitions assists the research psychologist involved in the planning and programming phases of the PPBS.

Growth

The third dynamic, growth, is typically transactional and transformational, and it can occur at various levels within any institution, organization, or unit. Transactional leadership is highly task-oriented, while transformational leadership is highly person-oriented (Bass, 1990). The combination of transactional and transformational leadership can establish meaningful trust, which in turn promotes synergy. Regardless of the personality factors infused in leadership style (McCormack and Mellor, 2002), elite warriors are highly responsive to growth in an organization that builds trust and promotes synergy (Westphal, Bonanno, and Bartone, 2008).

The research psychologist at USSOCOM facilitates the leadership's role in establishing the dynamic of growth, building trust, and promoting synergy. Assessments made by the research psychologist facilitate training and psychoeducation (Dixon et al., 2001) to enhance resilience throughout the headquarters and component levels, which establishes the groundwork for sharing individual and unit level reactions to operational demands. Resilience developed through training, psychoeducation, and operational demands fosters growth in Servicemembers and their families, instilling stronger purpose and adherence to SOF characteristics.

Leadership Perspectives

As leadership perspectives help the research psychologist apply SOF characteristics in program development, he or she should possess innovative leadership skills based on previous experience in deployable units. Though the command psychologist functions as a staff officer, the position requires strong familiarity with command-focused decision-making, field operations, and training.

Staff officers in the command are routinely engaged in airborne operations, other types of flight operations, and rotations into theater. Perspectives developed in the high-OPTEMPO USSOCOM environment reflect the capacity of the research psychologist to register the impact of OTBE from the bottom up and to responsively develop a command-directed program to mitigate that impact from the top down.

Social Learning

The delivery of leadership perspectives unique to SOF reinforces the unique ethos of warriors trained in the application of irregular warfare. Simple deliveries of leadership perspectives that keep organizations and individuals on target with purpose can be more meaningful than complex applications that diffuse efforts and distract from purpose. Social learning principles demonstrate the value of validation among warriors and further suggest that validation contributes to resilience (Campbell, Campbell, and Ness, 2008; Grossman, 1995). The intensity of the SOF selection and screening process provides a unique validation or sense of purpose to those accepted into the community. That sense of purpose contributes to a greater sense of coherence, or resilience, in adversity (Van Breda, 2001).

Grossman offers examples of how the trauma of warfare can be viewed in a social context that helps validate the aggressions and natural reactions involved. SOF warriors and families who thrive on a socially acknowledged sense of purpose train themselves to face adversity and overcome it. Grossman demonstrates, for example, how families that provide support to each other in ordinary crises tend to provide better support for returning warriors. The SOF family prepares, or trains, itself for adversity by adhering to its social commitment to purpose. The research psychologist assigned to USSOCOM should understand the importance of social validation for warriors and incorporate those principles into programs that provide command-directed support to SOF warriors and their families.

The Biopsychosocial Systems Approach

The research psychologist assigned to USSOCOM benefits from possessing a strong background in the biological sciences so that he or she can support the use of physiologic and genomic studies to augment psychological applications. Psychological and biological applications address potentially adverse OTBE in SOF and their families. Analysis of physiologic markers (such as neuropeptides and proteins) can be incorporated into studies that support programs designed to enhance resilience. From a BPS perspective (Engel, 1980), program development will benefit from a systems approach that spans from physiologic subsystems within the individual to macrosystems within society that influence health and behavior.

From a BPS perspective, the research psychologists at USSOCOM can track meaningful trends by gathering data from a continuum of factors, ranging from individual to social. Trained to recognize fragility in the force and formulate appropriate metrics, the research psychologist at USSOCOM offers a multidisciplinary approach to COSC that can increase the efficiency and productivity of SOF personnel and their families. The systems review methodologically supports confidential analyses of data (U.S. Department of Health and Human Services, 2003 and 2004). Conceptually, the systems review also supports an analysis of intrasystem changes associated with one's existence that influence the functional cognitive behavioral framework of what Antonovsky (1993) refers to as sense of coherence.

Biobehavioral Resilience

Leadership perspectives exercised at USSOCOM warrant the use of a research psychologist who can quantitatively assess OTBE and recommend training, protocols, and actions for preventing, interdicting, and mitigating COSI. The research psychologist's goal is to instill a sense of coherence and overall performance in individual SOF warriors and their family members through a program that supports resilience. Applications in COSC require clear distinctions between resilience and hardiness: resilience is a biobehavioral construct, while hardiness is a personality construct (Westphal, Bonanno, and Bartone, 2008). Hardiness is a component of resilience and is psychological in nature (Maddi, 2004). Resilience is both biological and psychological in nature (Lukey and Tepe, 2008).

Understanding the distinctions between biobehavioral resilience and hardiness can assist the USSOCOM research psychologist in tracking behavioral trends and potential causative or associative factors. When exploring the factors associated with PTSD or other behaviors, it is useful to know the measures of resilience and hardiness in SOF warriors. The biobehavioral construct, resilience, can be reinforced and strengthened to contribute to post-traumatic growth (Tedeschi and Calhoun, 2004). In contrast, one would not expect hardiness as a personality construct to be reinforced and strengthened based on the principles of trait-based personality theory (McCrae and Costa, 1997). Instead, it is a useful construct for operational psychologists engaged in selection and screening (Stephenson and Staal, 2007).

Consulting psychologists play a proactive role in preventing and managing COSR and COSI so as to ensure that SOF personnel are healthy and medically ready to deploy. This role is critical given the high OPTEMPO of SOF personnel and the contemporary operating environment. Research psychologists in USSOCOM also strengthen the commitment of SOF leadership to developing and supporting their elite personnel and families as they sustain and modernize the force.

Conclusion

In describing the observations of USSOCOM's first research psychologist as he employed new strategic and leadership perspectives in the application of military psychology, this treatise has emphasized several important lessons. First, the professional background of the research psychologist at USSOCOM affects his or her ability to carry out mission-essential duties, including psychology consultation and program development in accordance with the PPBS. Second, a successful USSOCOM research psychologist must possess a thorough appreciation of SOF's unique characteristics and, third, employ a distinct, leadership-focused strategic perspective.

The research psychologist assigned to USSOCOM receives direct guidance from the commander and command surgeon on the development of a centralized SOF-peculiar program with decentralized execution. Each shares a common breadth of command experience in highly deployable units, which helps to galvanize collective leadership focus and perspectives. The intent of the command-directed program development is to mitigate potential adverse effects stemming from OTBE in Servicemembers and their families. Furthermore, an understanding of the role of personality in the SOF screening/selection process and of how hardiness as a personality construct is a trait characteristic helps the USSOCOM research psychologist appreciate how important leadership focus is in the development of SOF resilience. Another important lesson is that in order to promote resilience, the USSOCOM research psychologist must employ a multidisciplinary approach while maintaining a mission-directed and interpersonal leadership focus. Finally, the research psychologist at USSOCOM must reinforce and strengthen biobehavioral resilience through the use of a BPS systems approach that supports social learning theory and the application of psychoeducation.

448 *THE 71F ADVANTAGE*

References

Antonovsky, A. (1993). The structure and properties of the sense of coherence scale. *Social Science Medicine* 36 (6), 725–733.

Bartone, P. (1999). Hardiness protects against war-related stress in Army reserve forces. *Consulting Psychology Journal 51*, 72–82.

Bass, B. (1990). *Bass and Stogdill's handbook of leadership* (3ᵈ ed.). New York: The Free Press.

Bernard, L., and Walsh, P. (2004). Socially desirable and non-purposeful responding on the neuroticism extraversion openness personality inventory-revised. *Counseling and Clinical Psychology Journal* 1 (1), 4–16.

Callister, J., King, R., Retzlaff, P., and Marsh, R. (1999). Revised NEO personality inventory profiles of male and female U.S. Air Force pilots. *Military Medicine* 164, 885–890.

Campbell, D., Campbell, K., and Ness, J. (2008). Resilience through leadership. In B. Lukey and V. Tepe (eds.), *Biobehavioral resilience to stress* (57–88). Boca Raton, FL: CRC Press.

Costa, P., and McCrae, R. (1997). Stability and change in personality assessment: The revised NEO personality inventory in the year 2000. *Journal of Personality Assessment* 68 (1), 86–94.

Dixon, L., McFarlane, W., Lefley, H., Lucksted, A., Cohen, M., Falloon, I., Mueser, K., Miklowitz, D., Solomon, P., and Sondheimer, D. (2001). Evidence-based practices for services to families of people with psychiatric disabilities. *Psychiatric Services* 52, 903–910.

Department of the Army. (2003). Field Manual 6–0. *Mission command: Command and control of Army forces*. Washington, DC: Headquarters Department of the Army.

Engel, G. (1980). The clinical application of the biopsychosocial model. *American Journal of Psychiatry* 137 (5), 535–544.

Eysenck, H., and Eysenck, M. (1985). *Personality and individual differences: A natural science approach*. New York: Plenum Publishers.

Grice, R., and Katz, L. (2006). *Personality profiles of experienced U.S. Army aviators across mission platforms* (Technical Report 1185). Arlington, VA: U.S. Army Research Institute for the Behavioral and Social Sciences.

Grossman, D. (1995). *On killing: The psychological cost of learning to kill in war and society*. Boston: Little, Brown and Company.

Hall, C., Lindzey, G., and Campbell, J. (1997). *Theories of personality* (4ᵗʰ ed.). New York: John Wiley and Sons, Inc.

Kennedy, C., and Zillmer, E. (eds.). (2006). *Military psychology: Clinical and operational applications*. New York: Guilford Press.

Lukey, B., and Tepe, V. (eds.). (2008). *Biobehavioral resilience to stress*. Boca Raton, FL: CRC Press.

Maddi, S. (2002). The story of hardiness: Twenty years of theorizing, research and practice. *Consulting Psychology Journal 54*, 173–185.

———. (2004). Hardiness: An operationalization of existential courage. *Journal of Humanistic Psychology* 44 (3), 279–298.

Mangelsdorff, A. (ed.). (2006). *Psychology in the service of national security*. Washington, DC: American Psychological Association.

McCormack, L., and Mellor, D. (2002). The role of personality in leadership: An application of the five-factor model in the Australian military. *Military Psychology* 14 (3), 179–197.

McCrae, R., and Costa, P. (1996). Toward a new generation of personality theories: Theoretical contexts for the five-factor model. In J.S. Wiggins (ed.), *The five-factor model of personality: Theoretical perspectives* (51–87). New York: Guilford Press.

———. (1997). Personality trait structure as a human universal. *American Psychologist* 52, 509–516.

Staal, M., and Stephenson, J. (2006). Operational psychology: An emerging subdiscipline. *Military Psychology* 18 (4), 269–282.

Stephenson, J., and Staal, M. (2007). An ethical decision-making model for operational psychology. *Ethics and Behavior* 17 (1), 61–82.

Tedeschi, R., and Calhoun, L. (2004). Posttraumatic growth: A new perspective on psychotraumatology. *Psychiatric Times* 21 (4).

U.S. Army Training and Doctrine Command. (2008). *The U.S. Army study of the human dimension in the future 2015-2024* (TRADOC Pamphlet 525-3-7-01). Fort Monroe, VA: TRADOC.

U.S. Department of Defense. (2008). *Irregular warfare*. (Department of Defense Directive 3000.07). Washington, DC: Office of the Under Secretary of Defense for Policy.

———. (2009). *DOD personnel and procurement statistics: Personnel and procurement reports and data files*. Statistical Information Analysis Division. Retrieved February 5, 2009, from <http://siadapp.dmdc.osd.mil/index.html>.

U.S. Department of Health and Human Services. (2003). *The health and insurance portability and accountability act (HIPAA) privacy rule (overview)*. Bethesda, MD: National Institutes of Health. Retrieved December 12, 2008, from <http://privacyruleandresearch.nih.gov/>.

———. (2004). *Protecting personal health information in research: Understanding HIPAA Privacy Rule*. Bethesda, MD: National Institutes of Health. Retrieved December 12, 2008, from <http://privacyruleandresearch.nih.gov/pr 02.asp>.

Van Breda, A.D. (2001). *Resilience theory: A literature review*. Pretoria, South Africa: South African Military Health Service. Retrieved from <www.vanbreda.org/adrian/resilience.htm>.

Wass de Czege, H. (2009). Systemic operational design: learning and adapting in complex missions. *Military Review*, January–February 2009, 2–12.

Westphal, M., Bonanno, G., and Bartone, P. (2008). Resilience and personality. In B. Lukey and V. Tepe (eds.), *Biobehavioral resilience to stress* (219–257). Boca Raton, FL: CRC Press.

Wiggins, J. (ed.). (1996). *The five-factor model of personality: Theoretical perspectives*. New York: Guilford Press.

Williams, T., Picano, J., Roland, R., and Banks, L. (2006). Introduction to operational psychology. In C. Kennedy and E. Zillmer (eds.), *Military psychology: Clinical and operational applications* (193–214). New York: Guilford Press.

Chapter 21

The Total Force and the Army Reserve Component: Contributions of Research Psychology

James E. Griffith

This chapter provides a brief history of the Army Reserve force and the changed nature of Reserve military service, particularly since the 1980s, and how such changes have introduced several issues relevant to research psychology. It briefly summarizes these issues and provides sources to obtain additional information. The chapter concludes by describing current and likely future concerns related to Reserve military service, in addition to opportunities for research psychologists in the Army Reserve.

What Is the Army Reserve?

At present, the U.S. Army is organized into two entities: the Active force, which consists of full-time Soldiers; and the Reserve force, which consists of part-time Soldiers. The Reserve force is a complex organization of several subentities representing different levels of Reservists' involvement in the military. Figure 21–1 depicts the organization of the U.S. Army Reserve Component.

The focus of this article is on the "Ready Reserve," the subentity that currently represents the most combat-ready Reservists (the dotted ellipse in figure 21–1). The Ready Reserve consists of three memberships: the Selected Reserve, the Individual Ready Reserve (IRR), and the Inactive National Guard. About three-quarters of the Ready Reserve force falls in the Selected Reserve, one-quarter in the IRR, and a small proportion in the Inactive National Guard. The former two represent Reservists most likely to be called on to support major military operations. The Selected Reserve consists of the Army Reserve and Army National Guard. Army Reservists make up the IRR, and Army National Guard members make up the Inactive National Guard. The Army Reserve represents about a third of the support organization of the entire

Figure 21–1. **Organization of Army Reserves**

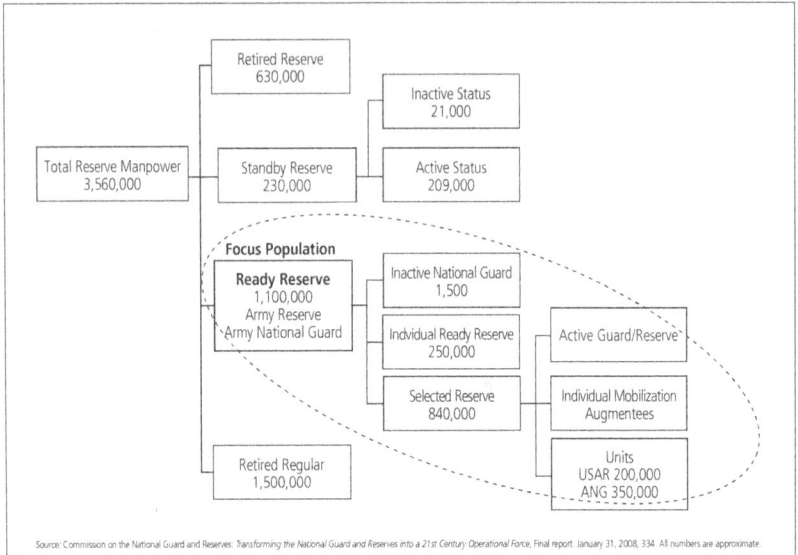

```
                        ┌──────────────────┐
                        │  Retired Reserve │
                        │      630,000     │
                        └──────────────────┘
                                          ┌──────────────────┐
                                          │  Inactive Status │
                                          │      21,000      │
                                          └──────────────────┘
┌──────────────────┐    ┌──────────────────┐
│Total Reserve     │    │  Standby Reserve │   ┌──────────────────┐
│Manpower          │    │     230,000      │   │  Active Status   │
│  3,560,000       │    └──────────────────┘   │     209,000      │
└──────────────────┘                           └──────────────────┘

       ┌─ Focus Population ────────────────────────────────────────────┐
       │  Ready Reserve        ┌──────────────────┐                     │
       │   1,100,000           │Inactive National │                    │
       │  Army Reserve         │Guard  1,500      │                    │
       │  Army National Guard  └──────────────────┘                    │
       │                       ┌──────────────────┐ ┌────────────────┐ │
       │                       │Individual Ready  │ │Active Guard/   │ │
       │                       │Reserve  250,000  │ │Reserve         │ │
       │                       └──────────────────┘ └────────────────┘ │
       │                       ┌──────────────────┐ ┌────────────────┐ │
       │                       │ Selected Reserve │ │Individual      │ │
       │                       │     840,000      │ │Mobilization    │ │
       │                       └──────────────────┘ │Augmentees      │ │
       │  ┌──────────────────┐                      └────────────────┘ │
       │  │ Retired Regular  │                      ┌────────────────┐ │
       │  │    1,500,000     │                      │Units           │ │
       │  └──────────────────┘                      │USAR 200,000    │ │
       └────────────────────────────────────────────│ANG 350,000 ────┘─┘
```

Source: Commission on the National Guard and Reserves. *Transforming the National Guard and Reserves into a 21st Century Operational Force*, Final report. January 31, 2008, 334. All numbers are approximate.

Army, and the Army National Guard represents about half of the Army's available combat force (Commission on the National Guard and Reserves, 2008). The Army Reserve now numbers about 200,000 drilling Soldiers, and its units primarily perform support functions to combat units. Reserve units mainly perform combat service support functions (for example, logistics, maintenance, transportation, and health services) and combat support functions (such as chemical warfare, intelligence, security, and communications). The Army National Guard now numbers about 350,000 drilling Soldiers and consists mainly of combat (infantry, armor, artillery, cavalry, air defense artillery) and combat support units.

Generally, Army Reserve Soldiers work part time, compared to Active-duty Soldiers who work full time. Army Reserve Soldiers usually perform training or duty one weekend a month and over two consecutive weeks (called annual training). These Reservists are often referred to as *drilling Reservists*, since they attend unit assemblies one weekend a month. Reservists generally serve in units that are within driving distance (50 to 100 miles), but some serve individually to augment either Reserve or Active Army units. Below is a description of the types of Reservists (for a more detailed discussion, see Commission on the National Guard and Reserves, 2008).

There are several general categories of part-time Reserve Soldiers representing progressively less involvement or participation in the Army (see table 21–1). The majority of members of most Reserve units serve one weekend a month plus 15 days annual training. These Soldiers are called *traditional, drilling, M-day* (mobilization asset), or *Troop Program Unit* (TPU) members. To manage administrative matters, many units are assigned a few full-time Reservists, called Active Guard/Reserve (AGR) and full-time, dual-status (M-day and Monday-through-Friday state or Federal employee) technicians. Some Reservists serve much like TPU members but with less fixed dates and times of monthly drills, and they usually support something other than a combat or combat support unit, such as a medical facility. They are called *Drilling Individual Mobilization Augmentees.* Reserve Soldiers who have an established relationship with or position in a unit but are not in a drilling status (one weekend a month) are called *Individual Mobilization Augmentees* (IMAs). These Soldiers are attached to specific units but usually drill for pay only during the 15-day annual training period. IRR Soldiers are not associated with any unit and do not perform monthly drills. These Soldiers can apply for annual training days to complete training required for their military occupational specialties.

There are three types of full-time Reserve Soldiers. First, the Active Guard/Reserve Soldier serves full time on Active-duty status in the Reserve Component in support of a Reserve unit. For example, a unit personnel specialist often serves full time to manage administrative affairs of the unit. Second, civilian positions are authorized in Reserve units that require the employee to be a member of the Army Reserve Component. These positions are called *technicians* and can be funded by Federal or state funds (in the case of the Guard). During the week, the employee serves as a civilian but usually wears the uniform, and during the drill weekend, the employee has similar status to the M-day Soldier. Third, some Reserve Soldiers are brought on Active duty for a short time (for example, 6 months) to perform a specific mission. These Soldiers are called *Active Duty for Special Work.*

All Soldiers in the U.S. Army incur an 8-year obligation that can be served in the capacities previously described. When joining, the initial contract specifies the amount of time to be served on Active duty (2, 3, or 4 years) and the remaining years are spent in Reserve military service. Soldiers who join only the Reserve force and enter the Army

Table 21–1. **Types of Reserve Military Service**

Type	Frequency in the Military Environment	Description	Location of Military Service
Part-time			
Troop Program Unit (TPU)	One weekend per month, 15 days annual training	Serves monthly in the Reserve military (one weekend a month and 15 days annual training). Also called *traditional Reservist* or *M-day* (mobilization asset). Comprises the greater part of members in Reserve units.	Unit
Drilling Individual Mobilization Augmentee (DIMA)	Certain number of drill days, 15 days annual training	Serves periodically, much like the TPU member, but with fewer fixed dates and times of monthly drills and usually in support of something other than a combat or combat support unit (for example, a medical facility).	Usually installation
Individual Mobilization Augmentee	15 days annual training	Serves periodically, much like a DIMA but less monthly frequency, usually having no or few drill days but having the 15-day annual training period. Often attached to specific units.	Usually installation
Individual Ready Reserve	15 days annual training	Serves infrequently, not associated with any unit, and does not perform monthly drills. Can apply for annual training days to complete training required for military occupational specialty.	No unit
Full-time			
Active Guard/ Reserve	Every workday	Serves full time on Active-duty status in the Reserve component in support of a Reserve unit (for example, unit personnel specialist often serves full time to manage administrative affairs of the unit).	Unit
Technician	Every workday	Serves full time as civilian "technician" in authorized positions in Reserve units, funded by Federal or state funds. Requires employee to be a member of the Army Reserve component. Usually wears uniform during the workweek, and during the drill weekend has similar status as the M-day soldier.	Unit
Active Duty for Special Work	Every workday for specified time period	Serves full time for short duration. Brought on Active duty to perform a specific mission	Unit

Reserve Component directly are called *non-prior service Reservists*. They are required to serve on Active or full-time military duty only during their introductory training (that is, basic training and advanced military occupational specialty training). Other Reservists serve an Active-duty term of 3 or 4 years and then enter Reserve military service, at which point they are considered prior service Reservists.

Historical Role of the Army Reserve

Both policy decisions and geopolitical events have determined what role the Reserve force has served. What follows is a cursory history of the Army Reserve and Army National Guard (taken from the Commission on the National Guard and Reserve, 2008), including what I see as their functional roles (see table 21–2). These roles are associated with different demands on individual Reservists, often giving rise to issues pertinent to research psychology.

Community Protection

The Army National Guard is the oldest military organization in the United States, dating back to the colonial militias. In 1636, militia companies were formally organized to protect the Jamestown colonists from the growing threat of the Pequot Indians. Lacking a formal military cadre, the colonies adopted the English militia system, which obligated all males between the ages of 16 and 60 to possess arms and participate in the defense of the community. These militias formed the basis for the modern-day Army National Guard, formally established by the Militia or Dick Act of 1903. This law aligned the militias with the Army Reserve and made the organization, pay, regulations, and equipment similar to those of the Active Army. The number of training days increased, and training was performed with the Active Army. The National Defense Act of 1933 made the National Guard part of the U.S. Army at all times. Prior to this time, the National Guard was a state's militia (Doubler, 2003). The act also established two Army Reserve entities: the Army National Guard and the U.S. Army Reserve.

Medical Support

The beginning of the Army Reserve can be traced to the need for medical care of Soldiers. In 1908, Congress passed a law authorizing a cadre of medical officers who would provide medical care to Soldiers

Table 21–2. **Army Reserve Component throughout History**

Time Period	Role	Precipitating Event
Early colonial times	Community protection	Need to protect colonists from growing threat of Pequot Indians
Civil War through start of 20th century	Medical support	Need for more formal medical care and support during early American wars
War of 1812 through Civil War, Spanish-American War, World Wars, Korean War into the Cold War	Strategic reserve of citizen-soldiers	Need to complement small Active-duty Army with great numbers of Soldiers to conduct major war effort to protect U.S. interests
1970 to 1990	Interdependency of Active and Reserve forces	—Need to reduce active military force and expense by optimizing Active-Reserve cooperation —Need to adopt policy that would integrate the Active and Reserve military organizations into a single force
1990s	Search for relevance	Need for new role in national security following the collapse of Soviet Union, strategic reserve questioned
2000 to present	Shift from strategic to operational force	Need to rethink use of Reserve force in the wake of required mobilizations and deployments of Army Reserve units for major wartime operations

and who could be ordered to Active Duty by the Secretary of War during emergencies (Currie and Crossland, 1977). Later, the National Defense Act of 1916 established the Officers Reserve Corps, the Enlisted Reserve Corps, and the Reserve Officers Training Corps.

Strategic Reserve of Citizen-Soldiers

Throughout U.S. history, meeting national security threats necessitated major military operations, including the War of 1812, the Civil War, the Spanish-American War, World Wars I and II, and the Korean War. Because of a lack of sufficient Active-duty personnel, Reservists were called up and youths were drafted to support these large-scale military operations. When operations ended, Reservists and draftees more often than not returned to society as civilians. This pattern changed after World War II. The United States had gained hegemony in world politics but faced perceived and real national security threats from both the Soviet Union and the People's Republic of China, thus requiring a large force to protect its interests (Knutsen, 1999). However, maintaining a large army was expensive. A small standing army, which would be complemented by Reservists and could serve as a Reserve force when threats escalated, was seen as less costly (Drew, 1994). Thus, the Reserve force became largely a strategic force, to be used for extraordinary circumstances.

Interdependency of Active and Reserve Forces

In 1970, Secretary of Defense Melvin Laird announced a new policy for the U.S. Armed Forces, called the Total Force concept (also known as the Abrams Doctrine). What prompted the policy change were the rising costs of maintaining a large military and growing discontent with the draft. Integrating the Active and Reserve force reduced the size of the U.S. military, especially the Active force. Efficiencies gained in the integration would mean fewer Soldiers, likely no draft, and preliminary steps toward the all-volunteer force. The policy called on the Army Reserve to assume many of the combat support and combat service support roles of Active force units and for the Army National Guard to augment several combat divisions. For any large-scale military operation, both the Army Reserve and Army National Guard would now have to be activated. As observed by Segal and Tiggle (1997), the Total Force "made the reserve components full partners with active forces in a structure that guaranteed that the active forces would not deploy without reserves." With the threat of thermonuclear warfare, the Reserve force continued to serve as a strategic deterrent from about 1970 through 1980 (Janowitz and Moskos, 1979). Thus, the Total Force would not be tested unless a major military operation was initiated, which was unlikely given that it might provoke a nuclear exchange.

Search for Relevance

The Soviet Union and many communist satellite nations collapsed in the 1990s. U.S. policymakers and the public were looking for peace dividends, which could easily be found in a large Reserve force no longer having to deter Soviet threats. During this time, the Army Reserve and Army National Guard leadership sought mission relevance, including becoming more like Active light infantry Soldiers (Osterberg, 1990); assuming new missions, such as overseas peacekeeping (Segal and

Tiggle, 1997; Segal, Reed, and Rohall, 1998); providing domestic support, such as medical assistance to the underprivileged; and conducting nationbuilding missions, such as several state National Guards developing relationships with former Soviet bloc countries. This search for relevance ended with the first Gulf War, during which the Total Force concept necessitated calling on Reserve forces.

Shift from Strategic to Operational Force

The reconfigured Total Force required the mobilization and deployment of Army Reserve units for major military operations. During Operation *Desert Storm* (ODS), more than 84,000 Army Reserve and 60,000 Army National Guard Soldiers were called to Active duty (Nelson et al., 2001). Later, operations in Iraq and Afghanistan would require that even more Reserve forces be called up and deployed overseas for periods of 1 to 1½ years. This marked the beginning of the shift from a strategic Reserve to an operational Reserve (Commission on the National Guard and Reserves, 2008; Wormuth, 2006), which introduced a new set of policy issues appropriate for research psychology study and evaluation.

Admittedly, this description of the changed roles of the Army Reserve Component is brief and simple. Even so, connections among the varied nature of threats to U.S. national security, changes in the Army Reserve Component, and its overall role in national defense policy are evident.

Reserve Policy Issues Pertinent to Research Psychology

Army Reserve issues having relevance to research psychology can be described in terms of three "Rs": retention, recruitment, and readiness. Each is described below.

Unscheduled Losses and Retention

An issue of historic interest (since the Vietnam War and during the all-volunteer Army in the early 1980s) has been retention, in particular, retaining non–prior service, first-term, junior-ranking enlistees (Perry, Griffith, and White, 1995). Typically, about one-third of these Soldiers leave either before the end of their current obligation or at the end of their 6-year obligation (White and Simpson, 1990). Soldiers who leave before the end of their obligation usually just stop attending drills. These losses are of great concern. The military invests about $80,000 in the recruitment and training of each new Soldier, so these unanticipated losses represent a significant fiscal drain in addition to manpower losses. This situation made retention a key research topic for the 1980s into the 1990s.

Historically, Reservists have reported leaving Reserve service due to poor training and leadership and the lack of engaging and meaningful work activities. This was especially evident among junior enlisted Soldiers. Troops often complained of doing little or nothing during weekend drills, though they desired to do something both meaningful and consistent with their military occupational specialty. Family affairs and civilian job conflicts have been of less importance to leaving Reserve service. However, with more frequent and prolonged deployments since 2005, Reservists have expressed less willingness to remain in Reserve service due to frequent and lengthy separations from families. For a recent summary of findings relating to retention and associated factors, see Griffith (2005). Losses before or at the end of Soldiers' obligations were, however, typically balanced by recruitment of more Soldiers. It seemed that units and commanders preferred to get new Soldiers than motivate those who had already joined but became disenchanted and stopped attending drills. Thus, a second research interest has been recruitment.

Recruitment to Replace Losses

Recruitment research has attempted to identify subpopulations having the greatest propensity to join the Army Reserve Component along with the incentives that would interest members of such subpopulations to join. As a result, many of the recruiting efforts in the 1980s and 1990s were directed at developing incentives for particular youth markets (Shavelson, Haggstrom, and Winkler, 1983). While other Services defined markets and developed successful recruitment methods and advertising (Barnes, Dempsey, Knapp, Lerro, and Schroyer, 1991; U.S. Army Research Institute, 1976), the Army, and in particular the Army Reserve, struggled.

Incentives to join the military, such as enlistment bonuses and educational scholarships, were offered. In particular, the rising cost of tuition and increased emphasis on college attendance in the 1970s and 1980s led the Army to offer young adults financial assistance for education. After some initial recruiting success, college-bound youth became the primary market for the Army Reserve Component (Hintze and Lehnus, 1996). The Army broadened efforts to develop educational and other monetary benefits as incentives to attract and recruit college-bound youth (Schmitz, 1990; Verdugo and Berliant, 1989). This market had the added advantage of providing higher aptitude Soldiers who represented the broad spectrum of the general population.

Offering material incentives, such as educational benefits, bonuses, and increased pay, has worked well for Reservists serving part-time (attending monthly weekend drill training and 2-week annual training). However, since dramatic changes in force organization, structure, and missions occurred under the Total Army concept in 2004, fewer college-bound high school students have joined the military. Nevertheless, offering bonuses in general and targeting certain military occupational specialties in particular continues to be an effective recruitment strategy, especially when bonus amounts are increased substantially. A sense of patriotism and the urge to serve the country have also been typical reasons Soldiers give for joining, although the Army's emphasis on such incentives has been less evident than the development of educational benefits and bonus money. For a report on the effects of ODS on enlistment motivations of Reservists, see Griffith and Perry (1993); and for a more recent analysis of motives, see Griffith (2008a).

Present, but Not Ready

ODS introduced an entirely new research issue: readiness. During ODS, more than 84,000 Army Reserve and 60,000 Army National Guard Soldiers were called to Active duty (Nelson et al., 2001). By some accounts (Kirby and Buddin, 1996; Orvis et al., 1996), many Army Reserve units during ODS could not be mobilized in a timely manner, due largely to insufficient skill-qualified Soldiers and training on mission-required equipment. These and other shortcomings—including insufficient individual job skills and combat preparation, utilization during weekend drills, noncommissioned officer and officer leadership, and collective training—were subsequently documented (General Accounting Office, 1992; Griffith, 1995). What ensued was a series of studies to examine the extent to which Reserve Soldiers were ready individually (medically, trained in a military occupational specialty, and so forth) and as units (based on ratings from Unit Status Reports). Findings from these studies spurred several initiatives to improve the readiness of Army Reserve Soldiers, including the Reserve Component Training Development Action Plan, aimed at improving individual Soldier skills and leader effectiveness, and the Leader Development Action Plan (U.S. Department of the Army, 1997), to better select, educate, and assess leaders. Other programs include the "Bold Shift" initiative (Sortor, Lippiatt, Polich, and Crowley, 1994), prescribing policies and procedures for combat maneuvers during peacetime, and Title XI (U.S. House of Representatives, 1993), to improve individual and unit readiness for mobilization and deployment. These initiatives, while better preparing Soldiers for future mobilizations and deployments, may well have had unintended negative effects on reenlistments by requiring Reservists to spend more time intermittently in their units, thus competing with civilian employment and family obligations (Sortor et al., 1994). Earlier studies showed the detrimental effects of extended training on retention rates (Kirby and Buddin, 1996). Detrimental effects of extended training and related negative attitudes of families and employers have been documented in analyses of the 1986 and 1990 Reserve Components Surveys (Grissmer, Buddin, and Kirby, 1989; Grissmer, Kirby, and Sze, 1992). Indeed, after ODS, proportionately fewer Soldiers remained in the Reserves (Griffith, 1995; Kirby, 1997).

Following September 11, 2001, about 300,000 of the 1.2 million National Guard and Reserve personnel were called to Active duty (General Accounting Office, 2003). Reservists combined with Active-duty members supported worldwide military operations, specifically in Afghanistan and Iraq. With the passing of several years since the last major callups, the Reserve Component was expected to have remedied many of the problems observed in ODS. Some studies showed improvements in mobilization of Reservists, whereas others pointed to persistent problems in Soldiers being trained and available for mobilizations (General Accounting Office, 2003; St. Laurent, 2004). Survey findings indicated that Soldiers' perceptions of individual and unit readiness were more positive and that Soldiers joined and stayed in Reserve service if properly trained and equipped, especially for their wartime deployments. Thus, readiness was tied to reasons for joining and reasons to stay. For a recent summary of Soldier reports of individual and unit readiness, see Griffith (2005).

The Three "Rs" Intertwined

Increasingly, recruitment, retention, and readiness have been recognized as being intertwined. Enough quality Soldiers had to be recruited to fill current positions and subsequently retained. If the Reserve

Component did not fill its strength allocations each year, Congress reexamined and often cut strength levels and units. To meet manpower requirements, those Soldiers who joined had to be retained, both during their obligations and after completing their obligations, through reenlistments. Research findings indicated that retained Soldiers have positive unit experiences, especially those indicating greater individual and unit readiness. Specifically, Soldiers who were more likely to remain in Reserve service were satisfied with their military duties, trained in a military occupational specialty, had challenging training, and had leaders who knew their jobs and showed interest in their Soldiers (Griffith, 2005; Perry, Griffith, and White, 1991). In other words, Soldiers who were "more ready" were more likely to want to stay in Reserve military service. For a recent summary of Reserve Soldiers' perceptions about readiness, see Griffith (2006). The focus of my own work has evolved from the issues of retention to recruitment, and finally to readiness. My recent work examines Reserve Soldier commitment in relation to both recruitment and reenlistment incentives (for example, monetary reward, meaningful training and work, positive unit relationships, and sense of obligation to others and the country), as well as perceptions about being combat-ready individually and as a unit (Griffith, 2009).

Current and Future Policy Issues and the Research Psychologist

What are the current and future issues concerning the Reserve force, and how might the research psychologist contribute? Below, I have identified some of these issues.

Recruiting and Retaining through Intangibles

The changed and uncertain nature of Reserve military service has had a negative effect on recruitment. The Army's response has been largely to increase the amount of monetary bonuses. Since 2000, the Army has more than tripled spending on enlistment bonuses, from $202 million to $639 million. The average bonus jumped from $7,900 to $16,500 (Associated Press, 2008). Such monetary bonuses are often double-edged. My recent work suggests this strategy has implications for retention and readiness. Soldiers who joined for material gain (educational benefits or money) are less likely to intend to remain in Reserve military service, perceive themselves as combat-ready, and report for duty out of a sense of obligation to others, belief in mission, and service to country (Griffith, 2008a). Less attention has been paid to what Eighmey (2006) has called "intangibles." These include faithfulness to goals and to the people who share them, willingness to make sacrifices, and desire to be near the family and have their approval. Such intangibles and their role in recruitment, retention, and especially readiness deserve more attention. Bondy (2004) has most recently called for a shift in strategy, advocating that Western militaries return to their past strengths and emphasize re-adoption of social forces associated with military effectiveness. He proposed that military ceremonies, symbols, explicit values, and institutional narratives be resurrected as part of unit experiences to reestablish Soldiers' trust of and commitment to fellow Soldiers, leaders, the unit, and the military organization. The extensive literature on social identity and groups (Ashforth and Mael, 1989) suggests several directions. Team experiences might be marketed as benefits of military service. Media messages might depict new recruits who gain positive self-identity through team experiences, such as cooperation and teamwork in accomplishing tasks and committed friendships with fellow Soldiers in their units. Through these and other group experiences, Soldiers may develop loyalty and a sense of obligation to others, mutual trust, and self-sacrifice. In other words, Soldiers experience a sense of "one for all" and "all for one."

Increasing Readiness through Soldier Identification

The changed nature of Reserve military service poses challenges that have not been fully addressed (Wallace, 1992). Because of the initial difficulties in deployment of Reserve forces, some have called for a reexamination of the viability of Reservists in a national security framework (Wormuth, 2006). Still others have argued for the importance of having citizens serving as Soldiers for positive civilian-military relationships (Cohen, 2001; Janowitz, 1975, 1979). The essential issue is whether citizen-Soldiers who participate intermittently in the military can be effective modern-day Soldiers. What can be done to better integrate Reservists as they move from being "partially included" to "fully included" without the benefit of increased time in the unit? From a research psychology perspective, it seems connections to Reserve military service will have to rely on social-psychological mechanisms. A promising area of research here includes social identity theory and perceptions of being a Soldier, the role of ideology in identifying with military service, and the relationship of recruitment and retention incentives and unit experiences to the Reservist's organizational commitment. Recent studies of ideologies in the military have related such attitudes to important military outcomes, such as morale (Goldberg and Kasirer, 1985), perceived legitimacy of assigned missions (Wyatt and Gal, 1990), unwillingness to serve as conscientious objectors (Linn, 1995), decisions to report for duty when called (Linn, 1995), and stress-buffering in combat (Shechner, Slone, and Bialik, 2007).

Commitment to Mission

In 1990 (less than a year before ODS), 80 percent of Army Reserve privates through corporals said it was "unlikely" or "uncertain" that they would be mobilized during their current obligations, normally 6 to 8 years in length. After ODS (a year later), the percentage decreased to 72 percent (Westat, Inc., 1990). After experiences in Iraq and Afghanistan, it might be expected that more Soldiers would predict being mobilized and deployed in the future. Such expectations would also likely bear on Reservists' commitment, especially to new, emerging, and often demanding missions. Part of the answer to bolstering commitment is perception of the importance and legitimacy of missions. However, complicating this process is that Reservists spend most of their time in nonmilitary or civilian contexts. For Reservists, the dominant context is the civilian environment, unlike the full-time Active-duty Soldiers whose dominant context is the military environment. Reserve Soldiers may broadly see themselves as Soldiers, but when not drilling one weekend a month or on annual training, they likely view themselves according to their civilian roles, such as students, workers, husbands or wives, or fathers or mothers. When the saliency of military service is low in everyday life, the identity as a Soldier is low. That is, Soldiers may not think about their unit and its mission, and focus instead on some enduring personal attributes associated with the military, such as helping others and national pride. Immersed in a civilian environment as opposed to a military environment makes Reservists more attuned to and influenced by public opinion (Goldberg and Cassirer, 1985). Studies of Reservists in the Israel Defense Forces have shown the dynamic interplay among reserve soldiers' political orientation, perceived legitimacy of assigned mission, and the degree of distress in its execution (Shechner et al., 2007). Commitment to missions relates to perceived legitimacy and ability to carry out missions (Linn, 1995a; Linn, 1995b; Wyatt and Gal, 1990), and as such, this issue deserves attention in future research, especially considering that Reservists are deployed to missions that may be controversial.

Post-traumatic Stress

An issue of growing concern for deployed Soldiers is the negative effects of war trauma on psychological well-being. The few studies including Reservists have shown them to have consistently higher rates of post-traumatic stress disorder (PTSD) and related symptoms than Active-duty members. These studies include Hourani et al. (2007), Martin (2007), Milliken et al. (2007), Jacobson et al. (2008), Smith et al. (2008), and Schell and Marshall (2008). There are various explanations for these findings (Griffith, 2008b). Few at-risk factors or background characteristics specific to Reservists have been associated with the occurrence of PTSD, such as marital status or age. However, the change from being part-time to full-time Soldiers likely introduces several stressors. Not expecting to be deployed (Griffith and Perry, 1993), not feeling adequately prepared and equipped (Erbes et al., 2008; Griffith, 1995), and being unaccustomed to prolonged family separations (Erbes et al., 2008) likely make deployments more difficult for Reservists than for Active-duty members. Additionally, Reservists have to negotiate their absences with civilian employers.

Evidence is mixed whether Reservists have more combat exposure than Active-duty Soldiers (Vogt et al., 2008; cf. Milliken et al., 2007). However, some studies suggest Reserve members are more adversely affected by it (Hourani et al., 2008; Jacobson et al., 2008; Vogt et al., 2007). Due to the intermittent nature of Reserve military service, Reservists may lack natural supports, such as the material and emotional support provided by unit cohesion that often mitigates the negative effects of stress on well-being (Brailey et al., 2007). Few studies have examined the stress-buffering effects of cohesion on PTSD symptoms among deployed and nondeployed samples of Reservists and Active-duty members. Finally, current demobilization practices may contribute to Reservists' overuse of mental health care, and accordingly, overidentification of PTSD symptoms.

Future studies must be conducted to determine which explanations have the most merit and to provide data to respond to important questions: What are the adverse psychological consequences of deployments on Reservists and their families? What are the common stressors throughout the deployment cycle? Given these stressors, what can be done to help Reservists and family members to recognize issues and develop initial coping strategies? Are there characteristics of individual Reservists and their families that increase the risk of traumatic stress after deployment? What are precipitating and/or contributing factors to post-deployment adaptation and post-traumatic stress in families of Reservists? What factors inhibit Reservists and their families in seeking mental health care, and how might they be reduced?

Opportunities in the Army Reserve as a Research Psychologist

Research psychologists in the Reserve component need not necessarily perform research relating to Reserve issues. However, as elaborated above, the marked changes in the organization, mission, and uses of Reserve forces have made several issues prominent and suitable for research. Indeed, before ODS, much of the research done by Reserve research psychologists related to issues of Active-duty Soldiers (Griffith, 1997, 2002; Griffith and Vaitkus, 1997). But as the nature of Reserve military service has changed, the Reserve research psychologist has become uniquely positioned to study emerging Reserve force questions of national security importance. For example, the research psychologist understands the distinctions and subtle differences among the various Reserve statuses. He also understands the competing demands of a civilian job and family. The research psychologist can draw on theories and concepts particularly appropriate for analyzing Reserve issues and recommending new strategies for improving recruitment, retention, and readiness.

Unfortunately, there are few opportunities to serve as a Reserve component research psychologist. As of this writing, there are no TPU positions for research psychologists in which the Reservist attends a weekend drill once a month with 15 days of annual training. Although not directly related to the specialty, research psychologists are qualified for TPU positions in Public Affairs and Psychological Operations units. While the primary mission of these units is to implement media campaigns, they often lack the methodological skills to evaluate campaign effectiveness, and thus research psychologists can assist them.

Alternatively, other opportunities do arise, as was the case for me. I have been a uniformed Army research psychologist since 1984, and after having served on Active duty for 3 years, I took a civilian research consulting position but continued my Army service in the Army Reserve. Because my interest was serving in an active Reserve status or as a TPU member, I explored alternatives. I was able to acquire a drilling or TPU position in the Maryland Army National Guard, slotted against a "branch immaterial" position. I was initially assigned to the Recruitment and Retention Branch to assist in efforts to retain Soldiers. I have served in the Maryland National Guard for 22 years and held such positions as the State Headquarters Detachment Executive Officer and subsequently the Commander. At present, most of my duties pertain to consulting on the design and data collection of evolving Reserve issues relating to psychology.

Other forms of Reserve service are somewhat limited. Several years ago, there were a few Individual Mobilization Augmentee positions supporting the Walter Reed Army Institute of Research and the Army Research Institute. Such positions entail serving a 2-week period at the unit or at the facility the Reservist is supporting. As of this writing, there is only one IMA position in the Army Reserve, at Fort Rucker, Alabama, and two psychologists are slotted against this position. Another possibility for the Reserve research psychologist is to be a member of the IRR. In that capacity, the research psychologist can enroll and complete Army correspondence courses to obtain retirement points and is also eligible to attend a 2-week training period at an Army school or to support a unit. The latter requires funding, if available, and coordination with the receiving unit.

Conclusion

This chapter provided a broad overview of the history and use of the Army Reserve Component. Changes in the structure and organization of the Reserve force in combination with geopolitical events have altered the role of the Reserve component, and in turn affected the nature of Reserve military service. Many changes have increased the demands on Reservists, in particular having to be deployed 1 to 1½ years every couple of years. These demands have adversely affected recruitment and retention. However, persistent problems in readiness have notably improved, and it seems that remedying problems in readiness (such as having better training and providing more meaningful duties) improves retention since Soldiers looked for such experiences when enlisting and see these experiences as better preparing them for deployment. They also likely relate to Reservists' ability to cope with deployment and combat stressors, and therefore the probability and severity of PTSD and related symptoms. It is important to assess the three "Rs"—recruitment, retention, and readiness—as the role of the Army Reserve Component evolves. In many ways, the research psychologist is well suited to assist in these assessments.

References

Ashforth, B.E., and Mael, F. (1989). Social identity and the organization. *Academy of Management Review* 14 (1), 20–39.

Associated Press (2008). Military recruiting bonuses grow by 25 percent. *International Herald Tribune*, October 2, 2008.

Barnes, J., Dempsey, J., Knapp, D., Lerro, P., and Schroyer, C. (1991). *Summary of military manpower market research studies: A technical report* (Report no. FR–PRD–91–08). Alexandria, VA: Human Resources Research Organization.

Bondy, H. (2004). Postmodernism and the source of military strength in the Anglo West. *Armed Forces and Society* 31 (1), 31–61.

Brailey, K., Vasterling, J.J., Proctor, S.P., Constans, J.I., and Friedman, M.J. (2007). PTSD symptoms, life events, and unit cohesion in U.S. soldiers: baseline findings from the neurocognition deployment health study. *Journal of Traumatic Stress* 20 (4), 495–503.

Cohen, E.A. (2001). Twilight of the citizen-soldier. *Parameters* 31 (2), 23–29.

Commission on the National Guard and Reserves (January 2008). *Transforming the National Guard and Reserves into a 21st century operational force*. Final report. Washington, DC: U.S. Congress.

Currie, J.T., and Crossland, R.B. (1997). *Twice the citizen: A history of the United States Army Reserve, 1908–1995* (2d ed.). Washington, DC: Office of the Chief, Army Reserve. See <www.militaryheritage.org/NationalGuardHistory.html>.

Doubler, M.D. (2003). *Civilian in peace, soldier in war, I am the Guard: A history of the Army National Guard, 1636–2000*. Lawrence, KS: University of Kansas Press.

Drew, N.S. (1994). *NSC–68: Forging the strategy of containment* (NSC–20/4, Sec. 21 (a), 31). Washington, DC: National Defense University Press.

Eighmey, J. (2006). Why do youth enlist? *Armed Forces and Society* 32 (2), 307–328.

Erbes, C.R., Arbisi, P.A., Courage, C., Polusny, M.A., Thuras, P., and Rath, M. (2008). *Contextual predictors of post-deployment symptoms in the RINGS study*. Presented at the annual American Psychological Association meeting, August 14, 2008.

General Accounting Office (1992). *Operation Desert Storm: Army Guard combat brigade war lessons reflect longstanding problems*. Testimony, GAO/T–NSIAD–92–36. Washington, DC: General Accounting Office.

Goldberg, G., and Kasirer, M. (1985). Voting patterns among Israel Defence Force soldiers who alternate between citizens and reserve soldiers called up for extended military duty. *Jewish Social Studies* 47 (1), 77–88.

Griffith, J. (2009). After 9/11, What kind of reserve soldier? Considerations given to emerging demands, organizational orientation, and individual commitment. *Armed Forces and Society* 35, 214–240.

———. (2008a). Institutional motives for serving in the U.S. Army National Guard: Implications for recruitment, retention, and readiness. *Armed Forces and Society* 34 (2), 230–258.

———. (2008b). *Citizens coping as soldiers: Prevalence of post-traumatic stress disorder symptoms among deployed reservists*. Paper prepared for publication, available from author.

———. (2006). What do the soldiers say? Needed ingredients for determining unit readiness. *Armed Forces and Society* 32 (3), 367–388.

————. (2005). Will citizens be soldiers? Examining retention of reserve component soldiers. *Armed Forces and Society* 31 (3), 353–383.

————. (2002). Multi-level effects of military cohesion on soldier well-being, army identification, group disintegration, and combat readiness. *Military Psychology* 14 (3), 217–240.

————. (1997). A test of a model incorporating stress, strain, and disintegration in the cohesion performance relation. *Journal of Applied Social Psychology* 27 (17), 1489–1526.

————. (1995). The Army Reserve soldier in Operation *Desert Storm*: Perceptions of being prepared for mobilization, deployment and combat readiness. *Armed Forces and Society* 21 (2), 195–215.

Griffith, J., and Perry, S. (1993). Wanting to soldier: Enlistment motivations of Army Reserve recruits before and after Operation *Desert Storm. Military Psychology* 5 (2), 127–139.

Griffith, J., and Vaitkus, M. (1999). Relating cohesion to stress, strain, disintegration, and performance: An organizing framework. *Military Psychology* 11 (1), 27–55.

Hintze, W., and Lehnus, J. (November 1996). *Recognition of military advertising slogans among American youth.* Paper presented at the 38th Annual Conference of the International Testing Association, San Antonio, TX.

Hourani, L.L. et al. (2007). *2006 Department of Defense survey of health related behaviors among Guard and Reserve force: A component of the Defense Lifestyle Assessment Program (DLAP).* Research Triangle Park, NC: RTI International.

Kirby, S.N., and Buddin, R. (1996). *Enlisted personnel trends in the selected reserve, 1986–1994* (Report No. MR–681/1). Santa Monica, CA: RAND Corporation.

Jacobson, I.G., Ryan, M.A., Hooper, T.I., Smith, T.C., Amoroso, P.J., Boyko, E.J., Gackstetter, G.D., Wells, T.S., and Bell, N.S. (2008). Alcohol use and alcohol-related problems before and after military combat exposure. *Journal of the American Medical Association* 300 (6), 663–675.

Janowitz, M. (1975). The all-volunteer military as a "sociopolitical" problem. *Social Problems* 22, 432–449.

————. (1979). The citizen soldier and national service. *Air University Review* 30 (1), November–December.

Janowitz, M., and Moskos, C.C. (1979). Five years of the all-volunteer force: 1973–1978. *Armed Forces and Society* 5 (2), 171–218.

Knutsen, T.L. (1999). *The rise and fall of world orders: The rise and demise of hegemonic powers in international politics.* Manchester, UK: Manchester University Press.

Linn, R.R. (1995a). The claim for moral maturity, consistency, and integrity among objecting Israeli Soldiers. *Journal of Applied Social Psychology* 25 (5), 399–417.

————. (1995b). Resistance and motivation—Moral, political, or personal? Israeli soldiers as selective conscientious objectors during the intifada. *Social Behavior and Personality* 23 (1), 35–44.

Martin, C.B. (2007). Routine screening and referrals for PTSD after returning from Operation *Iraqi Freedom* in 2005, U.S. Armed Forces. *MSMR: Medical Surveillance Monthly Report* 14 (6), 2–7.

Milliken, C.S., Auchterlonie, J.L., and Hoge, C.W. (2007). Longitudinal assessment of mental heath problems among active and reserve component soldiers returning from the Iraq War. *Journal of the American Medical Association* 298 (18), 2141–2148.

Nelson, H.W., Army, H., Jacobs, B., and Bluhm, R.K. (eds.). (2001). *The Army.* Westport, CT: Levin, Hugh, and Lauter Associates.

Osterberg, D.A. (1990). *Reserve component round out of light infantry divisions* (Report ADA226010). Fort Leavenworth, KS: Army Command and General Staff College, School of Advanced Military Studies.

Orvis, B.R., Shukiar, H.J., McDonald, L.L., Mattock, M.G., Kilburn, M.R., and Shanley, M.G. (1996). *Ensuring personnel readiness in the Army Reserve Components* (Report No. MR–659–A). Santa Monica, CA: RAND Corporation.

Perry, S., Griffith, J., and White, T. (1991). Retention of junior-ranking enlisted in the all-volunteer U.S. Army Reserve. *Armed Forces and Society* 18 (1), 111–133.

Schell, T.L., and Marshall, G.N. (2008). Survey of individuals previously deployed to OEF/OIF. In T. Tanielian and L.H. Jaycox (eds.). (2008). *Invisible wounds of war: Psychological and cognitive injuries, their consequences, and services to assist recovery* (87–161). Santa Monica, CA: RAND Corporation.

Schmitz, E.J. (1990). *The Army College Fund and military manpower: A review of existing research* (Technical Report 1572). Alexandria, VA: U.S. Army Research Institute for the Behavioral and Social Sciences.

Segal, D.R., and Tiggle, R.B. (1997). Attitudes of citizen-soldiers toward military missions in the post–Cold War. *Armed Forces and Society* 23 (3), 373–390.

Segal, D.R., Reed, B.J., and Rohall, D.E. (1998). Constabulary attitudes of National Guard and regular soldiers in the U.S. Army. *Armed Forces and Society* 24 (4), 535–548.

Shavelson, R.J., Haggstrom, G.W., and Winkler, J.D. (1983). *Potential for military recruiting from two-year colleges and postsecondary vocational schools* (Report no. N–1946–MRAL). Santa Monica, CA: RAND Corporation.

Shechner, T., Slone, M., and Bialik, G. (2007). Does political ideology moderate stress? The special case of soldiers conducting forced evacuation. *American Journal of Orthopsychiatry* 77 (2), 189–198.

Smith, T.C., Ryan, M.A.K., Wingard, D.L., Sallis, J.F., and Kritz-Silverstein, D. (2008). New onset and persistent symptoms of posttraumatic stress disorder self-reported after deployment and combat exposure: Prospective population-based U.S. military cohort study. *British Medical Journal* 336 (7640), 366–371.

Sortor, R.E., Lippiatt, T.F., Polich, J.M., and Crowley, J.C. (1994). *Training readiness in the Army Reserve components* (Report no. MR–474–A). Santa Monica, CA: RAND Corporation.

U.S. Army Research Institute for the Behavioral and Social Sciences (1976). *Research themes and technological base program in behavioral and social sciences for the U.S. Army* (Technical report ADA100830). Alexandria, VA: U.S. Army Research Institute for the Behavioral and Social Sciences.

U.S. Department of the Army (1997). *Institutional leader training and education* (TRADOC Regulation 351–10). Fort Monroe, VA: Headquarters, United States Army Training and Doctrine Command.

U.S. House of Representatives (1993). *Title XI: The Army National Guard combat readiness reform act in the national defense authorization act of fiscal year 1993* (House Report 102–966). Washington, DC: Government Printing Office.

Verdugo, N., and Berliant, K.R. (1989). *Estimating the Army's prime recruiting market* (Technical Report ARI TR 832). Alexandria, VA: U.S. Army Research Institute for the Behavioral and Social Sciences.

Walker, W.E. (1992). Comparing Army Reserve forces: A tale of multiple ironies, conflicting realities, and more certain prospects. *Armed Forces and Society* 18 (3), 303–323.

Westat, Inc. (1990). *1990 survey of United States Army Reserve (USAR) troop program unit (TPU) soldiers: Final report.* Washington, DC: Office of the Chief, Army Reserve.

White, T., and Simpson, J. (February, 1990). Why soldiers really leave the Army Reserve. *Military Review*, 76–77.

Wormuth, C.E. (2006). The future of the National Guard and Reserves, testimony before the Commission on the National Guard and Reserves, June 12, 2006.

Wyatt, T.C., and Gal, R. (1990). *Legitimacy and commitment in the military.* Westport, CT: Greenwood Press.

Epilogue

Research for the Soldier: The Story of an Army Research Psychologist

James A. Romano, Jr., as told to Richard V.N. Ginn

James A. (Jim) Romano entered the Army in 1978 and served with distinction for 28 years before retiring in 2006. His is a remarkable story, as told here to COL Richard Ginn, USA (Ret.). They talked on May 5, 2009, in Frederick, Maryland, at Dr. Romano's office with the Science Applications International Corporation.

A Jersey Boy

I grew up in Jersey City, New Jersey. I went to a Jesuit prep school there, then to a Jesuit College in Massachusetts, Holy Cross, followed by Fordham University in New York. At Holy Cross I majored in psychology, and at Fordham I did a Ph.D. in experimental psychology. I am married to Candy Grimes Romano—the best decision I ever made! To keep that in perspective, of all the many good decisions I made, that was the best one. We have three children: Candace, our oldest daughter, who's married and has two sons; Alicia, my middle daughter, who is married and has one son (and another on the way); and my son Jamie, who's in graduate school now.

My interest in psychology started when I was very young, and it is indirectly related to basketball. Some Medical Service Corps [MSC] officers and others who served with me know I was a basketball fanatic, even in my Active-duty days. As a high school player I would go to New York City to Greenwich Village to play. I played some games and then, of course, I'd get tired or I would lose, but more likely I would lose! I'd have time on my hands, so I would go to the bookstore in Greenwich Village, the 9th Street Bookstore. They had an enormous collection of psychology books and I would go there every week and start reading them. The bookstore owner took a shine to me and gave me a chair in the corner where I could read. By the time I finished high school and had gone on to college, I had read a considerable amount in psychology. To get accepted into the Psychology Department at Holy Cross, you had to interview and have a three-part oral exam. When I interviewed and they asked me questions, I was looking at the instructor's bookcase, and said, "Hey, I read that book!" He was astounded by the amount I had read as a youngster. I was accepted as a psychology major as a result of that, and that was a very good choice.

Discovering Army Research Psychology

I was at Fordham for a number of years before I finished my Ph.D., and while there I was hired into a permanent teaching position at Manhattan College. This couldn't happen today, to be hired into a permanent faculty position before finishing the Ph.D. But in the 1960s and '70s, this could happen! I finished my Ph.D. and then was promoted and awarded tenure in '76, but with tenure came a lot of administrative assignments at the university, committees and otherwise. So I reached out to a former colleague from college, Rick Manning, who at that time was Major Rick [Frederick J.] Manning at the Walter Reed Army Institute of Research. Rick put me in touch with some folks who would first of all see if I was suitable, and then would process me into the Army, and I was accessed as a direct commission into the Army after 10 years of university teaching.

In psychology you can go in a number of directions. I went in the direction of the scientific side of psychology, experimentation, rather than the more applied and clinical sides of psychology. I had an interest in the brain, its development, the development of the receiving characteristics of brain perception, and factors that influenced the development of the brain and its perceptual properties. I had a long-time interest in that and I exploited that in my first assignment in the Army, which was at Edgewood Arsenal, Maryland, where they were studying the impact of chemical warfare agents. My mission was to study the impact of the antidotes, the countermeasures to those agents, on human perception and performance.

The Early Years

So I entered the Army in 1978, completed the Officer Basic Course at Fort Sam Houston, Texas, and then went to Edgewood Arsenal as a research psychologist. I'm not one to ask for much. I'm very willing to accept the assignments given to me by any organization, so I didn't ask for anything other than Edgewood. I assumed, based on what

Rick Manning told me, the Army was a great place to work wherever you went, and now, after 30 years, I agree with that very much. Wherever you go, you can find the opportunity for great work; I firmly believe this. Now sometimes it's a little harder than others, but there are people, normally, at any assignment who will nurture you and help you to succeed and do good work in their organization if they believe that you're interested in doing that.

I was assigned to the Bio-Medical Research Lab, which later became the Institute of Chemical Defense [U.S. Army Medical Research Institute of Chemical Defense, or ICD]. Its name changed several times to keep people off the trail! But as with everything, it did actually change periodically. We were not in the Medical Department when I first reported there. One year later we were shifted by virtue of an Army staff study from something called DARCOM, or Department of the Army Materiel Command, into the Medical Department and the USAMRDC [U.S. Army Medical Research and Development Command] in, I believe, July of 1979. That was a major sea change for the laboratory. I don't want to denigrate any organization, but it did result in an improvement in the scientific capabilities of the lab. Perhaps, because we were in the Medical Department and not in the weapons development arm of the Army, so to speak, we were able to get good collaborations going with universities that would be responsive to the Surgeon General's defensive mission. This was very important, and I think that was a major factor in the improvement in the caliber of scientists both in the laboratory and those who were willing to collaborate with the scientists in the laboratories.

I was at Edgewood Arsenal for 5 years, and I got some additional duties that were very instrumental in my subsequent development. The Army has a way of giving you work to do that shapes the future work that you will do for them, and I was given the job as a custodian of surety materials. That sounds like you just owned the vault, but it meant that I had to learn all the rules governing surety materials and how to protect them, and how to pass the site assistance visits or inspections that accrue to surety materials. I learned a whole armful of Army regulations. It was something unique at the time, but the question of surety now is very much a significant issue in the Army Medical Department [AMEDD], since about 2001. So early on I was given that as an additional duty, and it was a good thing for me, because I carried that with me to a couple of future assignments where I was able to contribute that experience to policies and procedures, and to the writing of contract clauses that MRMC [USAMRDC was reorganized as the U.S. Army Medical Research and Materiel Command in 1994] put in place in the mid-1980s to enable universities to work with small amounts of agent, yet adhere to the broader principles of Army surety programs. That was sort of a compromise policy, but you could only do that if you understood the full policy. So that was a good additional duty that I got at the time.

In terms of research projects, we were very interested in the interaction of chemical warfare agents and the endogenous opiate system in the body. Some questions that come from that terminology might be the stress-enhanced toxicity of chemical warfare agents. Do the things that counteract stress perhaps have some impact in counteracting the chemical warfare agents and vice versa? I did a lot of that work and found some very interesting results, all publishable. I was fortunate to have worked with, at that time at ICD, Captain John McDonough [Major John H. McDonough, Jr., MSC, Ret.] and a civilian, Dr. Tony Shih. I published in some very reputable journals, which I also believe was an important factor in my subsequent vision of uniformed scientists, how they should be nurtured and developed at work.

While I certainly believe that it's important to publish early in your career, later in your career you take on the responsibilities of ensuring that 30 or 40 or 50 or 100 other people can publish effectively. If you're going to do that in the normal progression of leadership in the Army, you might need to know what it is to publish, how hard it is, what things obstruct it, and what things facilitate it, so you can better manage the process for other people.

An officer's job is to make 30 or 40 or 100 or 1,000 other people effective, and in research it's the same. You could be a great researcher and have a certain footprint that you make in your discipline, but you'll make a much more significant footprint if you can make 40 or 50 investigators more effective than they might have been otherwise, and to me that's the goal of the uniformed research scientist. This is not a totally accepted view, but my viewpoint is that you must lead. It's the analogy of, I'm 19 years old and if I'm in a squad, my defining goal is to become the squad leader, and when I make it there, my next goal is platoon sergeant. To me, that analogy holds true in research. If I'm an investigator, my next level is to become a branch chief (if that's the organizational makeup) for seven or eight investigators and make them more productive. Then the step beyond that is to make 30 or 40 investigators more productive for the benefit of Army missions. The Army works the same all the way across the spectrum.

So at Edgewood, I was just an investigator. My second assignment, after the AMEDD Officer Advanced Course, was to go to Brooks Air Force Base, where I was a liaison officer. We finished the Advanced Course in Christmas of '83. I went by way of the CAS³ [Combined Arms and Services Staff School] Course to Brooks Air Force Base in April, and there I had two assignments in one. That was a wonderful assignment. I was an investigator in a lab, and part of that job was to introduce the Air Force to what we called the Exempt Surety Program, as the Air Force wanted to build a chemical program at the School of Aerospace Medicine. Exempt surety basically means a compromise from the rules that might govern the stockpiling of 55-gallon drums of munitions chemical agents, to rules for using small amounts in a laboratory. It's a compromise program that allows you to work in the laboratory. For example, if you have chemical munitions and an alarm goes off, the regulation called for more than one respondent and they had to have crew-served weapons capability for further response.

Universities do not work like that, but universities might be required to if they had even a small vial of chemical agent. So I was asked to contribute to the development of that program there at Brooks as well as to conduct my own scientific work. Before 1982 or 1983, there was no middle program. The Medical Department was trying to recruit universities as collaborators to strengthen the Army program, but universities were not going to adhere to the requirements that governed chemical agents at the time. The Army had to adopt what I called a compromise program where the security requirement was commensurate with the risk. You had to do this without sacrificing the reputation of the Army and the safety and the security of the materials. The Army and the Medical Department had to do this, and a number of folks, myself being one of the main ones at the time, were asked to write a regulation that universities could live with and therefore work with the Army. This was very important to the Army Medical Department, and it was successful.

So I was working on that at Brooks as well as my own scientific investigations. Plus, I was also asked—and this was the odd thing—to spend half my time as the staff officer in the program manager's office for Air Force Medical Chemical Defense at the School of Aerospace Medicine. That was my first time dealing with staff work. I worked for a fellow who spent 7 years in North Vietnam in a prison camp and that

[work] was another great experience. Joe Milligan [Colonel Joseph E. Milligan, USAF] was released in the prisoner exchange of February 1973, and he threw out the first ball for the Mets in April 1983. He was a tremendous guy, as you might expect, and I learned a lot working with him. Of course, he was a true American hero. He was a graduate of the "Hanoi Hilton," and it was amazing to work with him. To say he was unflappable was putting it mildly. Joe shared a couple of anecdotes with me. He had a cockroach that was his pet for a while. He would give it bread crumbs in the afternoon, and it would climb up to the end of his bed. This showed the learned behavior of the cockroach. It's very interesting—a psychologist puts that twist on it. But yes, it was quite an experience to work with him. He still lives in San Antonio, and he's retired, of course, many years ago.

So I learned a lot from him. People joke, as I often do, about my approach to staff work. I have a rule that comes from working with Colonel Milligan. The first rule in staff work is: call the originating officer and ask for an extension. And you learn a lot by making that call; that was something that you just did. You learn how important it is.

My second rule was to have a cup of coffee or a Coke, take a pause, and think things over. And the third rule was to call the person who is the action officer listed at the bottom and talk with them. "What do you really want?" So sometimes I would appear to be too casual about approaches to suspense, but that was a learned behavior.

Moving Up the Ranks

The commander of the Institute of Chemical Defense came to Texas in the summer of '84, and he asked me how I liked my assignment. I learned something from him, too. I told him I loved it and it was the greatest assignment in the world: "Thanks so much for giving me this assignment." And he said, "Well, Jim, I'm bringing you back to the lab next April." So I learned, don't crow about your assignment, how great it is! I got a year and a half at Brooks. That's what happens when you have such a good assignment and you brag about it. So I came back to a branch chief's role at the Institute. The commander then undertook a number of reorganizations, and he created a branch called Plans, Operations, and Training [PO&T]. Now, a lot of people have trouble with this. What does that mean in a lab—plans, operations, and training?

There was training that went on there. We had a course, "Medical Management of Chemical Casualties," which at that time was more of a home-spun small course, one of the AMEDD-sponsored courses. It was one of four, I think, that were put in the MRMC as opposed to the AMEDD Center and School, because of the unique characteristics of the offering. So that would be included as well as the military training, but when the commander asked me to take that position I said, "No, I don't think so." And he said, "Well, I'll come back tomorrow and tell me what you think."

I went home and talked to my wife and she said, "Well, he's asked you to come back again, right? I think yes is the answer that you have to give, in my mind." My wife had, by this time, learned what's best for me. So I went back and said, "Yes, sir, I really meant to say yes." But that was an assignment that I dreaded. At the time when I took it I thought it was the furthest thing from scientific work. Well, it turned out to be one of the most valuable assignments in my career, and the reason for that was it was an integrator. It was a PO&T; we had everything but the kitchen sink in this branch. We were responsible for the "Medical Management of Chemical Casualties" course and the whole reporting requirement through the R&D [research and development] system, making sure our Institute's reports were correct. I learned about DTIC, the Defense Technical Information Center, and all the things that go along with that.

One of the "front burner" issues while I was in that job was a tasker we got from OTSG [Office of the Surgeon General] concerning a couple of Army IG [Inspector General] reports. One was on medical readiness for deployment, and one was about support of the demilitarization operations in the United States for reducing the stockpile of nuclear weapons. They both dinged the Army pretty seriously for not having good training programs for the medical support of either combat operations or demilitarization operations. But both of those reports contained almost the same sentence that said the one thing that seemed to be different was this course at Edgewood, and if it was made more available, then the Army might be able to address this deficiency. And here I am at Ground Zero.

The new commander, Colonel [Michael A.] Dunn, later General Dunn, comes in and says, "We're going to push hard to make this course be the go-to requirement for medical training to address this Army deficiency." I had a couple of jobs to do with respect to that. The first was to get the course elevated, and, as an incentive, to get it recognized to go on the ORB [Officer Record Brief] of the attending students. That was one of my tasks that I worked on with OTSG. General Dunn and I traveled to OTSG and we met Colonel [William T.] Tom Gray, who at that time was the Comptroller. We briefed Colonel Gray on what financial implications there would be to do this, and he took action. It wasn't a windfall but we got enough support to bring people TDY [temporary duty] in larger numbers. We were also successful in getting it labeled on the student's ORB, which, of course, is a pretty big incentive for someone. If they're going to take time away from a clinic, they've got to get some return for it that's tangible to them, so that was important.

We made a couple of other adjustments, too. I think the curriculum became more formalized and I believe we extended the training at the time. As to some of the other details, it's a pretty long time ago, but the point was we "plussed up" this course and made it responsive to those IG reports. That was probably one of my major contributions to ICD up to that point. However indirect it was from research, it was, nonetheless, a contribution, and General Dunn was supportive of me, very much, for that. Then I went up to Fort Detrick, in Frederick, Maryland, to be on the MRMC staff. This was around January 1988. I was a staff officer in one of the RADs [research area directors], the CHEM [chemical] RAD. I was there from '88 through '91, during the first Gulf War.

The RAD programs research efforts; it's like a controller in business. The RAD plans programs out for a few years, and it allocates money to various activities. It also monitors execution and assists the labs in reporting. The CHEM program had many labs that were playing in this area. As a result, I was able to learn a lot about the various other MRMC labs and even visit many of them.

The Gulf War Years

I was there in August of 1990, which you might remember was when Iraq invaded Kuwait. That was August 2, I believe, and on August 6, if I'm not mistaken, I was selected to go with General Russell [Major General Phillip H. Russell], the Commanding General of MRMC, to the CENTCOM [U.S. Central Command] command element to meet with the staff who were planning the response to possible chemical attacks by Saddam Hussein. The Army was, I think, accurately concerned about the potential use of chemical weapons. In the first week or so of August 1990 there was only a thin line of U.S. troops to prevent further invasion of the Iraqis should they have chosen to move in. There were real concerns about preparatory chemical fire being launched on our troops.

I was asked to discuss medical countermeasures to chemical weapons, and whether we would be able to fast-track them to get into the theater right away. I was asked to talk about three things, beginning with a skin decontamination kit. It was interesting because the NBC [nuclear, biological, and chemical] cell in their operations section was heavily Reservist. I remember talking to two fellows who worked for CVS and Rite-Aid and were activated really fast. I had brought a couple samples of the skin decontamination kit and a couple of the other countermeasures with me, and they were really appreciative of this information. The MRMC, to its credit, had just taken delivery of a large shipment of skin DECON kits not long before. They were in Pennsylvania, with just-in-time delivery. But this kit was so new that it wasn't in the training materials that a Reservist would have. This kit, the M291, had a number of significant benefits over the previous kit, the M258 Skin DECON Kit, which was not as effective. The new kit was a little better.

Because we seemed to be providing just-in-time delivery for a couple of things, there was also the CANA [Convulsant Antidote, Nerve Agent], the nerve agent antidote, which is Diazepam, 10 milligrams, IM with an auto-injector. The Army's Diazepam, of course, is a known anticonvulsant, but not very many of the routinely used anticonvulsants work against nerve agent seizures. Diazepam did, and the Army's contribution here was to show that in the auto-injector configuration, they can get good distribution of Diazepam throughout the body rapidly to counteract the nerve agent. We got that license approved by the FDA [Food and Drug Administration] between January and August of 1990 for this protection from a nerve agent—a poison—which is important. Fortunately, we did not have to use these countermeasures during the war.

There was some concern about putting Diazepam in the hands of Soldiers at the time, and the CENTCOM staff had to work through and deliberate this issue. My impression was that a really key person in the decision process was the CENTCOM surgeon in terms of policy for CENTCOM. I seem to remember it was a Colonel Belahar from the Air Force, to whom we had to present materials to a couple of times. He was open-minded, to his credit.

We were, in a sense, over-prepared in 1991, because chemical agents were not used against our troops. But it was good to be prepared, and it was based on a real concern about the chemical threat at the time. Because of that effort, General Russell took a positive view of me. I was a major at the time, and I wanted to go to the Walter Reed Army Institute of Research. But the general and a couple of other fellows would not go along with that plan, and I was really discouraged when they sent me back to the Institute of Chemical Defense. This was sort of like the Plans, Operations, and Training assignment, because I only recognized what the right answer was after a couple of attempts. I reported back there, and I was put in charge of the Drug Assessment Division. I remember General [Colonel at the time] Dunn when I came in. He said, "You're taking over a failing operation that may not last your time here." Soon thereafter, he changed out and Colonel [Ernest J.] Takafuji came in as the commander. However, when I left ICD at the completion of my tour, I remember that the incoming deputy commander said, "We need to do something to redress the imbalance that is now in favor of the drug assessment branch." That was because we were successful between 1991 and 1995 in terms of moving a couple of products through the acquisition channels to advanced development, which is what we were supposed to do. So along with that came resources.

In this job, I learned a lot. That was, again, another fabulous assignment, because I learned so much about medical product development and the importance of toxicology and pharmacology research in the drug development process. I sat for the American Board of Toxicology exam and became a board-certified toxicologist in 1994, because I really wanted to be able to talk with some credibility in terms of drug development to people I was working with who were toxicologists and pharmacologists. Ironically, that's my key credential for my day job today. So that was a good fortune that came along with that assignment.

Finding Fort Detrick

You don't always know what the reason is at the time, but there's a bigger plan. You can't anticipate this, and for me it was almost a personal thing. I remember one of my motivations occurred when I was speculating about doing the toxicology board, when the Institute's deputy commander said, "You shouldn't waste your time studying for that. You're a psychologist, you can't pass that exam." That was a red flag. I passed it eventually, although it is hard for a psychologist to pass that exam, I must say. But I now feel like I can go toe-to-toe with many toxicologists because of my experience and the crossroad that I took to do the self-study. So that was a great experience, too.

After that I was moved back to the headquarters at MRMC to be on the staff of Dr. Ana Johnson-Winegar in a combined CHEM/BIO [biological] RAD. She told me, "You can't just stay in your comfort zone of CHEM. You have to learn all about the BIO programs in the Army and what the big issues are." That was also another good development for me. She later became, at the Department of Defense [DOD] level, a very high level person in charge of the program for CHEM/BIO [Deputy Assistant Secretary of Defense for Chemical and Biological Defense Programs]. The CG [commanding general] then was [Brigadier] General [Russ] Zajtchuk. Now, as a lieutenant colonel in the Headquarters, you get bigger fish to fry, and for him I set up a number of international meetings and ran a number of trap lines—major programs where I had to brief him—and I guess I did okay.

For one of the international meetings, I selected the Xerox Center (as it was called at the time) in Leesburg, Virginia. General Zajtchuk couldn't find his way, and his aide let me down. His aide didn't show up; he needed to have that morning off. General Zajtchuk couldn't find his way, so that evening he had an after-action and he said—and this is so funny—"I couldn't find my way. Who picked this place?" Nobody's hand went up. I raised my hand and I said, "Sir, I picked this place." He looked at me for a moment and said, "Oh, Romano. I'm sorry." It was so funny. I was upset about his aide who was supposed to get him there, but in fact the CG had given him the morning off. So it was funny when he said, "I'm sorry." General Zajtchuk was a very nice fellow, and the meeting in the end was really successful. We in the CHEM/BIO program reassured the commanding general that we had our finger on everything—that one area of his domain was well covered, certainly by Dr. Johnson-Winegar, of course, but then she had a staff that seemed to be able to know what was going on and handle actions. He selected me to be the Deputy Commander at ICD. I don't know how they feel about me there. They feel a certain way about you when you're in the trenches, but when you get to be deputy and later, commander, they might feel differently because you're "management." Yes, you've crossed over the divide. But I still have friendships, of course, and fond memories of all the assignments. For me, every assignment was better than the previous one. I never expected that. Certainly, I never expected to have a better assignment after 1995 when I left the Drug Assessment Division, but every assignment kept getting better and better.

The Mantle of Command

So it was January 1, 1998, when I went back to ICD, and I worked under a great commander, Colonel Jim [James S.] Little. Oh, he's one of my heroes. He's a solid, all-around military officer in every domain in which you ask about an officer. I was deputy under him for 2 years. He was commander a long time, but then he transferred to CHPPM [U.S. Army Center for Health Promotion and Preventive Medicine], and I was selected by [Major] General [John S.] Parker to be the commander of ICD.

I was the commander for 3 years and that was also a wonderful assignment. It's hard to explain that experience to someone who hasn't been a commander. You have things that happen, parents of employees who pass away, or tragic incidents, and you as a commander have to represent the Institute to those people. There was one case where it was just a tragic family act of violence and I had to break the information to one of my Soldiers. For a few moments, I was representing the U.S. Army to that Soldier and I had to get it right. I got immediate assistance from the Chaplain and the travel office to help me assist that Soldier. You learn other things than just R&D and program management in command, and you'd never trade that experience for anything.

You said I might have had an ICD tattoo! By then, I was of course extraordinarily loyal to ICD, but as the commander, you could run the risk of thinking that everybody there works for you. I don't think that ever hit me once in my time there. Whenever I met anyone, I felt that I worked for them as the commander. I represented them. I was the proponent of that laboratory for its importance in the Army, importance to DOD, and importance to the Nation. For commanders, it's a very paternal or maternal feeling that you have toward your organization; you'd do anything for it. "Don't anyone come here and mess with my organization." You can tell me what you think is wrong with it and I'll take care of it, but don't mess with my organization and don't directly talk to my people and tell them things, because you might get them upset. I want my people working productively and fruitfully, and from my perspective they all do that. They would all do that whether or not anyone bugged them to do it. I mean, you could never lose the loyalty that you have to that organization no matter what happens afterward. If I'm still alive in 20 years, I will still be passionately loyal to that organization. I think that I'm beating a dead horse, but it's just unbelievable.

Looking Back

You know, I had a very, very fortunate career, almost as if I had been groomed from my first assignment for that job somehow, and many, many different players were along the way. There wasn't a plan for this, but I went from investigator to staff officer to sort of an inter-service assignment to branch chief to staff in the Army Medical Research and Materiel Command, to division chief and then staff and then Army lab command in a DOD program, and then MRMC Deputy Commander. It's almost as if I was being developed all along the way to take command of that organization.

Within the MRMC, there was sort of an active program of officer development management that may not have been codified in DA PAM [Department of the Army Pamphlet] 600–4 [*AMEDD Officer Development and Career Management*], although we tried to include it later, when I was the Consultant and an Assistant Corps Chief. My whole life experience was very much in my mind when we were relooking at 600–4 in terms of a scientific series officer, as to what that officer's career should be. You know, 600–4 is just a template, it's just a normative thing that says at this level you take this kind of assignment and you take this level of military training, and then at this level you take this kind of assignment and these are the corresponding courses that you should be looking at along the way. Many of those experiences helped me.

Q: So you're a part of a larger phenomenon that's starting to take hold—Army Medical Department branch-immaterial command. MRMC was really ahead of the curve.

A: Yes, MRMC was ahead of the Medical Department in that Colonel Little, I think, was such a strong and successful commander that it wasn't going to be hard for a qualified Medical Service Corps officer to come along after him to be nominated. I was followed by Colonel [Gennady E.] Platoff, who was a Medical Service Corps officer, and then Colonel [Brian J.] Lukey, who was also a Medical Service Corps officer. And then the string was broken when we had Colonel Tim [Timothy K.] Adams as commander, who is a Veterinary Corps officer, but who had similar experiences in terms of development. And now, of course, he's Brigadier General Adams.

Jim Little was an investigator, branch chief, and division chief. He went to command in Europe and then came back and commanded ICD. I think he might have been the ICD deputy briefly, under

Colonel Gary Hurst, but that pattern of progressive assignments in lab research leadership has been kind of holding true for a long time. I think the current commander, Harry Slife [Colonel Harry F. Slife, Jr.], also worked for me. He was an investigator, a branch chief, a division chief, the lab deputy, and now he's the commander; and he worked at the MRMC headquarters in the RAD shop.

One of the lessons I learned, by the way, in all of this, was that when I was asked by General Martinez [Major General Lester Martinez Lopez] to be the MRMC deputy commander, I felt that I couldn't do the job to replace Jeff Davies [COL Jeffrey W. Davies] as the deputy commander. So I said, "Okay, this seems to be way over my head, this job." And it was, but I said to myself, "If I'm going to go up there, I'm not going to fail with other people's personnel. I'm going to see if I have enough influence in the Medical Department to bring a team with me." And somehow I pulled it off. Now, I don't know if it was influence or pure good luck, but I brought five people with me who I had watched—two from my role as the 71F [Research Psychology] consultant: Mark Vaitkus [Lieutenant Colonel Mark A. Vaitkus] being one of them, and Melba Stetz, then a young captain, a very animated person, very energetic. I thought she'd be the perfect fit for the job in the RAD3, the soldier performance RAD. I brought Colonel Lukey [in] as the RAD3. Colonel Slife ultimately became the RAD4, which was the CHEM/BIO RAD, and Major Atchison [now Lieutenant Colonel Chessley Atchison, VC] worked in the Program Office. I brought those five people with me to Frederick. I did everything I could to make sure they were assigned where I was working, and if I was going to go down in flames, I'd go down with people I trust. They, of course, were wonderful people, and successful in those operations. Mark Vaitkus is now on the staff at NDU [National Defense University]. Interestingly, many 71Fs are very humble too, and may say "I can't do that job." I remembered telling General Martinez, "I don't think I can do this job." Yet at the same time I was reassuring other 71Fs that they can do the job, that I have all the confidence in the world in them, which is true, so there's an irony there. From ICD, I went to MRMC headquarters along with the five people mentioned above.

So in 2003, I was Deputy for General Martinez, and Colonel [Mary Regina] Gina Deutsch was Chief of Staff. The Deputy and the Chief at MRMC are joined at the hip in terms of supporting the CG, and he was a great boss. General Martinez trusted both Gina and me to keep his operation going on a daily basis. That, to me, was the biggest fulfillment of the managerial side of my career, that a CG would trust me to run his global operation day to day. Well, he trusted us, both Gina and me. She's an extraordinarily competent officer. She's a Kilo [70K, Health Services Materiel]. There's an MOS [military occupational specialty] or an AOC [area of concentration], and then there's the person. Sometimes the Army raises you above your AOC, and that's why branch-immaterial command, I think, is recognized, because through the assignment and education process it raises you above an AOC. I don't mean in the sense that you look down at the AOC; that's not it at all. In fact, it's quite the opposite. You have a great regard for all the AOCs and how they fit in the bigger picture of the Medical Service Corps.

Q: Going back to when you entered Active duty, did your expectations at that time as to what you would do as a research psychologist in the Army live up to what really happened?

A: You know, when I came in I had hoped that I would be a bench-level researcher for most of my career. Many of us come in with this idea. I thought I would work on great projects, that I would publish a lot and establish a reputation, and if I was lucky I would make lieutenant colonel in 20 years, and then I'd try to get a university position. If everything broke right, that would be my dream.

Well, much of it broke right in the beginning. The Army allows you to work hard in the lab and publish and develop yourself. But it puts other demands on you too, which have to do with the fact that you're not one individual, you are part of a team, and as a commissioned officer the Army expects you to lead. NCOs expect their Soldiers to want to lead, I mean, that's the Army. How can you join the Army and say, "I'm going to spend 20 years in the Army and never lead anybody?" You don't know that, though, when you go in. You don't realize that, "Oh, yes, I'm going to cultivate my expertise, but the Army's going to ask me to use it in ways that I never figured from my limited point of view as a university guy." Even if you're an ROTC graduate, you come in with a limited view of what you can do in the Army, and the Army keeps pushing that envelope and telling you that you can do more and more and more.

One of the things I always said was, what's the highest rank in the Army? It's general. They don't call them *specifics*, they call them *generals*, because they can address a wide range of functions. All of the generals that I ever served under or watched were unbelievably astute individuals. I could not believe the human intellect could morph that fast, or burrow that deeply into issues.

[Lieutenant] General [Eric B.] Schoomaker, oh my goodness, now there's a quick study. General Martinez, there's another quick study. These general officers are remarkable. General Zajtchuk, I thought, was brilliant. And General Parker, he's a little bit more deliberate than the others. Each was also an individual, but their analytic powers are amazing and quick. They are, by their assignment, asked to cover the waterfront, so to speak. Obviously, only a very small percentage of people are going be able to do this, but the Army tries to make everyone as broad and capable of being a general as they can through the assignment and education process. So I wound up sitting as deputy, and even briefly commanding. Between General Martinez and General Schoomaker, I commanded at MRMC. [Lieutenant] General [Kevin C.] Kiley turned to me and said, "You're not just an acting commander. I made you the *commander*." It was for about 10 or 12 weeks, but that was also a remarkable experience. Of course, many things were deferred until General Schoomaker came on board, but I had 1 or 2 days where I saw what it was like to be the CG. I don't know how they do it.

I remember one day when I got up at 4:00 in the morning and came to Fort Detrick. At about 4:45 we left (I had a driver—that was my big perk then). We drove down to a series of meetings in the Washington area, two meetings in the morning and an afternoon stop at Aberdeen. Then we hit the breast cancer meeting in Philadelphia at night, and I got home at 2:00 the next morning. Now, I only had to do that for one day. But General Martinez and General Schoomaker, I watched them doing it day after day after day, and being on their toes at every assemblage that they were at. I just don't know how they could do it. They were remarkable, but part of what their staff was supposed to do was to get them ready for those events. Oh my God, I can't believe that they could do that! I have so much respect for them, it's amazing. I mean, General Martinez and General Schoomaker, I worked closely with them, and their intellect is way out here. It's not in that ballpark, it's way out. I mean, they are remarkable.

I don't think the general public could possibly appreciate the scope of what they're asked to do because it's outside their range of experience. I mean, I couldn't appreciate it when I was a major. It's only when you stick close to them, and prep them and work with them on things that you see the level of energy that they have to bring to the table day after day after day. The Army gets it right, though, when they pick them.

I retired officially on 30 November 2006. Looking back over my career, what was my greatest personal satisfaction? Command, certainly, but there were many things, and they all seemed important at the time and then they seemed to pale as you got to bigger challenges. As Deputy Commander of MRMC, one of the things that I was working with under General Martinez, and a little bit under General Schoomaker, was the campus, the confederation at Fort Detrick [National Interagency Confederation for Biological Research and National Interagency Biodefense Campus]. To me, the fact that we got that off the ground was very, very significant. It remains to be seen, but it could be very, very significant on a national level.

At Fort Detrick (and this, of course, all spins off of 2001 and the national response to the 9/11 attack), since 2003 they have built a laboratory for the NIAID [National Institute of Allergy and Infectious Disease], which is another arm of the NIH [National Institutes of Health] separate from NCI [National Cancer Institute], but NCI, of course, is there as a large presence. There's also a lab built for the DHS [Department of Homeland Security], as well as plans for the construction of a new USAMRIID [U.S. Army Medical Research Institute of Infectious Diseases]. There's a medical logistics joint facility and then a few other lesser construction projects. What you have there, then, is not only joint, but enormous interagency pressures on a small Army installation. Then you have challenges of integration of the different agencies' business practices, so it remains to be seen if they'll be successful. That's not a 2- or 3-year process. The garrison at Fort Detrick takes the lead in this, and the CG of course. Congress has made it so, and put the money there to make it happen.

There is no Army installation that I know of quite like Fort Detrick. There is a campus with two Federal agencies, EPA [Environmental Protection Agency] and one of the arms of NIH, at the Research Triangle Park in North Carolina, and we looked to them for some lessons learned, but they were very different than the challenges here at Fort Detrick where the Army supplies the water and the utilities. Should the Army do that free for the DHS and for the NIH? Or should it ask for fair reimbursement? And you know, those agencies all appropriate their money very differently. It's remarkable to see that it's still afloat, that this is still working, this confederation at Fort Detrick.

Q: So it calls for a spirit of accommodation?

A: Yes, and some agencies are more accommodating than others, I can assure you. The Army has been very accommodating. The Army selected Colonel Deutsch after Colonel [John E.] Ball to be the garrison commander, another good call for having the right personality type to bring people to the table to talk. I mean, you can't be rigidly uncompromising in that situation or the whole initiative will fail. Again, I can assure you there are wide ranges in tolerance and flexibility between the agencies, as well as business practices, but so far it's working. Since I was in on the ground floor, that's probably one of my greatest feelings of accomplishment.

And command, of course, was very satisfying. Bringing the "Chemical Casualty Care" course to a broader customer base far exceeded anything I ever thought would happen after 2001, and to make it a course that trained Army physicians on a regular basis for their assignments, I think was important. I did that as a major. So at each point I had something that stood out to me as a significant accomplishment.

Now What?

After retiring in 2006, I came to work for SAIC. We have a contract with the NINDS [National Institute of Neurological Disorders and Stroke]. We are a prime systems contractor and we integrate the work conducted by a dozen or more subcontractors. Our purpose is to find a drug to treat spinal muscular atrophy, which is a disease of the newborn. There are various tiers of the disease, but Level 1, Type 1 is fatal usually in the first year or two, and then Type 2 is a little bit less severe, but usually fatal before age ten. Type 3 is a normal lifecycle, but it's accompanied by muscular weakness. It's a small population, so any drug that you might get or treatment for this disease would be called an orphan drug.

When I came into the project, for my government points of contact it was, "He's an Army guy." And that wasn't necessarily complimentary. Well, I might not be the most qualified guy, but everything that we developed in the Army for CHEM/BIO is an orphan drug by definition. There are no animal models that are really high fidelity, and you've got to do animal testing to get them approved. So you've got a lot of technical work that goes into supporting something that you can't test in a big population, which is the same thing as this disease.

So it turned out that my Army experience was, in fact, quite relevant to this project. I have a view of all of the possible pitfalls and pathways that you can encounter in development, so it was a pretty good match for me. Apparently I'm doing okay in the project for the government, but it's a good application of a subset of skills that I learned in the Army. It's a step backward in terms of program management size, and so on, but this disease is cruel to its victims, and I'd really like to be part of the group that solves it. I don't need to be in charge or in command anymore. I have a team of eight and that's hard enough for me; we're focused on this project and we want to make it work. The great thing is that everything I learned in the Army helps me on this job.

Do We Need Research Psychologists in Uniform?

Q: If we discuss the 71F research psychology field, I guess the first question could be, why have 71Fs in the Army? Why should research psychology be part of the Army's array of specialties?

A: That's a big question, actually, and it's a big question for me because I think the answer is easy. It's easy because there are so many reasons. So I'll try to just kind of take a couple of cracks at it. These are some things that I collected while I was consultant for the 71Fs. These are papers written by 71Fs; one of the papers is entitled, "The Psychological Resilience of Soldiers."

Q: This binder looks like it's at least 500 pages.

A: Oh, yes. These officers are very productive, but that isn't the rationale for retaining the AOC, because productivity is assumed. The title of this is revealing: "The Psychological Resilience of Soldiers." It doesn't say "Psychological Problems in Soldiers," and it doesn't say, "Post-traumatic Stress Disorder in Soldiers." It doesn't say, "Occupational Health and Related Problems in Soldiers." It says, "Psychological Resilience of Soldiers," and this is the important piece of it. 71Fs recognize, of course, all of these other issues and explore them, but in the end, uniformed psychologists have, perhaps, a very different view of the American Soldier than the media might portray, and that is that these are really awfully strong and psychologically healthy people that we have in the military. They look at those things that make them like that; they explore those things. Of course, they look at the things that might chip away at that: deployments, stressful experiences such as the Gander, Newfoundland, crash [July 20, 1989] and recovery of remains, and so on. They've explored all of the various domains of the things that nibble away at the psychological health of Soldiers, but always with this focus in mind, that the basic starting point is psychological health. These are aberrations, and how do we deal with that? How do we deal with these things in such a way as to mitigate their effect on the Soldier's behavioral health?

The other piece as to why we need them in uniform is that 71Fs recognize that the mitigating steps that one would take to, say, reduce deployment stress or impact PTSD [post-traumatic stress disorder] and so on, are really not medical steps. They're unit-based steps; that is, the training and events that go on in the unit actually are the most significant things. It's not really the work of the behavioral health system in the clinic; that's when things have gone too far. The basic process of keeping the integrity of the unit's mental health resides in the unit, and you can only understand that if you understand the relationship of the Soldier to the unit, the unit to the Soldier and his family, the commander to the unit, and the impact of a commander on the health of a unit. You need to understand that culture, and take the right frame of mind in approaching these problems. Otherwise you fall into the characterization of Soldiers that you've seen on TV—burned-out Vietnam War vets or Soldiers coming home from Iraq and then shooting up somebody. That's a disservice to our veterans, a big-time disservice. I think to keep the picture straight, you need psychologists, who understand the inner workings of the Army, to study the problems and then make recommendations about potential solutions. That's my primary reason for saying that you need them in uniform. In uniform they deploy with the units, or they go stateside to various places and stay with the units. They wear a uniform that doesn't differentiate them as Dr. So-and-So, but rather just as a fellow Soldier.

Q: The fall of the Soviet Unit coincided with various reductions of the Army, such as the series of SERBs [Selective Early Retirement

Boards]. As I recall, research psychology was being offered up for reductions pretty freely. Was that your impression at the time?

A: Well, I don't know how freely, but it was offered up, yes. But let me ask you, what year are you talking about?

Q: In the late 1980s, early 90s, a whole series of reductions were being pushed following the fall of the Soviet Union. They really were looking for cuts in the Army. Remember the "peace dividend"? It just struck me that there were sharp knives out for some of the scientific specialties of the MSC, and I wasn't hearing the kind of resistance that you would expect that would defend against those cuts.

A: Those are the same sharp knives that came out my last year on Active duty. They have a name for each of these rounds of cuts.

Q: Recently there was the MIL to CIV thing [Army program to convert military positions to civilian slots], which is actually part of the same process.

A: Yes, I had to trundle off in my last year to the Pentagon to talk about why you need uniformed scientists. How did it go? I was going to say it didn't go well, because the guy who asked for it didn't really want to hear it so he didn't show up, which I was told was going to happen. I took an NCO with me, and I had him talk about what he did while he was in his lab at USAMRIID in the period of 2001 to 2003, what he did when he was sent to Iraq, and what he did when he came back. What you saw in this very sharp NCO was his learning experience as he was actually leading a Navy unit in the very far north of Iraq doing environmental sampling and monitoring. He had a certain focus area in his training before he went to Iraq, and when he came back he knew that we needed to look for a lot more things than what were studied before. He knew what we needed to do with respect to another biological warfare threat, and he reported this back into the program. He modified the direction of the program based on his wartime experience.

I brought the NCO because NCOs are very credible to Pentagon executives and senior officers. NCOs are, in fact, where the rubber meets the road. I didn't speak to the fact that he had an officer who mentored him before he went, and that the officer who was with him in Iraq was the officer that he worked for at USAMRIID. In Iraq they were very thin, so he went this way and his officer went that way, and then he went back to Fort Detrick after the deployment.

So, rather than bringing in a Ph.D. officer who would talk in technical terms, I brought in his NCO. We were talking to bureaucrats. I didn't need to talk about the genome for this BW threat versus the genome for that threat, or the PCR [polymerase chain reaction] assay and how you need to make modifications. But I did stress that an officer learns things in a laboratory, then goes to a theater and comes back to apply what he learned in modifying the research that goes on in the laboratory, and that uniformed Soldiers have been doing this routinely in this war. It's more the case in this war than it has been in the past, at least since Vietnam, actually. My general theme was that because they can be deployed into theater, they can apply their expertise, which is not unidirectional, but bidirectional. When they come back, their experience shapes their research.

Of course, the challenge to that is, "Well, you can put a civilian or a contractor in there." Of course you could, but I just don't think you would have the flexibility or the appreciation for a unit organization as a uniformed scientist. I don't know how a civilian would fit into those very small, tight-knit deploying units, but I guess they do. The boundaries between contractors in theater and civilians in theater are really blurred.

But the boundaries between the stages of warfare have blurred pretty much too, haven't they? So I could be a company commander in a unit that could engage in humanitarian assistance operations in the morning, significant combat operations in the afternoon, and nationbuilding the rest of the day, and I would have to adjust my unit's protocols to all of them, which is an enormous challenge. This goes way beyond identifying friend or foe, it's just rapidly tuning up and down the military unit on a switch.

Just as the types of warfare have blurred, so have the demands on the individuals. Let me give you another example, and this goes along with what I was just saying. When I was Deputy Commander we had two guys awarded the Purple Heart from MRMC. Two guys awarded the Purple Heart from MRMC? What did they do? Well, one was a comptroller, [Colonel] Elias Nimmer. He was in a hotel that was bombed by a rocket attack and was seriously hurt. He was there to help the medical infrastructure in Iraq, to rebuild how they do appointment scheduling and ordering, billing, and collecting. He was a comptroller, but he was helping to rebuild and strengthen the Iraqi medical infrastructure.

And then we had another young fellow [Captain Brian J. Walrath] who was near Abu Ghraib and got the Purple Heart. He was a health facilities planner, which is a 70K 9I. What was he doing? Well, he was over there to design the conversion of a facility into a medical clinic to serve the Iraqi population, and he got a Purple Heart.

So you had these fellows in a clearly nationbuilding role, yet they're building nations in the midst of combat. Just as the phases of warfare have blended, so have the required military unit structures. I think one thing that you have to worry about is, are the organizations still accurate? This is just my view. The TO&E [Table of Organization and Equipment] doesn't have a comptroller or a health facilities planner in any of the units that these guys were assigned to, but the Army adapts and brings in the people that it needs to provide that expertise.

So I wouldn't ask the question anymore, "Why do we need them in uniform?" I would say this: We're way past that, and the question now in Iraq and Afghanistan is, "Where do we need them in uniform?" That's my view. Where could we best put these technical, administrative, and scientific experts? Research psychologists? I would have a research psychologist paired with the division psychologist. He's a healthcare provider, and a research psychologist is a data gatherer. They're very different roles, but they intersect at that point, and they would work, then, for the division surgeon. That's where I would put them. Now, I know that structure is changing too, if they still even have division surgeons—I presume they do. And what is in a BCT [Brigade Combat Team]? I think there is still a Division Surgeon.

Looking Ahead

Not long ago, I told a couple of young 71Fs that I wish I was starting over again as a research psychologist today. They have so many opportunities. For me, there was a certain frustration—not a great deal of frustration because my career was wonderful, it was the greatest—but I beat the drum for 30 years on PTSD and traumatic brain injury as under-recognized wounds. So I retire and, boom! Now those are probably the signature injuries of this current conflict. The recognition of the importance of these injuries signals another major change in military medicine over the last 30 years. It had already taken place probably 10 years ago, but now it's really on the front burner, and DOD can't get away with saying, "That's a VA [Veterans Administration] problem. That injury is a VA problem." That thinking is out of place today. The Army and the DOD are expected to apply the treatment that restores the person to full health even if they can't keep them in the Army. They're expected to ensure a continuity of care for the Servicemember as he or she transitions from Active duty to retirement. Furthermore, they're expected to provide the treatment that restores the person to full functioning, so that when they leave they are as close as they can be to when they came in. So the boundary (just like all other boundaries, I've learned) between where the Army lets go and the VA picks up is very much blurred. The general public doesn't care, and now that I'm a veteran I don't care anymore either. I cared when I was in the Army, but I don't care now, because caring about that boundary is fruitless.

The coming together is going to happen, and one of the things that might drive it is the unifying of the medical record. The Army has of course been using AHLTA [Armed Forces Health Longitudinal Technology Application—the DOD electronic health record] for some time. We wish we could find something better. But this problem is being worked on with vigor. The Veterans Administration of course has its own system. One of the things for the VA and the DOD is one of them has to give. Either the VA gives up its current system, which is built on an older platform, and rebuilds it from scratch so that it can marry up to the DOD system, or the DOD has to give. Psychologists are right in the middle of that. The injuries they study (for example, PTSD, TBI [traumatic brain injury]) are, in fact, right at the blurring boundary between the active force, DOD, and the VA.

One of the things that I have to qualify this with is that the 71Fs, as a group, are pretty good at studying the Active force, but they don't have accessibility to the Reserve force, and this is another thing that the DOD and VA have to resolve. The Reserve force is very distributed and it's hard for the 71Fs to capture that population and to do followup studies for redeploying Reservists. That's a challenge for all of DOD medicine. But the 71Fs do have pretty good ability now, more than ever, really, to look at the unfolding of problems in the Active force.

Q: Are the clinical psychologists and the research psychologists working as a combined arms team?

A: I don't believe so. If they are, that's news to me. That would have only happened in the last year or so. They're pretty much independent. As just one example, in the Medical Service Corps the research psychologists are grouped with the applied scientists, MFA [Medical Functional Area] 71, and the clinical psychologists are in MFA 73, the behavioral sciences, with social workers—they're healthcare providers.

Their assignment patterns are very different, and I don't believe there are any task forces that have necessarily united them. They have an occasional Medical Special Augmentation Team or the like that has joint membership, but there's no formal integrating mechanism in the Army for that. That may not be such a problem now that these problems are highly visible. In the past these behavioral health problems were not so visible, thus it was more difficult for the things that were written about by 71Fs to reach decisionmakers, especially since these decisionmakers always have other burning issues on their plate.

Let me give you a good example of a 71F statement that never really got traction. I asked you what year you were talking about with the fall of the Soviet Union. I think in the mid-1980s, Rick Manning wrote that the Cold War was ending and another hot war was just beginning and had already begun, and by that I mean the 1983 bombing of the Marine barracks in Lebanon, the attack on the USS *Cole* [October 12, 2000], and before that the Khobar Towers in 1996. He argued that it was going to be a lot harder on Soldiers when the Cold War ended, because we were no longer stationed and poised against another force as in Korea or Germany. We were going to be deploying all over the world, and that will turn out to be harder on the Soldiers. That was written I think, in '82 or '83 by Rick Manning. How right was that? How anticipatory was that?

The other thing is, you of course remember all of the hubbub, and rightly so, about Abu Ghraib? About 5 to 10 years before Abu Ghraib, research psychologists wrote that any deploying force that goes into an area and does any kind of peacekeeping mission will witness an Abu Ghraib–type phenomenon at a certain percentage rate. Two to six percent of any deploying group of soldiers from Denmark or the U.S. or Great Britain or Canada or Nigeria or elsewhere reported observing an Abu Ghraib phenomenon while they were deployed by the United Nations or NATO [North Atlantic Treaty Organization]. So 71Fs recognized the prevalence at a low level, but a persistent low level around the world of Abu Ghraib–type atrocities. Maybe if that had gotten more traction, more visibility, and more assessment, then possibly Abu Ghraib might have been anticipated and deflected.

So they touch on problems that are real, and sometimes when a problem goes awry, it's an embarrassment to the whole country. I think in the case of the 71Fs, it's not a question of should they be in uniform, but where should they be, and where are they best deployed? From my perspective as a 71F, you have to take that with a grain of salt, but I have watched what these people have been doing for 30 years, and as you see from this compilation of papers, they generally point out things that need to be addressed. That's what scientists are supposed to do, point out things, and verify that they are real in the sense that the science says this is a phenomenon that is actually validated. These are more descriptive models than predictive. Yet they are predictive when they try to recommend mitigation approaches. But those mitigation approaches are harder to validate.

In my last 5 years of Active duty, I was appointed by [Brigadier] General [Richard L.] Ursone to his senior leader team. Remember his deputies Jim Solomon and Dave Burns [Colonel James G. Solomon and Colonel David C. Burns]? General Ursone is tremendous, and together they were mentoring and developing a senior leader team. In Las Vegas, in November 2000, we set a target of what would be the primary purposes of the senior leader team, something that we would tackle over the next couple of years. Communication was number one, then competence; you probably have touched on this when looking at the Medical Service Corps. So we set forth a plan of what kind of activities would help to advance the Medical Service Corps in each of these areas.

I'll just kind of talk about the area of communication. We did a lot of study on that. We plussed up the material in the Corps Chief's letter, and tried to push hard to make it hit every Medical Service Corps officer. We then expanded the senior leader team to include all of the consultants. We tried to make sure that the consultants, as well as the senior leader team, would push out the news through their channels. Many of the consultants, me included, didn't have a newsletter before that. I got two young officers to do the newsletter for me to just say, more or less, where the Corps is going, what General Ursone is doing with the senior leader team, and so on. We also exploited use of Web pages more than in the past, keeping up with AKO [Army Knowledge Online].

Much of that was public service announcements, just keeping people informed of the activities of the Corps, such as a birthday ball and the various Silver Caduceus Societies. Later, we pushed through the annual report as a much more collective development that focused on young officers and their contributions. If you read some of that stuff, it was great. They did reach the young officers and trumpet them, not the colonels. I think there are many many officers whose careers were influenced by those communications in their view of the Medical Service Corps and the assignment process, and their experience was positively influenced by the annual report.

Another area where we tried to work, and this was a little bit harder, was what I think you can call relevance. This included the assignments process, making sure that the education that an officer completed, in particular their graduate training, when they came out of that educational experience, their skills were used in their next assignment. This required getting good linkage between the HRC [U.S. Army Human Resources Command] and the consultants in terms of the assignments process, a linkage that maybe wasn't as strong as it could have been before.

We even tried to influence the education experience by having a uniformed officer make contact with the officer in graduate school (I think we were trying to do this) to shape whatever their papers were that they had to do to accomplish their requirements so that they might pick a topic that was relevant to their next assignment. We worked on trying to shape that, and that's very hard to do. We tried to get more money for more opportunities for education, and I think some money came out. We tried to also push what they call continuing education, where you have an annual or biannual meeting of an AOC. It's called PPSC [postgraduate professional short course]. We'd try to get more money for that, and General Ursone was pretty successful. His successor, [Brigadier] General [Sheila R.] Baxter, adopted and expanded that with her deputy, Colonel Lisa

Weatherington, from my perspective. They talked about education and its value, and the instructions to promotion boards were influenced by the whole series of meetings of the senior leader team and the consultants.

The Corps chiefs were, in fact, very responsive to input from the Corps. It was remarkable how responsive they were despite enormous constraints within the Army system. I can't remember if it was General Ursone or General Baxter, but they then established a field leader team of majors who went to these meetings and then they were expected to go out and promulgate the news, "the Gospel." I think that was a good development when they brought them in. They have a very different viewpoint than colonels.

But that was good. It was good for the colonels to hear what they had to say. But communication to me was the major problem. It still is, but there were improvements, and I would suspect that over my last 5 years in the Medical Service Corps they made enormous improvements in the communication to the field. To me that was extremely important, because in my first few years I didn't hear much from the Corps. As a young officer, you hardly even knew it existed, and if it did exist you thought it was in San Antonio. So it was very different later and that's good, that's very good.

The Road to Success

Looking ahead to the future of 71Fs and the MSC, I would take a cue from General Martinez. He came and spoke to us at a Silver Caduceus meeting at Fort Detrick, and later, General Schoomaker said pretty much the same thing during another meeting at the Community Center. I want to repeat what they said because I respect them so much. They're not Medical Service Corps officers. They're physicians in the Medical Corps, but they certainly recognized the value and contributions of Medical Service Corps officers, no doubt in my mind about that.

I remember General Martinez talking about officers asking for an award and saying, "I did something, can I have an Army Achievement Medal or an ARCOM [Army Commendation Medal] for it?" I think that's not what an officer does, but officers have asked me for awards when I was more senior, and when they did that I think they lost something, in my view, I really do. So my answer is related to the flip side of that. It's all about service. Serve well. Serve the Army well; serve the country well. Serve with a passion, serve as if somebody is watching and thinks what you're doing is extremely important, because it's true. Somebody is watching you as an officer, and thinks that what you're doing is important. If you slip up or do something unethical or behave improperly, that sure as shooting will be noticed. But if you quietly serve with a passion and with devotion to the mission, that is also noticed and the rewards come on their own without asking. Rewards will come, and they may not be in the form of an ARCOM or an Army Achievement Medal. It may be your name that comes to mind when a challenging assignment is open, and a leader is asked if they have anyone who can step in and help us with some particularly important thing. Leaders tend to look to those who serve quietly with strength to fill those positions, and that's the reward for this service. At the end you could retire from the Army with just an MSM [Meritorious Service Medal] and an ARCOM, yet you have served the country tremendously. You could have served the country brilliantly and gotten nothing more than those awards.

But it's not about that. It's about service, and service brings recognition, not only to you but to your unit and to your AOC. If you're a 71F and you do a good job in a 71F position, and you point out things that go all the way up to the Surgeon General ("I have identified a problem here and I need to work on it") he or she may not remember the name of the officer who made the discovery, but will probably remember it was a psychologist who pointed this out. Then on the way back down, it's incumbent on those leaders to determine if there was any recognition and then give it as appropriate. General Schoomaker and General Martinez both said the same thing at these Silver Caduceus Society meetings, and I really believe that.

I was fortunate. People looked out for me all along the way. I think they looked out for me because I was trying. I certainly never got it all right in my assignments; I always made mistakes. Colonel Little once told me, "You made a mistake, but a sure-fire strategy to avoid a mistake was to not do anything." That was one of his mottos, that you try and you work, you explain what you're doing and it will come to you—good assignments, recognition if warranted, and rewards on a personal level that are really much more important than medals. I actually don't know where my awards are, but I'm very proud of my career and it was rewarding to me. I don't know where my medals are. They're somewhere in my basement. If I find them I would be very proud of them, but I don't need them to remind me that I had a wonderful career and had great blessings. They're just part of the periphery. If I got an award at a ceremony, to me the message was more to the people junior to me—that the organization will reward you for effort.

But that to me addresses, at the Corps level and the AOC level, the best strategy for making yourself relevant, which is to do the best darn job you can at the mission the Army gave you, and it will somehow get noticed and factored into the equation. You should be proud of everything that you do for the Army because the Army wouldn't have a 71F or a Medical Service Corps if somewhere along the way at the highest levels in the country someone didn't determine that they were needed. Why would an Army now currently fighting wars on two fronts have excess AOCs? Why would it have an excess Corps? They're there because along the way they've been revalidated time and time again. So if the question comes up in 2011, "What about 71F?" (and it probably will come up in 2011; there's no TO&E for those officers), if they say "Keep them," that's a national decision that says we still need them, and that's pretty significant and rewarding in and of itself. That an Army compelled to be lean and mean still recognizes the need for a 71F speaks volumes. And it also speaks volumes that an Army that is now providing healthcare in very demanding conditions recognizes that there's a Corps that is currently providing the best possible administrative, transportation, and logistical support to that healthcare system.

So, if the question comes up, "Should we combine the logisticians with the Army logisticians?" or "Should we combine the aviators with Army aviation?" or "Should we combine the administrative officers with Army administrative officers?"—yes, that's going to come up, just like the research people. It's not offensive. It's an Army trying to make sure that it's prepared to fight a war on multiple fronts with very limited resources. And every time they come down in your favor, that's a revalidation of your career and your mission.

It's a wonderful thing. I would have never expected this. Thirty years ago when I came into the Army, I was very selfish. I wanted to do research and publish papers. That was my goal. Now I look back on it and I say, "Gee, I did stuff that was important for the United States of America. Holy cow." I mean, that's wonderful. It's so wonderful. You don't want to wind me up and send me into a university because I'd be pitching it to the graduate students and the medical students as one of the greatest things that they can do in their life!

About the Contributors

Editors

Paul T. Bartone, COL (Ret.), USA, Ph.D., is a Senior Research Fellow in the Center for Technology and National Security Policy at the National Defense University (NDU). A Fulbright Scholar, Bartone has taught leadership at the Industrial College of the Armed Forces (ICAF) at NDU and at the U.S. Military Academy, West Point, where he also served as Director of the Leader Development Research Center. Bartone was the senior Research Psychologist in the U.S. Army and has served as Consultant to the Surgeon General for Research Psychology and as the Assistant Corps Chief for Medical Allied Sciences. He is also past President of the American Psychological Association's Division 19, Society for Military Psychology. He earned his Ph.D. in Psychology and Human Development from the University of Chicago.

Ross H. Pastel, LTC, USA, Ph.D., serves as an Assistant Professor in Medical and Clinical Psychology at the Uniformed Services University of the Health Sciences (USUHS). His previous assignments have included the Walter Reed Army Institute of Research (WRAIR), the U.S. Army Science and Technology Center, Far East (Japan), Armed Forces Radiobiology Research Institute (AFRRI), U.S. Army Medical Research Institute for Infectious Diseases, and U.S. Army Medical Research and Materiel Command (USAMRMC). He earned his Ph.D. in Neural and Endocrine Regulation at the Massachusetts Institute of Technology and did postdoctoral research at the Western Psychiatric Institute and Clinic in Pittsburgh, Pennsylvania.

Mark A. Vaitkus, COL, USA, Ph.D., is Associate Professor of Behavioral Science in the Strategic Leadership Department at ICAF. His prior assignments have included Deputy Technology Program Manager for the Future Force Warrior Soldier system, Deputy Director and Chief of Research at the Army Physical Fitness Research Institute at the U.S. Army War College, Sociology Program Director in the Department of Behavioral Sciences and Leadership at the United States Military Academy, Deputy Director at the U.S. Army Medical Research Unit in Heidelberg, Germany, and Research Psychologist at WRAIR. He holds a Ph.D. in sociology from the University of Michigan and an M.S. in National Resource

Strategy from ICAF. He is an author and co-editor of *The U.S. Army Guide to Executive Health and Fitness* and *The Total Trainer CD-ROM.*

Contributing Authors

Gina E. Adam, MAJ, USA, Ph.D., is assigned to the U.S. Army Research Institute of Environmental Medicine, where she investigates cognitive performance of the warfighter. Her research interests include the study of performance under dual- or multiple-task situations, the use of physiological measurements such as heart rate and eye tracking to understand the relationship of task performance to the stress of cognitive overload, and the impact of environmental stressors on higher order cognitive performance. Her recent work has involved analyzing marksmanship performance in cold and dehydration conditions, assessing judgment and decisionmaking in calorically deprived Soldier volunteers, measuring multiple task performance at sea level and at moderate and high altitudes, and assessing several domains of judgment and decisionmaking at sea level and high altitude.

Gary A. Adams, Ph.D., is a Professor at the University of Wisconsin–Oshkosh. His research interests include older workers and occupational stress and health. He has made over 50 professional presentations at national conferences such as those sponsored by the American Psychological Association, the Society for Industrial and Organizational Psychology, and the American Psychological Society. He has published two books, several book chapters, and over 25 articles in journals such as *Personnel Psychology, Journal of Applied Psychology, Journal of Occupational Health Psychology, Journal of Organizational Behavior*, and *Educational and Psychological Measurement.*

William F. Blakely has been a Staff Scientist with AFRRI for over 25 years. He is the Biodosimetry Research Group Advisor for the institute and the course director for USUHS's Radiation Biology graduate course. He also serves as a U.S. representative on the ISO TC85/SC2 (Radiation Protection) Working Group 18 (Performance Criteria for Service Laboratories Performing Biological Dosimetry by Cytogenetics), Chair of a NATO Research Study Group–Radiation Bioeffects and Countermeasures, was a founding member of AFRRI's Medical Radiobiology Advisory Team, and is on the National Council on Radiation Protection and Measurements.

Paul D. Bliese, LTC, USA, Ph.D., joined the WRAIR as a member of an interdisciplinary team studying stress and adaptation among Army Soldiers. While at WRAIR, he has studied Gulf War syndrome, the adaptation of Soldiers deployed on peacekeeping operations, the impact of new technology implementation, the effects of long-term activation on National Guard and Reserve units following September 11, and the health of Soldiers returning from Iraq. LTC Bliese's research interests include occupational stress, leadership, and multilevel methodology. As an outcome of these interests, he has published a number of theoretical papers examining the value of multilevel analyses in leadership and occupational stress research. He has written and maintains the multilevel package for the open source statistical language R.

Michael W. Boye, MAJ, USA, Ph.D., has been with the U.S. Army Medical Research Detachment since 2005, where he has been conducting research in the area of color vision and serving as the statistical consultant and reviewer for several research protocols. Previously, he was assigned to the U.S. Army Aeromedical Research Laboratory at Fort Rucker, Alabama, where he examined the relationships of job and life stress to the job performance of Army helicopter pilots. He received his Ph.D. from DePaul University with an emphasis in research and organizational psychology and has over 20 years of experience as a research psychologist and organizational consultant. MAJ Boye is a member of the American Psychological Association.

Thomas W. Britt, Ph.D., is a professor in the Department of Psychology at Clemson University. From 1994 to 1999, he was an Active-duty Research Psychologist for the U.S. Army and deployed in support of multiple military operations. Dr. Britt continues to do research on the determinants of resiliency among military personnel and the stigma of seeking treatment for psychological problems in the military.

Lolita Burrell, MAJ, USA, Ph.D., began her military career in March 1999 and is currently an assistant professor at the United States Military Academy. Prior to entering Active duty, MAJ Burrell earned her Ph.D. in Medical Psychology at the Uniformed Services University of the Health Sciences. MAJ Burrell's research experience includes conducting studies in the areas of stress, post-traumatic stress disorder, biofeedback, upper extremity disorders, Soldier performance and injuries, nutrition, and military family health.

Joel Carter is an Intelligence Analyst for the Department of Defense and is a graduate student in the Security Policy Studies Program at The George Washington University Elliott School. Mr. Carter served in the 82^d Airborne Division from 2000 to 2003, during which time he deployed to Afghanistan.

Carl Andrew Castro, COL, USA, Ph.D., is the Director of the Military Operational Medicine Research Program, Headquarters, USAMRMC. He began his military career as an infantryman in 1981 and joined the Medical Service Corps after obtaining his Ph.D. from the University of Colorado. Over the past 18 years, he has served in a variety of research positions, including as Chief of the Applied Pharmacology Branch at the U.S. Army Medical Research Institute of Chemical Defense (USAMRICD), Assistant Program Manager at AFRII, Director of the Primate Research Facility at WRAIR, and Commander of the U.S. Army Medical Research Unit–Europe, in Heidelberg, Germany. COL Castro's operational experience includes tours of duty in Iraq, Bosnia, and Kosovo. He has authored over 60 scientific articles and reports in numerous research areas.

Matthew G. Clark, MAJ, USA, Ph.D., obtained his Ph.D. in Behavioral and Neural Sciences at Rutgers University and is a certified defense acquisition professional. He has served in research and leadership positions at two Army medical research laboratories and commanded a medical company in Korea. Prior to his current position managing capability integration at the Army Asymmetric Warfare Office on the Army Staff, he served as the Deputy Commander of the U.S. Army Medical Materiel Development Activity. He also serves as a consultant on various deployment-related military medical issues and research.

Michael N. Dretsch, CPT, USA, Ph.D., is Chief of the Cognitive Assessment and Diagnostics Branch at the U.S. Army Aeromedical Research Laboratory. He is an experimental psychologist with a background in Cognitive Neuroscience techniques, and is currently involved in validating automated neurocognitive test batteries for the assessment of mild traumatic brain injury (mTBI) and identification and validation of behavioral, imaging, and serum biomarkers for both mTBI and post-traumatic stress disorder to improve diagnostic accuracy and develop return-to-duty guidelines.

Doris Durand, Ph.D., is retired from the Department of Military Psychiatry at WRAIR. Her specialty was Army families. Prior to her retirement, her major projects involved the study of deployment on Active duty, Reserve, and National Guard Soldiers and their families.

Edward N. Edens, CPT, USA, Ph.D., is a Research Psychologist in the Department of Military Psychiatry at WRAIR. Prior to joining the Army in 2007, CPT Edens spent over 10 years in corporate America as a senior research analyst for The Gallup Organization. Since joining the Army, CPT Edens has taken the lead with the WRAIR Land Combat Study in analyzing data and putting together briefs for the 160th and 5th Special Forces Group at Fort Bragg.

Sandra M. Escolas, LTC, USA, Ph.D., is the Assistant Dean for Research and Academic Affairs at the Academy for Health Sciences, Army Medical Department Center and School at Fort Sam Houston, Texas. LTC Escolas also serves as the Human Protections Administrator for the Army Medical Department Center and School and as a member of the Brooke Army Medical Center Institutional Review Board. She recently was on deployment orders for Iraq with a follow-on deployment to Afghanistan with the Institute of Surgical Research's Deployed Combat Casualty Research Team.

Richard V.N. Ginn is an Historian with the Army's Office of Medical History. He is a retired Army Medical Service Corps officer, health care company chief executive officer, and veteran of Vietnam, where he served with the 173d Airborne Brigade. Ginn is the author of *The History of the U.S. Army Medical Service Corps* (U.S. Army Center of Military History, 1997) and *In Their Own Words: The 498th Medical Company (Air Ambulance) in Iraq* (U.S. Army Office of Medical History, 2009).

James E. Griffith, COL, USA, Ph.D., serves as the Research Psychology Consultant (Reserve Component) to the Chief, Medical Service Corps, and provides research consultation to the Recruitment and Retention Branch, National Guard Bureau. His primary Army Reservist role is on the staff of the Joint Forces Headquarters, Maryland Army National Guard. While on Active duty, COL Griffith conducted research on military personnel manning systems and human dimensions issues at the Department of Military Psychiatry at WRAIR. He left Active duty in 1986 and entered the Maryland Army National Guard where he continues part-time military service. He received the "A" proficiency designator in research psychology from the U.S. Army and was given Fellow status by the American Psychological Association and American Education Research Association.

Charles W. Hoge, COL, USA, M.D., leads a psychiatric research program at WRAIR focused on mitigating the mental health impact of the current wars in Iraq and Afghanistan. He served in the U.S. Public

Health Service as an Epidemiology Intelligence Service officer at the Centers for Disease Control and Prevention before transferring to the Army in 1991, where he spent 6 years conducting field studies related to the treatment and prevention of tropical infections. He completed his residency in psychiatry at Walter Reed Army Medical Center in 2000. He has over 70 peer-reviewed publications, including a landmark article in the *New England Journal of Medicine* (January 2008) on mild traumatic brain injury among U.S. troops returning from Iraq.

Steven R. Hursh is the President of the Institutes for Behavior Resources and Professor of Behavioral Biology, Johns Hopkins University School of Medicine. Dr. Hursh is a retired Army research psychologist and served as Director of the Neuropsychiatry Division, Walter Reed Army Institute of Research, and as the 71F Consultant to the Surgeon General of the Army. Hursh is the author of over 65 articles, book chapters, and books. He holds the patent for the Sleep, Activity, Fatigue, and Task Effectiveness (SAFTE) Model used by the Federal Railroad Administration, the Federal Aviation Administration, and the Federal Motor Carrier Safety Administration to assess fatigue in transportation workers. He received his Ph.D. from the University of California.

William D.S. Killgore, MAJ, USA, Ph.D., is a Research Psychologist at McLean Hospital and Harvard Medical School, where he has been on the faculty for the past 10 years. Following the 9/11 attacks, Dr. Killgore transferred to Active duty status in the U.S. Army as a Research Psychologist and served as Chief of the Neurocognitive Performance Branch in the Department of Behavioral Biology at WRAIR. Most of his research has focused on the effects of sleep deprivation and combat stress on high-level cognitive capacities, including judgment, decision-making, and risk-taking propensity, as well as functional neuroimaging studies of affective disorders, anxiety, post-traumatic stress disorder, and brain development. He has published over 75 peer-reviewed manuscripts, 140 conference abstracts and presentations, and numerous book chapters.

Thomas A. Kolditz, COL, USA, Ph.D., is Professor and Head of the Department of Behavioral Sciences and Leadership at the United States Military Academy. He has served in an array of military tactical command and technical staff assignments worldwide, commanding troops through battalion level and acting as a leadership and human resources policy analyst in the Pentagon. COL Kolditz has published more than 40 articles in academic, military, and leadership trade journals and serves on the editorial and advisory boards of several academic journals. His most recent book, *In Extremis Leadership* (Jossey-Bass, 2007) focuses on leadership in dangerous contexts. He holds Master's and Doctoral degrees in Social Psychology from the University of Missouri.

Gerald P. Krueger, COL, USA (Ret.), Ph.D., spent 25 years in the Army, serving as a Human Factors Scientist at the Army Human Engineering Lab, Military Advisor for the Defense Advanced Research Projects Agency in Vietnam, Deputy Chief of Behavioral Biology at WRAIR, Staff Officer at the Army Medical Research and Development Command, and Division Director at the Army Aeromedical Research Laboratory. He culminated his career as Commander of the U.S. Army Research Institute of Environmental Medicine. Krueger is a recognized authority on sustained human performance, especially of equipment operators experiencing sleep deprivation and fatigue during sustained operations. Among his continuing activities, he is an adjunct associate professor of military psychology at the Uniformed Services University of the Health Sciences.

David M. Lam, M.D., retired from the Army in 2001 after 30 years of Active military service and now serves as Adjunct Associate Professor at the University of Maryland School of Medicine, from which position he has been seconded to the U.S. Army's Telemedicine and Advanced Technology Research Center (TATRC) in Fort Detrick, Maryland. He works primarily in the areas of telemedicine standardization, project development, and deployment of systems, mostly in the North Atlantic Treaty Organization (NATO) and Partnership for Peace nations. Currently, he serves as the TATRC European Liaison Officer, responsible for maintaining links to researchers in Europe, including with the NATO Medical Community, the European Space Agency, and the European Commission.

Ira H. Levine is a Senior Computer Scientist with AFRII and has served as head of the Information Technology Department for 20 years. He designed and programmed both the Biodosimetry Assessment Tool and First-responder Radiological Assessment Triage. Mr. Levine holds an M.S. in computer science from Johns Hopkins University and a B.S. in physics from the State University of New York–Stony Brook.

Philip Lewis, Ph.D., is a Professor Emeritus of Psychology at Auburn University, where he recently served as Acting Head of the Department of Music. He received his Ph.D. in Clinical Psychology from Syracuse

University and completed an internship in clinical psychology at Duke University Medical Center. His interests encompass adolescent and adult personality development, the assessment and development of managerial conceptual capability, and the role of individual differences in the workplace. Dr. Lewis was a co-investigator in a longitudinal study of West Point cadets and is currently involved in a longitudinal study of transformational change in Harvard M.B.A. students.

Sharon A. McBride, LTC, USA, Ph.D., is Chief of the Battlemind Transition Office at WRAIR. She received her Ph.D. in Experimental Psychology from American University, after which she was commissioned as a captain in the U.S. Army. She has had assignments at AFRII, the U.S. Army Research Institute of Environmental Medicine, and USAMRMC Congressionally Directed Medical Research Programs. She deployed to Afghanistan in 2007 as a member of the Mental Health Advisory Team to evaluate the mental health of deployed Soldiers.

Dennis McGurk, MAJ, USA, Ph.D., has been commander of the U.S. Army Medical Research Unit–Europe since 2007. In previous assignments, he was a Research Psychologist at WRAIR, the Senior Science Officer on the U.S. Army Surgeon General's Mental Health Advisory Team (MHAT) IIb in Afghanistan and MHAT IV in Iraq, and Assistant Department Chief in the Department of Military Psychiatry at WRAIR. He has published in peer-reviewed journals, authored book chapters, and presented at numerous scientific and military conferences around the world. He is a member of NATO Technical Team HFM 179 Moral Dilemmas and Mental Health Problems and NATO Exploratory Team 104 Mental Health Training.

Craig A. Myatt, LTC, USA, Ph.D., is the Command Psychologist assigned to the Command Surgeon's Office at the U.S. Special Operations Command, MacDill Air Force Base. Previous assignments include as Chief of the Information Management Division, Weed Army Community Hospital, Fort Irwin, California, and as Commander of the 145th Medical Battalion. He also was a Fellow at the University of Texas M.D. Anderson Cancer Center. In 2006, he obtained an individual tasking for deployment to Iraq as the Deputy G4, Task Force 3 Medical Command (Operation *Iraqi Freedom* 06–08) and resumed battalion command in September 2007 upon redeployment. He is the co-author of several neuroscience publications.

James Ness, LTC, USA, Ph.D., is currently assigned as an Engineer Scientist Exchange Program Officer to the *Streitkräfteamt*, Bonn, Germany.

Previously, he was an Associate Professor at the U.S. Military Academy, West Point. Throughout his career, LTC Ness has performed work ranging from researching the effects of directed energy on vision and visual performance to researching the social networks of terrorist groups. In recognition of his accomplishments across the breadth of the field of research psychology, the Army awarded LTC Ness the "A" proficiency designator.

Marek Rewers was a Research Assistant at the Center for Technology and National Security Policy. He is completing his M.A. in Security Policy Studies at The George Washington University, where he focuses on Political Psychology and Transnational Security.

Lyndon A. Riviere, Ph.D., has been a Soldier and Family Mental Health Research Scientist in the Department of Military Psychiatry at WRAIR since September 2005. He is the principal investigator of the Land Combat Study II protocol, "Impact of deployment and combat experiences on the mental health and well-being of military service members and their families." Dr. Riviere has presented his research findings and departmental training products at military and professional meetings and is currently working on book chapters on Army families. His research interests include National Guard Soldiers, military families, and mental health.

James A. Romano, Jr., COL, USA (Ret.), pursued a Ph.D. in Experimental Psychology from Fordham University and was in academia at Manhattan College before entering the U.S. Army in 1978. He attained the status of Diplomate in General Toxicology. He had several commands and served as the Surgeon General's Consultant in Research Psychology, where he worked as a recruiter, assigner, advisor, and proponent of 71F officers.

Joseph M. Rothberg, Ph.D., is an Adjunct Associate Professor of Psychiatry at USUHS. After serving as a draftee in the Army, he completed his undergraduate education at the Illinois Institute of Technology and his M.S. and Ph.D. at the University of Chicago. He retired as a Senior Scientist from WRAIR after 26 years with the Division of Neuropsychiatry.

Charles A. Salter, LTC, USA (Ret.), Ph.D., S.D., is Director of the Office of Scientific Management at USUHS. He served 9 years at AFRRI as Director of the Medical Effects of Ionizing Radiation course and also as senior military scientist on the biological dosimetry team. He holds a

Ph.D. in psychology from the University of Pennsylvania and a Doctor of Science in public health from Harvard University.

David J. Sandgren is a Research Biochemist in the Biodosimetry Research group at AFRRI and serves as the senior laboratory manager on three intramural protocols addressing basic science and applied biodosimetry applications. He obtained his B.S. and M.S. in Genetics from Iowa State University and previously worked at the U.S. Department of Agriculture in Ames, Iowa.

Maurice L. Sipos, LTC, USA, Ph.D., is Chief of the Research Division at USAMRICD, where he previously served as Director of the Behavioral Assessment Laboratory, Chief of the Advanced Assessment Branch, and Director of the Collaborative Research Program. Past assignments have included serving as a research psychologist at WRAIR and as the Deputy Director for Research at the Army Physical Fitness Research Institute at the U.S. Army War College. He obtained his Ph.D. from the Department of Psychology at Lehigh University.

Melba C. Stetz, MAJ, USA, Ph.D., is the Director of Research in the Department of Psychology, Tripler Army Medical Center, Hawaii. Her research area is combat stress. She has worked in both laboratory and managerial positions in places such as the WRAIR, USAMRMC Headquarters, and the U.S. Army Aeromedical Research Laboratory.

Jeffrey L. Thomas, MAJ, USA, Ph.D., is the Deputy Department Chief of Military Psychiatry at WRAIR and the Assistant Team Chief for the 6th Mental Health Advisory Team for Operation *Enduring Freedom*. His previous assignments include Executive Officer for the 5th Mental Health Advisory Team and Chief of Soldier Deployment Mental Health at WRAIR. His involvement in deployment mental health research includes work with U.S. Army Europe operations tempo studies, Army Suicide Evaluation Report, Psychological Screening Validation Studies, Psychological Debriefing Efficacy Studies, WRAIR Land Combat Studies, and the Battlemind Training Studies.

Michael D. Wood, CPT, USA, Ph.D., is assigned to the U.S. Army Medical Research Unit–Europe, in Heidelberg, Germany. Previous to that he was at the Walter Reed Army Institute of Research. He has been deployed to Iraq to support the Mental Health Advisory Team V. His research interests include how agents such as benefit finding and morale may buffer the stress of deployment and stigma to mental health care and lead to resilience.

Index